Understanding Evolution

Understanding Evolution

Earl D. Hanson

New York Oxford
OXFORD UNIVERSITY PRESS
1981

Library of Congress Cataloging in Publication Data

Hanson, Earl D
Understanding evolution.

Bibliography: p.
Includes index.
1. Evolution. I. Title.
QH366.2.H38 575 80-13037
ISBN 0-19-502784-1

Printing (last digit): 9 8 7 6 5 4 3 2 1

Printed in the United States of America

to Lyn
a special person

Preface

Evolution is biology's distinctive contribution to knowledge. The late Theodosius Dobzhansky, one of the leading evolutionists of this century, has said, "Nothing in biology makes sense except in the light of evolution." And beyond biology, among the questions that unendingly engage human curiosity—What am I? Where did I come from? Where am I going?—the concept of evolution is endlessly invoked for answers.

This text confines itself to the biological questions—What are living things? Where did they come from? What trends are apparent among them?—and uses Darwinian natural selection as the central theoretical explanation. Strictly speaking, natural selection says nothing directly about the nature of living things or where they came from in the sense of the origin of life. But given living things on this earth, natural selection is a powerful explanation for understanding why life is organized into well-adapted species and why there are species with varying degrees of similarities or differences. Indeed, the central effort of this book is to understand organismic diversity as a consequence of the action of natural selection. From that point of view, we must examine the nature of the living things affected by natural selection and also inquire into their primordial origins.

That forces us to realize two things about the study of evolution today: First, it is an exciting area of current research with its well-focused problems, its unique theoretical constructs and experimental methods, and its body of trained researchers and scholarly journals. Second, that unique body of research is continually spilling over and influencing ecology, physiology, systematics, cell biology, biochemistry and molecular biology, development, genetics, and all the other subdisciplines in biology, and extending also to anthropology, psychology, sociology, and beyond, and being influenced by them in return. Hence a textbook on evolution must not deal with evolution solely from the perspectives of population genetics or of speciation or of phylogeny. Such narrowly focused views minimize and downgrade the intellectual importance of evolution and are a disservice to those students whose minds and imagination deserve to be stretched and stimulated by the implications and applications of evolutionary thought.

With that in mind this book was written for biology majors, i.e., for a one-semester course for undergraduates concentrating on biology (or zoology or botany). A self-contained volume, it reviews organismic diversity, heredity, and selected portions of cell biology, development, molecular biology and ecology, and the history of evolutionary thought as needed background for the study of evolution itself. Needless to say, when such an array of facts and ideas are reviewed, there are bound to be differences among experts as to what is best included or excluded. The consistent focus here has been to provide those facts and ideas that will best bring the student into genuine contact with the problems and current answers of evolutionary biology. With that foundation, individual teachers are urged to take their classes further in those directions that best serve their own particular view of an exciting and rewarding experience in evolutionary studies. There is much in the areas of populational studies, molecular phylogenies, the evolution of behavior, human evolution, and so on that can usefully augment and extend this text.

Within the aims set forth above, the chapters written here have seemed to fall naturally into four parts. Part I presents the problem of explaining organismic diversity. It first surveys what is comprised by that diversity and next reviews Darwin's conceptual breakthrough, namely, the origin of species by natural selection. Additionally, with the advent of genetics, we have found a powerful new set of ideas from which to conceptualize evolutionary change through natural selection. Part II reviews what population genetics tells us about species. Genetics, as a science, was unavailable to Darwin. It has since transformed the older idea of a species into that of a reproductively isolated gene pool—the biological species concept—and thereby provided a new perspective on how adaptations are achieved and maintained. The concept recasts natural selection as the necessary force that molds chance mutations into organisms functionally adapted to their environment. Part III takes a modern look at the nub of Darwin's problem, the origin of species. Darwin's thinking has been revised and extended—a state of affairs of which Darwin could only approve. Additionally, this clearer vision on the mechanics of speciation provides more incentive for tracing out evolution beyond the species level. Today's vigorous interest in phylogeny attests to that. Part IV is concerned with phylogeny and the longer trends of evolution. The greatest evolutionary puzzles lurk in the phyletic relations among the larger groups of organisms—the phyla and kingdoms. Those problems are examined candidly, but for the newcomer to this area, the successes may seem minimal. Phylogeny—the history of life—is now only in its infancy. It is the great unfinished business of biology. How natural selection works is relatively clear. The nature of organismic diversity is extensively catalogued. We must put the two together so as to read and understand the history of life. It is the history of the biosphere, of that part of the earth we inhabit. It is also our history. Understanding evolution, as a process and as historical events, is integral to better understanding ourselves.

Among the many sources of criticisms, helpful ideas and encouragement I must acknowledge the most conspicuous. To my colleagues James Donady and Carol Lynch, both of Wesleyan University, with whom I have taught evolution, I owe special thanks. To my students, whose ideas and reactions have sharpened and clarified my thinking on many points, I have a deep debt of gratitude.

Special thanks are due Dr. Patrick Colin, of the Department of Marine Sciences, University of Puerto Rico, Mayaguez, for his help at the Marine Laboratory at La Parguera. The kind and able cooperation of the staff of that laboratory was extremely important in enabling the author to obtain needed illustrations of marine organisms. The manuscript was completed during a year's Fulbright Fellowship in India between lectures and laboratories on genetics and evolution. I am very grateful to Dr. A. Gnanam, Head, School of Biological Sciences, Madurai-Kamaraj University, Madurai, and Dr. Vinod C. Shah, Head, Department of Zoology, Gujarat University, Ahmadabad, for their support and the use of facilities at their universities. And very special thanks to my own university, the Biology Department office staff, and the science librarians for their constant help. Without their uncomplaining and cheerful efforts, the burden of completing this book would have been significantly heavier. I thank all of them.

Finally, special gratitude must be expressed to those of Oxford University Press, New York, who played major roles in the creation of this book. Robert J. Tilley, Editor, aroused in me the reality of doing this book and patiently (most of the time) let me work my way through its composition. Carol Miller did a superb job in rendering clearer what I tried to say. Joyce Berry was invaluable in locating effective illustrative materials. Margaret Joyner's book design is a creative realization of what I hoped for in this book. And to Margo Dittmer, Production Co-ordinator, I owe special thanks for her extraordinary efforts in producing this book. It was a joy to work with these and the other professionals on Oxford's staff. I do thank them.

Middletown, Connecticut E.D.H.
November 1980

Contents

CONTENTS

1

Organismic Diversity and the Concept of Evolution

Marine worm fossils in limestone

WE EXIST AS INDIVIDUALS, but our awareness extends beyond self to the world around us. We perceive many things in our world—some like us, others different; some living, others nonliving. We sense a tremendous variety, but in that variety we also perceive regularities or patterns. Such patterns intrigue us; we want to be able to explain them. Why do they exist? How did they arise? In particular, regarding the patterns we find among living things, we wonder why we share some characters with all living things, but share others with only a few. That is, there are some unifying aspects to life as well as aspects that distinguish one organism from another. Why?

One concept that biologists find crucial to understanding the unity and diversity of life is evolution. In Chapter 1 we will survey what is known about the unity and diversity of living things, in Chapter 2 we shall see how the idea of evolution helps us understand that unity and diversity. The major features of heredity are reviewed in Chapter 3, since heredity has a central role in evolutionary studies because it explains not only why and how organisms produce more organisms like themselves, but also, how variations or differences appear. In the rest of the book we will work out the consequences of the information and ideas found in these first three chapters. Overall, this book examines our understanding of evolution as an explanation of the unity and diversity of life.

ONE

Similarities and Differences among Organisms

THE DIVERSITY OF ORGANISMS arises from an underlying unity of life. Both unity and diversity are seen in functions and structures, which are complementary aspects of all living things.

The functional unity of life

Living things share three basic functions: (1) All of them obtain energy and materials from their environment; (2) all use energy and materials to maintain life; and (3) all reproduce themselves.

Energy from the environment. Different organisms obtain energy from different sources. Plants, as we know, carry out photosynthesis with the help of ultraviolet rays, from sunlight (radiant energy). Animals obtain energy from the food they eat (chemical energy). And among animals, different foods supply this energy: grazing animals obtain it from plants; meat eaters from other animals.

Most intriguing are the capabilities of bacteria. These microscopic organisms obtain energy from many different kinds of compounds, or even elements. In fact, it has been said that if there is an energy-yielding chemical reaction, one can find some kind of bacterium that utilizes it.

Hence, merely saying that organisms obtain energy from their environment obscures the fact that there are many different ways of obtaining that energy.

Materials from the environment. Plants, animals, and microorganisms require different kinds of molecular materials: first, because they vary in the chemical composition of body structures, although living things are more alike in *molecular* structure than they are different; second, because living things differ in their abilities to synthesize the molecules that constitute their cells.

Plants are the most able synthesizers of molecules. Starting with sunlight, water, and carbon dioxide plus small amounts of such heavy elements as iron, magnesium, calcium, and sulfur, they build more of themselves. Certain bacteria, like the plants, are adept at synthesizing, but most of them start with molecules that are larger and more complicated than water and carbon dioxide. Animals are the least able to synthesize essential molecules from simpler ones. They build their proteins from amino acids; their nucleic acids from nucleotides; and their carbohydrates and lipids from sugars and two-carbon compounds, respectively (Box 1–1).

These different abilities to synthesize essential molecular building blocks are reflected in the place different organisms have in nature. Plants are *producers,* obtaining their needs from non-living substances in their environments. Animals are *consumers,* eating plants and other animals. Many bacteria are *decomposers;* they utilize the products of animals and plants or directly attack the bodies of plants and animals to get the food they need.

Maintenance or homeostasis. All living organisms have an external and an internal environment, and the latter must always be kept more or less constant. This process is homeostasis. Thus, the fluid within the cell must be maintained at close to neutral pH (neither acid nor alkaline). Ionic (salt) concentrations are also carefully regulated, so that the structures essential to life continue to function.

Homeostasis is maintained by work (energy transfer), which is of several kinds. Enzymes, which are complex, protein-containing compounds, act as catalysts to regulate the reactions that take place in the cell. As a result of these reactions the structural order of living systems is maintained. In cells,

Box 1–1. Macromolecules and living systems.

The molecules most characteristic of living things are the organic, or carbon-containing, molecules. Although carbon-containing compounds exist in thousands of different forms, only a few are constituents of organisms. Among these, those whose molecular weights exceed a thousand daltons (units of molecular weight) fall into four important categories: proteins, carbohydrates, nucleic acids, and lipids.

A. *Proteins*. These large molecules are made up of amino acids, which form long chains called polypeptides. Proteins usually contain at least twenty different amino acids (Table 1–2, only five different amino acids are shown here). Each different polypeptide chain, with its specific sequence of amino acids, folds precisely; together the chains form a functional protein whose molecular weight can exceed one million daltons. Proteins can interact with each other or with carbohydrates or lipids as a cell synthesizes needed compounds (catabolism) and breaks down nutrients (anabolism).

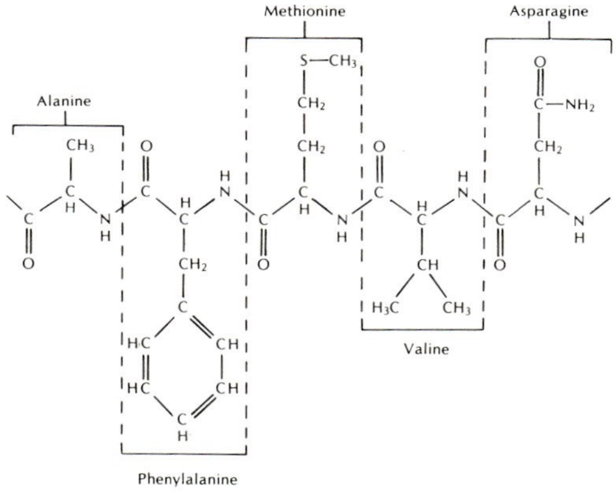

Methionine
Asparagine
Alanine
Valine
Phenylalanine

B. *Carbohydrates*. The basic unit of these carbon, hydrogen, and oxygen macromolecules is a six-carbon sugar molecule. (An example of a five-carbon sugar molecule is the ribose or deoxyribose sugar in a nucleotide. See p. 6.)

Long chains of these sugars make up starch, the carbohydrate found in plants. Glycogen, a very similar carbohydrate, is found in animals. The breakdown of these large molecules to their component sugars, and eventually to carbon dioxide and water, releases energy for cellular metabolism.

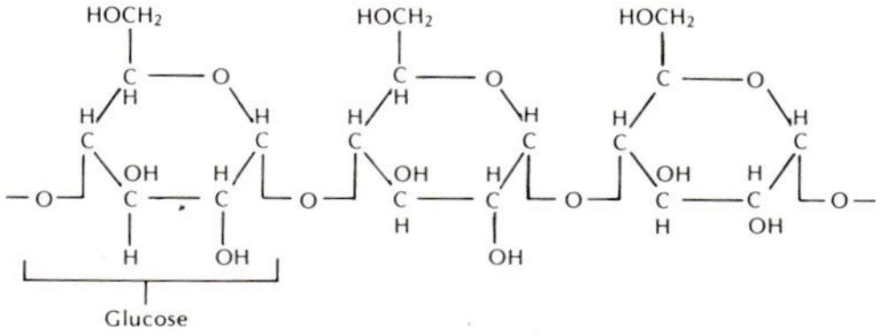

Glucose

(**Box 1–1** continues next page)

Box 1–1 (continued)

C. *Nucleic acids*. Deoxyribonucleic acid (DNA), a nucleic acid, contains four different nucleotides. They form two complementary polynucleotide chains, which intertwine as a double helix. Ribonucleic acid (RNA) is also made up of nucleotides, but (1) it does not form a double helix like DNA, (2) it contains uracil in place of thymine, and (3) it has one more oxygen on the ribose sugar of each nucleotide.

The role of DNA and RNA in protein synthesis and in heredity is further discussed in Chapter 3.

DNA double helix (diagram)	Atom distribution: Four paired nucleotides of the double helix

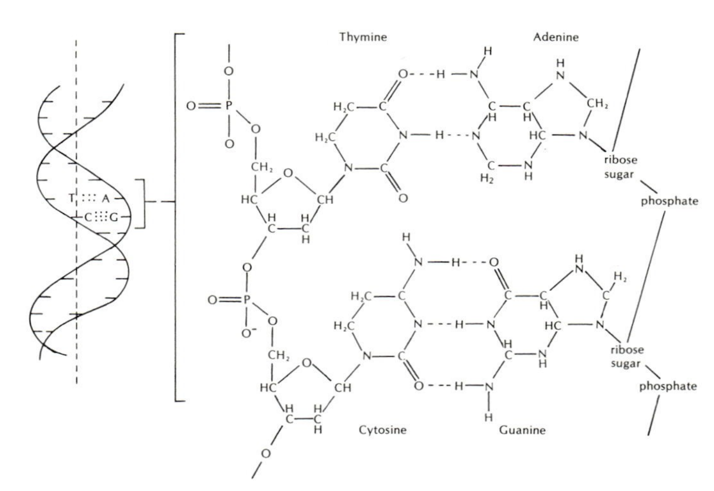

D. *Lipids*. Most lipids are made up of long chains of carbon atoms, with hydrogen and hydroxyl groups. The basic structure is one glycerol molecule, with three fatty acid molecules attached. This glycerol has one palmitic and two oleic acid molecules.

```
CH2— O — oleic acid
|
CH— O — oleic acid
|
CH2— O — palmitic acid
```

Oleic acid has the structure:

```
          H   H   H   H   H   H   H
— CH2— C — C — C — C — C — C — C — CH
          H   H   H   H   H   H   H    ||
 O                                     ||
   \\ H   H   H   H   H   H   H        ||
     C — C — C — C — C — C — C — CH
   /  H   H   H   H   H   H   H
HO
```

The bend in the middle of the chain and the double bond between the carbons at that point is characteristic of oleic acid.

These long carbon chains are synthesized by successively attaching acetic acid residues, each of which adds two carbon atoms. Breaking down these chains releases energy for cellular metabolism.

Other macromolecules vital to life include hormones, vitamins, and pigments.

some of the functions dependent on cellular structures are as follows:

the cell membrane protects the cell from its environment

and controls the flow of selected materials into and out of the cell;

the mitochondria generate molecules that release significant amounts of energy when they are broken down;

plastids, present in plant cells, carry out photosynthesis;

ribosomes aid in protein synthesis;

the nucleus contains the genetic material, as deoxyribonucleic acid (DNA);

and so on for each set of organelles in a cell.

In the final analysis, the activities of a cell come down to molecular interactions. This introduces one of the fundamental constraints on living things, the second law of thermodynamics. This law can be stated in several ways for living systems. The statement that concerns us is that whenever there is a reaction involving energy, some energy will be lost. Since all reactions involve energy, there is, then, a continuous energy loss, until no more energy can be lost. This is a state of maximum disorder or maximum entropy. In brief, the second law tells us there is a universal tendency toward disorder or increasing entropy.

To maintain order is to decrease entropy and that requires energy. Therefore, organisms must continually work to maintain the ordered structures and orderly functions that ensure life.

The plasma or cell membrane, as was just said, is involved in the transport of substances into and out of the cell. When a substance diffuses across the membrane because the concentration of it or another substitute is higher on one side of the membrane than on the other, energy is not required, and there is no evidence of work being done. But when a substance is moved against concentration gradients, work is being done and measurable amounts of energy are used.

Materials entering the cell are broken down to a size usable by the cell, if they are not already that size. Then the cell, using sunlight or chemical energy, which is ultimately derived from sunlight, synthesizes the molecules it needs. These include proteins, which are formed on the ribosomes using information stored in the nucleus. In the process some energy is used up and metabolic waste products are formed. These products are eliminated from the cell. They are among the materials recycled in our ecosystem.

Cells, therefore, continually form not only the molecules needed to maintain themselves, but also those necessary for growth and eventually reproduction.

Reproduction and variation. Reproduction has been called growth beyond the limits of the individual, that is, the production of new individuals. Most cells have the ability to form two cells. In some cases one cell can divide into many cells. In any case, reproduction is an increase in number of cells or of organisms.

There are two kinds of reproduction, asexual and sexual. The former involves only one parent. That is, only one organism produces more organisms. In contrast, sexual reproduction, except for cases of self-fertilization, always involves two parents. But even in self-fertilization, the new organism or offspring, is the result of the union of two distinct cells, or more basically, of two nuclei. Such a union implies a sexual process. The distinctions between sexual and asexual processes will be examined in greater detail in Chapter 4. Here, we need only make sure the basic concepts are clear in terms of reproduction being a mode of increase common to all living things.

Reproduction ensures survival of the species. Organisms are relatively fragile systems; being complex and highly ordered they are liable to a variety of malfunctions. Heat, drought, too much or too little light, a variety of everyday occurrences can severely damage, and even kill, living systems. Survival wins out over extinction of a species when organisms reproduce faster than old ones die. But this can, on occasion, lead to overpopulation. When that happens, death rates often catch up again to birth rates, and the population goes back to a size that can be supported by the environment.

The appearance of the organism depends upon (1) the genetic material passed on by the parents and (2) factors in the environment. In asexual reproduction, the genetic material of the parent is passed intact, except for rare mutations, to the progeny. This is a very conservative way of reproducing. In sexual reproduction, only one-half the genetic material of each parent is passed to the progeny. (See Chapter 3 for a review of the basic features of heredity.) This results in a combination of genetic material different from that of each parent.

In both asexual and sexual reproduction, the environment can affect the expression of the genetic material, so that the phenotype (the outward appearance) of the progeny varies. Overall, then, it is rare that progeny are perfect replicas of their parents, and thus reproduction with variation is a common characteristic of all living things.

From the foregoing discussion we can now define an organism in functional terms. It is *a system that takes matter and energy from its environment to maintain itself and to reproduce itself with variations.*

So far, little has been said about the structures that make the three basic functions possible. Although bacteria, plants, and animals bear little resemblance to each other a certain structural unity is shared by all of them.

The structural unity of life

One very obvious structural characteristic shared by living things is that they are cellular, although viruses are an exception. We need to say something more about cellular organization, and then we can take a more meaningful look at viruses. All cells (Fig. 1–1) have three major parts: (1) the outermost cell or plasma membrane, (2) the cytoplasm surrounded by the cell membrane, and (3) the chromosomes within the cytoplasm, with or without a nuclear membrane.

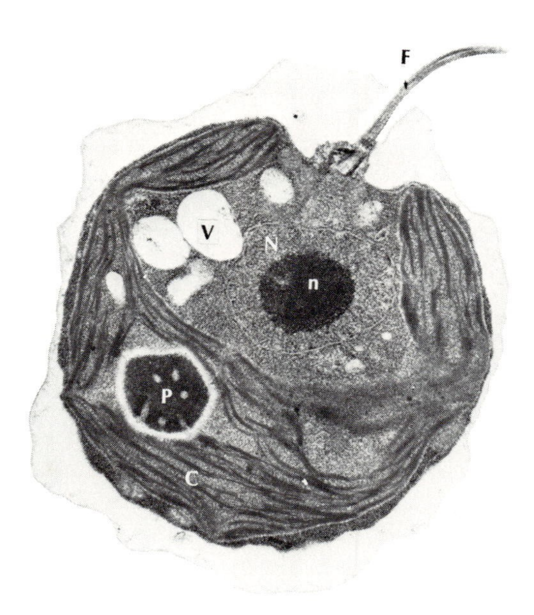

Fig. 1-1: An electron micrograph of the unicellular green alga *Chlamydomonas* (cross section). The nucleus (N) contains a nucleolus (n); the cup-shaped, parallel membranes of a chloroplast (C) extend around the nucleus; within the chloroplast lies a pyrenoid (P). Vacuoles (V) lie outside the nucleus; the flagellum (F) is used for locomotion.

Fig. 1-2: Major features of plant and animal cells. A diagram of selected features of a specialized bacterial cell (A), the pleuropneumonia-like organism (PPLO); a plant cell (B); and an animal cell (C).

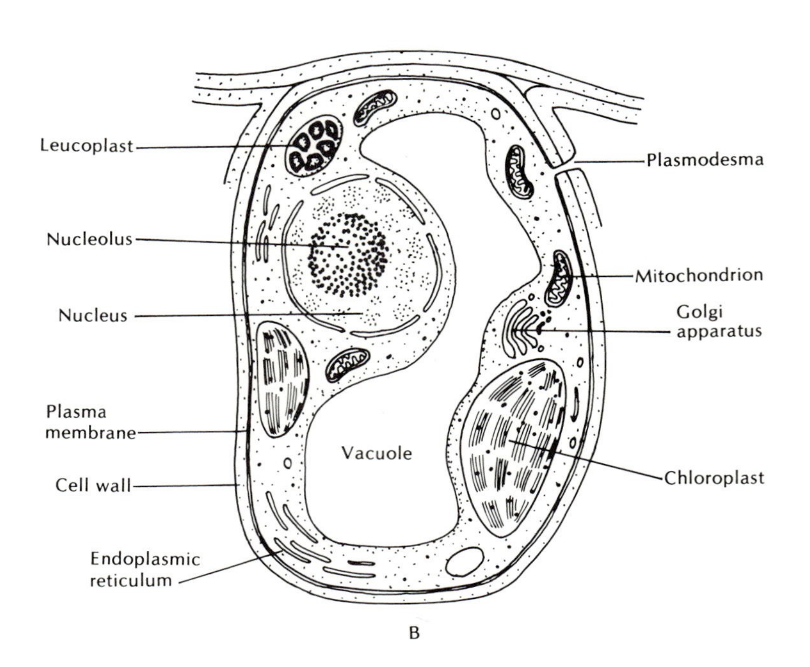

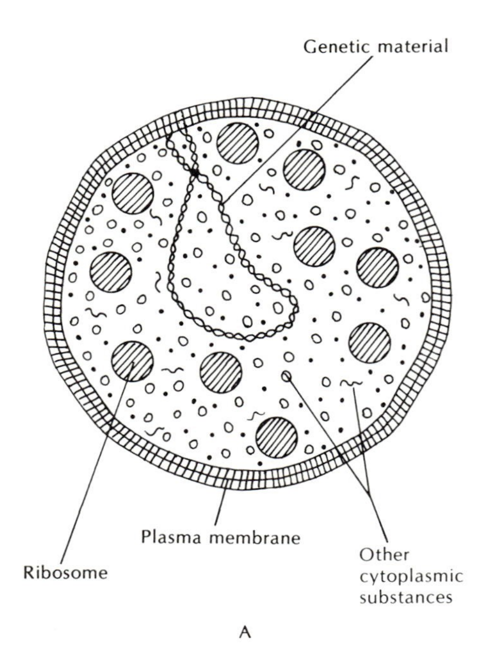

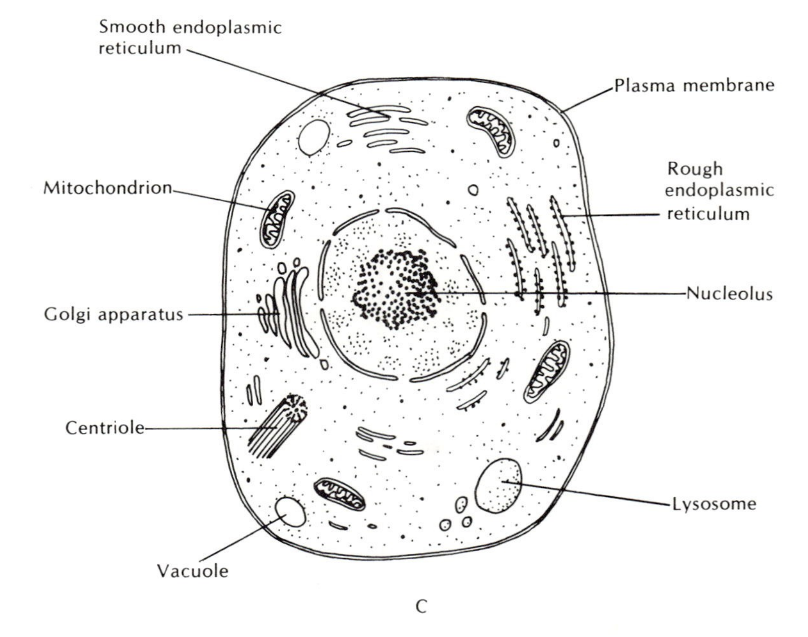

There are two kinds of cells: the prokaryote and the eukaryote. A major difference between them is the absence of a nuclear membrane in prokaryote cells and its presence in eukaryote cells. There are also other differences between these two kinds of cells, and within the eukaryotes there are certain differences between those found in plants and those found in animals (see Table 1–1 and Fig. 1–2).

The other structural characteristic shared by all living things is the genetic code. This code transmits the information that specifies the amino acid sequence of proteins. Different proteins, of course, contain different amounts and combinations of the essential twenty amino acids. But the information that specifies (codes for) a certain amino acid is the same for all organisms. It consists of a sequence of three nucleotides, called a codon (Table 1–2). Because the genetic code is universal, it is our most convincing evidence for the unity of living things.

Viruses have always been a stumbling block to any definition of an organism or of life, in general. It is important to

Table 1–1. Characters found in animal, plant, and bacterial cells.

Cell structure	Animal	Cell type * Plant	Bacterium
cell wall	–	+	usually +
cell or plasma membrane	+	+	+
ribosomes	+	+	+
endoplasmic reticulum	+	+	–
mitochondria	+	+	–
plastids	–	+	–
Golgi apparatus	+	rare	–
lysosomes	+	rare	–
peroxisomes	+	–	–
cilium with basal body	+	only in certain gametes	–
simple flagellum	–	–	+
vacuoles	+	+	–
centrioles	+	–	–
spindle during division	+	+	–
nuclear membrane	+	+	–
chromosome of DNA, RNA, and protein	+	+	–
circular chromosome of DNA	–	–	+
nucleolus	+	+	–

* + = present; – = absent.

Table 1–2. Dictionary relating DNA codons (capital letters) and RNA codons (italic capital letters) to specific amino acids.

AAA	*UUU*	Phe	AGA	*UCU*	Ser	ATA	*UAU*	Tyr	ACA	*UGU*	Cys
AAG	*UUC*		AGG	*UCC*		ATG	*UAC*		ACG	*UGC*	
AAT	*UUA*	Leu	AGT	*UCA*		ATT	*UAA*	Stop	ACT	*UGA*	Stop
AAC	*UUG*		AGC	*UCG*		ATC	*UAG*		ACC	*UGG*	Tip
GAA	*CUU*	Leu	GGA	*CCU*	Pro	GTA	*CAU*	His	GCA	*CGU*	Arg
GAG	*CUC*		GGG	*CCC*		GTG	*CAC*		GCG	*CGC*	
GAT	*CUA*		GGT	*CCA*		GTT	*CAA*	Gln	GCT	*CGA*	
GAC	*CUG*		GGC	*CCG*		GTC	*CAG*		GCC	*CGG*	
TAA	*AUU*	Ile	TGA	*ACU*	Thr	TTA	*AAU*	Asn	TCA	*AGU*	Ser
TAG	*AUC*		TGG	*ACC*		TTG	*AAC*		TCG	*AGC*	
TAT	*AUA*		TGT	*ACA*		TTT	*AAA*	Lys	TCT	*AGA*	Arg
TAC	*AUG*	Met	TGC	*ACG*		TTC	*AAG*		TCC	*AGG*	
CAA	*GUU*	Val	CGA	*GCU*	Ala	CTA	*GAU*	Asp	CCA	*GGU*	Gly
CAG	*GUC*		CGG	*GCC*		CTG	*GAC*		CCG	*GGC*	
CAT	*GUA*		CGT	*GCA*		CTT	*GAA*	Glu	CCT	*GGA*	
CAC	*GUG*		CGC	*GCG*		CTC	*GAG*		CCC	*GGG*	

This code, as stated in the text, occurs in all organisms. One exception has been reported: The UGA codon in the DNA of yeast mitochondria is thought to code for tryptophan rather than being a terminator (Macino et al., 1979). Each codon has three nucleotide bases (A, adenine; C, cytosine; G, guanine; and T, thymine for DNA, or U, uridine for RNA). The sequence stored in DNA is transcribed into RNA, whose bases are complementary to those of DNA; for example, the DNA codon GTA specifies the RNA codon CAU and is translated as the amino acid histidine. (The twenty amino acids in the table are abbreviated as follows: Ala, alanine; Asn, asparagine; Asp, aspartic acid; Arg, arginine; Cys, cysteine; Gln, glutamine; Glu, glutamic acid; Gly, glycine; His, histidine; Ile, isoleucine; Leu, leucine; Lys, lysine; Met, methionine; Phe, phenylalanine; Pro, proline; Ser, serine; Thr, threonine; Try, tryptophan; Tyr, tryosine; Val, valine.)

recall that viruses live *only* inside cells of other organisms, which can be bacteria, plants, or animals. Obviously, then, the environment exploited by viruses for matter and energy is the cell that surrounds them. They both maintain themselves and reproduce there. Among the virus progeny there can be variant viruses (mutant viruses) that also reproduce. In these terms, there is no problem in calling a virus an organism.

The problem with defining viruses as organisms comes when they are outside of their host cells. Under such conditions they can be handled chemically, much as if they were lifeless molecules. Wendell Stanley was the first to discover that viruses could be put into solution and then recovered, as viral particles, by recrystallizing them from the solvent. Further, he discovered that the dissolved particles could not infect and grow in a host whereas the recrystallized ones could. From this the conclusion is suggested that viruses are living in one state (particles inside the cell), and are not living in another state (dissolved outside the cell) and that these two states are reversible.

That conclusion does not throw out the definition of organism proposed above, but it is not what most people accept as a characteristic of living organisms. The fact is, viruses are not typical organisms (Fig. 1–3). Their genetic code is the same as that used by other organisms, but some virus particles store that code in ribonucleic acid (RNA) rather than DNA. Viruses synthesize some of their essential enzymes from amino acids from the host cell, and some of the metabolic activity that supports a virus is carried out by the host cell. Metabolic pathways leading to the formation of viral proteins and nucleic acids occur outside the virus particle, in the host cell surrounding it. And finally, a virus particle is not a cell. Although it contains nucleic acids that form a chromosome, which is surrounded by a protein coat in the mature particle, that coat is not a plasma membrane and there is no cytoplasm within the protein coat. Thus structurally, viruses are quite different from other living organisms.

This emphasizes the need to define organisms functionally rather than structurally. It also demonstrates that certain molecules, namely viral proteins and nucleic acids, are right at the transition between living and nonliving systems. The lesson from viruses is that they are at the "threshold of life," as one of the greatest viral researchers, Dr. Fraenkel-Conrat, has put it. Depending on their organizational state, they may or may not show evidence of being alive.

Finally, we can anticipate one of the questions to be discussed later when the origin of life will be considered. What was the first living thing? Was it a virus? The answer will be no, because viruses depend on cells to live. This means we must first find out how cells appeared. When we come to that topic we will see that even though our functional definition of an organism still holds, the early living systems need not have been organized the way organisms are today.

Diversity and classification

Organisms, even within the same species, are diverse. If we look at any group of humans we see evidence of that fact. Except for twins, there are easily distinguished differences between individuals. They differ in height and weight, age and sex, color of skin, hair, and eyes, and facial features and other details.

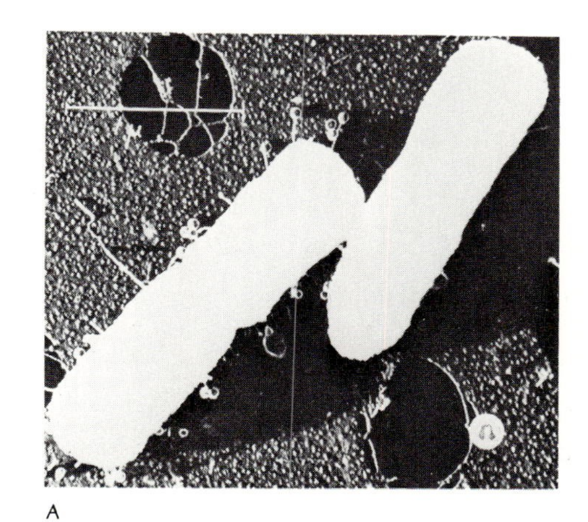

A

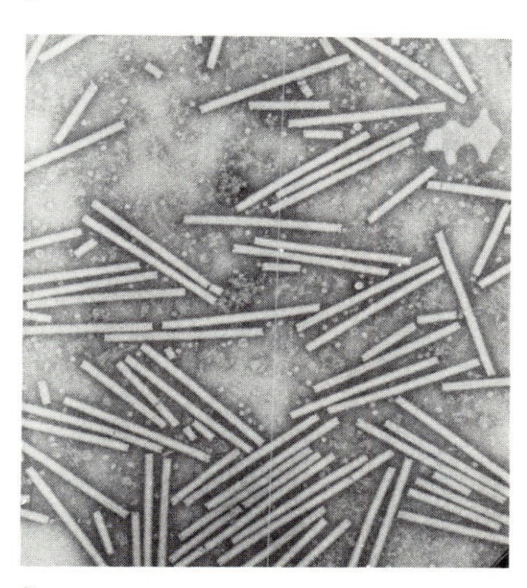

B

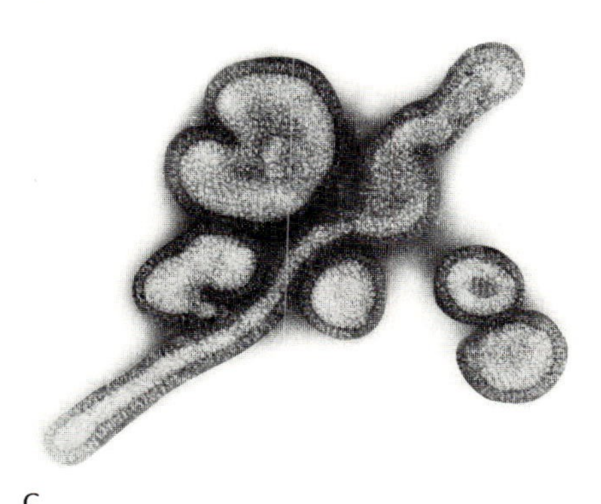

C

Fig. 1-3: Selected viruses: bacterial (A), plant (B), and animal (C).

Other organisms also show variations of the same sort. In some plants, for example, the age differences are very striking as when we realize that a tree may have one kind of leaf as a young plant, another very different kind of leaf as an adult. But in other organisms such as earthworms, differences are much more difficult to see.

In addition to looking for variations within a species, there are enormous differences between species. The smallest organisms—viruses—are 20 nm across; the largest—giant redwoods—are over 100 meters high and weigh some thousands of tons. (See Table 1-3 for metric units of length.) Between these extremes there are, literally, millions of other species. And they are found around the earth—from the north pole to the south pole, from ocean floors to mountain peaks, and from deserts to rain forests.

To be able to keep track of and understand this extraordinary array of life we have to have a system of classification. Such a system must include, at the least, basic descriptions of organisms, storage of those descriptions, and the means for recovering the stored descriptions when needed. The eighteenth-century Swedish biologist Carolus Linnaeus produced, in his *Systema Natura,* a system for classification that is still in use in a modified form.

One of Linnaeus's great contributions was to nomencla-

Table 1–3. Metric units of measurement.

Unit of Measure (Symbol)	Length in Meters	Equivalent in English system
kilometer (km)	1,000	0.62 mile
meter (m)	1.0	39.4 inches
centimeter (cm)	0.01	0.39 inch
millimeter (mm)	0.001	
micrometer * µm)	0.000001	
nanometer (nm)	0.000000001	
angstrom (Å)	0.0000000001	

1 m = 10^2 cm = 10^3 mm = 10^6 µm = 10^9 nm = 10^{10} Å

* Micrometers were previously termed microns.
The meter is the basic unit of length in the metric system.

Geographic distances are commonly given in kilometers. The heights of large organisms are given in meters, but the size of cells and their organelles are usually given in micrometers. For example, a human red blood cell is about 7 µm in diameter, a paramecium is about 120 µm in length.

Subcellular fine structure is often given in nanometers. A ribosome is about 25 nm in diameter, a cell membrane about 10 nm in diameter. Component parts of such substructures, seen in an electron miscrope, can be measured in angstroms, which also are used for macromolecules. For example, the diameter of a DNA double helix is about 20 Å.

Table 1–4. The more commonly recognized taxa and the classification of two animals and a plant.

Taxon	Human	Dog	Sugar maple
Kingdom	Animalia **	Animalia **	Plantae **
Phylum *	Chordata	Chordata	Anthophyta
Subphylum	Vertebrata	Vertebrata	
Class	Mammalia	Mammalia	Dicotyledonae
Order	Primates	Carnivora	Sapindales
Family	Hominidae	Canidae	Aceraceae
Genus	*Homo*	*Canis*	*Acer*
Species	*sapiens*	*domesticus*	*saccharum*

Other divisions sometimes used are cohort (between class and order) and tribe (between family and genus). Very commonly finer subdividing is achieved by using the prefixes super- or sub- with the appropriate taxon, e.g., see Subphylum, above. (See also Appendix I.)

* The term phylum is often replaced by division in plant taxonomy.
** These are Linnaean terms. See also Metazoa and Metaphyta in Fig. 1–21 and in Chapters 11, 12, and 13.

ture, when he devised rules for naming organisms. Linnaeus started with varieties and then went to species. At this basic level (excluding varieties for the time being) each distinctive type of organism was given two names. For example, humans were called *Homo sapiens*. The first name, *Homo* is the genus to which our species belongs, the particular species being indicated by *sapiens*. Different genera (the plural of genus) contained different species. Speaking broadly, the species within a genus resemble each other more closely than they resemble species in other genera, the genera within a family resemble each other more closely than they resemble genera in other families, and so on.

This system of degrees of resemblance is still largely used to determine the membership of any classificatory group. Linnaeus recognized six such groups: in addition to variety, species, and genus he also recognized order, class, and kingdom. Today we recognize certain additional groups or taxa. These are given in Table 1–4. As can be seen from the table, each taxon fits into the one above it. And since membership in any taxon is determined by the similarity of its members, the whole scheme is based on degrees of similarity or shared characteristics. This means that species within the same genus have more characteristics in common than do families within the same order.

One of the problems of a classifier or a taxonomist is to decide what key characteristics to classify.

At one time animals were classified as to whether they swam in water, ran on land, or flew in the air. That was, obviously, one way to describe organisms, store that information, and recover it from the categories swimmer, runner, or flier. Problems came when, for example, it was realized that the wings of a butterfly were quite different from those of a bird or that some swimming animals (e.g., seals) have backbones like running animals (e.g., dogs). In fact birds, too, have backbones. And what do we do with plants, which neither swim, run, nor fly, but yet live in water or on land and extend into the air?

Linnaeus called his classification a "natural" one. He believed, as was common in his day, that species were divinely created and that they persisted unchanged in their original forms. The presence or absence of certain characteristics allowed their groupings into more or less similar taxa. In particular, his descriptions of a species centered on the concept of an ideal type. Organisms closely similar in form to a certain ideal or type were then, by definition, members of a certain species. The type concept ignored variation in a species, which as we will see later on, is now considered of central importance. Though Linnaeus never made clear what he really meant by a "natural classification," his wide experience with organisms and his system of nomenclature gave us a taxonomy much more sophisticated and useful than any that preceded it. He used quite detailed information on the structure of organisms to determine the key characteristics for his classification, rather than relying on such simple categories as swimmer, runner, or flier.

The major taxa: Animals

A look at the major groups of organisms is useful now for two reasons: (1) It makes familiar the names and characteristics of organisms we will be discussing throughout this book. (2) It allows us see more clearly the diversity we hope to explain when we come to talk about natural selection and evolution. Since most of us are probably more familiar with animals than with plants, we will start with animals, but it will soon be apparent that many animal phyla contain forms that are not part of our everyday experience. The same will be true for plants, as we go from the more familiar to the less familiar ones. Nonetheless, we must try to explain the *whole* array of living things when we look for patterns in the diversity of living things. (See Appendix I for more details on classification and illustrations.)

Box 1–2. Nomenclature.

Some simple rules apply to nomenclature. Their proper usage, as with rules of grammar, aids considerably in the conveying of information.

1. The names of all genera and all species are italicized. The generic name is capitalized, the specific name is never capitalized. (Examples: *Homo sapiens, Amoeba proteus, Felis leo, Escherichia coli,* which refer, respectively, to humans, a species of ameba, African lions, and a common intestinal bacterium.)
2. The scientific name of the other taxa are capitalized. They are used as proper nouns, in regular Roman letters. The plural is never used, since each taxon is unique. (Examples: Chordata, Vertebrata, Mammalia, Primates, Hominidae, *Homo*. These refer to a phylum, subphylum, class, order, family, and genus, respectively.)
3. Certain conventions are used when scientific names are used as common nouns. They are no longer capitalized. (Examples: The phylum Chordata becomes chordates, the order Primates becomes primates—no change in spelling but, for purists, there can be a change in pronunciation—and the genus *Amoeba* becomes ameba. This last example indicates the complications that can arise as one transposes Latin into English. Note that the *o* has been dropped from *Amoeba*. The plural is either amebae or amebas.)
4. To avoid repetition, after the genus and species of an organism is given once, the generic name is abbreviated when that genus and species is repeated. (For example: *Escherichia coli* becomes *E. coli*.)

THE CHORDATES

The phylum Chordata (Box 1–2) contains many familiar animals; all have a backbone or its structural equivalent. The structural equivalent of the backbone, a long, round piece of cartilage, is called a notochord (Fig. 1–4). Not only is it found in certain of the less complex chordates, it is also found in the embryos of other chordates. In these it is replaced during development by the bony vertebrae that make up the backbone. Where this occurs the organisms are placed in the subphylum Vertebrata; hence they are vertebrates.

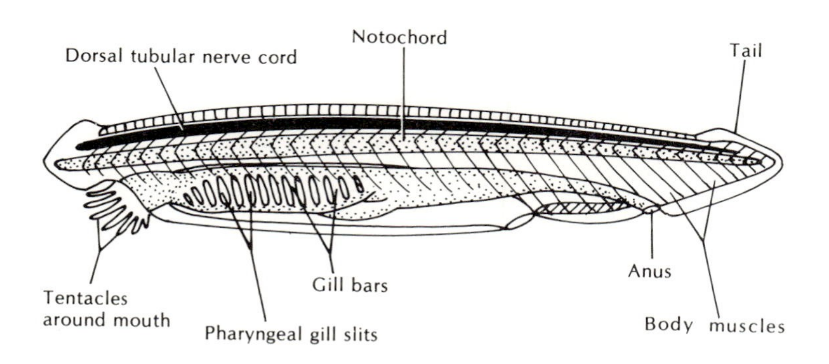

Fig. 1-4: A drawing of *Amphioxus* shows the notochord.

There are other key characteristics that describe the chordates and their major subgroup, the vertebrates. All these animals possess a dorsal, hollow nerve cord; they have gills or gill-like structures some time in their life history; and they are all bilaterally symmetrical. Figure 1–4 shows these characteristics in more detail.

The hollow nerve cord is, of course, the spinal cord found in ourselves and other vertebrates. It extends posteriorly from the brain and is surrounded by the bony vertebrae. Although the brain is variously formed in different groups of vertebrates, and the spinal nerves extend from the cord in different ways in different vertebrates, in all vertebrates and in the other chordates there is always a dorsal, hollow nerve cord.

Gills are most prominent in the simpler chordates and the fishes. They are found in young and adult forms. In the amphibians, gills are important in the young forms, but not in the adults. In reptiles, birds, and mammals the gills are only partially developed, even in embryonic forms (Fig. 1–5).

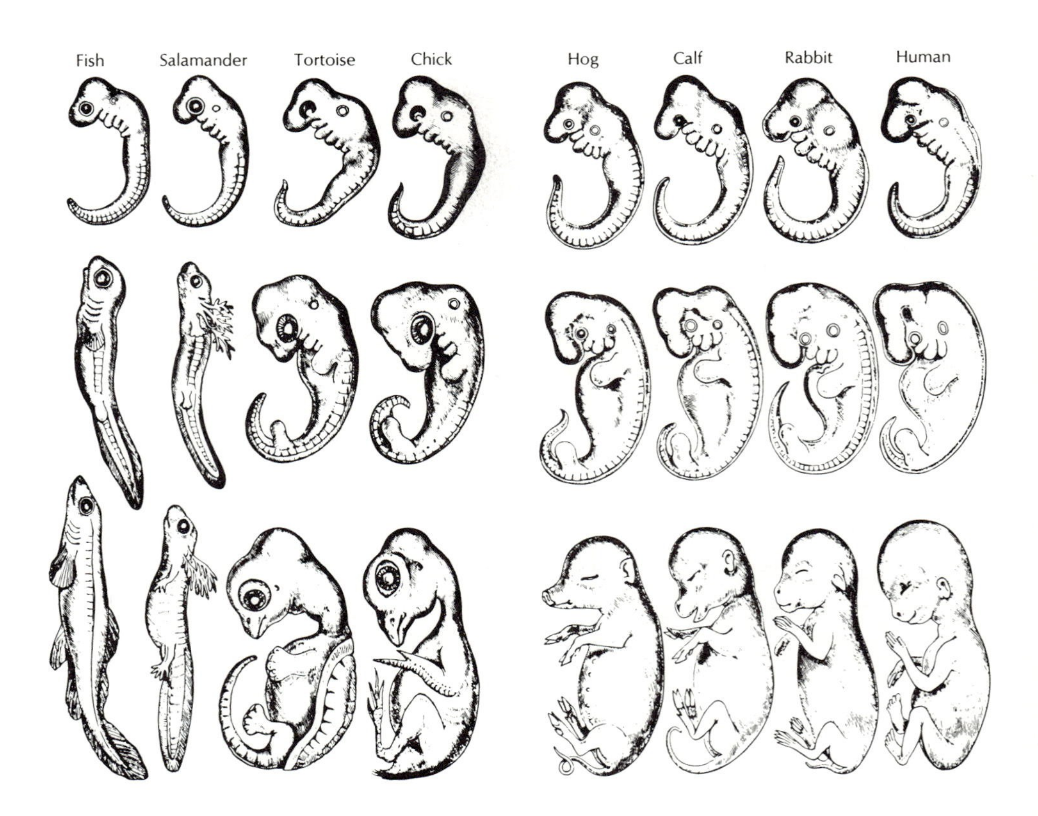

Fig. 1-5: Some vertebrate embryos. These were drawn by Ernst Haeckel, an ardent evolutionist, in the nineteenth century, and contain certain errors, but they do illustrate various important common features.

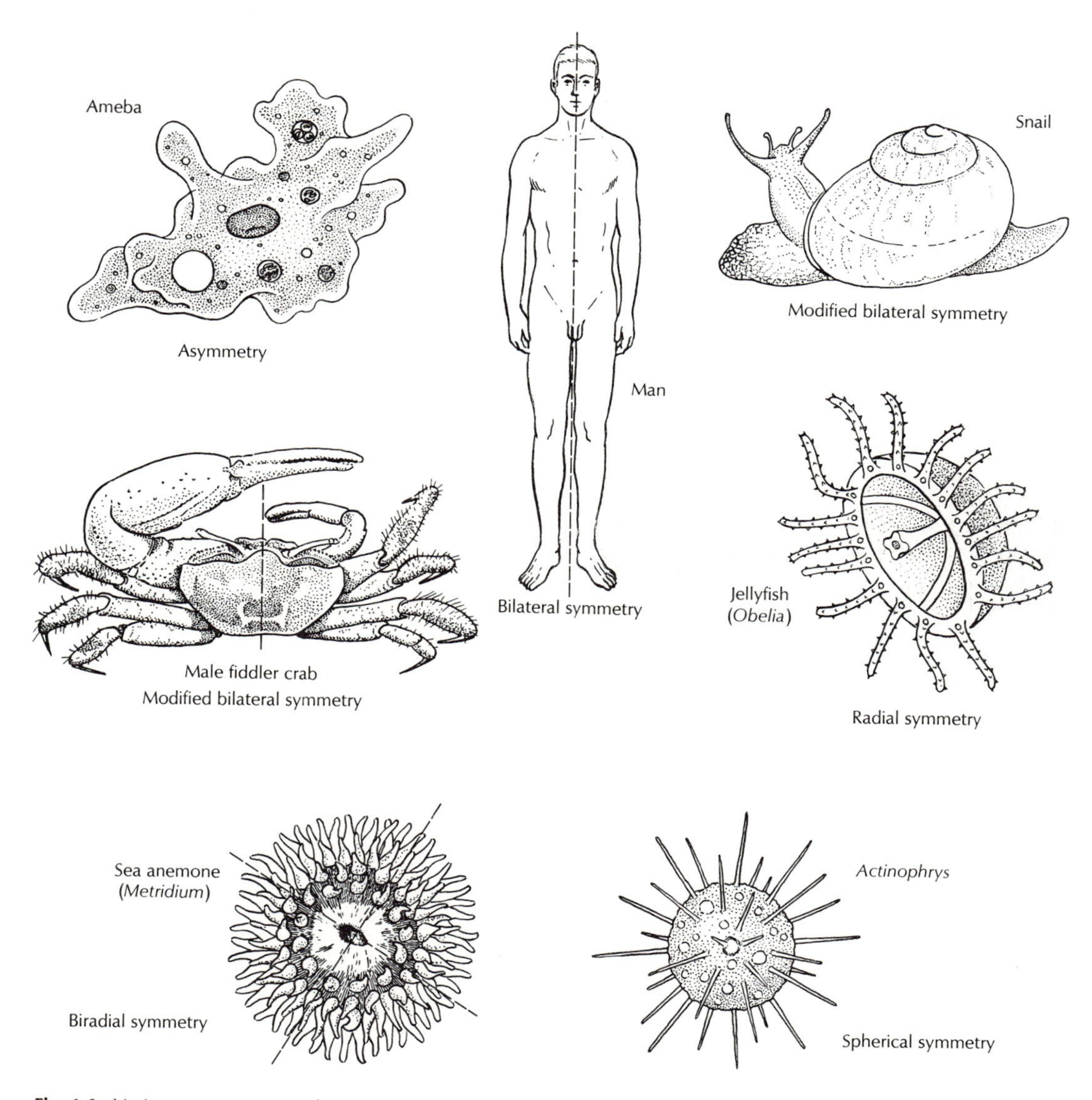

Fig. 1-6: Various types of animal symmetry.

In referring to the symmetry in living things we imagine a flat surface or plane that divides the organism into mirror halves. Bilateral symmetry means that only one plane can provide mirror halves. These are right and left halves, since the plane passes through the organism from its dorsal to ventral side and from its head to its tail (Fig. 1–6). There are other kinds of symmetry in living things, and some organisms are not symmetrical. Also, it must be kept in mind that symmetry is a general description; in practice, animals are not totally symmetrical. For example, the heart in humans is somewhat to the left of the midline of the body, which is the plane of bilateral symmetry. Therefore, people are not exactly bilaterally symmetrical, but bilateral symmetry comes closest to describing our form and is therefore a useful term.

The chordates have been intensively studied. This is largely the result of interests that go beyond evolutionary concerns. We are curious about forms that look like us, namely the monkeys. And we are concerned about our pets—dogs and cats—and about our domestic animals—cattle, sheep, hogs, and poultry. Also of interest are fish, an important source of food. There are about 50,000 well-described species of these free-living vertebrates in the oceans and in freshwater.

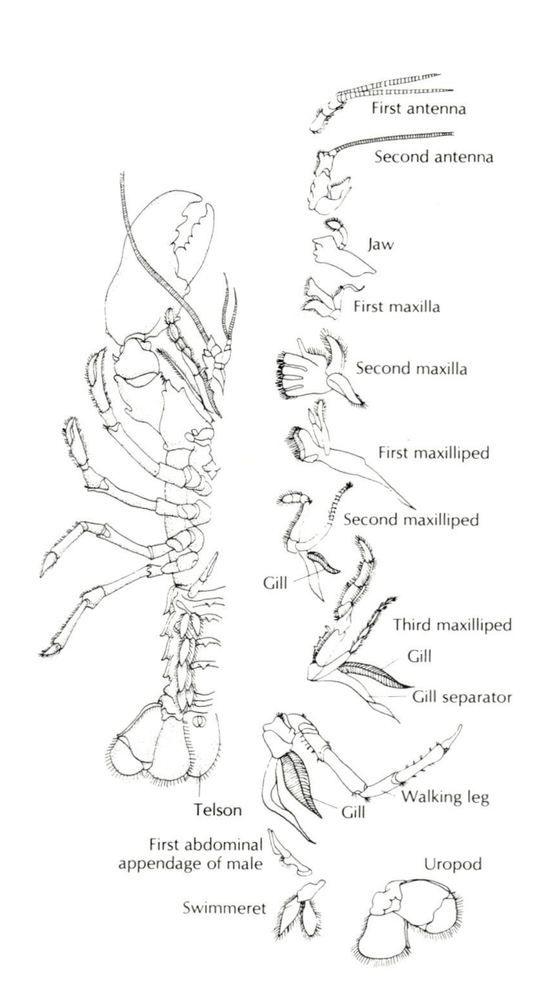

Fig. 1-7: A lobster and its appendages.

THE ARTHROPODS

The phylum Arthropoda is a huge one in terms of numbers of different species. Its members range in size from tiny insects to large crustaceans whose legs might be as much as eight feet across, as in the king crab of the northern Pacific Ocean. This variety has presented quite a problem to taxonomists over the years. As we will see later, recent work proposes that this taxon is best seen as a group of several different phyla.

But there are certain characters shared by all the arthropods. Two key ones are the presence of an exoskeleton and of jointed legs and other appendages that are jointed. The exoskeleton is a hard outermost covering to the body that (1) protects the soft inner parts and (2) provides sites for muscle attachment, so that the appendages and the body itself can be moved.

The appendages of the lobster (Fig. 1–7) have a variety of functions. Some appendages are sensory, i.e., the antennae. Others are for ingestion, and still others for locomotion (walking or swimming). Think of insects you know—flies, bees, butterflies, dragonflies—all have jointed legs. Their wings are appendages too, but they are not jointed.

All arthropods are bilaterally symmetrical organisms with a ventral nerve cord. They are segmented, but in many cases, as in crabs, the segments have fused and are difficult to see.

There are perhaps as many as 1 million species described; the insects alone account for over 800,000 of them. Along with insects, crabs, and lobsters, this group also includes spiders, centipedes, and millipedes. They are land-dwellers, especially the insects, and aquatic forms.

THE MOLLUSCS

The phylum Mollusca contains forms that, basically, are bilaterally symmetrical. Some forms, however, such as the snails with their coiled shells and octopuses with their numerous flexible tentacles, make it hard to see that symmetry. The one characteristic that unites all the molluscs is the presence of a muscular foot, as it is called (Fig. 1–8). In snails, the foot is the flat portion of the body on which the snail glides. In clams

Fig. 1-8: The molluscan foot. A. Drawings of a snail and a squid show the broad muscular food of the former and the tentaculated foot of the latter. B. The foot of a marile snail (ventral view). The ovoid foot (center) is attached to the glass wall of an aquarium.

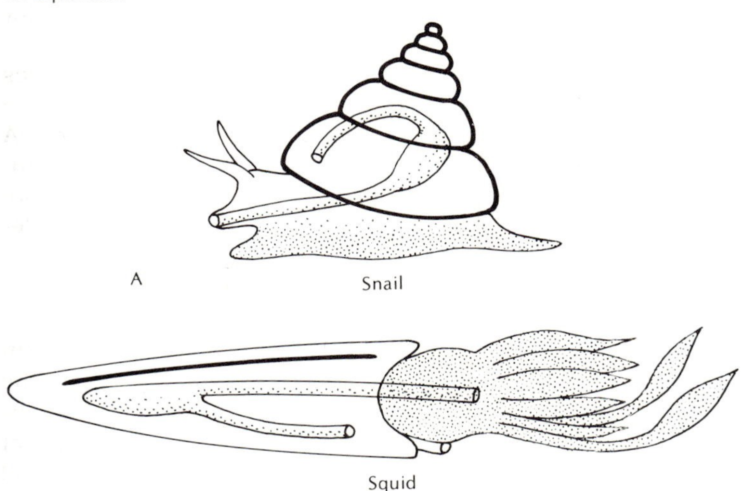

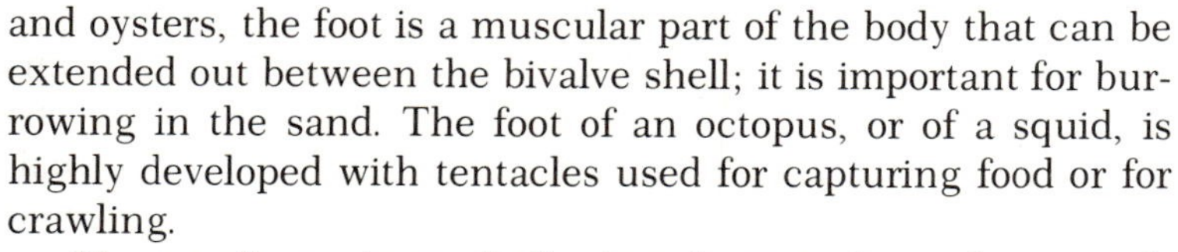

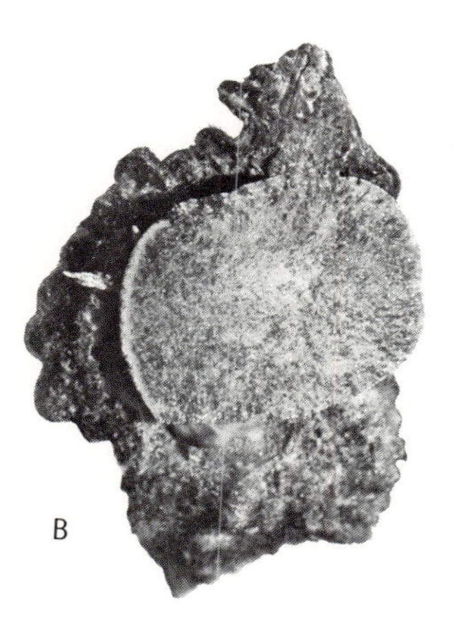

and oysters, the foot is a muscular part of the body that can be extended out between the bivalve shell; it is important for burrowing in the sand. The foot of an octopus, or of a squid, is highly developed with tentacles used for capturing food or for crawling.

Most molluscs have shells, but those without them, such as certain land snails and squids and octopuses, are thought to have lost them. That is, their ancestors had shells many millions of years ago. The loss of such a complex body part that is still important to most other members of this phylum is an intriguing aspect of their evolutionary history, since we often think of evolution as a series of changes leading to the appear-

Fig. 1-9: Segmentation in earthworms. The segments are clearer in the more extended worm.

Fig. 1-10: Pentaradiate symmetry in a starfish (ventral view through the glass of an aquarium). The round structures, sucker-like tube feet, are used for attachment.

ance of new characters. This shows that the reverse situation is also of evolutionary significance. Many mollusc species have been found only as fossils.

There are many species of molluscs—over 80,000 are known. They occur in the oceans, in freshwater, and on land.

THE ANNELIDS

Segmentation is the key characteristic of the phylum Annelida. In segmentation, each rather similar body part along the long axis of the body is marked off from the preceding and following part, as in the earthworm (Fig. 1–9). In the annelid worms, this segmentation is very obvious, the segments changing slightly in size and form from head to tail.

Annelids possess a ventral nerve cord and are bilaterally symmetrical. Appendages are often present, being used for locomotion (crawling or swimming) and as gills (breathing).

In addition to terrestrial earthworms, this phylum includes many marine worms, such as clam worms, and leeches. These latter are often called parasites, but are really symbionts. A symbiont is an organism living in close association with another of a different species. We will discuss symbionts in more detail later. About 9,000 species of annelids have been described.

THE ECHINODERMS

Starfish are probably the best-known examples of the phylum Echinodermata. They, and another group in this phylum, the sea urchins, make clear why the name of the phylum comes from the Latin for "spiny skin." In certain other members of the phylum, the skin is not spiny—it is simply tough and leathery or embedded with smooth plates.

Adult echinoderms show a radial symmetry. That is, there are several planes, all on one axis of the body, that result in mirror halves. In these animals that axis is an oral-aboral one, passing through the mouth and out the opposite side of the animal. Furthermore, since the radial symmetry is based on the presence of five repeated parts (for example, a starfish usually has five arms), there are five planes of symmetry (Fig. 1–10).

Another characteristic, found only in this phylum, is a water vascular system. This is a system of tubes that extends the tube feet, used for locomotion or for holding prey, by controlling the water pressure within them (Fig. 1–10).

All echinoderms live in ocean waters, with over 4,000 species being described. They are well represented as fossils.

THE ASCHELMINTHS

The phylum Aschelminthes is a very diverse collection of animals. Many taxonomists recognize several different phyla here. For our purpose it is easiest to discuss these forms as one group. These bilaterally symmetrical animals live in water or as internal symbionts or parasites. If the organisms damage their host, then they are true parasites; the relationship is *parasitism*. If the host and its symbiont both benefit, then the relationship is *mutualism*. And if no damage or benefit occurs then the relationship is *commensalism*.

A roundworm, such as *Ascaris* (Fig. 1–11), is a parasite found in the intestine of mammals. Its internal organization shows a significant simplification over that in all the preceding phyla, with the intestine lying freely in the body cavity (the pseudocoel).

Fig. 1-11: A parasitic roundworm (×142½).

The number of species of aschelminths depends upon what assumptions we make about the number of species of roundworms. Those that are not roundworms come to about 3,500 different species. But when we include the roundworms, one educated guess is that there are at least 500,000 different species. This figure is arrived at as follows: every vertebrate species has at least one (usually more than one) kind of roundworm in it. Since there are over 45,000 species of vertebrates, there probably are 100,000 species of roundworms. Roundworms are also parasitic in plants, molluscs, and most of the arthropods, including insects. Furthermore, free-living roundworms (in water and in soil) seem to be more numerous than the parasites. Therefore, 500,000 would be a conservative estimate. There are, however, good descriptions for only some thousands of roundworm species.

THE NEMERTINES

The ribbon worms included in the phylum Nemertina show a further simplification of body organization beyond that found in the roundworms. The nemertines have no body cavity. Apart from the digestive cavity and a cavity for the special feeding apparatus, the rest of the body is a solid mass of tissue.

Most of these bilaterally symmetrical worms are marine organisms.

There are under 1,000 species in this phylum, but they are worth considering because of the role their ancestors may have played in the evolution of animals. We will discuss this later.

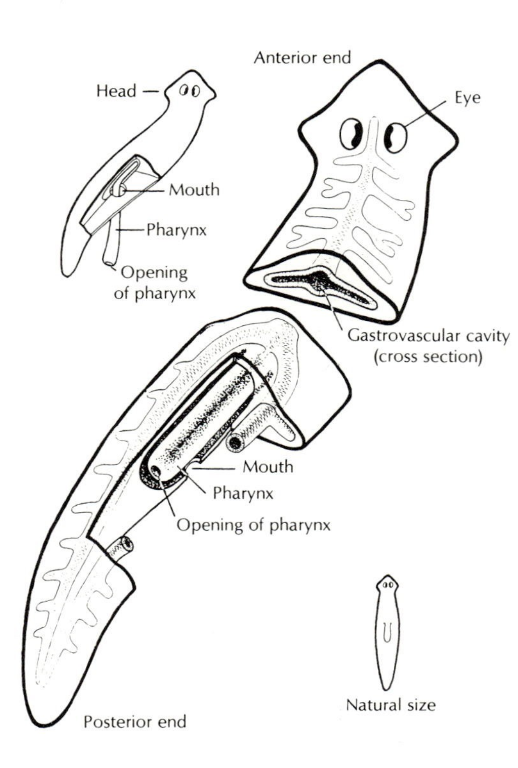

Fig. 1-12: Cutaway diagrams of internal structure in a planarian. The digestive system is seen as a branching gastrovascular cavity.

THE PLATYHELMINTHS

This phylum, the Platyhelminthes, contains bilaterally symmetrical worms, which are usually flattened in appearance, hence their common name flatworm. Flatworms, like the nemertines, have a simpler internal organization than the aschelminths; there are no internal body cavities except for the digestive cavity. And this cavity is so simply constructed that it has only one opening, which serves both as a mouth (ingestion) and as an anus (elimination). In planaria-type forms, the digestive cavity is highly branched (Fig. 1–12). The side of the worm next to the ground is ciliated. (Cilia are microscopic, hair-like structures that beat together.) In planaria the cilia are used in locomotion.

Planaria are rather large flatworms, and can be up to 1 centimeter in length. Other flatworms are much smaller, some being only a few hundred micrometers long, and are totally covered by cilia. Though planaria and many other flatworms are free living (some live on land, but most are aquatic, living in both fresh and salt water), others are symbiotic. For example, the tapeworm is a parasite that can be found in the human intestine.

Around 6,000 species of flatworms have been described.

THE CNIDARIANS OR COELENTERATES

The phylum Cnidaria has two common names, a result of once being grouped in the phylum Coelenterata. That taxon was eliminated when its two major subtaxa, the Cnidaria and Ctenophora, were described as separate phyla. We will have little occasion to refer again to the ctenophores, but we will return again to the cnidarians.

The cnidarians are radially symmetrical forms with two different but related body forms, the medusa and the polyp (Fig. 1–13). Ocean jellyfish (there is one freshwater species) are all medusae, the corals and sea anemones (the majority marine species) are all polyps. There are no body cavities except for the digestive cavity. Both body forms have tentacles with special cells that contain nematocysts. These structures can be suddenly extruded for defense or food capture. Many contain a poison whose effect on humans ranges from a mild irritation to a very painful sting.

About 10,000 species of cnidarians have been described.

THE SPONGES

Sponges belong to the phylum Porifera. More often than not, a

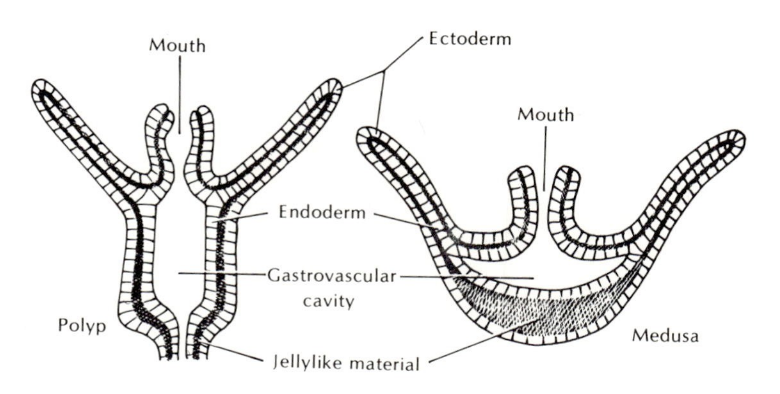

Fig. 1-13: Cutaway diagrams of polyp and medusa bodies.

sponge appears to be a blob, showing little evidence of being alive, but closer inspection reveals holes in the sponge body. And if one follows what is happening to the nearby water—all sponges are aquatic—it is seen to be circulating through the sponge. It goes in through the many smaller holes in the sponge body and comes out, through the few larger ones. Within the sponge, water enters a chamber in which flagella (structures like cilia, but longer) beat to circulate the water through the chamber, where food is ingested. These radially symmetric flagellated chambers (Fig. 1–14) and the channels

Fig. 1-14: Pattern of water flow through three types of sponges. A single sponge (A); a complex sponge (B); a very complex sponge (C), such as a bath sponge of the genus *Spongia*.

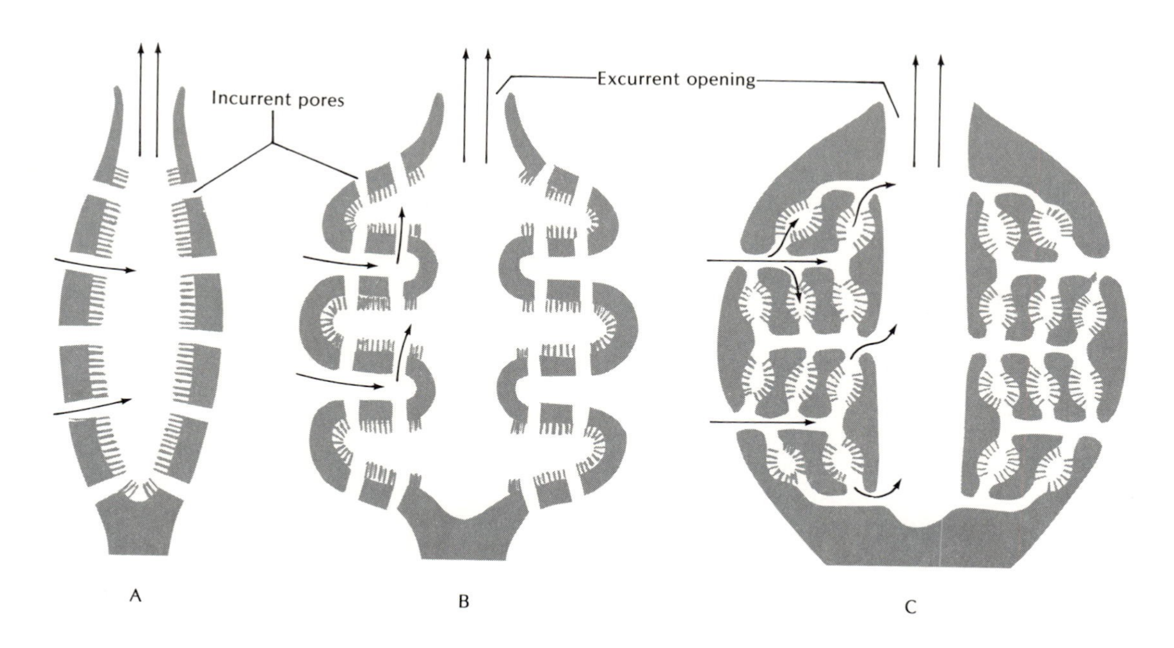

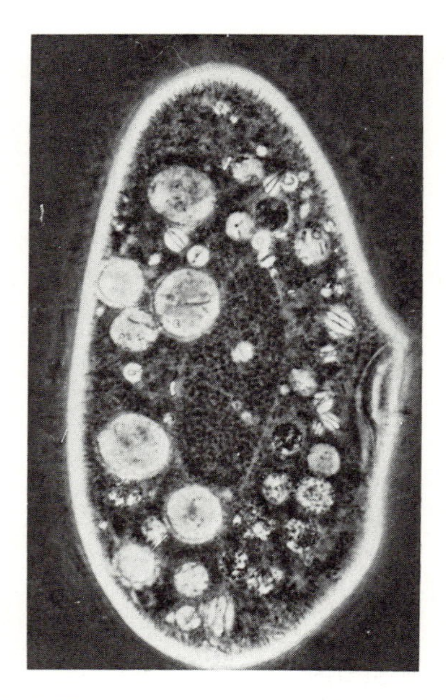

Fig. 1-15: A ciliated protozoan, here, a live *Paramecium aurelia.* Note the feeding structure (right) and the large, oval macronucleus (center). Vesicles and other cytoplasmic organelles are seen between the macronucleus and the outer cell membrane.

leading to them and from them are the only cavities in sponges.

The great majority of the 3,000 or so sponges described are marine.

THE PROTOZOANS

These unicellular animals are so diverse that few experts are satisfied with their placement in just one phylum, the Protozoa. Though each organism is a single cell, that cell can be large enough and complex enough to prey on certain multicellular animals (metazoans). (All animals except the protozoa are metazoans.) Other protozoa are tiny symbionts, which can be as small as 5 μm in length. Furthermore, protozoans can move about by cilia or flagella or variously shaped extensions of the cell body, the pseudopodia (Fig. 1–15).

Protozoa feed by taking food particles directly into their cells. A vacuole, formed by a single continuous membrane, which initially was part of the cell membrane (Fig. 1–16), forms around these particles. The vacuole is a food vacuole, the process phagocytosis.

There are about 30,000 described species of protozoans.

With the protozoans, we conclude our survey of phyla

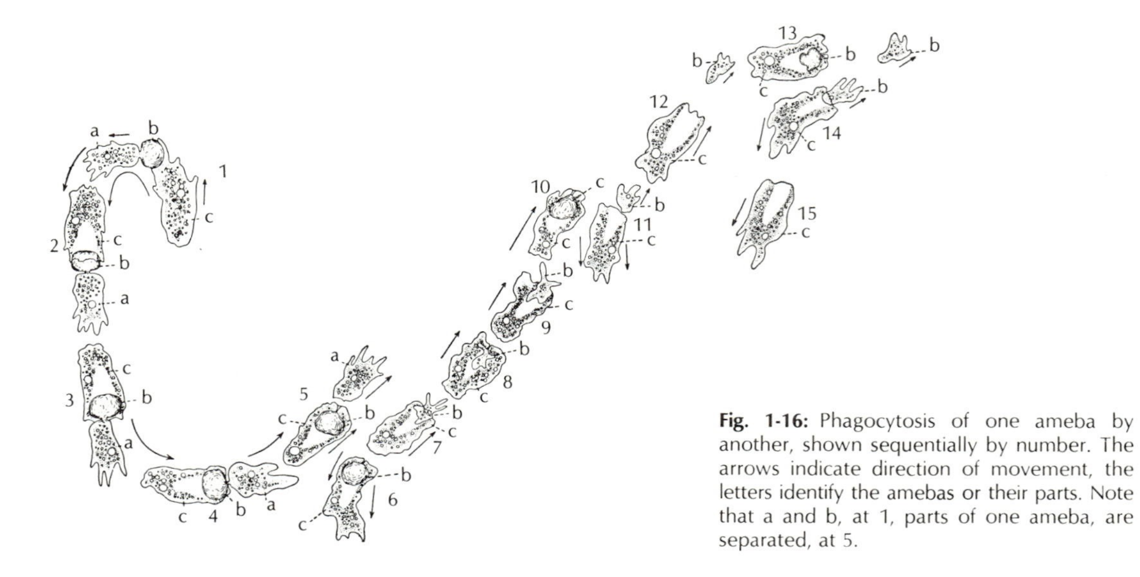

Fig. 1-16: Phagocytosis of one ameba by another, shown sequentially by number. The arrows indicate direction of movement, the letters identify the amebas or their parts. Note that a and b, at 1, parts of one ameba, are separated, at 5.

that characterize well over 80% of the known species of animals. The other known species lie in another six or more phyla. The actual number of phyla depends on the system of classification used. For example, as discussed, the protozoans, aschelminths, and arthropods are considered by many experts to contain several phyla apiece. (See Appendix I.)

The major taxa: Plants

We will next examine the major plant phyla, or divisions, as they are also called. In addition we will also look at fungi and bacteria. These latter groups of organisms assimilate or absorb their food as molecules taken largely from the breakdown of plant or animal materials around them. They are decomposers.

THE TRACHEOPHYTES

The phylum, or division, Tracheophyta contains the plants best known to us. These include trees, shrubs, herbaceous plants, grasses, and ferns. The term *tracheophyte* means plants with tubes. The tubes referred to are special cells that help transport substances from one part of the plant to another. This means of transport is also called a vascular system, and hence, the tracheophytes are also called vascular plants. Like the arthropods, aschelminths, and protozoans in the animal kingdom, this group has been subdivided into separate phyla. We will retain the grouping as a diverse phylum, but describe its members in terms of certain subgroups.

The angiosperms are the flowering plants. This subgroup has the greatest number of species: over 250,000 have been described. Flowers can be viewed as specialized structures derived from leaves and containing reproductive organs within which seeds develop. This view of flowers is accepted by most botanists. But there are many different opinions as to just how leaves evolved into flowers.

The gymnosperms include the vascular plants we often call evergreens. Their reproductive structures are found in cones. Within the female cones seeds are formed; the male cones produce pollen. In contrast, most flowers produce both pollen and the female parts from which seeds are formed (Fig. 1–17).

Both the angiosperms and gymnosperms are seed plants. Other tracheophytes lack true seeds.

In many ferns, for example, the spores lie in special structures on the underside of the leaves or fronds. These spores develop into a tiny gametophyte, which produces eggs and

Fig. 1-17: Reproductive structures of certain tracheophytes. A. The flower of this mountain laurel (*Kalmia latifolia*) has all its petals fused. The style (center) extends outward, terminating in a sticky tip (stigma). Ten filaments radiate out from the base of the style, each tipped with a dark anther that will release pollen upon maturation. Note the unopened flower buds, top center. B. A pollen-bearing pine cone is seen at the base of a larger female cone. The airborne pollen lands on the female cone, where it falls into the spaces between the scales. At the base of each scale lies an immature seed. Upon fertilization, a seed develops into a pine nut.

Fig. 1-18: A fern sporophyte with two types of fronds. The central, taller darker fronds are spore-bearers; the outer fronds, which will grow taller, carry out photosynthesis for the entire plant. Such separation of function is not typical of ferns, although it occurs in this cinnamon fern.

Fig. 1-19: Moss: The typical, mossy areas are gametophytes, the capsule-like structures on stalks, the sporophytes.

sperm. When these unite, the fertilized egg develops into a typical fern plant or sporophyte (Fig. 1–18). This type of life cycle, with its separate sporophytic and gametophytic plants, is absent in the gymnosperms and angiosperms. However, as we will see later, there is a remnant of it as seen in the way pollen and egg develop in cones and in flowers.

THE BRYOPHYTES

The phylum Bryophyta contains the mosses and another, less-known group, the liverworts. These plants do not have vascular tissue. They are always small—never over a few centimeters (or inches) high—and occur in wet or very moist areas.

The gametophytes of these plants grow out of the ground and rhizoids (root-like structures) are used to obtain water and nutrients from the soil. The gametophyte carries either male or female reproductive cells. Sperm from the male plants fertilize eggs in the female plants and from this the sporophyte develops (Fig. 1–19). The sporophyte therefore depends on the gametophyte for its survival. This is the opposite of what occurs in the tracheophytes.

The spores of the sporophyte, when released, develop into a new male or female gametophyte.

There are about 9,000 known species of liverworts and 14,000 of mosses.

THE ALGAE

This group consists of several phyla or divisions. The Phaeophyta (or as they are also called, the Phaeophycophyta) include the brown algae; the Rhodophyta (or Rhodophycophyta) are the red algae; and the Chlorophyta (or Chlorophycophyta) includes the green algae. There are other algal phyla, too.

The three phyla just mentioned all contain multicellular forms. The brown algae contain the kelps; the largest ones, from the Pacific coast of North America, can be up to 70 m in length. The red algal seaweeds are medium to small in size, and usually measured in centimeters. There are few freshwater forms in the brown and red algae. They are the dominant seaweeds.

The green algae occur in marine environments, but are most common in freshwater. A few terrestrial species are found on rocks or tree trunks. These algae show a great variety of forms, being unicellular and organized into a variety of colonial forms. In some of the latter, the members of the colonies are so highly integrated that the colony is recognized by many biologists as a kind of superorganism or, simply, as a multicellular organism. No unicellular organisms are known among the brown algae, but their gametes are unicellular. The red algae have only a few unicellular species. More intriguing is the fact that the red algae have no flagella, even on their male gametes.

Photosynthetic pigments determine the colors of these algae. These pigments, as we shall see, offer important clues to the evolutionary histories of these phyla. For example, the green algae have the same photosynthetic pigments as the bryophytes and tracheophytes, and the red algae have some of the same pigments as the blue-green algae. The blue-green algae, as we shall see shortly, belong in a very different group.

In the three phyla of the brown, red, and green algae there are about 16,000 species. The multicellular ones often attach to the bottom of the sea or other bodies of water by special structures called hold-fasts. The depth of attachment is limited by the sunlight available for photosynthesis. The rest of the plant extends passively in the water. The unicellular forms and a few of the colonial forms are free swimming.

THE FUNGI

In this group we again find a diversity that justifies classification into several phyla. Two of the most common fungi, the Basidiomycota and Ascomycota, contain many familiar toadstools, mushrooms, and molds. These are not photosynthetic organisms, although there is good reason to think that they evolved from plants. This is another case of the loss of a major characteristic.

The fungi contain some unicellular forms, like yeast, and many multicellular ones, like mushrooms. Details of the structure in the latter show them to be made of many fine filaments, made up of cells joined end-to-end. The true mushroom and toadstool parts are actually special reproductive structures. They arise from a mat of filaments (a mycelium), which is continually growing and feeding (Fig. 1–20).

Fig. 1-20: The mycelium, the white, thread-like structure below the mushroom cap (to which a leaf has stuck), of this fungus has been revealed by careful digging. It penetrates the earth and at points sends up its spore-forming structures, what we know as mushrooms.

Fungi are decomposers. They feed by taking into the mycelial filaments material obtained by breaking down plants and animals to their constituent molecules.

There are over 45,000 species of different fungi. They occur just about everywhere, being aquatic, terrestrial, and symbiotic.

THE BACTERIA AND BLUE-GREEN ALGAE

These two groups are considered together because their cellular structure differs from that found in all other living things.

All the groups mentioned so far are eukaryotic, whereas bacteria and blue-green algae are prokaryotic. Some of the differences between these two types of cells are shown in Table 1–1.

The blue-green algae and certain bacteria carry out photosynthesis and are, therefore, producers. Most bacteria, however, are decomposers, like the fungi. But such similarities are of limited value, as we will find out, when we try to understand how eukaryotic cells evolved.

When we inquire into the origin of prokaryotic cells we come up against one of the great questions in biology, that is, How did life originate? Except for the viruses, these cells are the simplest living forms, and this is especially puzzling when we think of the prokaryotes as producers and decomposers. Prokaryotic producers use sunlight (energy) and simple molecules (carbon dioxide and water) to build more of themselves. Is that how life started? To build more of oneself from large preformed molecules, as the decomposers do, is much easier. But where did the preformed molecules come from? There are answers to these questions, but no one answer is completely satisfying.

There are a few thousand known species of prokaryotes. However, because of their small size and their worldwide occurrence, their number in terms of individual cells is enormous.

The last organisms we will mention here, the *viruses*, are the simplest organisms we know, in terms of structure and function. They are not thought to be of much value in explaining the origin of life, and they are not usefully classified as belonging to any taxon because they are symbiotic. They live inside bacteria, fungi, plants, and animals. Our present view of their evolutionary history is that they are highly modified genes escaped from their original hosts. Hence, they evolved after their hosts and cannot represent the first living things. And being descended from many different hosts, they are best classified with them rather than being arbitrarily collected into one taxon. New viruses are being described all the time. Theoretically, it seems there can be as many different viruses as there are species of organisms.

Even though the foregoing survey encompasses a large majority of the known species of organisms into seventeen groups—largely phyla or divisions—there is still great diversity among those seventeen groups, from viruses to redwood trees and blue whales, from organisms living deep in the ocean to those in rain forests and deserts. This diversity includes many kinds of symbionts. It hints at the stability of key characteristics, such as photosynthetic pigments, as one looks at different groups of plants, but loss of those pigments seems possible when one considers the fungi and protozoa.

One of the easiest ways to demonstrate a sense of order in this diversity is to look for a moment at how organisms relate to each other. This brings us to ecological considerations.

Kingdoms as ecological groupings

Ecosystems are groups of interacting organisms and the environment in which they live. Ecosystems are not entirely independent of each other. For example, a pond ecosystem will receive matter from the forest or prairie that may surround it. The ecosystem of a saltwater marsh is connected to that of the nearby sea and to that of the land on its other margin. Nonetheless, ecologists recognize that a more intimate interaction occurs among the living things in a pond than between them and neighboring forest or prairie dwellers. And certainly marsh dwellers interact among themselves more than they do with organisms living in the sea or on the land.

What are these interactions? They are most easily seen by pointing out that any ecosystem has four major components. First, there is the nonliving or abiotic component. This includes water, air, and various inorganic and organic compounds that make up the soil or are suspended or dissolved in air and water. It also includes temperature and light. The other three components are living or biotic ones. They are the organisms that are producers, consumers, or decomposers. We have already briefly described these forms in the seventeen groups surveyed. Here we are talking about their interrelationships.

Producers are photosynthetic forms; they are plants. They utilize the abiotic components of the ecosystem. Plants, by and large, take carbon dioxide, water, and certain elements from the environment and, obtaining energy from sunlight, maintain and reproduce themselves.

Consumers ingest their food; they are animals. They eat the plants in their ecosystem, if they are herbivores, or other animals, if they are carnivores. The animals that eat both are omnivores.

Decomposers assimilate or absorb their food directly. Fungi and certain bacteria make up this final biotic component of an ecosystem. They breakdown or decompose the bodies of plants and animals to nutrients they can absorb.

In any ecosystem, then, producers, or plants, depend on nonliving material and energy sources in that environment. Consumers, or animals, depend on plants and each other. And decomposers depend on both plants and animals. The products of all this biological activity are, in the long run, returned to the nonliving part of the ecosystem where they are reused by the producers.

In addition to these important ecological groupings there are certain other key characteristics found among our seventeen groups. Many forms are multicellular, and some are unicellular. Many are eukaryotic, and some are prokaryotic. If we consider these characteristics with ecological characteristics, we can place our seventeen groups into five large groups called kingdoms—the highest taxon (Table 1–4). This classification is seen in Fig. 1–21. Multicellular producers are all placed in the kingdom Metaphyta; multicellular consumers, in the Metazoa; and multicellular decomposers, in the Fungi. All the unicellular eukaryotes are placed in the kingdom Protista.

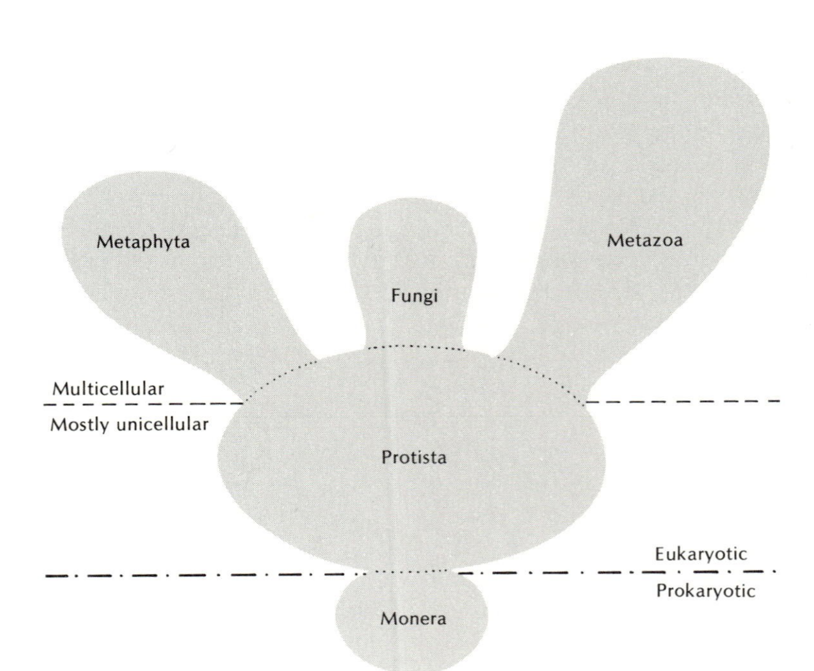

Fig. 1-21: The five-kingdom scheme of classification.

This includes certain algae, which are unicellular producers (multicellular algae are in the Metaphyta). These unicellular algae are sometimes designated as protophyta. The Protista also contains unicellular consumers in the form of protozoa. This leaves as members of the fifth kingdom—the Monera—the prokaryotic forms that include bacteria (unicellular decomposers for the most part) and photosynthetic blue-green algae.

The forms we call plants and animals were placed by Linnaeus in just two kingdoms, the Plantae and the Animalia. His kingdom Plantae included the Metaphyta, the Fungi, the protophyta from the Protista, and the Monera. His Animalia included the Metazoa and the protozoa from the Protista. Today we find it more accurate and useful to use the term plants as a common name for all producers and apparently related forms, regardless of the kingdom to which they belong. When the term is used this loosely, the fungi, bacteria, and blue-green algae are also sometimes covered by it; in the same way, the term animals covers the metazoans and protozoans. Thus the terms plants and animals are useful as general terms, but are confusing when considered as taxonomic groupings.

Look again at Fig. 1–21 and recall that, without exception today, all organisms come from pre-existing ones. What, then, is the origin of the first organisms, i.e., What is the origin of life? There are other questions. How did the diversity we see today come about? In what ways do today's organisms resemble and in what ways do they differ from the first organisms? How did producers, consumers, and decomposers differentiate? Why are there both unicellular and multicellular organisms? Why are there both prokaryotes and eukaryotes? These are among the most profound questions in all of biology.

These are among the questions we shall address in this book. We want to know what successes biologists have had in trying to understand the variety of life around us. Can we explain organismic diversity? In this chapter we have briefly described that diversity. In the following chapters we shall try to explain how it came about.

References

Blackwelder, R. E., 1967. *Taxonomy: A Text and Reference Book.* John Wiley and Sons, New York.

Bold, H., 1973. *The Morphology of Plants.* 3rd ed. Harper & Row, New York.

Buchsbaum, R., 1948. *Animals without Backbones.* University of Chicago Press, Chicago.

Fraenkel-Conrat, H., 1962. *Design and Function at the Threshold of Life: The Viruses.* Academic Press, New York.

Grassé, P. P. (ed.), 1952–1977. *Traité de Zoologie.* Masson et Cie, Paris.

Macino, G. et al. 1979. *Proceedings of the National Academy, USA. 76:* 3784–85.

Mayr, E., 1969. *Principles of Systematic Zoology.* McGraw-Hill, New York.

Morowitz, H., 1978. *Foundations of Bioenergetics.* Academic Press, New York.

Whittaker, R. H., 1963. New concepts of kingdoms of organisms. *Science 163:* 150–160.

Wolfe, S. L., 1972. *Biology of the Cell.* Wadsworth, Belmont, Calif.

TWO

Explaining Organismic Diversity

Because the diversity of living things is so great, the ambition to explain it might seem foolishly ambitious. But that really depends on how we seek our explanation. Sticking to questions that science can answer, we find ourselves simply asking: What causes the diversity of the living things we see around us? And how we answer such questions in science can be outlined quite simply. But the actual scientific inquiry is a mixture of excitement and tedium, of many frustrations and deeply satisfying moments, of intuitive leaps and rational analysis. Finally, there is a presentation of the essential knowledge obtained from all our work.

Darwin's work on evolution typifies this view of scientific inquiry. It can be clearly outlined, but it contained its share of ups and downs, its moments of discouragement and of insight. The book *The Origin of Species*, published in 1859, was but a summary, as Darwin referred to it, of all he had analyzed during his study. Nevertheless, Darwin's work followed the classical outlines of problem-solving in science. Initially, he defined the problem of organismic diversity. Then he looked for and found a solution for it. And, finally he carefully tested that solution. Darwin was not the first to try to explain organismic diversity (Box 2–1), but he was the first to offer an explanation that has lasted.

Box 2–1. Darwin's predecessors.

In the Western World, with its intellectual traditions derived in large measure from the Greeks, there has always been great interest in the variety of living forms. These include attempts to describe and classify that variety as well as attempts to explain it. Among those scientists close to Darwin's time, three Frenchmen stand out as students of organismic diversity.

Georges Louis Leclerc (1701–1788), the Comte de Buffon, known more simply as Buffon, was a naturalist trained originally as a physicist. He held an influential position as director of the Museum of Natural History in Paris and he published one of the great encyclopedias on natural history. It was vividly written and grossly inaccurate in places, but it was infused with the idea that there is unity in nature. Perhaps because of his education as a physicist, Buffon looked for unifying laws, and especially causal laws, that could explain nature. He did not find any, but he lead many people to think of nature as dynamic and even changeable. This helped scientists of his day break away from the Greek tradition of a static and fixed world.

Georges Cuvier (1769–1832) was almost the opposite of Buffon in scientific outlook. He was an inspector general for education under Napoleon and a professor of anatomy at Buffon's Museum of Natural History. A careful observer, his imaginative powers lead him to the theory of catastrophism as an explanation of the sequence of different forms over time as observed in the fossil record. He stayed with the biblical notion of divine creation and argued that the world had been visited by successive catastrophes, after which obscure species became widespread and so appeared in the fossil record. There was no real change.

Jean Baptiste de Lamarck (1744–1829) was also a professor at the Museum of Natural History, and a contemporary there of Cuvier. He was strongly influenced by Buffon's encyclopedia, but due to the chaos caused by the French Revolution, he did not start his major studies on invertebrate zoology until late in life. One of

his most important works, *La Philosophie Zoologique,* was published in 1809, the year of Darwin's birth. Lamarck was an evolutionist in the sense that he believed in change, but his theory was very different from Darwin's natural selection. Lamarck took a romantic view of nature and believed living things contained within them a perfecting principle. By exercising their drive to improve their abilities to live in their environments, they acquired certain habits. These habits resulted in bodily changes, which were then passed on to their progeny. There is, however, no biological evidence supporting this theory of the inheritance of acquired characters.

No other scientists really anticipated Darwin's theory of natural selection, though some did believe that organisms could change.

Patterns of diversity and natural selection

An outline of scientific inquiry sketches out the major features of such inquiry, but it is not a How-To-Do-It set of directions. Because good inquiry is an art, a step-by-step description of scientific problem-solving is not an instructional manual. Problem-solving requires experience, disciplined skill, and intuitive insights. The products of such efforts are typically laid out as no-nonsense, objective facts and the logically rigorous, plausible ideas arising from them.

Our look at the steps of inquiry has, therefore, the purpose of examining the nature of the evidence Darwin used to explain organismic diversity. In attempting to recreate the structure of his inquiry, we can see where some of his ideas came from and so evaluate them more clearly. Additionally, we can see better what work remains to be done—something Darwin continually emphasized. For example, in his chapter on laws of variation, Darwin said, "Our ignorance of the laws of variation is profound. Not in one case out of a hundred can we pretend to assign any reason why this or that part differs, more or less, from the same part in the parents." Much of that ignorance has been dispelled today, but we need to see how it has been dispelled and how our current ideas affect ongoing studies of natural variation in populations.

THE PROBLEM

A natural phenomenon may be first observed as a regularity or a pattern. That then raises the questions, "Why should that regularity exist? or How can I explain such a pattern?" Darwin was quite explicit in formulating his problem, and his formulation was much more precise than the general one we stated above (What causes organismic diversity?). Darwin noted that certain structures in closely related species occurred regularly. This came from his observations as a naturalist on the *H.M.S. Beagle*. He wrote

> During the voyage of the *Beagle*, I had been deeply impressed discovering in the Pampean formation great fossil animals covered with armour like that on existing armadillos; secondly, by the manner in which closely allied animals replace one another in proceeding southwards and over the [South American] Continent; and thirdly, by the South American character of most of the productions of the Galapagos Archipelago, more especially by the manner in which they differ slightly on each

> island of the group, none of the islands appearing to be very ancient in a geological sense.
>
> It was evident that such facts as these, as well as many others, could only be explained on the supposition that species gradually became modified; and the subject haunted me. (C. Darwin, *Autobiography*, p. 57.)

Darwin observed similar patterns in the forms of fossil and living armadillos (Fig. 2–1) and in animals living on the South American continent and on the Galapagos Islands. He was, as he said, intrigued by the fact that many of these forms differed from one another, to a degree. Specifically, fossil and live armadillos are recognizably armadillos, but they differ in certain ways. The forms found in northern South America are like those further south, but they also differ. And those on the Galapagos Islands were similar to the ones he saw on the South American mainland, but they too differed. And those on one island differed from those he observed on other islands of the Galapagos Archipelago (Fig. 2–2).

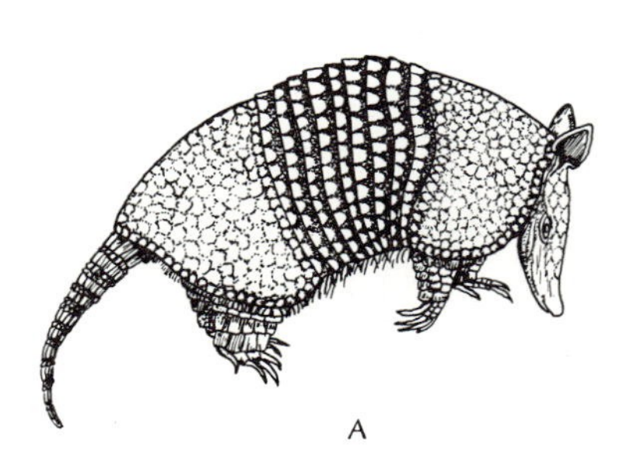

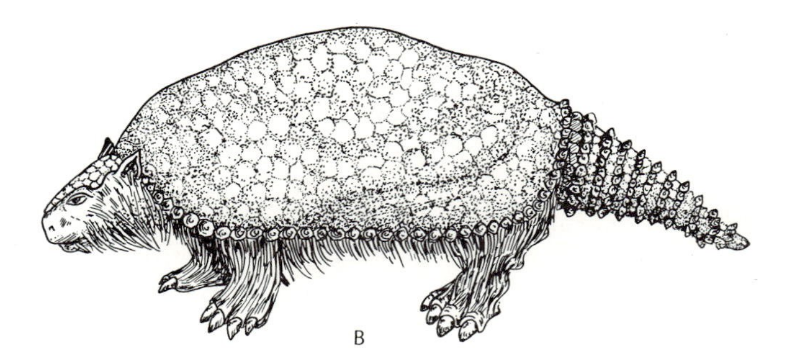

Fig. 2-1: Examples of living (A) and fossil (B) armadillos.

At this point we should remind ourselves that Darwin's voyage on the *Beagle* occurred from 1831–1836 (Box 2–2). Furthermore, he accepted Linnaeus's view of taxonomy, that species were described in terms of a single type, whereby variations in the members of the species were ignored. And finally, Darwin was brought up believing in a Christian view of creation. He went to Cambridge University to become a clergyman before he embarked on the *Beagle*. Darwin was able to say of himself at that time ". . . I did not then in the least doubt the strict and literal truth of every word of the Bible, I soon persuaded myself that our Creed must fully be accepted." In other words, Darwin was fully convinced that species were divinely created as separate entities. Hence, when he felt

Fig. 2-2: Darwin's finches. Eight of the fourteen paired (male and female) species, all from the Galapagos Islands, are essentially bush- and tree-dwellers; the other six are ground-dwellers. The sizes and shapes of their bills indicate their food preferences, which range from insects to fruits, nuts, and seeds.

forced to the opposite view ". . . that species gradually became modified . . . ," it was a revolution in his own way of thinking. No wonder, then, in a letter to a colleague in 1844, he wrote, "I have read heaps of agricultural books and have never ceased collecting facts. At last gleams of light have come, and I am almost convinced (quite contrary to the opinion I started with) that species are not (it is like confessing to a murder) immutable."

THE EXPLANATION

However, it is one thing to suppose that species can change and another thing to explain how that change comes about. Darwin had to move from his statement of the problem,

Box 2–2. The voyage of the *Beagle* and the Galapagos Islands.

By Darwin's own admission, the trip on the *Beagle* was of singular importance to him. In the *Autobiography* he writes,

> The voyage of the *Beagle* has been by far the important event in my life, and has determined my whole career. . . . I have always felt that I owe to the voyage the first real training on education of my mind. . . .

Charles Darwin at age 40

And then half-humorously he also remarked,

> That my mind became developed through my pursuit during the voyage is rendered probable by a remark made by my father, who was the most acute observer whom I ever saw, of a skeptical disposition, and far from being a believer in phrenology; for on first seeing me after the voyage, he turned round to my sisters and exclaimed, "Why, the shape of his head is quite altered."

The route of the voyage is shown on the accompanying map.

The most famous stop was at the Galapagos Islands, an archipelago lying about six hundred miles off the west coast of South America. Here, on different islands, he found the various species (see Fig. 2-2, for example) that led him to the idea that species could change. At the time of his visit he wrote in one of his notebooks the following significant lines

> . . . when I see these Islands in sight of each other and possessed of but a scanty stock of animals, tenanted by these birds but slightly different in structure and filling the same place in nature, I must suspect they are only varieties. . . . If there is the slightest foundation for these remarks, the Zoology of archipelagoes will be well worth examining for such facts would undermine the stability of species.

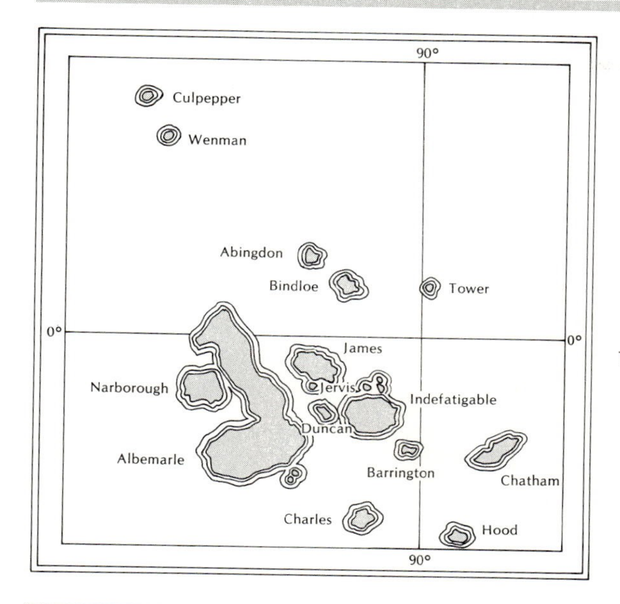

The Galapagos Islands

The route of the *Beagle*

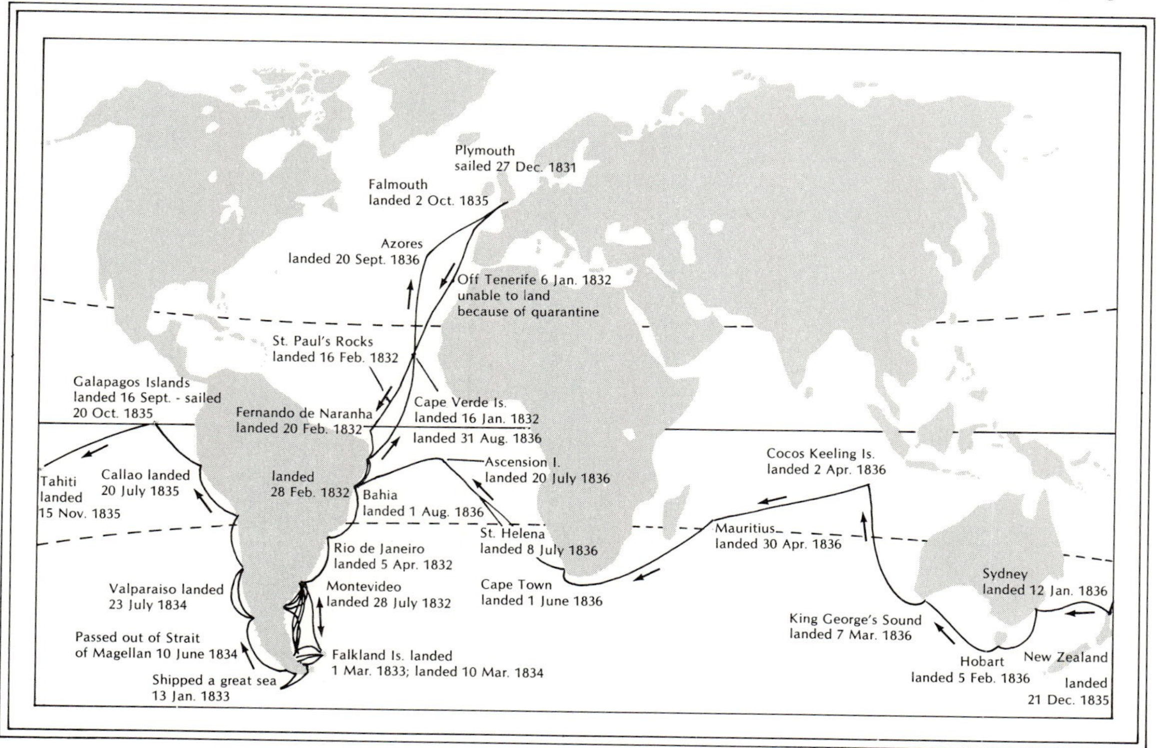

namely, the pattern of differences among related species must mean they are mutable, to an explanation of that mutability.

His explanation was the *theory of natural selection.*

What Darwin meant by natural selection can be expressed and illustrated the following way:

1. All species may reproduce to the extent that overpopulation results. Many observations from our everyday experience support this. Think of the number of eggs a hen can lay in her lifetime. Look at the number of seeds a single plant produces year after year. What is happening to the human population right now? Perhaps the most dramatic example of reproductive potential is seen in bacteria. Some prokaryotes can reproduce once every twenty minutes. If they were able to reproduce at that rate for 48 hours, there would be approximately 2^{144} or 2.2×10^{43} bacteria whose total mass would be 2.2×10^{31} grams. (Each cell weighs about 10^{-12} grams.) That is about 24×10^{24} tons or 4,000 times the weight of the earth.

2. Despite the potential for overproduction of their kind, species remain rather constant in population size over a period of time. We are not overrun by chickens or oak trees or bacteria. We are, however, running into severe problems because of the increase in the human population. That represents a special problem to which we will return later. In general, we observe that for the vast majority of species, their population sizes remain relatively constant.

3. From these two sets of observations one concludes there is a "struggle for life," resulting in a "survival of the fittest," as Darwin described it. From among all those organisms born, only a few survive to perpetuate the species. The question immediately arises: Which ones survive? Is it by chance or is there something else? Darwin answers those questions, but we must first make one more observation.

4. Heritable variations occur. Darwin did not know how these variations occurred, since he had never heard of Mendel and his ideas on heredity at that time. (Mendel published his study on hereditary factors in 1865 and Darwin died in 1886, but Mendel's work was not recognized until 1900, and so it never helped Darwin.) Nonetheless, it was known that variations occur and that at least some of them could be inherited. Darwin collected extensive evidence on this point from plant and animal breeders of his day.

5. Heritable variations that benefit the organism in the struggle for survival will be preserved. This is Darwin's theory

of *natural selection*. Organisms that win the struggle for survival do not win by chance; they are selected. They have some feature that gives them an advantage, however slight, in exploiting their environment to maintain and reproduce themselves and at least some of their variations.

So viewed it can be seen that these selected forms will tend to produce more progeny than those in the population not having the advantage or advantages in question. This means differential reproduction will occur. In this way, members of the species will change slightly over time as more organisms carry the new beneficial or selectively advantageous character. And the process can occur again and again. In this way changes accumulate and give rise to the differences in anatomy and distribution of organisms observed.

Thus natural selection leads to organismic diversity.

TESTING

Now comes the third step in scientific inquiry—testing the explanation.

Ideally, testing involves two steps. First, the explanation, or theory, is used to predict observable results. Then, observations are made, often by performing experiments, to see if what is predicted actually occurs. If the prediction is fulfilled, the explanation is supported. If the prediction is not fulfilled, then the explanation is clearly placed in doubt or rejected.

Prediction using the theory of natural selection, as formulated by Darwin, was and still is very difficult, since the most obvious experiments necessitate long stretches of time so that changes can accumulate and generate species differences. Darwin realized this and took a different approach. Sensing that natural selection is going on around us, all the time, Darwin wanted to intercept the process as it occurs naturally. He wanted to be able to say that, if natural selection occurs, we should expect to see those phenomena that result from natural selection. Chapter IV of *The Origin of Species* is devoted to natural selection, and especially in the latter half of that chapter we find the kinds of things Darwin was looking for, as well as the problems he had finding them. But no one was more aware of these problems than Darwin himself. In Chapters VI, VII, and VIII, he brings together these problems and tries to solve them. In effect, Darwin tests his theory. What conclusions can we draw regarding such tests?

Darwin discusses six major tests of natural selection. If

natural selection occurs then we should expect to find:

1. Some forms lose in the struggle for survival and become extinct.
2. In small areas organisms will diverge as a result of the struggle for survival going on among them.
3. Transitional forms, which are on the way to forming distinct species, will appear.
4. New physical characters will emerge, including very complex ones, such as complex eyes.
5. New behaviors or instincts, as Darwin called them, will also appear.
6. Variations within species will be seen to be of the same sort as those that give rise to species themselves.

These six points are expressed here somewhat differently than Darwin put them. He actually saw the first two as tests of his theory. The last four he spoke of as "Difficulties on Theory." These latter were criticisms sure to be directed against his theory by others. For example: Are there transitional forms? Can new characters arise through natural selection? Can "instincts" be a product of natural selection? Why are hybrids within a species fertile, but hybrids between species often sterile? Darwin, wisely, anticipated his critics by raising these questions. His answers to them, however, turned out to strengthen the theory and in that sense were a test of natural selection.

A careful look at this way of going about testing a theory raises certain questions, so let us look critically at each one of these "tests." Certainly the first two tests represent facts known to Darwin and his contemporaries (Box 2–3) before the *Beagle* voyage. He was an enthusiastic amateur geologist, and he was aware of the fossil record being compiled by geologists of his day. That record included the fact that certain organisms that once existed on this earth were no longer around. Such facts describe but do not explain extinction. Also, as an enthusiastic naturalist, Darwin was very familiar with the diversity of life that can be found on limited plots of earth. Again, this was descriptive information, unless divine creation were to be accepted as an explanation. (Initially, as we have seen, Darwin did accept this.)

The important point regarding testing is this: Darwin was aware of these facts *before* he advanced the theory of natural selection. He could just as well have cited them as observations leading to the idea that species are not immutable, along with his observations from the voyage of the *Beagle*. Strictly

Box 2–3. Darwin's contemporaries.

It is worth noting that Darwin (1809–1882) depended on the help of his fellow scientists. Though chronically ill (more recently, it has been suggested that this illness was caused by a parasite picked up during the voyage of the *Beagle*), he stayed close to his country house, but nonetheless interacted vigorously with others in the scientific community.

Sir Charles Lyell (1797–1875) was the acknowledged leader in British geology. His theory of uniformitarianism had greatly influenced Darwin's willingness to accept change in nature. In Lyell's view, uniformitarianism referred to the historical continuity of such observable, ongoing geological changes, as erosion, deposition of sediments, and weathering of rocks. In his three-volume classic *Principles of Geology,* Lyell established, once and for all, that existing forces in nature, given enough time, change the nature of our physical environment. Darwin took the first volume with him on the *Beagle* and received the second volume by mail, in Montevideo. The sense of enormous time in geological history and of continuous change provided Darwin with a perspective he never lost, a perspective that ripened his ideas of biological change over time and in the face of a continually changing environment.

Sir Charles Lyell

Sir Joseph Hooker (1814–1879) was a botanist of wide learning. For years he was the only scientist with whom Darwin shared his ideas on natural selection [". . . that species are not (it is like confessing to a murder) immutable," was in a letter to Hooker.] Hooker encouraged Darwin and was responsible for Darwin's first publication of his ideas in 1858. In that year, another naturalist, Alfred Russell Wallace (1823–1913), submitted a paper to the Linnaean Society in London, putting forth ideas on natural selection that were essentially the same as Darwin's. Hooker, the President of that Society, prevailed on Darwin to present a summary of his views concurrently with the reading of Wallace's paper. This finally brought natural selection to public view. Wallace, the younger of the two men, recognized Darwin's superior grasp of the issues and a mutually supportive and friendly relation persisted between both men until Darwin's death.

Thomas Henry Huxley (1825–1895) has been called Darwin's bulldog. Where Darwin was retiring, soft-spoken, and given to careful pondering of his own and his opponents views, Huxley was vigorous and unabashed, with a diamond bright intellect, and thought quickly on his feet. He is reported to have said, when he first read Darwin's book on natural selection, "How extremely stupid not to have thought of that!" Huxley, too, was a capable and experienced student of natural history. In the subsequent public debates on the question of evolution by natural selection, it was Huxley who carried the day for the Darwinian point of view.

Thomas Henry Huxley

speaking, we expect tests to be predictions of *new* observations—of observations yet to be made. They are usually not hindsight explanations of *old* observations. But that is how Darwin generated these two "tests."

What about the four remaining points? The story is somewhat the same for three of them. Though Darwin and other biologists before him had seen varying degrees of difference (or similarities) between species, and had noticed different degrees of complexity in structures and in behaviors considered to be instinctual, it was the idea of natural selection that made them understandable. Before natural selection, these problems were the curiosities and the puzzles of natural history. Again, therefore, Darwin explained by hindsight rather than foresight or prediction.

Many people feel that explanation by hindsight is a major limitation on Darwin's theory. It really is not a predictive theory, such as we often associate with science. We shall return to this point after discussing the final test of natural selection, which Darwin referred to as "hybridism."

One incisive criticism of Darwin's viewpoint on hybrids appeared in 1867. William North Rice, then a graduate student at Yale University and subsequently Professor of Geology and Natural History at Wesleyan University, wrote

> The most decisive argument against the doctrine of Transmutation [mutability of species], is that which is drawn from the phenomena of hybridism. It appears to be a general law that, within the same species, a union between individuals, as diverse as possible, is most favorable to fertility. . . . When different species are crossed, the result is directly the contrary. In most cases the result is no issue [progeny] whatever. . . . Here, then, we have the law that divergence of character, within the limits of a specific type [species], tends to increase fertility, but beyond this limit tends as surely to diminish fertility or to produce absolute sterility.
>
> In the face of these facts, is it not absurd to claim that varieties and species differ only in degree? Are we not forced to the conclusion that there is between them a radical difference in nature and origin? [From William North Rice "Darwinian Theory of the Origin of Species" *New Englander* (1867). Reprinted in *Darwinism Comes to America* ed. George Daniels, Blaisdell Publishing Company, Waltham, Mass. pp. 55–56.]

Darwin had of course, claimed that varieties and species differ only in degree. In *The Origin of Species* he says

> Nevertheless, according to my view, varieties are species in

> the process of formation, or are, as I have called them, incipient species. (Chapter IV.)
>
> . . . there is no fundamental distinction between species and varieties. (Chapter VIII.)

On this point, Darwin was both right and wrong. The variations within a species are, indeed, the kind of differences out of which species are formed. But, as Rice insisted, when we come to species, organisms are unable to mate effectively with each other, and, therefore, differences between species cannot be altogether like those within a species. In a later chapter of this book the nature of this puzzle will be made explicit. At present we need to note simply that Darwin's argument for natural selection as the basis of hybridism is not what he thought it to be. This test does not support his theory the way he thought it did.

Overall, this appears to leave us with an uncomfortable view of testing Darwinian natural selection. Can this widely respected theory explain only by hindsight? A straightforward answer is not yet possible. The chief problem is that natural selection is more complex than even Darwin thought it to be. This will become clearer when we examine the relation between natural selection and evolution.

Evolution and natural selection

Evolution is the product of natural selection. Too often the two ideas are confused and treated as if they were the same. When natural selection acts for a long time on a population there may be accumulated a significant series of changes in the members of a population. Those changes, collectively called evolution, are caused by natural selection. A cause and its effects are never the same thing.

In fact, some changes are independent of selection. Such changes, for example, those called genetic drift, are also considered to be part of evolution. But for the moment let us consider only the changes resulting from natural selection.

Without exception, in his six tests of natural selection Darwin was looking for evolutionary changes—the long-term effects of selection. He was right in saying the changes should be observable. But as we have seen, he was looking at phenomena that had already occurred rather than at phenomena that were occurring as the result of an experiment. Furthermore, we now know today that such changes depend on the interactions of several factors in addition to selection. They include mutation, migration of subpopulations or other populations into and out of a population, the size of a population, and breeding patterns. Also, these interactions are played out on the stage provided by the environment—itself complex and often unpredictable.

It is no wonder then that Darwin did not specify new phenomena as tests for natural selection. Looking at specific, already familiar problems was really the only alternative he had and he used it well. However, that must also mean that the burden of the testing is now fully on our shoulders. If we know so much more about how natural selection works than Darwin did, why don't we test the theory experimentally? The answer is clear; that is precisely what modern biologists are trying to do.

However, it must be understood that what we test is the theory of natural selection and not evolution. We test the cause, not its effect. There will be instances when we will use a kind of short cut to say that evolution explains certain facts of embryology, physiology, or behavior. That kind of discussion really means that through natural selection certain changes have accumulated and such evolutionary changes can account for what we observe in the development or function of certain organisms. Therefore, natural selection is still invoked as the causal agent to explain the diversity of organismic form and function, even though we consider selection from the perspective of its result, namely, evolution.

Natural selection and evolution after Darwin

Before turning to the testing of Darwinian natural selection, we need a clearer picture of the advances in biology that have been made since Darwin's day.

As the nineteenth century drew to a close, all of biology reflected the concept of evolutionary change through natural selection. All, that is, except genetics. Genetics came alive as an important biological field only in 1900.

In comparative anatomy, no longer was there just an encyclopedic accumulation of descriptive facts. These facts made new sense as the record of life. They suggested a history stretching back far beyond human records, to pre-human times. Fossils were now seen as pages from this historical record. The clue to reading them was provided by Darwin's evolutionary outlook. Similar forms often had a common evolutionary ancestor. Very similar forms often had a recent ancestor—there being relatively little time to accumulate differences; and less similar forms often had ancient ancestors—significant differences having accumulated over relatively long periods of time. Comparisons of mammalian skeletons revealed clearly the possibility of common ancestors among these animals. Further comparisons were made, between mammals and birds, and then between mammals and birds and reptiles, amphibians, and fishes (Fig. 2–3). Comparable studies were begun among the invertebrates and among

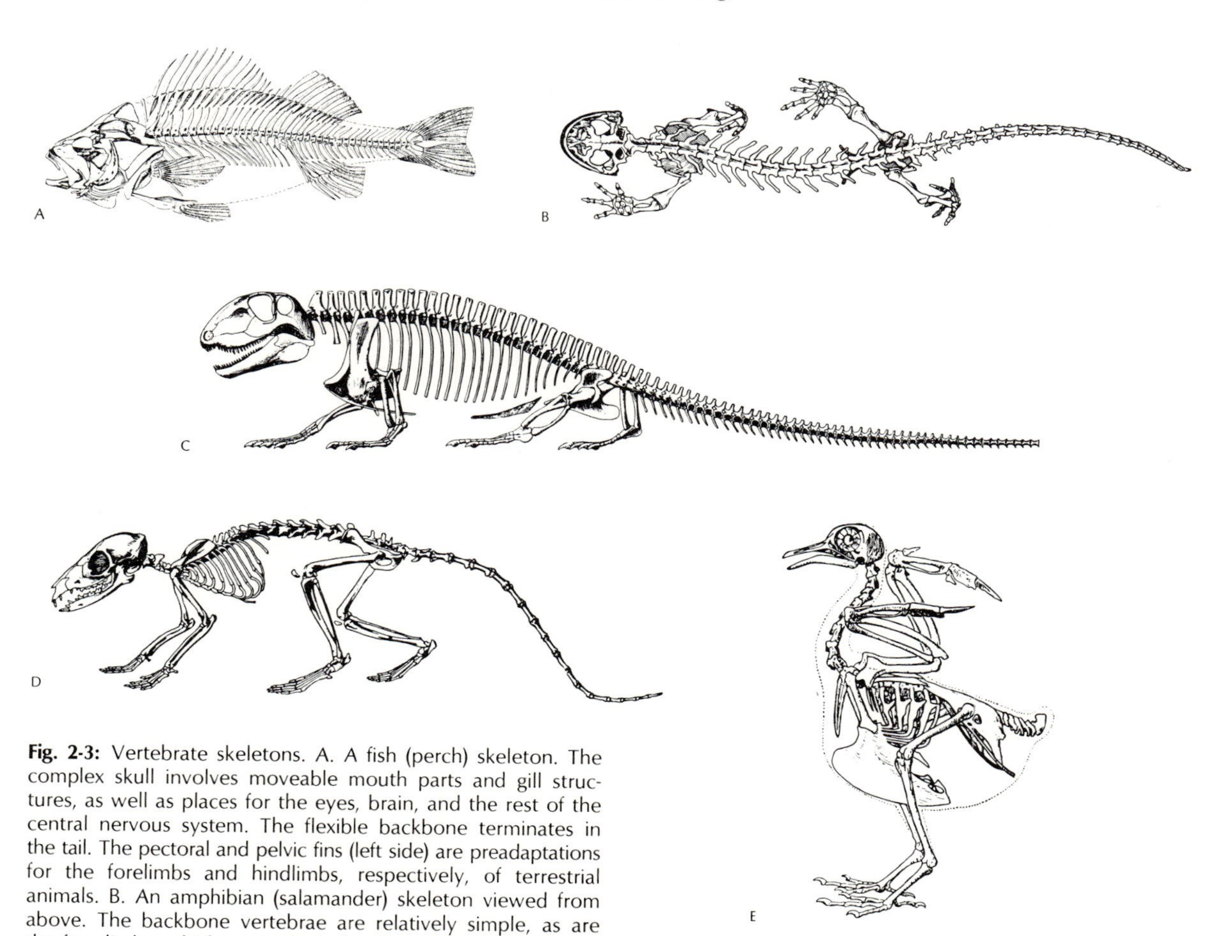

Fig. 2-3: Vertebrate skeletons. A. A fish (perch) skeleton. The complex skull involves moveable mouth parts and gill structures, as well as places for the eyes, brain, and the rest of the central nervous system. The flexible backbone terminates in the tail. The pectoral and pelvic fins (left side) are preadaptations for the forelimbs and hindlimbs, respectively, of terrestrial animals. B. An amphibian (salamander) skeleton viewed from above. The backbone vertebrae are relatively simple, as are the four limbs, which are typical of such terrestrial vertebrates. C. A reptilian (*Polypterus*) skeleton. The body of this early reptile is carried close to the ground. D. A mammalian (tree shrew) skeleton. The body is carried off the ground on well-developed limbs. E. A bird (pigeon) skeleton. Here, as above, the major features of the skull, backbone, and appendages are apparent and highly adapted to their special functions.

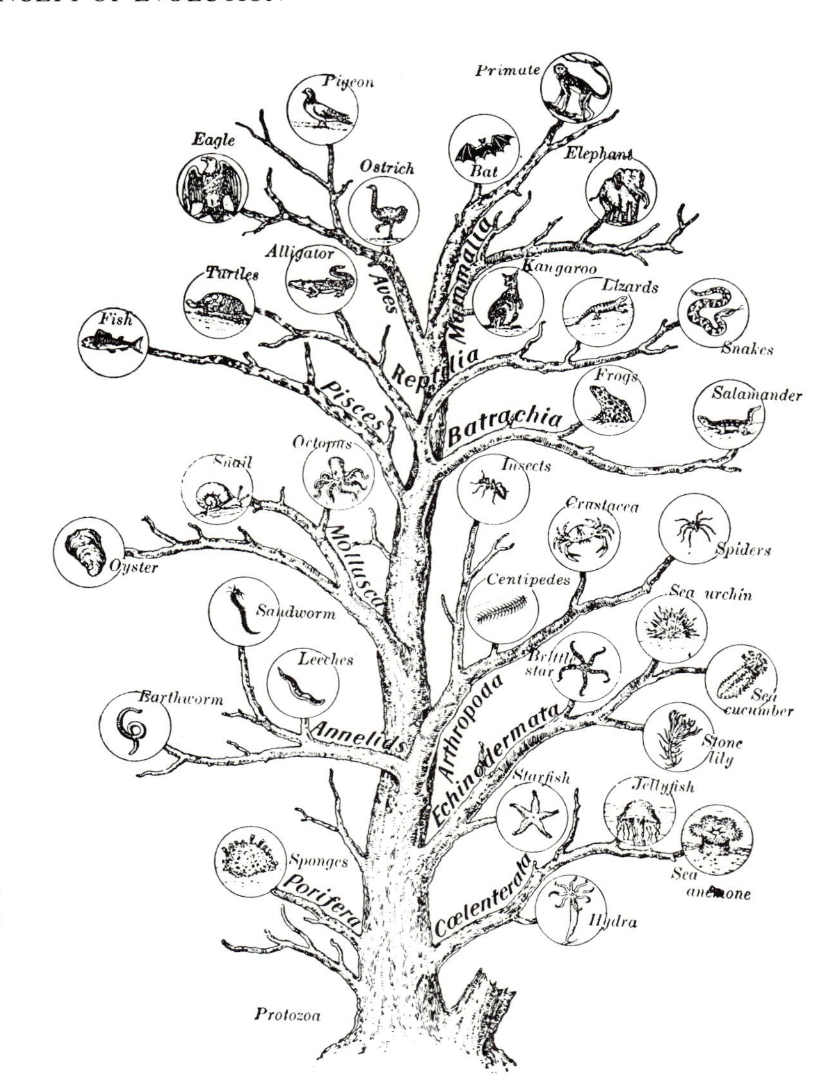

Fig. 2-4: This very fanciful phylogenetic tree proposes to show animal evolution from the Protozoa. Other, more accurate trees, or dendograms, are given in Part 4.

plants. The whole study of the structure and organization of living things was transformed. Simple collections of different species became the challenge of unraveling the history of life.

Such a viewpoint was contagious. It also infected taxonomy. Taxonomy has taken on a new name, i.e., systematics, and many twentieth-century systematists continue the enthusiastic work of their nineteenth-century forebears but with a new insight. Now they know why species fit into a genus and why different genera were placed into a family, and so on. Taxonomy became phylogeny, the study of evolutionary history. Some students of phylogenetics are overenthusiastic (Fig. 2–4). The tree of life, or dendrogram, which summarizes these speculative reconstructions of the history of life has been presented in a variety of forms, and there is still serious disagreement as to how living forms evolved. As these differences surfaced over the last hundred years or so, it became progressively clearer that reading the historical record takes very special skills. Some researchers said, in effect, that until we know how to read the historical record let us stop trying to read it. Let us not clutter the scientific literature with mistaken notions. The fundamental business of taxonomy, in any case, is to describe and classify the diversity of life, and we can do that without speculating, probably errroneously, on the course of evolutionary history. Others said that only by continued effort will the techniques for the correct reading of history become apparent. Efforts, even mistaken ones, must continue for they will eventually pave the road to success. Out of this has come a lively debate that has only become livelier as biochemists have found ways to decode the evolutionary history of such molecules as proteins and nucleic acids. This debate will be examined in some detail in later chapters of this book.

Embryologists with an evolutionary perspective made a startling discovery. Development itself seemed to be an abbreviation of at least certain parts of the fossil record (Fig. 1–5). Early in development vertebrate embryos showed very similar features, which diverged later on. Gill slits, so important in the classification of all chordates, were found in all vertebrate embryos. The conclusion seemed obvious that they were inherited from a common ancestor, but in time, some vertebrates developed other means of exchanging gases with the environment. Nevertheless, their past history of having gills persisted as a historical relic. Thus, embryonic development, called ontogeny, was a kind of repetition of historical developments, called phylogeny. The great nineteenth-century German biologist Ernst Haeckel summed it up as *ontogeny recapitulates phylogeny*. It does not literally work that way, but there is more than a germ of truth in that phrase.

Physiology, when viewed as adaptive functions, took on a whole new set of problems. Not only were organisms seen as integrated organ systems, cooperatively maintaining life, but the nature of the systems were seen as the results of evolutionary events. It became obvious that parts of organisms could evolve in different ways depending on adaptive needs.

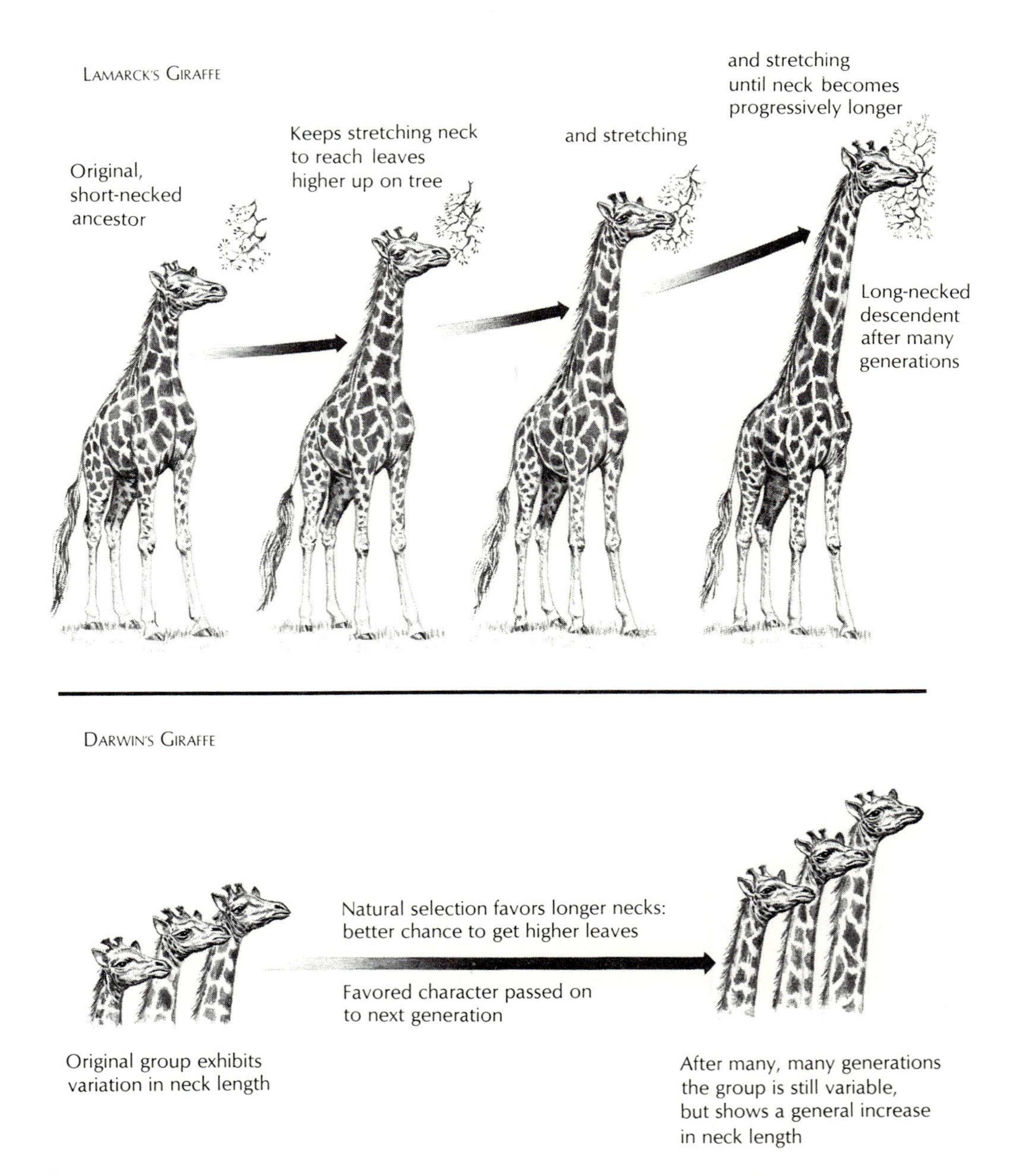

Fig. 2-5: Two views, Lamarckian and Darwinian, of the evolution of the giraffe's neck.

Such needs were not the result of an organism striving for some sort of perfection, as was proposed by the French biologist Lamarck in the early nineteenth century. Nor was the fulfilling of these needs the result of divine creation, as proposed by Christian theologists. Rather, such needs were met by natural selection preserving the heritable variations that improved the functioning of the organism. In fact, in its most precise formulation, natural selection really acts only on the functions of an organism.

Though we may refer to selection acting on giraffes to produce longer necks, what we really mean is better stated another way. Giraffes are browsers and whatever aids them in exploiting their environments for leafy forage will be selected for. Selection acts on the function of food-getting, which in the case of giraffes, involves, among others, those structures making up their necks. Lamarck simply proposed that the habit of stretching for leaves results in the perfection of longer necks. Darwin, writing in one of his numerous letters to a friend, said, "Heaven forfend me from the nonsense that is Lamarck." He, of course, preferred the explanation that the accumulation of variations resulting in a slightly longer neck eventually achieved the long necks of giraffes as we know them (Fig. 2–5).

But how are these variations inherited? Darwin insisted that it was *heritable* variations that count. Changes due to accidental injury or disease were not inherited, even though they might be beneficial. For a change to be meaningful in evolutionary terms it must be transmitted from one generation to another.

Darwin was so concerned with this question that he developed his own theory of pangenesis. In this explanation of heredity, each part of the body released particles into the bloodstream, which were finally collected by the ovaries and the testes and were incorporated into the ova and the sperm, respectively. These particles expressed are the characteristics of the body parts from which they came. So when fertilization occurred and a new organism developed, it would have body parts like those in its parents. This view of heredity is only of historical interest now, for the work of Mendel has given us a quite different view.

In fact, Mendel's work and that of the genetic researchers of the twentieth century is so basic that we shall devote the next chapter to reviewing the principles of genetics. We shall then go directly to the genetics underlying the study of evolution by natural selection.

References

Appleman, P., 1979. *Darwin: A Norton Critical Edition.* 2nd ed. W. W. Norton, New York.

Daniels, G. (ed.), 1968. *Darwinism Comes to America.* Blaisdell, Waltham, Mass.

Darwin, C., 1964. *The Origin of Species.* A Facsimile of the First Edition. Introduction by E. Mayr. Harvard University Press, Cambridge, Mass.

Darwin, C., 1979. *The Illustrated Origin of Species.* Abridged and introduced by R. E. Leakey. Hill & Wang, New York.

Gillispie, C. C., 1960. *The Edge of Objectivity.* Princeton University Press, Princeton, N.J.

Huxley, J., and H. B. D. Kettlewell, 1965. *Charles Darwin and His World.* Viking Press, New York.

Irvine, W., 1955. *Apes, Angels and Victorians.* McGraw-Hill, New York.

Moorhead, A. 1969. *Darwin and the Beagle.* Harper & Row.

Reese, M., 1979. *The Darwinian Revolution. Science Red in Tooth and Claw.* University of Chicago Press, Chicago.

Turner, F. M., 1978. The Victorian conflict between science and religion: A professional dimension. *Isis 69:* 356–376.

THREE

The Principles of Genetics

THE SIMPLEST SUMMARY of heredity is that like produces like. There is no exception to this generalization. A tree produces a tree like itself. Only humans make more humans and dogs produce dogs and cats produce more cats. No peacock ever came from a bacterium or vice versa. No whale ever produced a fern, a fungus, or even a dolphin. In fact blue whales produce only blue whales and killer whales, killer whales. All of genetics is aimed at elucidating how it is that like produces like, for genetics is the study of heredity.

Mendelian genetics

The Austrian monk Gregor Mendel advanced an explicit and widely accepted explanation of heredity in 1865, publishing a precise description of how certain characters are transmitted from one generation to the next; he then proposed a theory to explain the observed patterns of inheritance and also tested that theory in a convincing manner. Rarely in science, and especially in the biological sciences, does it fall to one person to formulate a major problem, to explain it, and also to test it. Usually scientific inquiry is a stop and go process in which many people contribute bits and pieces and finally the whole picture falls into place. This was certainly true regarding our understanding of deoxyribonucleic acid (DNA) as genetic material. That story will come after we look more closely at Mendel's magnificent analysis.

MENDEL'S ANALYSIS

Mendel modestly described his problem as that of trying to find the laws governing the formation of hybrids. By "hybrids" he meant the progeny of a cross between two pure lines. And by "pure lines" he meant strains that always produced the character in question, generation after generation. From these ideas of pure lines and hybrids it is clear that, in Mendel's time, people knew that like produced like even to the point of making controlled matings to ensure the perpetuation of desired characters. In Mendel's case, he worked with seven pairs of characters in the garden pea *Pisum sativum*. He knew how to ensure that tall plants always produced tall plants and short plants produced short ones, and so on, for the other six pairs of characters. In brief, he knew how to self-fertilize flowers to protect against cross-breeding and how to perpetuate a pure line. He could also cross-breed two pure lines to form the hybrids he desired.

The pattern of transmission of traits. Let us follow what Mendel found when he crossed a plant with round seeds to one with wrinkled seeds. These results are given in Box 3–1. In the first filial generation (F_1), Mendel found what others before had also noticed: one character disappears. But whereas many other workers had then given up, Mendel persisted and made a further cross by self-fertilizing the F_1 generation. The progeny from that cross, the F_2 generation, again had wrinkled seeds but only one-third as often as the round seeds. (There were 5,474 round: 1,850 wrinkled, which is a 2.96:1 ratio, obviously close to 3:1.) This apparent loss of wrinkled seeds in the F_1 and its reappearance in the F_2 generation led Mendel to refer to wrinkled as a recessive character and to round as dominant—a purely descriptive statement at this point. Mendel also noted that this same relation of dominant to recessive held for all seven pairs of traits he studied and that in no case was the recessive trait ever modified in any way as a result of its temporary disappearance, in the F_1 plants.

We can summarize the pattern of transmission of characters in garden peas as follows: F_1 show only the dominant trait, F_2 show the dominant trait three times as often as the reces-

Box 3–1. Mendel's first crosses.

The results of crossing a pure line of *Pisum sativum,* which produces round seeds, with a different pure line, which produces wrinkled seeds: P_1 is the first parental generation, P_2 and P_3 are the second and third parental generations, respectively; F_1 is the first filial generation, or the progeny of the P_1, F_2, and F_3 are the progeny of the P_2 and P_3 plants, respectively.

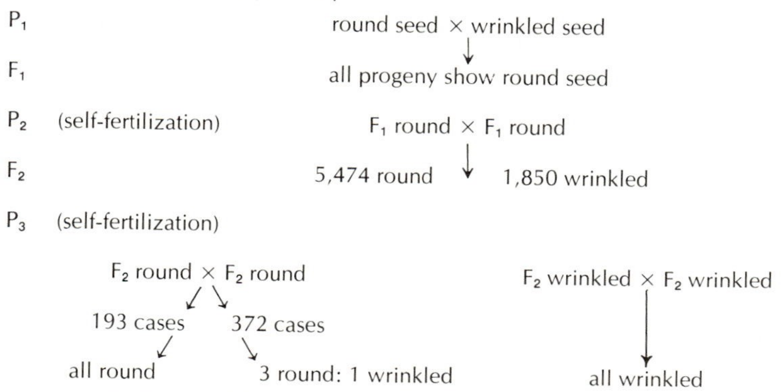

Only in P_1 did Mendel cross-breed two different lines. This was necessary to produce the hybrids shown in F_1. In P_2 and P_3, he self-fertilized the F_1 and F_2 plants, respectively.

Also look carefully at the results in F_3. The self-fertilization of the F_2 wrinkled plants gave only progeny that produced wrinkled seed. Mendel does not tell us how many cases he observed. But he obtained two types of results when he self-fertilized the F_2 round plants. In 193 cases of self-fertilization, the F_3 plants always produced round seed. But in 372 cases, the F_3 progeny produced either round or wrinkled seed and there were three of the round for every one of the wrinkled.

sive one (3:1 ratio); and in F_3 the recessive trait breeds true to type while the dominant trait shows one plant breeding true to type for every two that show 3:1 ratios of dominant to recessive progeny. Actually, the F_3 results tell us that there are three different types of plants in the F_3, i.e., pure dominants, those that can be called mixed dominants, and pure recessives and that they occur in a 1:2:1 ratio.

Clearly, this curiously regular pattern of inheritance required an explanation.

Mendelian factors. Unfortunately, Mendel's explanation is, at this point, too often reduced to what is called the law of segregation. Mendel himself is partly responsible for this confusion, since he overemphasized the one point of segregation (or assortment as it is also called). Mendel introduced the concept of factors here to explain the pattern of hereditary transmission and his factors involve five points, not one.

1. For each trait or character there is a factor.
2. Each factor is paired.
3. Only one factor of a pair is transmitted to a progeny organism. (This is the law of segregation, i.e., the paired factors segregate or separate from each other.)
4. Each factor of a pair has an equal probability of being transmitted to a progeny organism.
5. There is a dominant/recessive relationship between the pairs of factors.

Using these five points, we can explain as follows. Let us use the symbol R for the dominant factor determining round seed and r for the recessive one determining wrinkled seed. This tells us we can describe a pure line of round seed plants as RR and a pure line for wrinkled seed as rr. Using these symbols, we can diagram Mendel's crosses as shown in Box 3–2. We will go through this diagram step by step.

The use of RR and rr to denote the round and wrinkled pure lines used for the P_1 is straightforward. Being pure lines they, logically, have *only* the factors either for round or for wrinkled. On the other hand, the F_1 plants must have at least one R and one r. Each parent contributes one factor and the contributions are of each kind. Here the dominant/recessive relationship will account for our seeing only round seeds, even though the factor for wrinkled seeds is present. Further, when these F_1 plants are selfed (as they are in the P_2), we now see where the factors for wrinkled are coming from to give the F_2 wrinkled plants. They were present, though unexpressed, in

Box 3–2. A summary of Mendelian inheritance in terms of factors.

Here the factors are R for round seed and r for wrinkled seed. Note how the transmission of the factors follows the five points needed to determine the behavior of Mendelian factors.

P_1		RR		$\times$		rr	
F_1				Rr			
P_2		Rr		$\times$		Rr	(self-fertilization)
F_2	1 RR	:	2 Rr	:	1 rr		
P_3	$RR \times RR$		$Rr \times Rr$		$rr \times rr$		(self-fertilization)
F_3	RR		1 RR:2 Rr:1 rr		rr		

the parent (P_2) plants. What is not so clear is the numerical ratio of 1 *RR*:2 *Rr*:1 *rr* among the F_2. Before considering this ratio, we can point out that both the RR and Rr plants have round seeds. Therefore, in terms of traits rather than factors, F_2 show 3 round to 1 wrinkled, as described earlier (Box 3–1). The 1:2:1 ratio comes from Mendel's fourth point regarding his factors, i.e., the equal probability of each factor of a pair being transmitted to a progeny. This is shown most precisely by the familiar device called a Punnett square. This shows how parental factors can combine to form pairs of factors in the progeny.

	R	*r*
R	*RR*	*Rr*
r	*Rr*	*rr*

Across the top of this square are given the factors provided by pollen, for example, and at the left are those provided by the eggs. Remember we are dealing with self-fertilization here in the P_2. Hence Mendel uses the pollen of a single flower to fertilize the eggs lying in the ovule of the same flower. According to Mendel's explanation, the factors in the pollen must have segregated so that there is *R*, or *r*, in each pollen grain. Therefore, at the top of the square we find *R* and *r* and the implication is that they occur at the same frequency. This also applies to *R* and *r* at the left of the square in reference to the factors in the egg. Then, looking at all possible fertilizations of eggs by pollen we have the four combinations of paired factors. We find 1 *RR* to 2 *Rr* to 1 *rr,* as pairs of factors. This tells us that we must, in accordance with Mendel's thinking, expect 1 *RR*:2 *Rr*:1 *rr* progeny. This, then, explains the results observed in F_2.

It also explains the results from selfing the F_2 plants, as they are seen in the F_3. Each F_2 *RR* plant and each F_2 *rr* plant can only produce more *RR* and *rr* plants, respectively. And the F_2 *Rr* plants, when selfed, behave just like F_1 *Rr* plants; that is, they produce 3 round to 1 wrinkled progeny.

Testing the explanation of Mendelian factors. Any problem in science has, at least theoretically, an infinite number of explanations. Since they cannot all be right, it is therefore necessary to find the right one by testing. Even though Mendel proposed only one explanation, he knew that it was not necessarily the right one. He had to test it. In principle, testing is straightforward: the explanation is used to generate a prediction and then an experiment is performed to see if the prediction is confirmed or not.

Mendel's test was to describe a cross he had not studied earlier and predict what the progeny would be like from it. He chose as parents an F_1 plant, whose factors were indicated as *Rr*, and a pure wrinkled plant, with factors *rr*. Mendel predicted that the progeny should be half round and half wrinkled, i.e., 1 *Rr*:1 *rr*. He made the cross and found 106 round: 102 wrinkled. This was so close to his predicted 1:1 ratio that he justifiably argued that his prediction was confirmed.

Literally thousands of genetics students repeat these Mendelian experiments each year and all corroborate Mendel's work. His analysis stands as a monument of clear, convincing scientific inquiry. However, paradoxically, Mendel's analysis was ahead of its time. No one took his work seriously until 1900 when it was, in effect, rescued from oblivion. The biological world was then finally ready for it, even though Mendel himself died in 1884, relatively unknown. Mendel's explanation of a problem in heredity immediately raised a host of questions. These included such questions as What are these factors? Where are they located? Before turning to these new questions, we need to add a bit more from Mendel's contributions and to put it all in a more modern perspective.

The dihybrid cross. The essential further point that Mendel demonstrated with regard to his factors is that they segregate, or assort, independently of each other. (This is his so-called second law, the Law of Independent Assortment.) It is better seen as a sixth point to be added to the earlier five. Another way to express it is to say that the behavior of one pair of factors is not influenced by, or is independent of, that of any other pair. Mendel proposed this further point to explain the 9:3:3:1 ratios he found when he made crosses between pure lines that differed with regard to two pairs of characters (Box 3–3). Mendel called this type of a cross a *dihybrid cross,* since it involved two pairs of characters that normally bred true to type. The result of such a cross were offspring carrying two different sets of factors.

Terminology. In the foregoing discussion, we have purposely used only those terms proposed by Mendel. There has been no mention of genes and alleles, none of homozygous or heterozygous genotypes, and so on. All such terms came after Mendel, indeed they came after the rediscovery of his work in 1900. We will introduce them now since they are necessary to further define and use genetic ideas. This can be done by comparing Mendelian terms with more modern terms.

Box 3–3. Independent assortment of Mendelian factors.

The behavior of factors when lines differing in two pairs of characters are crossed is shown in the following diagram. Here the two pairs of characters are seed shape (round vs. wrinkled) and seed color (yellow vs. green).

P_1 round *RR* yellow *YY* × wrinkled *rr* green *yy*

F_1 all round yellow *Rr Yy*

P_2 (self-fertilization)
F_1 round yellow *Rr Yy* × F_1 round yellow *Rr Yy*

F_2 315 round yellow
108 round green
101 wrinkled yellow
32 wrinkled green

The factors in the F_2 plants are not given above, in all other cases the characters and their factors are given. It is instructive to treat this situation as if we were making a prediction from Mendel's explanation, including, of course, his sixth point. Then let us see if the prediction is confirmed. (Mendel did not present it this way; he obtained the results given here and then concluded the sixth point was necessary. He tested his sixth point by an appropriate prediction and experiment, as seen below.)

A Punnett square for the P_2 plants is as follows:

Factors in ova \ Factors in pollen	*RY*	*Ry*	*rY*	*ry*
RY	*RRYY*	*RRYy*	*RrYY*	*RrYy*
Ry	*RRYy*	*RRyy*	*RrYy*	*Rryy*
rY	*RrYY*	*RrYy*	*rrYY*	*rrYy*
ry	*RrYy*	*Rryy*	*rrYy*	*rryy*

Note that the four possible types of pollen (above the square) and of eggs (to the left of the square) depend upon the other pair. Thus one *R* factor occurs as often

(**Box 3–3** continues next page)

Mendelian	*Modern*
factor	gene
different paired factors	alleles
factors present	genotype
characters present	phenotype
pure line	homozygous genotype
hybrid	heterozygous genotype

Now we can return to the questions arising from Mendel's work and ask, What are genes? Where are they located?

Box 3–3 (continued)

with *Y* as with *y*, and *r* occurs as often with *Y* as with *y*, the ratio of each type of pollen or ova being 1 *RY* : 1 *Ry* : 1 *rY* : 1 ry. Then all possible combinations of these four types of eggs and pollen, all occurring in equal frequencies, give us the contents of the sixteen boxes within the square. If we remember dominance and recessive relationships, we can work out the characteristics that will appear in each of the sixteen individuals. We can also determine the frequency at which they occur, i.e., how many out of sixteen are round yellow or round green or wrinkled yellow or wrinkled green. (These are the only possible combinations of characteristics.)

Characters	Factors	Numbers	
Round yellow	*RRYY*	1	9
	RRYy	2	
	RrYY	2	
	RrYy	4	
Round green	*RRyy*	1	3
	Rryy	2	
Wrinkled yellow	*rrYY*	1	3
	rrYy	2	
Wrinkled green	*rryy*	1	

You will have noticed that the characters predicted here are the same as those observed to occur in the F_2 by Mendel. They are predicted, by the above reasoning, to occur in a 9:3:3:1 ratio. Compare this to the 315:108:101:32 ratio found by Mendel. The assumption of independent assortment does explain the results of this dihybrid cross.

Mendel tested this proposal of independent assortment by predicting the results of a cross where the F_1 round yellow was crossed to a wrinkled green plant. He predicted and found a 1:1:1:1 segregation of characters in the progeny. All the seven pairs of Mendelian factors were found to assort independently of each other. Mendel was clearly ready to say any pair of factors whatsoever can be expected to behave independently of any other pair. As we shall see this is where he would have been wrong, had he pressed this conclusion.

Chromosomal theory of heredity.

In 1902, two biologists—the distinguished German Theodor Boveri and a young American graduate student, Walter Sutton—both proposed that Mendel's factors lie on chromosomes. (The term gene, genotype, and phenotype were coined a few years later, but we shall use them here for convenience.) Clearly, Boveri and Sutton were telling us where genes are located. As it turned out, the answer to that question led very rapidly to much more information on the related question regarding the nature of a gene.

Boveri's and Sutton's explanation of gene location really evolved from a desire to understand why Mendel's factors behaved the way they did. The clue to this came from studying

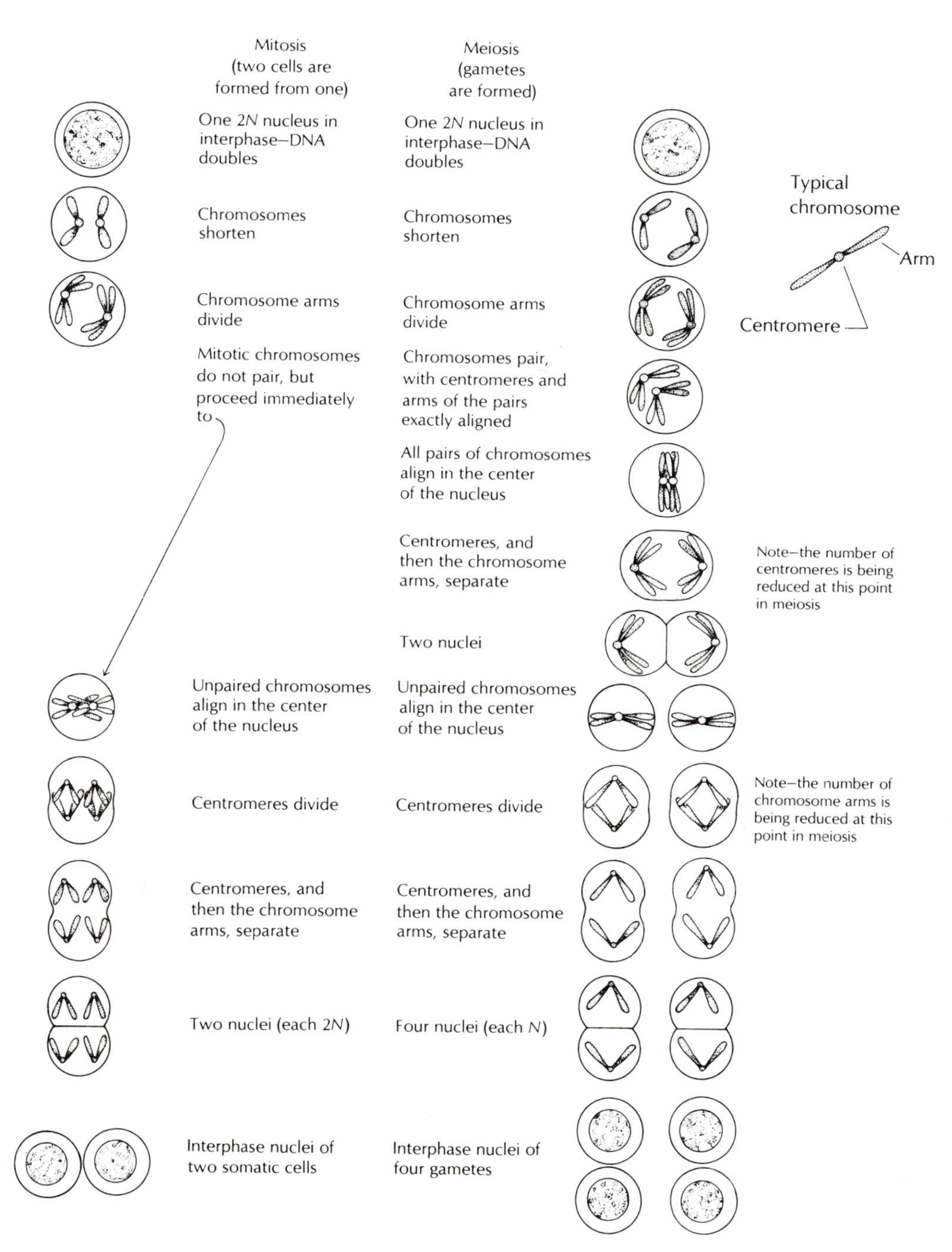

Fig. 3-1: This diagrammatic summary compares mitosis and meiosis.

reduction divisions of chromosomes. It was known that the cells that form gametes—pollen and eggs in plants, sperm and eggs in animals—divide to reduce their chromosome numbers by one-half (meiosis). The behavior of meiotic chromosomes lead to the following conclusions among others (Fig. 3–1).

1. Chromosomes differ from each other.
2. Chromosomes pair during meiosis, i.e., synapsis occurs.
3. The pairs then separate and one chromosome of each pair enters a gamete.
4. The way one pair of chromosomes separates has no apparent bearing on the way another pair separates.

To Sutton and Boveri, there was a peculiar and intriguing ring to the foregoing points. They could be reformulated as follows:

1. There are individual chromosomes.
2. However, they exist in pairs.
3. The chromosomes of a pair segregate.
4. The segregation of one pair of chromosomes is independent of segregation of any other pair.

If we could also say that the presence of any one chromosome in a gamete does not affect the chances of that gamete partaking in fertilization, we would see that we have a parallel between the above points and five of Mendel's six points regarding the behavior of factors. (Mendel's ideas on dominant and recessive factors, however, have no parallel here.) It was, therefore, proposed that factors behave as they do because they occur on chromosomes. This led to the chromosomal theory of heredity, which simultaneously explained genic behavior and genic location. And like any good theory it suggested various predictions whereby it could be tested. An important prediction was that the behavior of a gene and of the chromosome on which it was located must be perfectly correlated. That is, the hereditary transmission of the two must be identical.

With the use of the fruit fly *Drosophila melanogaster* as a laboratory organism, the means for testing the chromosome theory experimentally was at hand.

SEX LINKAGE

Around the time the fruit fly investigations were getting underway it was discovered, in another insect, that the inheritance of sexual differences correlated perfectly with the presence of certain chromosomes. When this discovery was applied to fruit flies, it was found that females have two of what were called X chromosomes and males have one X and one Y chromosome (Fig. 3–2). Clearly the inheritance of these chromosomes, if they carried whatever genes controlled sexual differences, would correlate perfectly with the inheritance of male and female phenotypes.

How would a test of the chromosomal theory use this situation of the sex chromosomes and the inheritance of male and female phenotypes? If genes are on chromosomes, we would predict that the transmission of a gene must follow exactly that of the chromosome on which it lies. But recall that we cannot *see* genes on chromosomes; in fact, we only know about genes from the characters they determine. So what we are really predicting is that the character controlled by a gene will be inherited in a pattern identical to the chromosome on which the gene lies. So, in the case of sex determination, we predict that the characters of maleness or femaleness will be inherited in a pattern that follows exactly that of the chromosomes that carry the genes for being a male or female fruit fly. But though there is a perfect correlation between XX chromosomes and the female sex and between XY chromosomes and male flies, where are the genes in question?

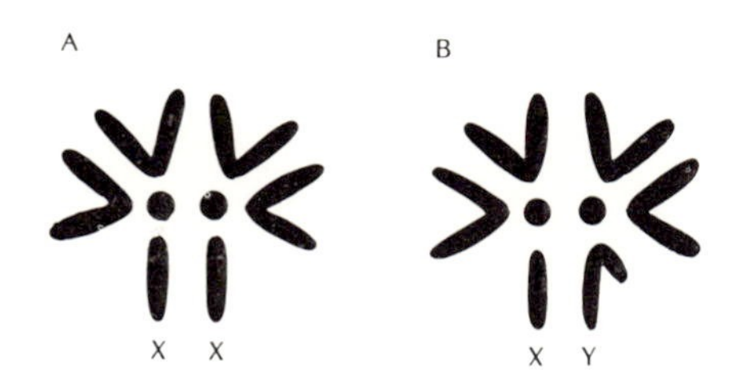

Fig. 3-2: The four pairs of chromosomes of the male and female *Drosophila melanogaster* are from cells in metaphase. The female (A) has two X chromosomes; the male (B), one X and one Y chromosome.

To make a rather complex story short, there is no such thing in *Drosophila* as a gene for a male or a gene for a female, since in this insect, sex turns out to be determined largely by the ratio of X chromosomes to the other chromosomes present. It involves many as yet unidentified genes and their complex interactions. If one X is present, the fly is a male (even if the Y chromosome is absent). If two X chromosomes are present the fly is a female. In other words this test of the chromosomal theory of inheritance was not rigorous. It neither proved nor disproved the explanation.

A convincing test was made when a single trait was found such that we could conclude it was determined by a single gene, which controlled eye color, located on the X chromosome. The unusual Y chromosome turned out not to have any allele for this gene. This study is summarized in Box 3–4. Here, then, was a convincing example of a trait, white eye color, whose inheritance could only be explained if its gene lay on the X chromosome. Clearly, then, the inheritance of the X chromosome correlated with that of the white eye trait and its gene. Hence, this prediction of the chromosomal theory of inheritance was confirmed.

Other tests of this major theory in genetics were carried out, and in all cases they supported the theory. One especially convincing example was the inheritance of attached X chro-

Box 3–4 Sex linkage in fruit flies.

The inheritance of eye color, a sex-linked character, in *D. melanogaster*. The first two crosses are reciprocal crosses, wherein the phenotypes of the male and female parents are switched. Notice, especially, the distribution of eye color and sex in the progeny of the two crosses.

1. P_1 white eye ♂ × red eye ♀
   F_1 all ♂ and ♀ with red eyes
   P_2 F_1 red eye ♂ × F_1 red eye ♀
   F_2 2 red eye ♀ : 1 red eye ♂ : 1 white eye ♂
2. P_1 red eye ♂ × white eye ♀
   F_1 all ♂ white eye; all ♀ red eye
   P_2 F_1 white eye ♂ × F_1 red eye ♀
   F_2 1 red eye ♀ : 1 white eye ♂ : 1 red eye ♀ : 1 white eye ♀

This distribution of phenotypes can be explained if we assume that the gene for white eye exists on the X chromosome along with its normal allele for red eye, i.e., red eye is X^R and white eye is X^r, and that no eye color gene is found on the Y chromosome.

P_1	X^rY	×	X^RX^R	X^RY	×	X^rX^r
F_1	X^RY		X^RX^r	X^rY		X^RX^r
P_2	X^RY	×	X^RX^r	X^rY	×	X^RX^r

	X^R	Y
X^R	X^RX^R	X^RY
X^r	X^RX^r	X^rY

	X^r	Y
X^R	X^RX^r	X^RY
X^r	X^rX^r	X^rY

F_2 $X^RX^R{:}X^rX^r{:}X^RY{:}X^rY$ $X^RY{:}X^rY{:}X^RX^r{:}X^rX^r$

In particular note how the predicted genotypes, above, explain the observed phenotypic results. Moreover, inheritance of the phenotypes follows a specific pattern of inheritance of different X chromosomes—different in terms of the allele, *R* or *r*, they are thought to carry.

A purist might at this point say that this test of the chromosomal theory of heredity is still not rigorous because we cannot really follow one particular X chromosome, for example, the X^R one as distinct from the X^r chromosome. That is true. And that is why, in the accompanying discussion we have referred to Bridge's work with non-disjunction in which particular X chromosomes were followed. The interested student is urged to examine that important piece of research in any good genetics text.

mosomes. The double occurrence of this chromosome led to very specific predictions regarding the sex and the eye color of the flies involved. This analysis is often cited as the most convincing of the early experiments testing the chromosomal theory of heredity.

LINEAR ARRAY OF GENES ON CHROMOSOMES

Although it seems obvious that genes should lie in a row along a chromosome, nothing in the work described so far really necessitates such a conclusion. All we can say from the foregoing is that genes must be somewhere on chromosomes. But how do we find out just how they are arranged? In fact, how many genes are on a chromosome? Again we turn to the work on *D. melanogaster*. During the early decades of this century, the genetics laboratory of Thomas Hunt Morgan and his students, especially H. J. Muller, J. Sturtevant, and C. Bridges, at Columbia University in New York City, produced an extraordinary flow of significant ideas and experiments. After Mendel, it was this group that put genetics in the forefront of scientific research.

Now consider the number of genes on a chromosome. First the number of genes we find in an organism depends on the number of gene-controlled characters identified. The location of these genes on particular chromosomes are identified and are then to be located on chromosomes. Quite obviously, there will be more genes than chromosomes and so chromosomes can be expected to carry many genes. But in no case, except that of certain bacteria and viruses, do we have a clear idea of the number of genes an organism has. In the fruit fly it is estimated, conservatively, that there are 10,000 genes in the haploid set of chromosomes, but fewer than one-tenth of these genes have been associated with specific traits or characters. Nevertheless, it must be concluded that at least hundreds of genes can lie on a single chromosome. Now we can return to the question of how they are arranged on that chromosome. To start, it will be necessary to identify a group of genes that are linked as a result of lying together chromosomally. Next, we can try to sort out their positions relative to each other. The first problem is called *linkage*, the second *mapping*.

Chromosomal linkage groups. The genes on a particular chromosome must be linked. We have seen that inheritance of white eye color is linked to the determination of sex—hence, sex linkage occurs. An example of autosomal linkage is given in Box 3–5. A little pondering on the matter makes it clear that any time the number of genes exceeds the haploid number of chromosomes in an organism, we must expect to find linkage. At least two of the genes will have to lie on the same chromosome; they will be linked. Of course, more than two can be linked, since it all depends on which chromosome or chromosomes they happen to lie on.

As work continues on a given organism, its catalog of genes grows accordingly. Today in the fruit fly, for example, we know that there are about 12 genes in the small fourth chromosome, and on each of the larger third, second, and X

Box 3–5. Autosomal linkage and recombination of linked genes.

In the fruit fly the following genes have been identified: *B* determines normal or gray body color and its allele *b* black body color; *V* determines normal or long wings, its allele *v* vestigial wings. The following breeding analysis was made:

P_1	*bb VV* ♂	×	*BB vv* ♀
F_1	all *Bb Vv*		
P_2	*Bb Vv* ♂	×	*bb vv* ♀

Now if the locus for body color lies on a different chromosome from that for wing length we should expect the following result:

F_2					
	Genotype	*Bb Vv*	*Bb vv*	*bb Vv*	*bb vv*
	Phenotype	gray long	gray vestigial	black long	black vestigial
	Ratio	1 :	1 :	1 :	1

However, if the two loci lie on the same chromosome, that is if they are linked, then we should expect the results given below because the genotypes of the gametes from the male P_2 parent are *bV* and *Bv* and those from the female are all *bv*.

F_2			
	Genotype	*bb VV*	*Bb vv*
	Phenotype	black long	gray vestigial
	Ratio	1 :	1

Since black long and gray vestigial flies occurred in equal numbers, these loci are linked, but not sex-linked (Box 3–4). They lie on an autosome (Fig. 3–3).

Now consider another cross with these same alleles. (The cross is written to acknowledge the fact of linkage by representing the single chromosome carrying the genes as a solid line. Actually, there are two homologous chromosomes here.) Notice that the double heterozygote is now the female parent rather than the male, as in the preceding cross.

$$P_3 \quad \frac{b\ v}{b\ v}\ ♂ \quad \times \quad \frac{B\ v}{b\ V}\ ♀$$

F_3 Observed progeny phenotypes and their frequencies

Black long	Gray vestigial	Black vestigial	Gray long
41%	41%	9%	9%

Here we find the four phenotypes that would be expected if the loci were unlinked, but their ratio is not the 1 : 1 : 1 : 1 that goes with unlinked loci. What has happened? Apparently the genes of the female parent have somehow recombined to provide 9% of the eggs with a *bv* autosome and another 9% with *BV*. This recombination is the result of crossing-over, which is explained in Fig. 3–4. We need to question why the recombinants resulting from crossing-over occurred in the P_2 cross and not in the P_3. The answer is an example of Haldane's Rule, which says crossing-over is suppressed in the heterogametic sex. In fruit flies the male is heterogametic (XY) and the female is homogametic (XX). Recombinations between X and Y chromosomes could upset the mechanism of sex determination, which is not true for recombinations between two X chromosomes. Hence, the results in the F_3 females. They also probably occurred in the P_2 females, but could not be detected because they were all homozygous; recombinants would have had the same genotypes as unrecombined chromosomes.

chromosomes there are over 150 genes. On the puzzling Y chromosome we have found around 12 genes (Fig. 3–3).

It can be pointed out now that chromosomal linkage groups are another test of the chromosomal theory of heredity. If that theory is correct, then we should expect that (1) the number of linkage groups is equal to the haploid number of chromosomes and (2) the size of the linkage group should reflect the size of the individual chromosomes. Both predictions are confirmed. (Compare Figs. 3–2 and 3–3.)

Incidentally, the haploid number of chromosomes in Mendel's garden pea, *Pisum sativum,* is seven. How many independently segregating pairs of characters did Mendel analyze? Suppose he had eight pairs of characters, what would have happened to his sixth point regarding independent segregation of traits and their factors? What are the chances of finding seven characters in *P. sativium* and having each one lie on a different chromosome? Those who are curious should consult a sound genetics text.

Mapping. The genes in Fig. 3–3 are all arrayed linearly on their chromosomes. What is the evidence for this arrangement? It comes from the discovery of *genic combination* through *chromosomal crossing-over.*

Genic recombination was discovered in Morgan's laboratory when a second sex-linked gene, miniature wing, was discovered. In this work it was seen that the two marker genes miniature wing and white eye do not always remain linked. They are incompletely linked, and as a consequence they recombine to form genotypes not present in the parent fruit flies (Fig. 3–4). This recombination immediately raised a question as to the mechanics of the process. Various explanations were advanced and today we are left with the point of view that somehow—we don't yet know exactly how—from four homologous chromatids only two exchange parts at a given place during meiotic prophase. This explanation of chromosomal crossing-over is supported by visual evidence of exchanged parts and by the fact that genetic recombination never involves more than one-half the progeny.

Long before the debate over the mechanisms of crossing-over was resolved in favor of chromosomal exchanges, geneticists were using the fact of recombination as the basic technique for mapping the location of genes on a chromosome. The principle is quite straightforward and a comparison to road maps can illustrate what is involved here. Let us assume

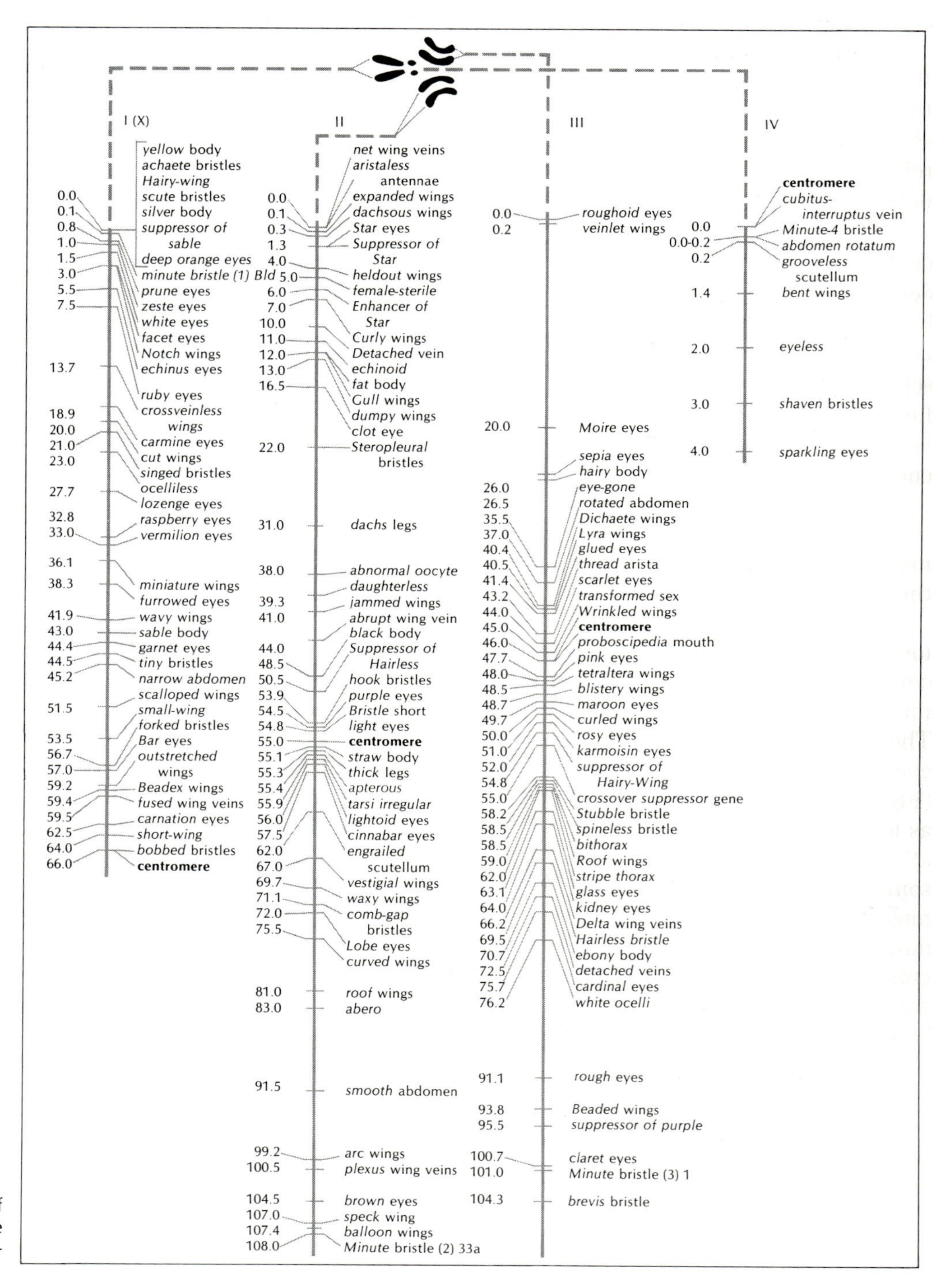

Fig. 3-3: A map of some of the important genes in the four chromosomes of *Drosophila melanogaster.*

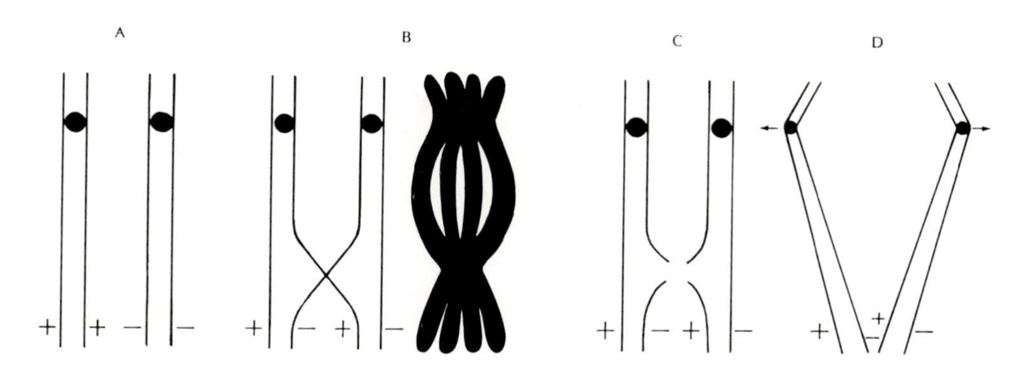

Fig. 3-4: An interpretative diagram of crossing-over between two of the four chromatids present in a prophase chromosome. A. The homologous chromosomes have synapsed and each member of the pair has copied itself, to form two chromatids each. Alleles are indicated as + and − on the chromatids. The centromeres (dark circles) have not replicated yet. B. The inner chromatids have crossed-over, as can be indirectly inferred from the chiasma just to the right of the crossing-over diagram. Frequency of chiasmata occurrence and frequency of genetic recombination show a high correlation. C. A break is assumed to occur in each of the inner chromatids at the site of the chiasma. The mechanism that causes the apparent repulsion of chromosomes at this stage is unknown. D. The broken ends join, reciprocally, thus recombining the genes on the two chromatids that crossed-over. Note the fate of the + and − markers. The centromeres now migrate to opposite poles of the spindle.

that a person unfamiliar with the United States is trying to figure out the relative positions of New York, Washington, D.C., and Miami on the east coast. If that person is provided with the frequency of auto accidents on the highways joining those three cities, what can be concluded? Obviously, the cities with the longest stretch of highway between them will have the greatest number of accidents, if we assume that the probability of an accident occurring is the same for each mile of the highways. Hence New York and Miami will be farthest apart and Washington will be in between. Now if we consider genic recombination to result from some kind of chance event, or accident, between homologous chromatids, then it will occur more often between genes further apart on these chromatids than between those closer together. In fact, *the frequency of recombination is a measure of the distance between linked genes*. This holds, however, only within certain limits. If genes are very close together, recombination can be so rare as to be negligible, and then the two markers may seem to occupy the same location or locus on a chromosome. On the other hand, if the markers are so far apart that recombination occurs frequently, it will appear that random assortment is occurring between the markers, which would imply no linkage. Thus mapping distances are most reliable when values of 10 to 30% recombination are found—this avoids the extreme values that lead to confusing results.

The analysis of many different markers in the fruit fly has produced chromosomal maps (see Fig. 3–3) that depend upon the conclusion that a single linear array of genes is the best interpretation of the recombination data. This applies not only to *Drosophila*, but to all organisms, since recombination has been observed in all living things studied thus far.

Properties of genetic material

What general conclusions can we come to regarding genetic materials?

Genes are sources of information; they are programmed to specify the structure and to regulate the function of an organism. Considered this way, genes have the following capabilities: (1) they store information, (2) they replicate that information, (3) they express the stored information, (4) the stored information can be changed (mutation), and (5) the information can be recombined. The obvious next question is How do genes carry out these five functions? To answer, we must at some point look at genes from a biochemical and molecular viewpoint. Before doing that let us examine expression and change in genetic material in more detail. How genes express themselves is basic to our ideas of what they are and how they function. Since evolution is the result of accumulated heritable changes, the reason why changes occur in the first place is basic to our studies of evolution. Let us look first at changes in a general way, later we will discuss genic changes in chemical terms.

CHANGES IN GENETIC MATERIAL

There are, broadly speaking, three categories of changes to be considered: (1) genic mutations, (2) chromosomal changes, and (3) genomic changes.

Genic mutations. These changes have been intensively studied in the fruit fly. Muller, in particular, pursued this problem, and in 1927 he showed that X-ray irradiation could increase by 15,000% the natural rate of change of genetic material. This was only one of Muller's many fundamental contributions to genetics. He realized most clearly the importance of mutation to evolution. Mutations in genes are also called point mutations in that they identify changes within individual genes. The precise nature of these changes is clearer now that we know more about the biochemistry of genes. This will be examined shortly.

Chromosomal changes. These changes, occurring within or between chromosomes, are rearrangements of genes. Whereas point mutations are qualitatively new genes, chromosomal changes are new combinations of genes already present. Box 3–6 summarizes the kinds of recombinational changes

Box 3–6. Various kinds of chromosomal changes.

In the symbols used below, the letters all refer to genes, the solid lines to chromosomes. The causes of these changes always involves chromosomal breakage and subsequent repair. Beyond that general similarity, various differences account for different kinds of changes. These will be discussed later when the evolutionary implications of the changes are examined.

1. *Addition* by duplication of a gene.

   A B C D E → A B B C D E

2. *Deletion* by loss of a gene.

   A B C D E → A C D E

3. *Inversion* of a chromosomal segment containing various genes.

   A B C D E → A D C B E

4. *Crossing-over* by exchange of chromosomal parts between homologous chromosomes.

   A B C D E / a b c d e → A B c d e / a b C D E

5. *Reciprocal translocation* between non-homologous chromosomes.

   A B C D E / M N O P → A B C O P / M N D E

that have been described by various workers for a great variety of eukaryotic organisms.

Genomic changes. Changes of this sort involve chromosomal number. In aneuploidy, a chromosome has been added or subtracted to otherwise complete sets of chromosomes. For example, in humans an extra chromosome-21 results in trisomy or three copies of that chromosome. This causes an abnormal phenotype, called Down's syndrome, which includes a swollen tongue, a short body with stubby fingers, eyelid folds, and mental retardation.

In polyploidy whole sets of chromosomes are added to existing sets. Very often two diploid sets combine to form a tetraploid. Cases of triploidy and of higher polyploid series, such as hexaploids or octoploids, are found, or they can be created experimentally. Tetraploids are probably the most common example of polyploidy and occur naturally in many species of flowering plants. The reason why polyploidy succeeds in plants and does not succeed in animals is thought to be related to the mechanisms of sex determination. In animals sex determination is genotypic; it depends on the interaction of X and Y chromosomes with each other and with autosomes. In plants, sex determination is phenotypic; in many plants both pollen and eggs are produced in the flowers. Obviously, differences in sex chromosomes do not explain the situation in plants, since all cells contain the same chromosomes. Differentiation, therefore, is imposed on reproductive cells from the surrounding tissues to produce different gametic phenotypes. Not being dependent on the number of X or Y chromosomes or the number of autosomes, plants can thus change chromosomal numbers without affecting sexual differentiation.

PROBLEMS RELATING TO GENE EXPRESSION

Ideally, geneticists would like to start with the characters found in the adult organism and trace each character back to the gene or genes that determine it. That has turned out to be exceedingly complex for at least three reasons: (1) Genes, for the most part, lie in the nucleus of a cell, and cytoplasmic events bring them to expression, depending upon the capabilities of the individual cell. Cytoplasmic mediation of gene expression is still being worked out, although remarkable progress has been made. (2) Genes, for the most part, interact to determine characters. We now know that the simple Mendelian relationship of one factor to one trait is the exception rather than the rule. Gene interactions affect both individual cells and tissues. The adult organism is not the sum of the expression of its individual genes. It is a complex product of different genes acting at different times in development to produce orderly structures and functions as the organism matures. This has been referred to in Chapters 4 and 5 as the expression of the epigenotype. (3) Genes determine the potential of an organism, but it is the environment that determines what aspect of that potential will be expressed. Most humans have the pigment (coded for by genes) that allows tanning upon exposure to ultraviolet light, but it takes that exposure for tanning to occur. This also applies to many other characters, ranging from the molecular level (induction of enzyme in the presence of a particular substrate) to embryonic inductions in developing tissues, and physiological adaptations of organisms in response to nutrients and to complex behaviors aroused by the proximity of certain other organisms.

Behavior, in fact, is one of the most difficult problems for geneticists today. Some behaviors, such as nest-building among parakeets, clearly show genic control, others, including the furiously controversial debates in this decade over the genetic bases, or lack of them, for human intelligence, are unresolved. Much of the behavior we most cherish as humans seems, on present evidence, to be socially determined. That is, a social environment can bring a very complex potential for musical creativity or linguistic abilities to expression. We are far from understanding what our genetic potential has to do with the control, if any, of complex behaviors whose expression is apparently largely controlled by the environment.

We need to examine in a bit more detail these complex topics of nucleocytoplasmic interaction, genic interaction, and environmental control of gene expression.

Nucleocytoplasmic genetic interaction. We have stated that the genes "for the most part, lie in the nucleus." For the other part, some genes lie in the cytoplasm. They have been identified with certainty in plastids, mitochondria, and certain other particles. In all these cases DNA is the genetic material. This association in one cell of nuclear and cytoplasmic genes is one model of nucleocytoplasmic genetic interaction. Another model involves the role of differentiated cytoplasmic organelles, with no evidence of any cytoplasmic DNA involvement, in the perpetuation of cellular phenotypes. In a third model of cytoplasmic hereditary factor involvement we can bring together a heterogeneous grouping, such as pollen sterility, cellular antigens, or long-lasting physiological adaptations of cells to higher than normal temperatures or to otherwise toxic concentrations of chemicals, for example.

The occurrence of DNA in the cytoplasm is often explained as the result of a symbiotic association between the host organism and some smaller, often microbial or viral invader. That seems to be clearly the case in certain well-studied instances. For example, in the sex-ratio character in *Drosophila melanogaster*, a spirochaete-like entity is the causative agent; for the killer character in *Paramecium aurelia*, the kappa factor is now thought to be a highly specialized bacterial symbiont. It may be that, if the endosymbiont theory of the origin of eukaryotes is correct (see Chapter 11), mitochondria and plastids also represent types of symbionts. Whatever the origin of the associations of nuclear and cytoplasmic DNA, their interactions have certain features in common. The cytoplasmic DNA is associated with a distinct cytoplasmic structure, e.g., the plastid, mitochondrion, or other membrane-bound structure; the DNA is a circular Watson-Crick double helix; the cytoplasmic entity depends on nuclear genes; and sometimes the cell as a whole depends on the cytoplasmic factor.

In the case of mitochondria and plastids, the host cell relies on their presence for survival. Their loss is lethal in the case of mitochondria and often lethal in the case of plastids, although colorless algae are not uncommon among the unicellular protophyta (see Chapter 11). In the case of other cytoplasmic factors, loss often involves no hardship for the erstwhile host, but in terms of gene expression, the presence or absence of the cytoplasmic component determines the presence or absence of certain phenotypes. The nuclear genes are often only involved in maintaining the cytoplasmic partner. They may also produce changes in that partner, as when a nuclear gene in the green alga *Chlamydomonas reinhardi* mutates to change the plastid phenotype. But the important point here is that cytoplasmic factors can change independently of nuclear genes, and these changes determine the cellular phenotype. Such changes can be mutations of the cytoplasmic DNA, quantitative changes in the kinds or amounts of the cytoplasmic factors, or functional changes (responses) in the cytoplasmic factors. Hence many changes in characters can be expressed just through the action of the cytoplasm without any detectable change occurring in the nuclear genotype.

The cytoplasmic DNA is apparently not involved in the control of the cellular phenotype in cytoplasmic differentiation. Here, many factors have been intensively studied using the highly structured cortex of *Paramecium aurelia, Tetrahymena pyriformis, Stentor coeruleus,* and other ciliated protozoans. And somewhat similar stories are found, as in the role of the egg cortex in the development of such animals as insects and amphibians. The essential point is that preformed structures, not always identifiable as specific organelles, determine the formation of other structures. The new structures may or may not resemble the pre-existing structures. A certain row of cilia or a feeding organelle may determine another comparable ciliary row or feeding organelle. But larger topographical relations seem to determine the placement and location of differentiating structures during regeneration.

The role of nuclear genes seems to be that of supplying building materials (specific proteins in all probability) for this or that cortical organelle. But, the critical point is that the demand for the building material and the phase and type of assembly of the structure are cytoplasmically controlled. Genic potential is expressed through cytoplasmic control. In all probability various genes are called upon for products that interact at predetermined cytoplasmic locations to produce certain cellular phenotypical structures.

Other cases of nucleocytoplasmic interaction are not as well studied as these two models, but they too are instances in which the cytoplasm, not the nucleus, controls gene expression.

Genic interaction. Genic interaction was apparent in Mendel's work with the occurrence of the dominant and recessive genes, the dominant allele suppressing expression of the recessive one. Furthermore, this is not the only type of interaction. In codominance in a heterozygote, both alleles are expressed. This is seen in sickle cell anemia in which all three genotypes, i.e., Hb^AHb^A, Hb^AHb^S, and Hb^SHb^S, are expressed as three different phenotypes, i.e., normal, mild anemia, and severe anemia, respectively (see Chapter 5). In another case, two different loci can affect the same character. Here a dominant allele at either locus or at both loci gives one phenotype; the alternate phenotype occurs only with the double recessive homozygote. (In such a dihybrid cross, the F_1 heterozygote crossed to the F_1 heterozygote results in a 1:2:1 ratio of the two possible phenotypes.)

In brief, these interactions modify the classical Mendelian ratios of 3:1 in a monohybrid cross and 9:3:3:1 in a dihybrid one. How the ratios will be modified depends on the kind of interaction. None of this negates the principles of Mendelian heredity. On the contrary, Mendel is supported, but a broader understanding of how genes interact to express phenotypes is required.

Multiple factor interaction can become so complex that it is almost impossible to isolate the contributions of individual genes. A simple illustration of the consequences of what is often called quantitative inheritance is given in Fig. 3–5. Here phenotypes controlled by one locus with two alleles are diagrammatically compared with another phenotype in which many loci and alleles are involved.

One of the most interesting instances of genic interaction is that referred to as hybrid vigor. This describes the increased physical vigor and the often correlated increase in size that commonly occurs when members of different species are crossed and offspring are produced. The mule, a commonly cited example, is the result of a mating between a horse and a donkey. Though not always larger than the horse parent, a mule is larger than the donkey parent, and it is significantly

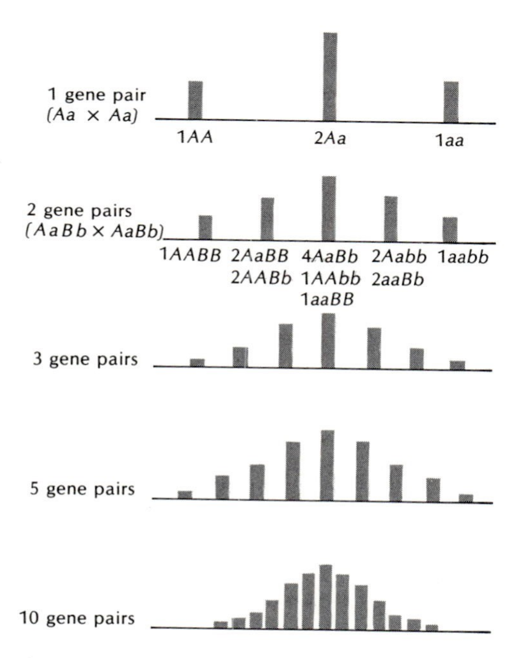

Fig. 3-5: Genotype distribution when different numbers of unlinked gene pairs are involved in crosses between heterozygous individuals. The height of the genotype histograms represents the proportion of individuals with that genotype, compared to other individuals in the population.

more hardy than either parent. Mules, however, are sterile, a common occurrence in such interspecies crosses.

Hybrid vigor has been turned to significant agricultural advantage, notably in the case of hybrid corn. Here the crosses are not between different species of plants, but between varieties of one species, *Zea mays*. This corn is designated a hybrid because there is significant heterozygosity in the offspring plants. We do not know why heterozygous genotypes result in more vigorous phenotypes. Current thinking favors the idea that the products of different alleles somehow produce a physiologically more active organism than if the alleles are all alike. In any case, hybrid vigor is a real phenomenon and a genetically and evolutionarily important example of genic interactions.

Genes and the environment. Organisms depend on their environment for food, light, water, nesting sites, individual and social territories, and so on. That fact underlies the responses of genotypes to the environment in expressing adapted phenotypes. Some features of an organism are so basic, such as its genetic code, its cellular organization, and its enzymatically regulated metabolic pathways, that they are a constant feature of life regardless of environmental exigencies. But superimposed on these features are the specializations expressed through epigenetic development that produce specific adult phenotypes, and superimposed on the latter—starting from early development onward—are the responses of the developing system to its environment. Out of all this there emerges a functional phenotype that assures survival. Some features of that phenotype are, as we have just said, relatively invariant, but others are more variable, especially those that respond to environmental pressures. As was mentioned earlier, such pressures range from enzyme induction to learned choices in social contexts. The spectrum is broad and complex and the role of gene action and interaction, though quite clear in the case of enzyme induction, is not understood in such cases as learned behaviors.

The following categories of environmental control of gene expression are known: Single allele differences, in conjunction with the environment, can, for example, determine the color of the fat in certain strains of rabbits or whether the gene for phenylketonuria (PKU) is expressed in humans. In normal rabbits, an enzyme breaks down xanthophyll, a yellow plant pigment, and the fat is white. But when the gene for that enzyme is absent or nonfunctional the pigment accumulates, and the fat is yellow. Diets low in xanthophyll can produce white fat in rabbits without the normal allele. Thus the phenotype can be environmentally controlled. A comparable situation holds for PKU. Here the inability to metabolize phenylpyruvic acid (due to the absence of a gene-controlled enzyme) can lead to the extreme phenotype of severe mental retardation. Controlling the diet of affected individuals at an early age can modify the undesirable phenotype. Here environmental control of the availability of phenylalanine affects the phenotype of individuals lacking the enzyme.

Environmental interactions with characters controlled by many genes are also known. Such physical factors as temperature, light, and availability of adequate oxygen and food affect living systems to produce abnormal phenotypes. Hence it is always useful and often necessary to describe growth or culture conditions along with the phenotype. This is a tacit recognition that the same genotype can, under different conditions, produce a different phenotype. Instances such as these illustrate the range and complexity of genotype-phenotype-environment interactions.

The chemical basis of heredity

We now turn to a question that has intrigued generations of geneticists: What is the genetic material; What is this mutable material that stores, replicates, expresses, and recombines hereditary information? The answer is nucleic acid. In most organisms it is deoxyribonucleic acid (DNA), but in certain viruses it is ribonucleic acid (RNA).

Initially, especially in the years before 1944, there was vigorous debate over whether proteins or nucleic acids were the genetic material. Many favored proteins for two excellent reasons. First, proteins occurred in the nucleus and cytoplasm, whereas DNA was restricted to chromosomes. Cytoplasmic DNA was unknown then, but the importance of the cytoplasm in determining cellular phenotypes was an accepted, though still unrefined concept. Therefore the presence of nuclear and cytoplasmic proteins suggested a basis for expressing stored genetic programs. Second, the structure of DNA, as it was then known, did not seem to be complex enough to provide the specificity needed for genetic programming. Geneticists were unaware of the unique array of bases that characterize DNA and RNA, whereas they were familiar with the unique arrays of amino acids in proteins.

All of this changed as a result of work done in 1944 by three scientists from the Rockefeller Center in New York City, O. T. Avery, C. M. MacLeod, and M. McCarty. They showed that in one species of bacterium the DNA from one strain could change the phenotype of a second strain into that of the first one. And the changed phenotype was heritable. (See also Box 11–1.) From that point on, evidence accumulated rapidly for the essential genetic role of DNA in most organisms, the only exceptions being those viruses in which RNA is the genetic material. In all organisms, then, without exception, nucleic acids are the chemical basis of heredity.

The next problem was to explain how DNA could account for the storage, replication, expression, mutation, and recombination of genetic information. The answers depended on the chemical structure of DNA (Box 1–1) and when that was elucidated by James Watson and Frances Crick, in 1953 at the Cavendish Laboratories in England, further refinement of the gene concept, in particular, and of heredity, in general, was made possible.

DNA AND THE PROPERTIES OF GENETIC MATERIAL

Let us start with gene expression and then turn to the other properties, but we will return to gene expression from time to time.

One gene-one polypeptide. Prior to the Watson-Crick model of DNA as a double helix with complementary purine and pyrimidine bases, the relation between a gene and the character it determines was that of each gene controlling the formation of an enzyme. This came about through the work of Beadle and Tatum, at the California Institute of Technology, on the red bread mold *Neurospora crassa*. This mold was almost ideal for such studies. It grew on a semisolid agar medium that contained sugar (glucose), a vitamin (biotin), and various salts as nutrients. From this it made everything—including proteins and nucleic acids—needed to grow and reproduce. It reproduced asexually, so that large amounts of genetically stable material were available, and it also underwent a sexual process, followed by meiosis, which made Mendelian crosses and genetic recombination remarkably easy to analyze. In a series of unique experiments, Beadle and Tatum obtained convincing evidence that genes control phenotypes through their ability to form enzymes, hence the concept one gene-one enzyme (Box 3–7).

We have now refined that concept. All enzymes are proteins. Some proteins are compound structures, consisting of two or more proteins that combine to form the final functional product. But not all proteins are enzymes; they also are structural components of cell membranes, ribosomes, microtubules, microfilaments, and other organelles. What we now realize is that the linear array of nucleotides (molecules made up of phosphoric acid, deoxyribose sugar, and purine or pyrimidines and thus comprising one helix of DNA) is what specifies the linear array of amino acids that ultimately forms one protein (Box 1–1). Such an amino acid array is, strictly speaking, a polypeptide. When it folds into a functional configuration, whether it combines with other macromolecules or not, it is a protein. So our idea of the genotype to phenotype relationship now stands as one gene-one polypeptide. Actually, the gene here is a unit of function, a cistron. One, therefore, hears of one cistron-one polypeptide as the ultimate refinement of Mendel's initial concept of one factor-one character.

This elucidation of the genotype-phenotype relationship allows us to restate the problem of gene expression. We now must ask how various linear arrays of nucleotides can specify various linear arrangements of the twenty essential amino acids (Box 1–1 and Table 1–2). We can only answer that question after we look at the problems of information storage and protein synthesis.

Information storage and protein synthesis. When it was realized that nucleotides containing adenine, guanine, and thymine or cytosine were not randomly distributed along the ribose sugar and phosphoric acid backbone of a DNA helix, but were distributed in a sequence that amounted to a code for amino acid sequences, the question of information storage presented itself. Biochemical geneticists, notably Severo Ochoa, of New York University, and Marshall Nirenberg, of the National Institutes of Health, led the way in answering this question.

Before this question could be answered, certain other important points had to be elucidated. Thus it was found that DNA was not directly translated into polypeptides, but, rather, first transcribed into RNA, which was then translated into polypeptides. The discovery of this key RNA, known as messenger RNA or simply *m*RNA, was necessary to crack the secret of the genetic code: The transcription of DNA to RNA was a direct transfer of information (specific nucleotide sequences) by means of complementary base-pairing. In DNA the adenine (A) on one helix was bound to the thymine (T) on the other helix; guanine (G) was bound to cytosine (C). These

A–T and G–C pairings were among the clues used by Watson and Crick to develop their idea of the double helix. As regards transcription, it was assumed, and later experimentally verified, that the G, C, and T in DNA paired with the C, G, and A, respectively, in RNA, and that the A from DNA paired with the uracil (U) in RNA. [RNA contains no thymine; uracil replaces it (Table 1–2).] So a complementary set of RNA bases contained the information present in one strand of DNA.

Furthermore, it was also learned that in translating *m*RNA into a polypeptide, two other types of RNA, enzymes, and energy were needed. Translation was a precise, but complex process. The other RNA molecules were called transfer or *t*RNA and ribosomal or *r*RNA. Their interactions in forming specific polypeptides are summarized diagrammatically in Box 3–8. The most critical point in that diagram, in terms of information

Box 3–7. The one gene-one enzyme theory.

The genetic control of nutritional mutants or auxotrophs was first demonstrated in *Neurospora crassa,* the red bread mold. A. This fungus can be grown on a minimal medium. Agar, to which glucose, biotin, and certain salts have been added, is sufficient to support growth of the normal genotype. The organism, which reproduces asexually, can also produce spore-like structures called conidia. B. After the conidia have been harvested, they are irradiated to induce mutations.

The rest of the diagram shows procedures for finding nutritional mutants or auxotrophs. These are mutants unable to synthesize one or more chemicals needed for growth. (The normal plant possessing a complete array of synthetic capabilities is called a prototroph.) C. The irradiated conidia are placed on a complete medium where they germinate and grow. A complete medium is a minimal medium to which has been added vitamins, essential amino acids, and the four different nucleotides. This assures growth of the auxotroph. D. The auxotroph is then crossed to a prototroph of the complementary mating type and fruiting bodies are formed within which meiosis occurs in each ascus. E. In a single ascus from the fruiting bodies there are eight ascospores. An adjacent pair of ascospores is derived from one chromatid. Therefore they represent the four products of meiosis produced by diploid cells formed from the cross being investigated. F. Each ascospore is isolated as a culture and grown separately on the complete medium. G. Each culture is subcultured on minimal medium. If the suspected auxotroph does contain a mutated gene that blocks growth, this will segregate from its normal allele (in the prototroph) and show a 4:4 or 1:1 segregation of growing and non-growing cultures. H. The nature of the nutritional mutation is determined by culturing on media to which one or another of the constituent of the complete medium, i.e., individual vitamins, amino acids, and nucleotides, has been added. When the culture responds by growth it can be concluded that some enzyme leading to the biosynthesis of the nutrient in question is missing because the gene necessary for its formation is inactive or forms an inactive product. These results are expected on the basis of the one gene-one enzyme theory.

transfer, is the site of *m*RNA, *t*RNA, and ribosome interaction.

It had long been clear that a one nucleotide to one amino acid relationship was an inadequate model. At most it would allow the four different nucleotides to specify four different amino acids. But 20 amino acids had to be specified. Combinations of two nucleotides were also inadequate, since that would specify only 16 amino acids. But combinations of three of them ($4^3 = 64$) would more than cover all twenty amino acids. Hence Nirenberg and Ochoa synthesized RNA molecules that contained three bases, in all possible combinations, and demonstrated that when these were attached to ribosomes, specific *t*RNA molecules and their amino acids then attached to the bound triplets of *m*RNA. In this way a dictionary was compiled that matched one or more nucleotide triplet to each amino acid (Table 1–2). Each RNA triplet or codon was complementary to

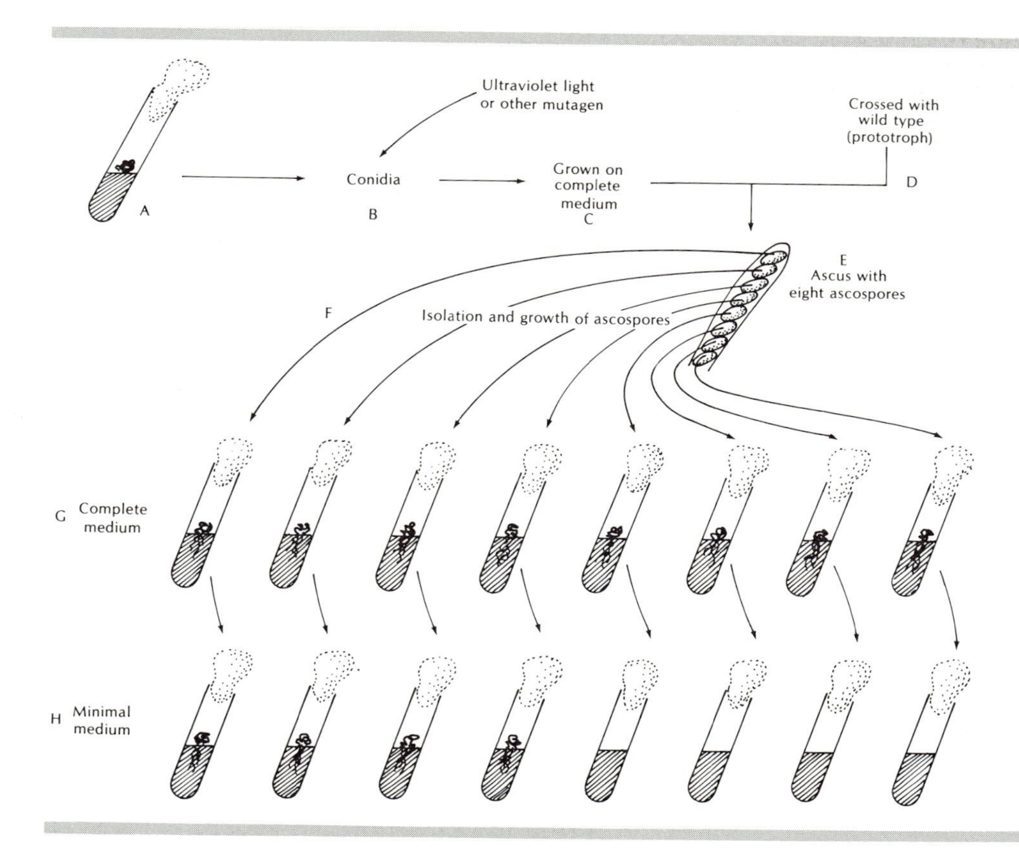

Box 3–8. The synthesis of a polypeptide illustrating the transfer of genetic information from DNA to a polypeptide.

A. Of the two polynucleotide strands in a DNA molecules, one is the sense strand, and other the nonsense strand. Thus only one strand, the sense strand, is transcribed into *m*RNA. B. Transcription occurs when a segment of the double helix unwinds and the two strands separate. Onto the sense strand an RNA polymerase attaches; this enzyme is largely responsible for the production of the *m*RNA, whose bases are complementary to those in the DNA sense strand. C. The *m*RNA moves to the cytoplasm and there attaches to a ribosome. The ribosome of (protein and *r*RNA) is the site at which translation occurs. Here *t*RNA molecules become active. First a sequence of these nucleotides, complementary to an *m*RNA codon, and called an anticodon, attaches temporarily to the *m*RNA. Second, an amino acid, at another point on the *t*RNA, is picked up. The introduction of a specific anticodon and a specific amino acid is essential in the translating of codons into amino acids, i.e., in translating a polynucleotide into a polypeptide. D. The completed polypeptide can then fold into its functional conformation as an independent molecule or in conjunction with other molecules. The ribosome again attaches to another *m*RNA molecule and the *t*RNA molecule picks up its respective amino acids. The *m*RNA may be again translated or destroyed; unless protected these molecules have a short half-life.

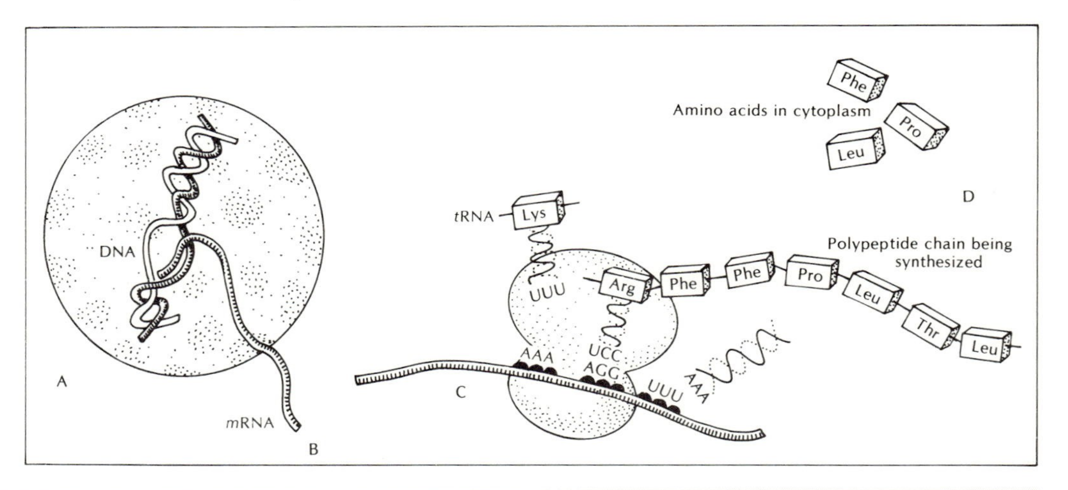

a DNA codon. It now became clear how the information in a DNA polynucleotide could be expressed as a polypeptide. Information storage depends, then, on specific nucleotide sequences. But those sequences are first organized as codons—three nucleotides that specify, via *m*RNA codons and *t*RNA anticodons, a certain amino acid—and then groups of codons, corresponding to a whole polypeptide, which make up a cistron or gene.

DNA and replication, mutation, and recombination. In

their first paper (1954), Watson and Crick wrote, "It has not escaped our notice that the specific pairing we have postulated [i.e., A–T and G–C] immediately suggests a possible copying mechanism for the genetic material." This has come to be known as the semi-conservative mode of replication of DNA (Fig. 3–6). The name refers to the understanding that each new double helix contains one polynucleotide strand from the parent double helix and one newly synthesized polynucleotide strand. In terms of preserving the genetic information present in the parent double helix, the semi-conservative mode of replication is totally and precisely conservative, except for very rare mistakes.

Those mistakes are, of course, mutations. Such mutations, also called point mutations, result from a change in nucleotide sequence. In terms of information in the genetic code, these point mutations are codon changes. The *m*RNA codon for phenylalanine, UUU, is transcribed from an AAA codon in DNA. A change of the last nucleotide will result in a new DNA codon, AAG. This will transcribe to UUC. But note that since more than one codon can specify the same amino acid (this is called degeneracy of the code), the mutated code will still specify phenylalanine (Table 1–2). This mutation has no phenotypic effect. Now suppose there is a mutation that changes AAA to CAA. This leads to an *m*RNA codon, GUU, which will specify valine. This may or may not have a phenotypic effect, depending on whether the new amino acid affects folding of the polypeptide. If there is an effect, a protein with

Fig. 3-6: The semi-conservative mode of DNA replication. In synthesis, which is always in the 5′ to 3′ direction, the complementary parental strands unwind. In the 3′-5′ strand (left, each diagram), uninterrupted copying is largely by a DNA polymerase; in the antiparallel 5′-3′ strand, the DNA polymerase is inserted to make 5′→3′ segments, joined by ligases, as space becomes available. The resulting pairs of double helices are each a perfect copy, except for the rare mistakes in complementary pairing of nucleotides, of the original double helix. (See also Box 1-1).

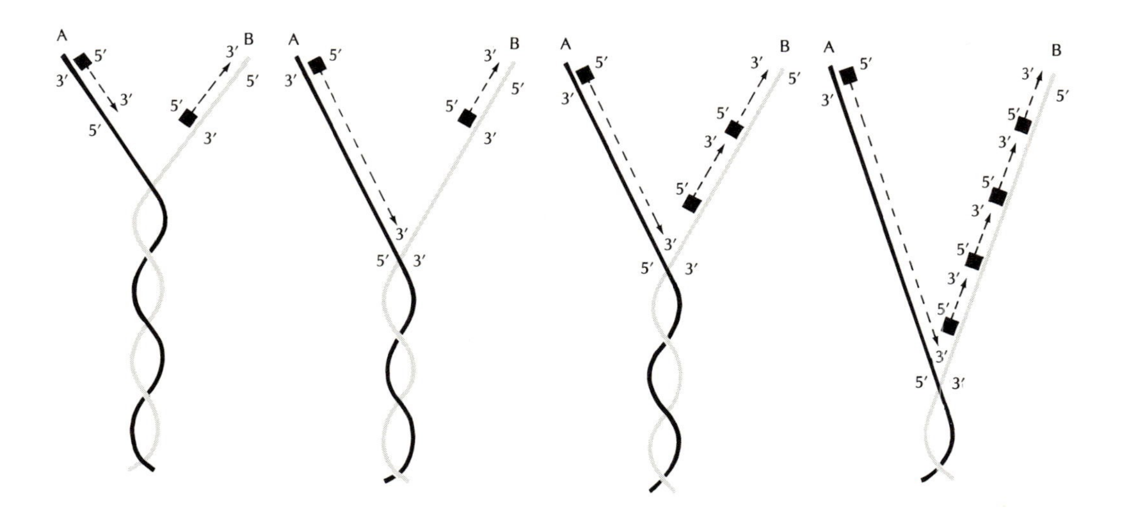

changed properties can result and then there may be a detectable phenotypic effect.

Such nucleotide changes illustrate two important points. First, any nucleotide in a gene, or cistron, can change. Such a change is the theoretically smallest mutable unit, or muton. Second most of these point mutations will not be detected. Twenty amino acids are specified by 61 codons. (Three of the 64 are used for punctuating the code.) Because of this degeneracy, changes can occur in about two-thirds of the codons without being expressed as amino acid changes. And, further, not all amino acid changes affect protein function. (In the case of hemoglobin, a change in one amino acid can be important. In other proteins many changes in or even losses of amino acids have no effect on protein function.)

The second point, above, tells us we are certainly *under*estimating mutation rates. We are probably detecting less than one-third of the nucleotide changes in DNA.

The effect of changes that affect *m*RNA codons, as compared to those affecting *t*RNA anticodons, is important in DNA point mutations. In the first instance, if we assume that the change results in amino acid change, the result will affect *just* that polypeptide (and protein) specified by the cistron containing the mutated nucleotide. The situation can be quite different for an anticodon change. There is a precise relation between an anticodon and the amino acid that attaches to the *t*RNA with that anticodon. This is what translates codons into specific amino acids. A change in the anticodon can, therefore, insert into the translation a different amino acid than that specified by the codon. For example, if the *m*RNA codon is UUU the usual *t*RNA has AAA in its anticodon and specifies phenylalanine for the polypeptide. If, however, the anticodon for phenylalanine *t*RNA has mutated to CAA, it will respond to an *m*RNA codon of GUU. Instead of supplying valine in response to GUU, it inserts phenylalanine. So far this just seems to be another amino acid substitution, which may or may not have a phenotypic effect. But it is much more than that. This mistake will occur in *every* protein that calls for valine in its synthesis. Mutations affecting *t*RNA anticodons can affect all or nearly all proteins, whereas *m*RNA mutations affect *only* the protein translated from a single mutated cistron.

Of even greater significance, mutations of anticodon sites are the equivalent of changing the dictionary of the code. Such changes introduce much nonsense into previous conventions of the code. It is like arbitrarily changing a letter in our alphabet. Take this sentence and introduce the letter x in place of every e. (Takx this sxntxncx and introducx thx lxttxr x in placx of xvxry x.) Compare that with the same sentence where only one e is replaced by an x. The difference in the effect on the sentence is comparable to anticodon mutations and isolated *m*RNA point mutations. It says, further, that once the genetic code evolved it became frozen because tampering with it could be lethal. This may well be the reason why we have a universal genetic code, and it is also our strongest argument for the unitary origin of all living forms.

Finally, let us briefly examine recombination as it relates to the model of DNA as a double helix. We can postulate that there is at least one such helical molecule running the length of a chromosome. Various experimental studies have supported this view of chromosomes, except for polytene chromosomes. [These latter are chromosomes, such as those in the salivary glands of the fruit fly, made up of many (up to 1,024) DNA double helices.] In prophase chromosomes undergoing meiosis, each chromatid is generally thought to contain a single double helix. Hence recombination, by crossing-over, involves the breaking of two homologous double helices and their reciprocal rejoining. One view of how this may occur is given in Fig. 3–7. This figure is based on the known action of such enzymes as the nicking enzyme (endonuclease), the strand digestion enzyme (exonuclease), repair enzymes (polymerases), and joining enzyme (ligase). Whether this is really the way they interact during recombination is not known. More generally, we still cannot explain how homologous chromosomes pair in synapsis. This is, of course, a necessary prerequisite to the events proposed in both Figs. 3–4 and 3–7. Nor do we know how chiasmata form, why breaks (nicking) first occur where they do, and why only two of four chromatids recombine at any one site along the chromatids. Obviously, much remains to be learned regarding recombination.

NUCLEIC ACIDS AND GENOMIC ORGANIZATION

Lastly we need to look at the whole genome in the light of present knowledge regarding the chemical basis of heredity.

Viral genome. In two instances we know the nucleotide sequence for *all* the DNA in an organism. Predictably, these organisms are viruses. One of these is the bacterial virus called ϕX174; in its genome there are 5,386 nucleotides that specify ten proteins (Fig. 3–8). This work was reported in 1978 by a team headed by Michael Smith, of the University of British

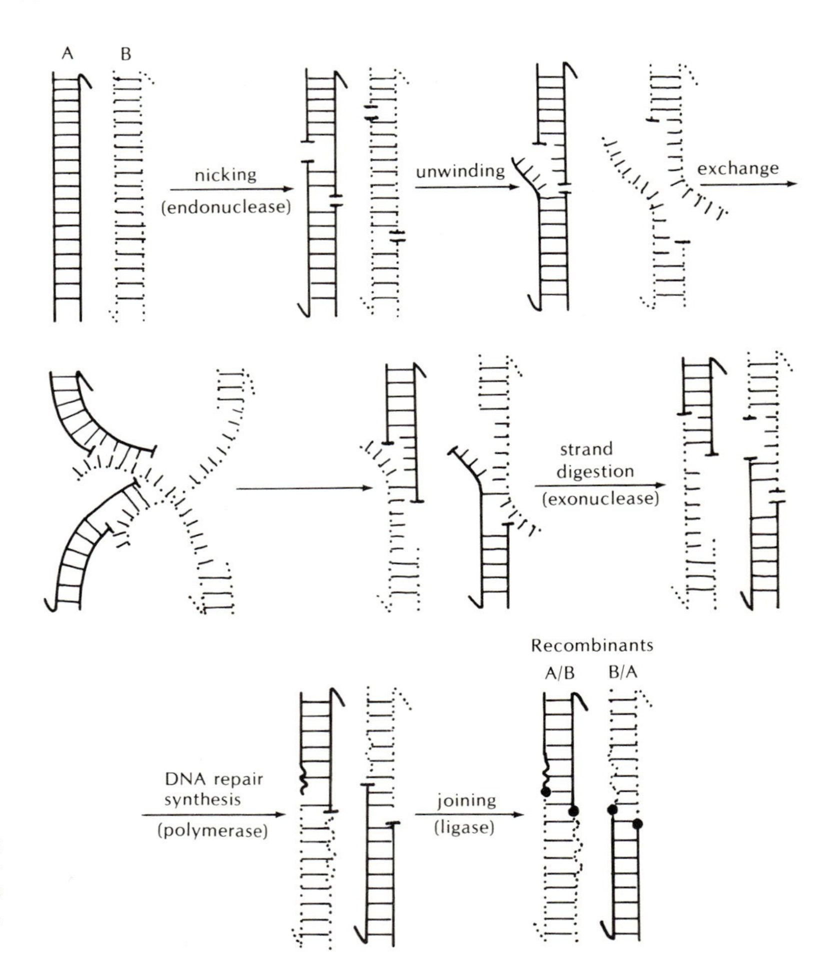

Fig. 3-7: Possible recombination between two DNA molecules (A and B). Our knowledge of DNA structure and of the enzymes involved in DNA breakage, repair, synthesis, and joining is applied here.

Columbia. But much of the work was done in the laboratory of Dr. F. Sanger, a pioneer in the sequencing of large proteins (Box 4–3). Two interesting points emerge from this study of ϕX174. First, there are only 217 nucleotides in the non-coding sequences between the genes; all the rest code for proteins. Second, there is overlap in the code. For example, gene A overlaps all of gene B and part of gene K. (In the other virus for which the nucleotide sequence is known, the simian virus SV40, there is also significant overlap.) Although overlapping seems an efficient way to use genetic information (Fig. 3–9), intriguing questions arise regarding its role in controlling the timing of protein production, for example. Its role is under investigation now.

The smallest genome known is that of the RNA virus of the

bacterium *Escherichia coli*. This virus, MS2, has three genes. One specifies the protein coat that surrounds the mature virus; another determines the assembly protein that brings the RNA and coat protein together; and the last specifies a replicase needed for RNA genome replication. In MS2 there are about 3,000 nucleotides. Compare this with the human, in whom there are approximately 3.2×10^{29} nucleotide pairs in the haploid set of 23 chromosomes.

Prokaryote genome. The bacteria and blue-green algae comprise the prokaryotes, and their genomes, like the genome of ϕX174, are invariably one continuous Watson-Crick double helix of DNA. Though often referred to as circular molecules, they are far from being circular since they lie in a highly compacted mass in their cells. Genetically their genes map out to a linear array with no end. These circular genetic maps have been confirmed cytologically. Special isolation techniques and electron microscopy demonstrate delicate strands that are continuous threads of DNA. In addition to this nuclear DNA, other genetic material is also found in bacterial cells. This includes viruses, which we have already mentioned and whose genetic material can be DNA or occasionally RNA, and plasmids, which are always DNA. In bacteria a virus can attach to the bacterial chromosome and persist there indefinitely, dividing every time the host chromosome divides. Or the virus can remain in the host cytoplasm, taking over portions of the host cell's synthetic capabilities, and form more viral particles. This kills the host cell, but during the process, which takes about 20 minutes, a hundred or more new viruses are formed. Plasmids are smaller than the host chromosomes. They remain in the cytoplasm, but replicate in synchrony with DNA replication in the rest of the cell. Plasmids can be transferred from one bacterial cell to another when pairing occurs and they can also be transferred when a host cell breaks up and another cell takes up the free plasmid.

From intensive work on bacteria, and especially on *Escherichia coli,* there has been described an important pattern of organization of genes in the prokaryotes. This is the operon. In the case of *E. coli* this involves what are called structural and regulatory genes (Box 3–9). The three structural genes, or cistrons, each control the translation of part of one long *m*RNA. This *m*RNA is, in turn, translated into three separate polypeptides, which fold into three different enzymes. The three regulatory genes consist of a regulator, a promotor, and an operator locus. They control transcription of the cistronal, or structural,

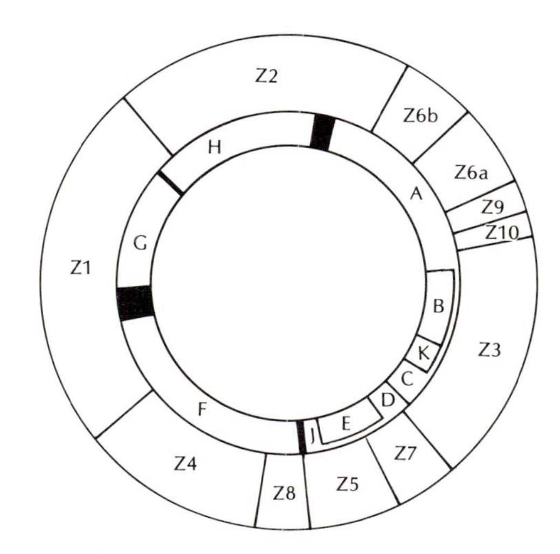

Fig. 3-8: Gene map of the circular chromosome of ϕX174.

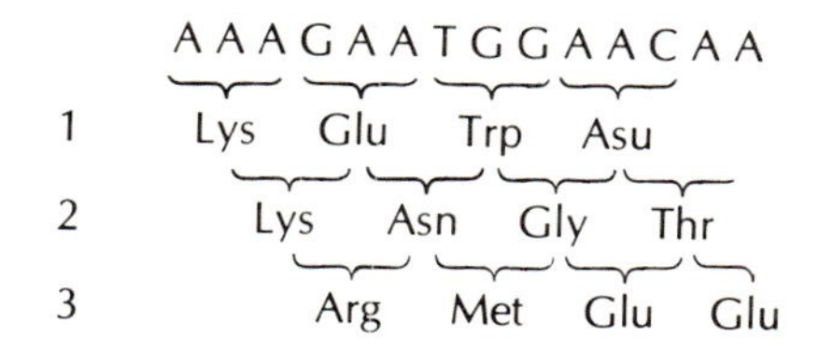

Fig. 3-9: Part of the overlapping genetic code in ϕX174. The same sequence of DNA nucleotides can specify three different amino acid sequences: Sequence 1 is part of protein A, sequence 3 of protein B. Sequence 2 is not translated because the needed read-out signal is missing.

Box 3–9. The operon in bacteria.

In *Escherichia coli,* the best-studied operon controls the formation of the enzyme β-galactosidase. The DNA consists of an array of six genes—the regulator, promoter, and operator are regulatory in function, the z, y, and a loci are structural genes. The z locus controls the formation of β-galactosidase, the enzyme that breaks down lactose, a sugar of the β-galactoside form, to galactose and glucose; the y locus controls the formation of permease, an enzyme that controls the rate of entry of lactose into the cell; and the a locus controls the production of transacetylase, an enzyme whose role is as yet unclear.

A. In the absence of lactose, the regulator locus produces (via *m*RNA) a small repressor protein that blocks the operator locus. This blockage apparently prevents RNA polymerase, which attaches to the promoter locus, from transcribing the structural genes.

B. When lactose is present, it acts to induce transcription by combining with the repressor and preventing repressor interaction with the operator. Now the RNA polymerase—a very complex protein made up of seven smaller proteins—proceeds to transcribe the structural genes. The product is one long RNA message. It, however, is translated in three separate polypeptides that fold into β-galactosidase, permease, and transacetylase. In this example of the regulation of the expression of three structural genes we see how the cell containing the operon is regulated by the presence of lactose. When lactose is present, the residual permease allows it to enter the cell. There it combines with the suppressor, and transcription proceeds. The β-galactosidase acts on the lactose to produce sugars utilizable by the cell. More permease is produced, so that more lactose enters the cell; this acts as a positive feedback system for the production of more enzymes. When the environmental lactose is used up, the repressor again asserts itself at the operator locus and enzyme synthesis abates.

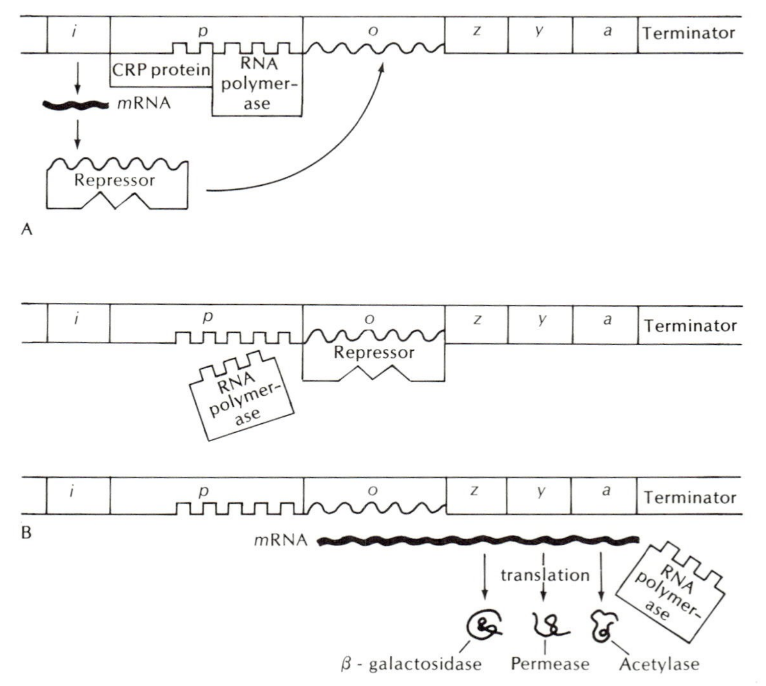

genes in response to the availability of substrate for the enzyme β-galactosidase, produced by the first cistron. Thus the operon represents an interaction of six different loci lying together in the bacterial genome. Such tandem arrangements of functionally related genes are found in other bacteria, too.

The operon is an important model of gene expression in these cells. It advanced our ideas on gene interaction and the effect of environmental factors (here, the availability of substrate for β-galactosidase) on gene action. Although the search for operons in eukaryotic cells has not been successful, such a model of how genes can be organized to interact has stimulated work in gene expression.

Eukaryote genomes. Although no operon has been identified in eukaryote cells, there is strong evidence that both structural and regulatory genes are present. Perhaps as much as one-half of the genome may involve regulatory genes.

The eukaryote genome is made up of different segments of double-ended DNA molecules, that is, chromosomes. The DNA is intimately associated with proteins, especially histones, and RNA. As mentioned, some chromosomes have more than one strand of DNA, although all are identical except for rare mistakes in replication. Most are thought to have only one double helix per chromosome or chromatid (when the chromosome has replicated). But a haploid genome can have as few as one chromosome (a special case is the nematode *Ascaris megacephalus;* the mosquito *Culex pipiens* has $n = 3$) or as many as 255 (the fern *Ophioglossum petiolatum*). The average haploid number is usually between 6 and 25.

Because of the molecular complexity of chromosomes, and the amount of DNA and the number of genic sites within them, we still do not have a clear view of intrachromosomal organization. We do have certain basic ideas about their structure—genes are linearly arranged; each chromosome has two ends and a centromere for attaching to mitotic and meiotic spindles; chromosomes are highly coiled during mitosis and meiosis and uncoil to varying degrees in interphase—but these emphasize how little we know about their end-to-end organization. We are, however, beginning to determine the locations of regulatory genes on chromosomes by such elegant techniques as those developed by Joe Gall, of Yale University, and his coworkers. *Drosphila* chromosomes are placed on a microscope slide, protein is removed, and the paired polynucleotide strands of the DNA double helix are gently heated to separate them. Then RNA, labeled with a radioisotope, is added to the slides. This radioactive RNA, transcribed from regulatory gene

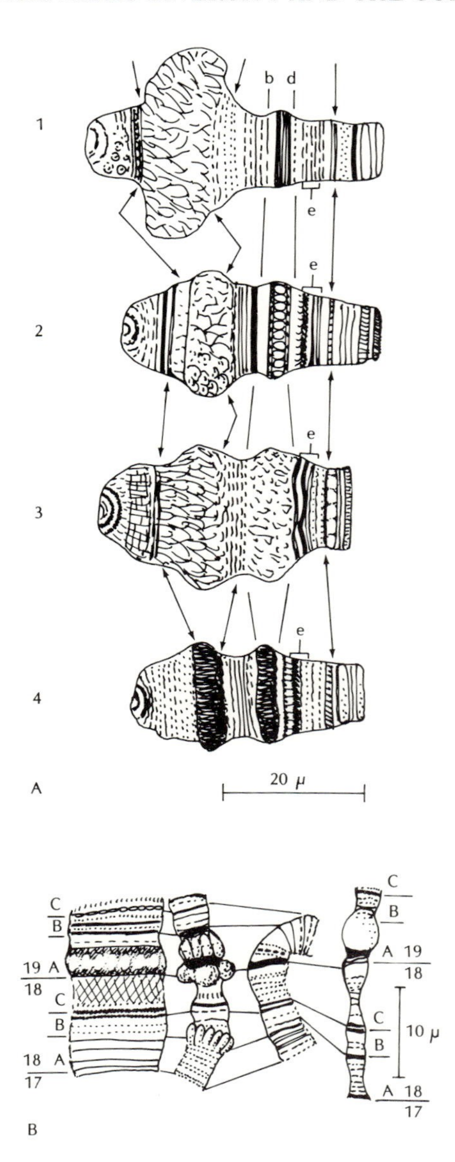

Fig. 3-10: Visual evidence for differential gene action. A. In the chromosomes of the fungus gnat *Rhynocosciara angelae,* segments 1 through 4 show the same chromosome in different stages of larval development. The puffing, indicative of RNA transcription, is most pronounced in segments 1 and 3. B. Segments 1 through 4 are of the same chromosome in *Chironomus tentans,* at the same stage of development, but from cells of four different tissues.

DNA isolated by special techniques, joins to the segments of chromosomal DNA complementary to it. The radioactivity along the DNA chromosomes then labels the sites of regulatory genes. This type of ingenious cytological mapping is one approach to unraveling some of the mysteries of the eukaryote genome.

One of the most intriguing of these mysteries, which is now being unraveled, is that of gene expression in eukaryotes. For decades the basic problem has been how different cells arise in a multicellular plant or animal when they all start out with the same genes. The genetic endowment of a multicellular organism is determined by its zygote, and all cells of the organism are derived from that zygote. Therefore, there is genotypic identity in all cells, but yet we find brain cells and liver cells in a developing animal and cells for photosynthesis in leaves and cells for transport of metabolites in stems in a developing plant. The answer is that although different cells in an organism all have the same genes, these genes do or do not produce their products depending on the cell in which they lie. This is known as differential gene action. It has been shown most clearly in the large chromosomes of the fungus gnat *Chironomus tentans* (Fig. 3–10). In this gnat the localized swelling of a chromosome into puffs means that *m*RNA transcription is occurring, that is, structural genes or cistroms are active. (There are probably some regulatory genes acting, too, but what they actually do is not known.) At different times in development the same area on a chromosome has different puff patterns, that is, different activities. In different tissues, at the same time in development, the same area on a corresponding chromosome also has different puff patterns. Both sets of observations support the idea of differential gene activity as the basis of gene expression during development.

This chapter is a summary of an extraordinary page in the history of science. From Mendel, in 1865, to the present, we have moved from a concept of heredity as involving unseen, hypothetical factors to a molecular description of genes. We still have a great more to do in explaining how the genes of a whole genome transform a zygote in a congenial environment into a functional, adult organism. Nevertheless, our knowledge of how genes are transmitted from one generation to the next, of how they mutate and recombine, and of how they are arranged on the chromosome is now so advanced that we can turn to evolutionary questions vastly better equipped than was Charles Darwin when he first struggled with the problems of organismic diversity.

References

Avery, D. T., C. M. MacLeod, and M. McCarthy, 1944. Studies on the chemical nature of the substance including transformation of pneumococcal types. Induction of transformation by deoxyribonucleic acid fraction isolated from pneumococcus type III. *Journal of Experimental Medicine 79:*137–158. (Also reprinted in Peters, 1959.)

Carlson, E. A., 1968. *The Gene: A Critical History.* Saunders, Philadelphia.

Goodenough, U., 1977. *Genetics.* 2nd ed. Holt, Rinehart and Winston, New York.

Hexter, W. M. and I. Yost, 1977. *The Science of Genetics.* Prentice-Hall, Englewood Cliffs, N.J.

Lai, E., S. Woo, A. Dungaiczyk, J. Catterall, and B. O'Malley, 1978. The ovalbumin gene. Structural sequences in native chicken DNA are not contiguous. *Proceedings of the National Academy of Science, U.S.A., 75:*2205.

Mendel, Gregor, 1865. *Experiments in Plant Hybridization.* Reprinted in translation in Peters, 1959.

Min Jou, W., G. Haegeman, M. Ysebaert, and W. Fiers, 1972. Nucleotide sequence of the gene coding for bacteriophage MS2 coat protein. *Nature 237:*82–88.

Muller, H. J., 1927. Artificial transmutation of the gene. *Science 66:* 84–87.

Pardue, M. L., and J. G. Gall, 1970. Chromosomal localization of mouse satellite DNA. *Science 168:*1356–1358.

Peters, J. A. (ed.), 1959. *Classic Papers in Genetics.* Prentice-Hall, Englewood Cliffs, N.J.

Smith, M., 1979. The first complete nucleotide sequencing of an organism's DNA. *American Scientist 67:*57–67.

Srb, A., R. Owen, and R. S. Edgar, 1965. *General Genetics.* 2nd ed. W. H. Freeman, San Francisco.

Stern, C., and E. R. Sherwood (eds.), 1966. *The Origin of Genetics: A Mendel Source Book.* W. H. Freeman, San Francisco. (Includes a discussion of Mendel's ratios by R. A. Fisher and S. Wright.)

Strickberger, M. W., 1976. *Genetics.* 2nd ed. Macmillan, New York.

Watson, J. D. and F.H.C. Crick, 1953. Molecular structure of nucleic acids. A structure for deoxyribose nucleic acid. *Nature 171:*737–738.

Williamson, B., 1977. Globin hnRNA = mRNA. *Nature 269:*9.

2

Adaptation in Reproductive Communities

Polar bear family in ice den with air vent

SPECIES ARE REPRODUCTIVE COMMUNITIES: groups of individuals among whom sexual reproduction occurs successfully. This means that when otherwise fertile individuals attempt, unsuccessfully, to reproduce, they are probably members of different species. This biological isolation of reproductive communities is one of the fundamental facts of organistic diversity.

Every organism, *without exception,* belongs to one species or another. Why is this so? The answer must lie in the function of a species. Hence we must ask what a species does as reproductive community. One of the foremost students of evolution, Professor Ernst Mayr, answers that question. He says that a species has three sets of attributes, "those that (1) adapt species to their physical environments, (2) enable species to coexist with potential competitors, and (3) permit species to maintain reproductive isolation from other species . . ." (from E. Mayr, 1963, *Animal Species and Evolution*. Belknap Press of Harvard University Press, p. 60).

In Part 2, we will study how a species achieves these attributes. The arguments will be developed as follows: In the next chapter (Chapter 4), we examine genetic variation within a species and how we measure that variation. In Chapter 5, we study selection and its effects on the genetic resources available to a species. In Chapter 6, the roles of certain other factors as they affect those same genetic resources are examined, and we pull together all these factors as they interact to achieve well-adapted reproductive communities.

Overall, we shall see that species are often referred to as the fundamental units of evolution because they are the means for generating and preserving well-adapted organisms. That is why species are the universal context for organisms.

FOUR

Genetic Variation within a Species

WHY START OUR STUDY of species as adapted reproductive communities by looking at their genetic resources and, in particular, at the variation in those resources? The first and obvious answer is that an understanding of genetic resources is essential to an understanding of biological potentialities. And since potentialities set the limits to capabilities, we can thus see what a species is capable of doing as an adapted reproductive community. To really understand genetic resources would require that we identify and characterize the thousands of genes present in the genome of a species. But such an insight into the heredity of a given species is not possible today, although it remains an ideal to be pursued. Just how close we are to realizing it will become clearer later in this chapter.

Another, less ambitious reason for studying the genetic variation of a species is to settle once and for all how we see a species, in genetic terms. One view, now called the *classical theory of population structure,* is that members of a population are characterized by certain highly adapted genes. That is, at each locus the wild-type allele is the predominant one. Other, less well-adapted alleles occur, but at very low frequencies. In contrast to this view there is the *balance theory of population structure.* This proposes that an *adaptive norm* or a certain phenotype is the expression of various different genotypes. Hence there is no predominant set of genes, but several genotypes, and they produce an ideal phenotype in the sense of an optimally adapted one. If we can choose between these models—and we can by looking at natural populations and determining their genetic variability—then we are in a position to go on, in subsequent chapters, to see how selection and other factors act on the genetic material to ensure survival.

Our first task, then, is not that of generating a full catalog of all the genes present in a species; it is the more modest one of seeing how much variability there is at selected loci. To do this we will take a rather detailed look at gene or point mutations so as to fully understand how new alleles arise and how they might be detected. Next we want to see just how allelic variation in populations is measured. And, finally, we will see how alleles can be recombined through the processes of asexual and sexual reproduction. That introduces us to the Mendelian population, which is the geneticist's view of a reproductive community.

Further aspects of point mutation

In Chapter 3 we said simply that point mutations are changes in the nucleotide sequences of a gene and that they most commonly arose spontaneously as mistakes in replication of genic DNA. A further point to be emphasized is that such changes are not always detected. This will be clear from looking at the two chief kinds of base sequence changes, namely, base substitutions and frameshifts.

BASE SUBSTITUTIONS

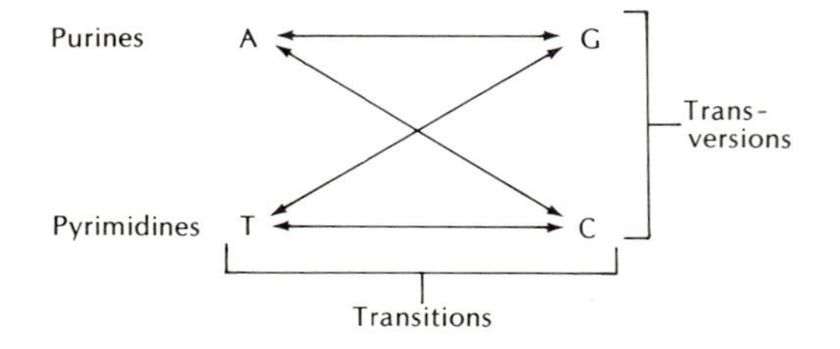

Fig. 4-1: Transition and transversion mutations.

As the name implies, one base is replaced by another. Transition substitutions are most common. In them one purine (adenine or guanine) is substituted for the other, or one pyrimidine (cytosine or thymine) is substituted for the other. Less common are transversions wherein a purine substitutes for a pyrimidine or vice versa (Fig. 4–1).

As regards the phenotypic effects of these changes, two questions must be kept in mind: Does the mutation result in a codon change such that a different amino acid is specified by that codon? If a different amino acid is specified, is that change detectable?

To answer the first question look again at Table 1–2, for example, at the codons that specify leucine. The DNA codons

Table 4–1. Amino acid substitutions due to a change in a single base.

Amino acid	Possible substitutes												
ala	asp	glu	gly	pro	ser	thr	val						
arg	cys	gln	gly	his	ile	leu	lys	met	non	pro	ser	thr	trp
asn	asp	his	ile	lys	ser	thr	tyr						
asp	ala	asn	glu	gly	his	tyr	val						
cys	arg	gly	non	phe	ser	trp	tyr						
gln	arg	glu	his	lcu	lys	non	pro						
glu	ala	asp	gln	gly	lys	non	val						
gly	ala	arg	asp	cys	glu	non	ser	trp	val				
his	arg	asn	asp	gln	leu	pro	tyr						
ile	arg	asn	leu	lys	met	phe	ser	thr	val				
leu	arg	gln	his	ile	met	non	phe	pro	ser	trp	val		
lys	arg	asn	gln	glu	ile	met	non	thr					
met	arg	ile	leu	lys	thr	val							
non	arg	cys	gln	glu	gly	leu	lys	ser	trp	tyr			
phe	cys	ilu	leu	ser	tyr	val							
pro	ala	arg	gln	his	leu	ser	thr						
ser	ala	arg	asn	cys	gly	ile	leu	non	phe	pro	thr	trp	tyr
thr	ala	arg	asn	ile	lys	met	pro	ser					
trp	arg	cys	gly	leu	non	ser							
tyr	asn	asp	cys	his	non	phe	ser						
val	ala	asp	glu	gly	ile	leu	met	phe					

See Table 1–2 for the genetic code for these changes.

are GAA, GAG, GAT, and GAC. A transitional mutation can change GAA to GAG and the codon will still specify leucine. And a transversional change of GAG to GAT will also still specify leucine. Hence these mutations will have no phenotypic effect. Many other similar examples can also be found in Table 1–2.

On the other hand, many mutations do change an amino acid (Table 4–1). Many of these changes, however, do not result in a detectable phenotypic effect, whereas others do. The case of the hemoglobins is worth looking at here. Hemoglobin is the blood protein involved in oxygen transport. Normal vertebrate beta hemoglobin has glutamic acid at position 6, but hemoglobin S has valine in that position (Fig. 4–2B). In this case substitution of a single amino acid has a significant phenotypic effect.

There are other cases, however, in which changes in amino acids cannot be expected to change the phenotype. Actual cases will be hard to find precisely because it may not be possible to detect phenotypic differences between a normal and a mutant protein. Very often the differences we look for are different functional capabilities, as in the case of the different hemoglobins and their capacity to bind oxygen or dif-

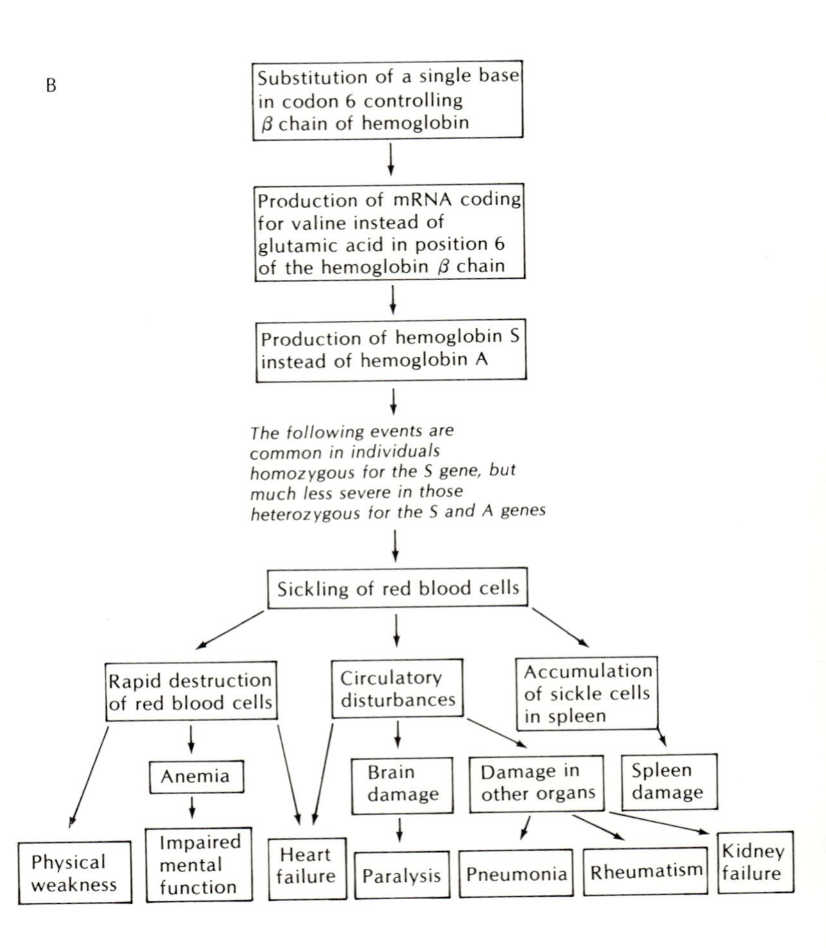

Fig. 4-2: The molecular basis for sickle-cell anemia is shown (A). Here, part of the sequence of amino acids in the beta polypeptide chain of normal hemoglobin is compared with the same part of sickle-cell anemia hemoglobin (hemoglobin S). The latter chain has valine at position 6, the former glutamic acid. The codons specifying these amino acids could be any of the relevant ones shown in Table 1-2. For example, if glutamic acid is specified by the DNA codon CTT, a transversional mutuation of the middle nucleotide thymine to adenine would produce the codon CAT, to specify valine. B. The phenotypic effects of hemoglobin S.

ferent proteins and their enzymatic activities. The enzyme papain can have 80 of its 180 amino acid residues removed and it will still act in its usual way. Or the adrenocorticotropic hormone (ACTH) appears to require only 16 or perhaps 24 of

Box 4–1. The neutralist-selectionist controversy.

This controversy revolves around this one question: What role does selection play in determining variation present in natural populations? We will have occasion to return to the question when selection is discussed in more detail. Here the issue is raised because of its obvious connection to the genetic structure of natural populations.

If variation is the result of selection, i.e., only those alleles persist that are of selective advantage, then we might predict results consistent with the classic theory of population structure. If, however, variants arise that were neither selected for nor against, then we might expect results consistent with the balance theory. However, the proponents of the balance theory make a strong argument that their view of natural populations is the consequence of natural selection, that the adaptive norm generated by several phenotypes is a perfectly good strategy for survival by natural selection. The reasons for that will become clearer in the next chapters. The first question to be answered now is why do gene mutations and their phenotypic effects sometimes not affect natural selection?

The first answer comes down to gene mutations as changes in base sequences. Because code degeneracy exists (Table 1–2) base changes may not change the amino acid the codon specifies. Therefore, the point mutation in question has no phenotypic effect and cannot be acted upon by natural selection.

At the next level of possible change, namely the polypeptide, we know that certain amino acid changes have pronouned phenotypic effects (Fig. 4–2), whereas many others have not. The latter make up another category of neutral changes.

Finally, there are many examples of detectable phenotypic changes at the organismic level—structures and functions—that have no known selective advantage or disadvantage. Certainly, in many cases, the absence of selective value reflects our ignorance rather than being a true measure of the presence or absence of selection pressures. But certain evolutionists are making a strong case today that much of the variation we see around us has no relation to natural selection. That is the neutralist position, which is opposed by selectionists. Of course, there is room for an intermediate position, too. Whatever the points of view in this selectionist-neutralist controversy, researchers go on trying carefully, case by case, to establish what the selective advantage or disadvantage may be or whether a given phenotypic change is neutral. Thus far there are examples to support each viewpoint: selectionist, neutralist, and a combination of both.

its 39 amino acids for hormonal activity. In such cases, it is possible that various amino acid substitutions could occur without producing detectable phenotypic effects. Mutations causing these changes are described as silent, or more accurately, as occurring in the *silent areas* of DNA.

What these silent areas mean in evolution is simply not known. They could, of course, not be silent, since silence could simply be a synonym for our ignorance. But if these areas really are silent and have no phenotypic function, why are they there? How can an organism afford the luxury of nonfunctional molecular structures? It takes nutrients and energy to build and maintain these molecules and this would be hard to explain in the light of natural selection (Box 4–1).

The explanations of protein function depend on the intricate foldings of the polypeptides. In hemoglobin, certain amino acids on the surface define the configuration that binds the molecule to oxygen. Presumably that site cannot mutate without affecting the oxygen-binding capacities of hemoglobin. And it also seems clear that such a site would be quite conservative. Of the 141 amino acid residues in the globin polypeptides, only eleven were first reported always to be present. Presumably these bound to the heme molecule, which in turn, assured binding to oxygen. But more recently another mutant hemoglobin has been found, Hemoglobin Gun Hill, which lacks both the heme groups and also five amino acids near the site of attachment of the heme. This reduces the always present amino acids or invariant sites to six or 4% of the total. Looking at another protein, cytochrome c, which is important in cellular respiration, we find a somewhat different story. Here, about 30% (34/113) of the amino acid residues seem invariant.

Invariant sites can be expected for a number of reasons. They are needed by enzymes for joining with substrates that are essentially invariant. They represent attachment sites to membranes (this has been suggested for most of the invariant sites in cytochrome c). And there are many intermolecular attachments that need to be preserved, such as those between proteins that interact to form enzymes; between proteins and sugars that form cell walls; or between proteins and lipids that form membranes. In general, invariant amino acid sites can range from one, as in a certain dehydrogenase, to dozens. Apart from these sites there seems to be significant areas for mutational changes, the phenotypic effects of which will be minimal or even undetected.

FRAMESHIFTS

Frameshift changes result from additions or deletions to the nucleotide sequence of a gene. In most cases known today, the transcription or reading of a gene is unidirectional. Consider what a deletion could do in such a situation. Suppose we have the following sequence of bases (codons are emphasized by added space between them):

TAC CAT CAT CAT CAT . . . .

The reading starts at the left codon, which is a codon for read-out initiation. The next four codons specify the amino acid sequence.

valine - valine - valine - valine . . .

Suppose the thymine nucleotide of the first valine codon is deleted. The codon sequence and its polypeptide product then become

TAC CAC ATC ATC AT . . .
valine - (stop) - (stop) . . .

In other words, no polypeptide is produced by this gene.

One can, similarly, determine the consequences of two deletions, either close together or at different places in the DNA molecule, as well as the consequences of the addition of one or two bases.

Note, however, if three bases are added or deleted as a block, the gene product may or may not be affected. Normally it will only add or subtract a single amino acid. The location of that added amino acid in the protein will determine the effect. A change in one of the amino acids essential for protein function—a change in an invariant site—will have a detectable phenotypic effect. A change elsewhere in the molecule may not have a detectable effect.

Finally, the position of frameshift changes is important. If the change occurs near the beginning of the transcriptional process, then all subsequent codons are changed. If the change comes later, it may have little effect. This is easily illustrated by looking at the result of the deletion or addition of a letter to *this* sentence. Insert the letter m, for instance, in the first word, i.e., change This to Thim. Then shift the s of This to the next word and take that next word's final letter and put it at the beginning of the next word. And so on, making the final letter of one word the initial letter of the next word, throughout the sentence. In a literal sense, because of this frameshift, the sentence becomes unintelligible. It reads

> Thim si seasil yillustrate db ylookin ga tth eresul to fth edeletio no radditio no falette rt othi ssentenc e.

Such a mutant sentence may or may not be decipherable as a result of this kind of frameshift. Certainly the change would cause less difficulty if it had occurred only in the last word of the sentence, and it might even be undetected by most readers. In brief, shifts in the reading frame of codon triplets, like nucleotide substitutions, may or may not cause a detectable mutation.

In any case, the fact remains that gene mutations, comprised of base substitutions, deletions, and additions, are the raw material of evolutionary change. They are the source of new genes and, as a consequence, the source of new phenotypic effects.

MUTATION RATES

There is now evidence that most single base changes occur spontaneously as mistakes in DNA replication. They can, of course, be induced experimentally by various physical agents or chemical or even biological mutagens. Physical agents include X-rays and ultraviolet irradiation. Chemical mutagens include a variety of substances ranging from relatively simple chemicals, such as formaldehyde (HCHO), to more complex ones, such as mustard gas ($ClCH_2CH_2SCH_2CH_2Cl$). Especially efficient are the molecular analogs of the bases naturally found

Table 4–2. Mutation rates for different genetic loci.

Organism	Trait	Mutation per 100,000 cells (gametes in most cases)
bacteria (from many sources)		
Escherichia coli (K 12)	to streptomycin resistance	.00004
	to phage Tl resistance	.003
	to leucine independence (*leu*$^-$→*leu*$^+$)	.00007
	to arginine independence	.0004
	to tryptophan independence	.006
	to arabinose dependence (*ara*$^+$→*ara*$^-$)	.2
Neurospora crassa	to adenine independence	.0008-.029
	to inositol independence	.001-.010
	(one *inos* allele. JH5202	1.5)
Drosophila melanogaster males	y^+ to *yellow*	12
	bw^+ to *brown*	3
	e^+ to *ebony*	2
	ey^+ to *eyeless*	6
corn	*Wx* to *waxy*	.00
	Sh to *shrunken*	.12
	C to *colorless*	.23
	Su to *sugary*	.24
	Pr to *purple*	1.10
	I to *i*	10.60
	R^r to r^r	49.20

Adapted from M. W. Strickberger (1976).

in DNA. These compounds are similar enough to the natural purines and pyrimidines to substitute for them in DNA replication. However, they do not substitute for them when it comes to transcription or further replication. Some of these analogs are shown in Fig. 4–3. Biological mutagens have been found as mutator genes. In corn, for example, the presence of certain genes results in more frequent mutations at other loci in the corn genome.

Our basic concern in these evolutionary studies is spontaneous mutation rates. Such rates determine how often gene mutations occur in nature. Such rates are determined by careful observation of changes in phenotypic characters known to be controlled by a single locus. Table 4–2 lists the results of such studies on different loci in a wide variety of organisms.

mouse	a^+ to *nonagouti*	2.97
	b^+ to *brown*	.39
	c^+ to *albino*	1.02
	d^+ to *dilute*	1.25
	ln^+ to *leaden*	.80
	reverse mutations for above genes	.27
man	retinoblastoma	
	England (Philip and Sorsby)	1.2
	U.S.A., Michigan (Neel and Falls)	2.3
	U.S.A., Ohio (Macklin)	1.8
	Germany (Vogel)	1.7
	Switzerland (Böhringer)	2.1
	Japan (Matsunaga and Ogvu)	2.1
	achondroplasia (chondrodystrophy)	
	Denmark (Mørch)	4.2
	North Ireland (Stevenson)	14.3
	Sweden (Böök)	7.0
	Japan (Neel, Schull, and Takeshima)	12.2
	partial albinism with deafness	
	Holland (Waardenburg)	.4
	Pelger's anomaly	
	Germany (Nachtsheim)	2.7
	Japan (Handa)	1.7
	neurofibromatosis	
	U.S.A., Michigan (Crowe, Schull, and Neal)	13.0–25.0
	Huntington's chorea	
	U.S.A., Michigan	.5

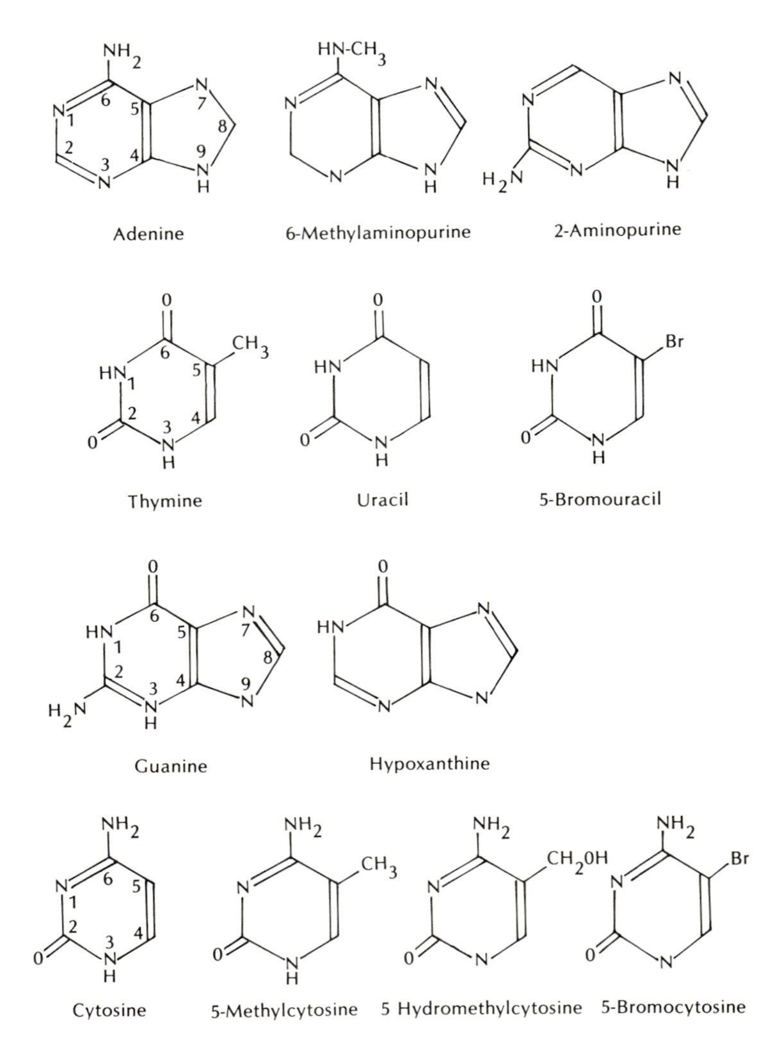

Fig. 4-3: Normal DNA bases (left) and their analogs (right).

Spontaneous mutations can vary a million-fold in frequency. Resistance to the antibiotic streptomycin is controlled by a very stable gene in *Escherichia coli;* its mutation rate is about once in 10^{11} cells. By comparison the nervous tissue tumor neurofibromatosis, in humans, and yellow body color, in the fruit fly, show relatively high mutation rates. On the average, the mutation rate of a given locus seems to be about once in every 10^5 or 10^6 cells, which is about once in 10^5 or 10^6 cell generations. Thus mutations at a given locus are quite rare.

But equally important is the number of mutable loci in a given cell. Estimates differ as to what that number is. In many viruses and in certain bacteria, the mutable loci might number up to a hundred or a thousand or more. In multicellular organisms, such as the fruit fly, there are as many as 10,000 loci in the haploid set of genes. Humans are thought to have even more loci. But regarding mutation rates and mutable loci, consider the consequences of the presence of thousands of loci in a population of some thousands of individuals. There are $10{,}000 \times 10{,}000$ (or 10^8) loci at which mutations can occur. (These assumptions of 10,000 genes in a haploid genome and a population of 10,000 individuals could easily apply to fruit flies, for example.) Even with a mutation rate of 10^{-5} or 10^{-6}, a significant number of mutational changes, namely 100 to 1,000 new mutations each generation, will occur in the population.

It must be remembered, however, that these gene mutations occur independently of the survival needs of the organism. They represent a process of continuing accidental change.

Measuring genic variation

How much genic variability actually occurs in an organism? To answer this question we must first decide on our approach. The ideal approach, as we said earlier, would be to catalog all the genes in the organisms we want to study; since that is not technically possible, what do we do? How do we measure genic variation within a species?

First of all we must decide which organisms we want to study. We shall first concentrate on organisms that reproduce sexually, a group of individuals who can interbreed successfully. In short, a Mendelian population. But remember that a Mendelian population can be very large. There are more than four billion humans, for example, so it is obvious that no observer can study all human beings. A sampling technique must be used.

A variety of techniques have been used for estimating genotypic variability by observing phenotypic variability. But what phenotypic effects are best studied? We could look at the whole organism and observe differences in size, color, or other characters (Fig. 4–4). However, most of these traits are polygenic; they are determined by several genes acting together. Very often the number of these interacting genes is

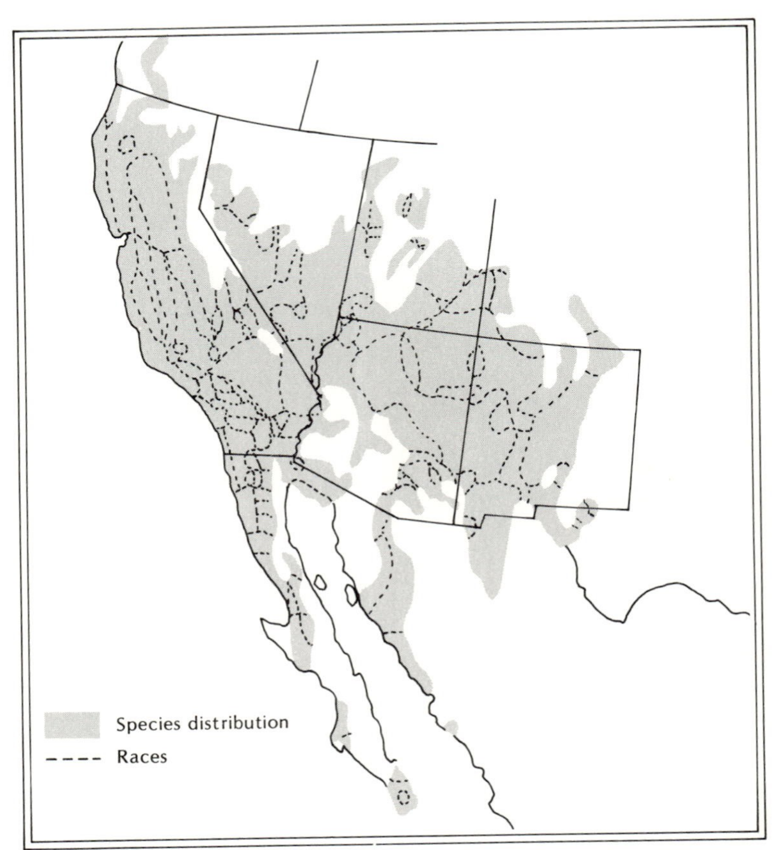

Fig. 4-4: Phenotypic variation in the pocket gopher *Thomomys bottae*. Visible variations are coat color, body size, and body (skeletal) proportions. Because similar phenotypes (races) occur in ecologically similar areas, e.g., similar soil types, though separated by hundreds of miles, variations may be adaptations to local conditions.

unknown. Additionally, there are environmental effects that are not hereditary, but that clearly produce a measurable variation. For these reasons, in analyses of variation in Mendelian populations, we try to stay away from phenotypic traits for which the genetic control is poorly understood. To obtain precise information regarding the genic control of certain traits is a long, painstaking job when approached by classical Mendelian techniques, namely, breeding analyses. It is often difficult to breed organisms in captivity, which must be done to achieve the necessary controlled matings. When successful, such an approach provides useful information on animals or plants that are easily cultured, such as fruit flies (genus *Drosophila*) or evening primroses (genus *Oenothera*). But there is no real genetic information on most deep-sea fishes or on many tropical plants, for example. In fact, fewer than 1% of the known species have been analyzed genetically.

In recent years a technical breakthrough has occurred. This new approach, which bypasses all breeding studies, also avoids the problems of polygenic characters and environmental effects. It is a technique that looks at proteins, the translational products of genes (Box 3–7). These proteins determine the phenotypes studied. In fact, depending on the exact technique used, proteins demonstrate variations between the various alleles found at a single gene locus or variations at different loci. The first is best shown by determining the sequence of amino acids in a given protein. The second depends on the behavior of proteins in an electrical field, with a technique called electrophoresis; this involves the movement of proteins in a medium, the rate of movement depending upon their size, shape, and electrical charge. By this technique soluble proteins from different organisms—proteins readily isolated in salt solutions—are identified and compared.

Let us look at the techniques of breeding and protein analysis in more detail to get a clearer understanding of what kind of data we can obtain from them.

BREEDING ANALYSES

Lethality is the trait most readily studied, since mutations that kill organisms are most common and most easily studied. A simple analogy makes clear why that is so. Take a complex mechanism, such as that of a watch, and introduce a change in it. For example, jam a screwdriver into it or bang it or drop it. What is the likelihood that the randomly introduced change will allow the watch to go on functioning? Very, very small. Mutations are random changes in the genes of organisms, and since organisms are very complex systems, it is likely that most random changes will be harmful, even to the point of being lethal.

For lethal genes to persist they must, of course, be recessive and occur in the heterozygous condition. For them to be detected, they must then become homozygous and the loss of organisms resulting from the expression of the lethal character must be detectable.

Box 4–2. The detection of recessive lethal genes.

This analysis is a refinement of the one first used by H. J. Muller, in 1927, to demonstrate the production of mutations by X-rays in *Drosophila melanogaster* (see Chapter 3). The basic stock for the flies used involves two lethal genes lying at different positions on what is called the *second chromosome*. They are arranged to achieve a balanced lethal stock. In this stock only those flies survive that are heterozygous for the two lethal markers. One marker is the dominant gene *Cy*, for curly wings. Flies homozygous for this gene do not survive, nor do those with the other marker, which is dominant for plum eye-color, i.e., *Pm*. The elimination of homozygotes and the survival of the heterozygote is shown in the following cross. (Each chromosome contains large inversions that prevent crossing-over between them or between them and normal chromosomes.)

$$P_1 \qquad \dfrac{Cy \quad +}{+ \quad Pm} \times \dfrac{Cy \quad +}{+ \quad Pm}$$

$$F_1 \qquad \underset{\text{(dies)}}{\dfrac{Cy \quad +}{Cy \quad +}} \qquad \dfrac{Cy \quad +}{+ \quad Pm} \qquad \underset{\text{(dies)}}{\dfrac{+ \quad Pm}{+ \quad Pm}}$$

To look for lethal genes on the second chromosome using the curly/plum balanced lethal stock, the following breeding analysis was made by the Japanese geneticist T. Mukai, in 1964. It starts with a cross between a curly/plum fly and a normal appearing fly. The aim is to determine whether or not there is a lethal on one of the second chromosomes of the normal fly. The chromosomes under analysis are indicated as $\sim\sim\sim$ for the purpose of identification.

$$P_1 \qquad \dfrac{Cy \quad +}{+ \quad Pm} \times \dfrac{\widetilde{+ \quad +}}{+ \quad +}$$

$$F_1 \qquad \dfrac{Cy \quad +}{\widetilde{+ \quad +}} \qquad \dfrac{Cy \quad +}{+ \quad +} \qquad \dfrac{+ \quad Pm}{\widetilde{+ \quad +}} \qquad \dfrac{+ \quad Pm}{+ \quad +}$$

$$P_2 \qquad \dfrac{Cy \quad +}{+ \quad Pm} \times \dfrac{+ \quad Pm}{\widetilde{+ \quad +}}$$

$$F_2 \qquad \dfrac{+ \quad Pm}{\widetilde{+ \quad +}} \qquad \dfrac{Cy \quad +}{\widetilde{+ \quad +}} \qquad \dfrac{Cy \quad +}{+ \quad Pm} \qquad \underset{\text{(dies)}}{\dfrac{+ \quad Cy}{+ \quad Pm}}$$

$$P_3 \qquad \dfrac{+ \quad Pm}{\widetilde{+ \quad +}} \times \dfrac{+ \quad Pm}{\widetilde{+ \quad +}}$$

$$F_3 \qquad \underset{\text{(dies)}}{\dfrac{+ \quad Pm}{+ \quad Pm}} \qquad \underset{2}{\dfrac{+ \quad Pm}{\widetilde{+ \quad +}}} \quad : \quad \underset{1}{\dfrac{\widetilde{+ \quad +}}{+ \quad +}}$$

In the P_2, a curly/plum fly is crossed to one of the F_1 progeny. It really does not matter which genotype is used here, since each F_1 fly carries one of the second chromosomes being tested. The point now is to make that chromosome homozygous so that if any recessive lethal is present it will be expressed in the F_3. The P_3 is made by crossing males and females of the same phenotype and the desired homozygous individuals are then obtained in the F_3. If no lethals are present on the chromosome, then normal flies will appear in the F_3. Plum flies will also survive, in a 2:1 ratio relative to the normals. If a lethal is present it will eliminate all the normal flies. Their absence demonstrates the presence of a recessive lethal gene.

The results of Mukai's analysis for lethals and detrimentals are summarized in the following table:

lethal lines	15
deleterious lines	2
normal appearing lines	84
missing lines	3
total lines studied	104
total number of flies examined in these lines	70,456

The study described in Box 4–2 is an excellent example of how this can be done. Here the occurrence of a single lethal gene on one fruit fly chromosome is being detected. Actually, the analysis can go even further than that shown in Box 4–2. If no lethals are present, the expected ratio is two plum eye-color flies for each red eye-color fly. And as Box 4–2 shows, if a lethal is present then only plum eye-color flies are expected. Now suppose the data fall between the two outcomes just described. One interpretation of this result would be that there is a detrimental gene present on the chromosome being analyzed. It is not a lethal because some homozygotes survive, but it is not normal, since fewer than the expected number of normal phenotypes occur.

This type of cross can, therefore, detect both lethal and detrimental mutants. With this technique, a significant number of such mutants was found. And of course, such analyses are not restricted to one chromosome. Given a system of balanced lethals, other chromosomes can also be analyzed.

It is very important to understand what these studies tell us about hidden variability in general. The most important point is that the organism, in the diploid condition, stores much more hereditary material than is expressed. In the heterozygote, the effect of recessive genes on the phenotype is minimized. This allows lethals, detrimentals, and other nonbeneficial genes, as well as beneficial genes, to be stored. The presence of beneficial genes is not surprising, since such genes would be favored by natural selection if expressed. But the occurrence of deleterious genes is significant. We will see later why the availability of a wide range of genetic material may be of adaptive benefit to a species. The fact that this range of genetic material exists is all we can document at this point in our discussion. It is an important outcome of these breeding analyses and strongly favors the balance theory of population structure rather than the classical theory, wherein we would expect one normal or wild-type allele to predominate with other alleles being very rare.

PROTEIN ANALYSES

These studies provide answers to two different, but related questions. (1) What is the genetic variation at one chromosomal locus? (2) What is the genetic variation among different loci? The first question can be answered two different ways. We can compare the sequences of amino acids in proteins produced by different alleles. Or, more simply, we can look for differences that can be attributed to amino acid differences. The latter technique is less precise than the former, but it can be much easier to do and it can give us the essential information we want, i.e., allelic proteins are present. This is done by electrophoresis.

The difference between different loci can also be examined electrophoretically. Here the proteins being studied are so different that they are reasonably presumed to be controlled by different loci. This difference is demonstrated by the necessity of using different extraction procedures and also by wide differences in mobility on the gel medium of the electrophoresis apparatus.

Amino acid sequences. Let us look again at hemoglobin, the oxygen-binding protein molecule of the blood, since it is an informative example of differences among alleles. Vertebrate hemoglobin consists of two different pairs of polypeptides, with each type coded for by a different gene locus. For example, hemoglobin A, the most common hemoglobin in humans after birth, consists of two alpha and two beta chains of amino acids. The alpha polypeptide chain is coded by one gene and the beta chain by another gene at a different locus. The alpha chain contains 141 amino acids, the beta 146 amino acids; the exact sequence of these subunits is known for each chain. As we saw earlier (Fig. 4–2), normal beta chains have glutamic acid in the sixth amino acid position, but in hemoglobin S, formed by a mutant allele of the gene for the normal beta chain, the abnormal beta chain has valine in position six. In another mutant of this allele, found in the beta chain of hemoglobin C, lysine is substituted in the sixth position. This, too, can be accounted for by a change in a single DNA nucleotide.

When we compare differences between different genes, such as those coding for alpha and for beta polypeptides, we find more extensive differences. Amino acid sequencing tells us that the two chains differ in 21 amino acids out of those present. There are other kinds of hemoglobin, too, with yet other polypeptide chains: gamma, delta, and epsilon, and they differ significantly in their amino acids. Box 4–3 outlines a chromatographic procedure for sequence determination and gives details on another widely studied protein, cytochrome c.

Electrophoretic studies. Polypeptides with different amino acids have different properties. One such property is electrical charge. Valine is electrically neutral, but glutamic acid has a net negative charge. When polypeptide molecules with different amounts of valine or glutamic acid or both are placed in an electrical field, they move at different rates.

Proteins moving on a starch gel (Box 4–4) are usually treated chemically, to color them and, thus, increase their visibility. Hence a given protein and its varients are detectable when they occupy even slightly different positions on the gel. These are usually interpreted as products of one set of alleles. One such study is shown in Box 4–5.

Proteins formed by allelic genes differ relatively little in size, hence their electrophoretic mobility depends almost entirely on electrical charge. But different combinations of amino acids in different polypeptides can have the same charge, since glycine and alanine, among other amino acids, are also neutral. Glutamic acid and aspartic acid have negative charges; lysine and arginine positive charges. Given these facts, and the way the genetic code determines the sequencing of different amino acids, it is generally accepted that only one-third of all amino acid differences between genetically related proteins are detectable electrophoretically. So although experiments using electrophoresis are much easier than experiments using amino acid sequencing, electrophoresis is a less sensitive measure of genetic variation.

Proteins formed at different loci can also be studied. This can be done either of two ways. First, and obvious from the foregoing discussion, one can examine a number of enzymes from each organism in a sampling of organisms. Second, one can isolate a number of soluble proteins from various organisms. Since soluble proteins apparently exist in solution in the cells of an organism, they can be separated, run on gels, and visualized by a technique that identifies them as proteins. This method does not specify the nature of each protein. We do not know if they have an enzymatic or some other function; whether they are made up of one or more constituent polypeptide chains; or exactly how many different genetic loci they represent. In short, all we learn from this technique is that soluble proteins can be obtained from different organisms and that they differ from each other, in varying degrees, as shown by electrophoresis.

The kind of data obtained by this last approach is shown in Table 4–3, the kind obtained by surveying a variety of identifiable enzymes in Table 4–4. The data presented in both these tables indicate the significant genetically determined variation in the proteins studied. In Table 4–3, eighteen dif-

Box 4–3. Amino acid sequence determination in proteins.

The British biochemist Dr. F. Sanger received a Nobel prize for his elucidation of the sequence of amino acids in the protein insulin. First he identified the end of a polypeptide chain that terminates in the amino group, the N terminal group. Second, he broke the polypeptide chain at known sites by using different enzymes to obtain fragments of different sizes. And third, he isolated these fragments by

A, alanine	I, isoleucine	R, arginine
C, cysteine	K, lysine	S, serine
D, aspartic acid	L, leucine	T, threonine
E, glutamic acid	M, methionine	V, valine
F, phenylalanine	N, asparagine	W, tryptophan
G, glycine	P, proline	Y, tyrosine
H, histidine	Q, glutamine	

	1–9	10–19	20–29	30–39	40–49	50–59
human	– – – – – – – – G	D V E K G K K I F I	M K C S Q C H T V E	K G G K H K T G P N	L H G L F G R K T G	Q A P G Y S Y T A A
rhesus monkey	– – – – – – – – G	D V E K G K K I F I	M K C S Q C H T V E	K G G K H K T G P N	L H G L F G R K T G	Q A P G Y S Y T A A
horse	– – – – – – – – G	D V E K G K K I F V	Q K C A Q C H T V E	K G G K H K T G P N	L H G L F G R K T G	Q A P G F T Y T D A
pig, bovine, sheep	– – – – – – – – G	D V E K G K K I F V	Q K C A Q C H T V E	K G G K H K T G P N	L H G L F G R K T G	Q A P G F S Y T D A
dog	– – – – – – – – G	D V E K G K K I F V	Q K C A Q C H T V E	K G G K H K T G P N	L H G L F G R K T G	Q A P G F S Y T D A
gray whale	– – – – – – – – G	D V E K G K K I F V	Q K C A Q C H T V E	K G G K H K T G P N	L H G L F G R K T G	Q A V G F S Y T D A
rabbit	– – – – – – – – G	D V E K G K K I F V	Q K C A Q C H T V E	K G G K H K T G P N	L H G L F G R K T G	Q A V G F S Y T D A
kangaroo	– – – – – – – – G	D V E K G K K I F V	Q K C A Q C H T V E	K G G K H K T G P N	L N G I F G R K T G	Q A P G F T Y T D A
chicken, turkey	– – – – – – – – G	D I E K G K K I F V	Q K C S Q C H T V E	K G G K H K T G P N	L H G L F G R K T G	Q A E G F S Y T D A
penguin	– – – – – – – – G	D I E K G K K I F V	Q K C S Q C H T V E	K G G K H K T G P N	L H G I F G R K T G	Q A E G F S Y T D A
Pekin duck	– – – – – – – – G	D V E K G K K I F V	Q K C S Q C H T V E	K G G K H K T G P N	L H G L F G R K T G	Q A E G F S Y T D A
snapping turtle	– – – – – – – – G	D V E K G K K I F V	Q K C A Q C H T V E	K G G K H K T G P N	L N G L I G R K T G	Q A E G F S Y T E A
bullfrog	– – – – – – – – G	D V E K G K K I F V	Q K C A Q C H T C E	K G G K H K V G P N	L Y G L I G R K T G	Q A A G F S Y T D A
tuna	– – – – – – – – G	D V A K G K K T F V	Q K C A Q C H T V E	N G G K H K V G P N	L W G L F G R K T G	Q A E G Y S Y T D A
screwworm fly	– – – – G V P A G	D V E K G K K I F V	Q R C A Q C H T V E	A G G K H K V G P N	L H G L F G R K T G	Q A A G F A Y T N A
silkworm moth	– – – – G V P A G	N A E N G K K I F V	Q R C A Q C H T V E	A G G K H K V G P N	L H G F Y G R K T G	Q A P G F S Y S N A
wheat	A S F S E A P P G	N P D A G A K I F K	T K C A Q C H T V D	A G A G H K Q G P N	L H G L F G R Q S G	T T A G Y S Y S A A
fungus (*Neurospora*)	– – – – G F S A G	D S K K G A N L F K	T R C A E C H G E G	G N L T Q K I G P A	L H G L F G R K T G	S V D G Y A Y T D A
fungus (baker's yeast)	– – – T E F K A G	S A K K G A T L F K	T R C E L C H T V E	K G G P H K V G P N	L H G I F G R H S G	Q A Q G Y S Y T D A
fungus (*Candida*)	– – P A P F E Q G	S A K K G A T L F K	T R C A E C H T I E	A G G P H K V G P N	L H G I F S R H S G	Q A Q G Y S Y T D A

ferent loci were identified in five different populations of wild *Drosophila pseudoobscura*. About one-third of the loci in each population were found to have more than one allele present. (That is, as determined by electrophoretic mobility, more than one protein was produced at a given locus.) Eight of these loci were known to control the formation of certain enzymes, the other ten formed unidentified proteins. It is difficult to say whether this sample of proteins and their genetic loci represented the whole genome from which they were taken. Bearing that in mind, as well as the estimate that only one-third of the amino acid changes were detectable, we realize that we must be seeing the lower limit of genetic variability of those structural genes that form soluble proteins.

chromatography, a technique in which a solvent is added to a sheet of special paper on which the fragments become isolated. The different polypeptides move varying degrees depending on the nature of the fragment. Next follows a sophisticated game of deducing the sequence of the amino acids in the various fragments from the data provided by known N-terminal groups, known enzymatic breaks, and chromatographic positions. The goal is to find the one sequence of amino acids that can account for all the data. Since Sanger's pioneering analysis, much further work has been done, and a large number of proteins have had their amino acid sequence elucidated.

As an example of the comparisons possible by Sanger's method, see the tabulated list of the amino acid sequences in cytochrome c, a protein important in electron transport, from twenty different species. The few invariant sites are shaded in gray. Note how similar cytochrome c molecules are in humans and rhesus monkeys, i.e., 103 of 104 amino acid residues are shared. (From Dobzhansky et al., Fig. 9–18.)

	60–69 (0123456789)	70–79 (0123456789)	80–89 (0123456789)	90–99 (0123456789)	100–109 (0123456789)	110–112 (012)
human	NKNKGIIWGE	DTLMEYLENP	KKYIPGTKMI	FVGIKKKEER	ADLIAYLKKA	TNE
rhesus monkey	NKNKGITWGE	DTLMEYLENP	KKYIPGTKMI	FVGIKKKEER	ADLIAYLKKA	TNE
horse	NKNKGITWKE	ETLMEYLENP	KKYIPGTKMI	FAGIKKKTER	EDLIAYLKKA	TNE
pig, bovine, sheep	NKNKGITWGE	ETLMEYLENP	KKYIPGTKMI	FAGIKKKGER	EDLIAYLKKA	TNE
dog	NKNKGITWGE	ETLMEYLENP	KKYIPGTKMI	FAGIKKTGER	ADLIAYLKKA	TKE
gray whale	NKNKGITWGE	ETLMEYLFNP	KKYIPGTKMI	FAGIKKKGER	ADLIAYLKKA	TNE
rabbit	NKNKGITWGE	DTLMEYLENP	KKYIPGTKMI	FAGIKKKDER	ADLIAYLKKA	TNE
kangaroo	NKNKGIIWGE	DTLMEYLENP	KKYIPGTKMI	FAGIKKKGER	ADLIAYLKKA	TNE
chicken, turkey	NKNKGITWGE	DTLMEYLENP	KKYIPGTKMI	FAGIKKKSER	VDLIAYLKDA	TSK
penguin	NKNKGITWGE	DTLMEYLENP	KKYIPGTKMI	FAGIKKKSER	ADLIAYLKDA	TSK
Pekin duck	NKNKGITWGE	DTLMEYLENP	KKYIPGTKMI	FAGIKKKSER	ADLIAYLKDA	TAK
snapping turtle	NKNKGITWGE	ETLMEYLENP	KKYIPGTKMI	FAGIKKKAER	ADLIAYLKDA	TSK
bullfrog	NKNKGITWGE	DTLMEYLENP	KKYIPGTKMI	FAGIKKKGER	QDLIAYLKSA	CSK
tuna	NKSKGIVWNN	DTLMEYLENP	KKYIPGTKMI	FAGIKKKGER	QDLVAYLKSA	TS–
screwworm fly	NKAKGITWQD	DTLFEYLENP	KKYIPGTKMI	FAGLKKPNER	GDLIAYLKSA	TK–
silkworm moth	NKAKGITWGD	DTLFEYLENP	KKYIPGTKMV	FAGLKKANER	ADLIAYLKES	TK–
wheat	NKNKAVEWEE	NTLYDYLLNP	KKYIPGTKMV	FPGLKKPQDR	ADLIAYLKKA	TSS
fungus (*Neurospora*)	NKQKGITWDE	NTLFEYLENP	KKYIPGTKMA	FGGLKKDKDR	NDIITFMKEA	TA–
fungus (baker's yeast)	NIKKNVLWDE	NNMSEYLTNP	KKYIPGTKMA	FGGLKKEKDR	NDLITYLKKA	CE–
fungus (*Candida*)	NKRAGVEWAE	PTMSDYLENP	KKYIPGTKMA	FGGLKKAKDR	NDLVTYMLEA	SK–

Box 4–4. Identification of different proteins by electrophoresis.

Proteins with different surface charges, and of different sizes and shapes, move differentially in an electric field. These different mobilities separate many proteins from each other. One extremely useful variant of this electrophoretic separation is to use starch gels in a buffer solution. The following diagram illustrates how proteins from different samples are separated and identified. A. Samples are placed in a row in the middle of the gel. B. An electrical current is passed through the gel for several hours. C. The distance the proteins in the sample move varies with their net electrical charges, size, and shape. They are visualized by adding chemicals, which form colors in conjunction with the protein.

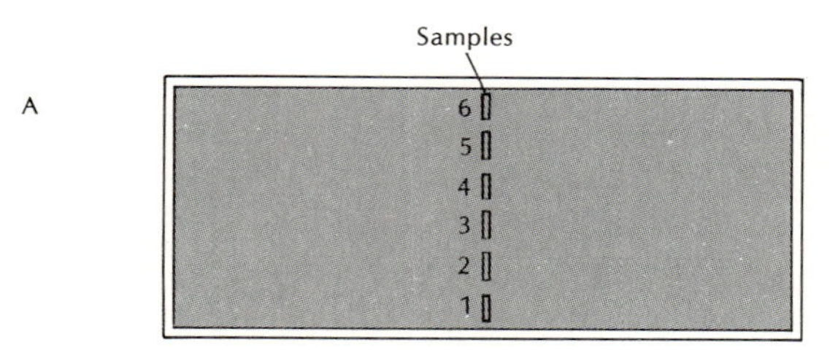

Top view

B

Electrode
Contact wick
Sample slot
Gel
+
−
Buffer tank

Side view

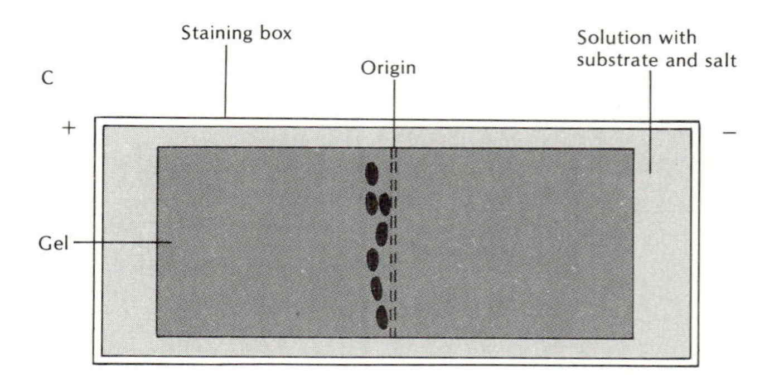

Top view

Nevertheless, because these estimates represent some of the most direct evidence we have regarding genetic variation, these techniques have been extensively used. Table 4–4 presents an analysis of known enzymes. In this study it was found that in 53% of the populations studied a given locus carried two (or more) alleles. This is direct evidence of significant genetic diversity in natural populations and supports the balance theory of population structure.

Putting together analyses of the sorts just discussed, we can summarize what we know about allelic variation within

Box 4–5. Electrophoretic phenotypes of a cross between wild butterflies.

The phenotypes are expressed as localization of protein esterases on gels. The genotypes of the parents are given at the left as 100-85 × 120-80, an interpretation of the observed protein bands. The three bands for each parental phenotype indicate that the parents are heterozygous and the protein in question is a dimer, i.e., it is formed by the union of two separate polypeptide chains. This accounts for the three-band phenotype: one band is the product of two polypeptides from

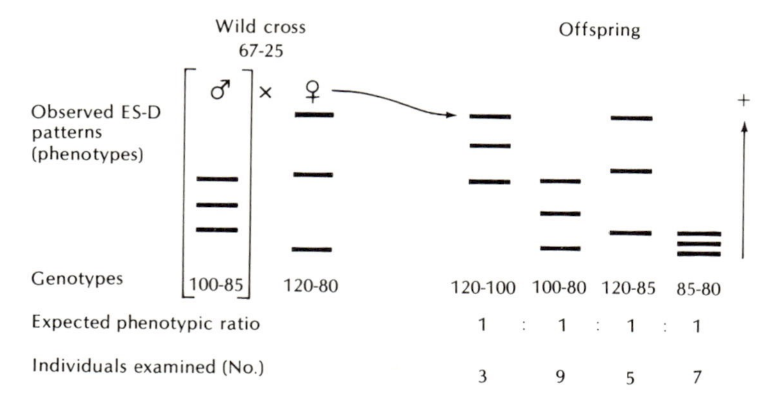

the 100-locus, one from two polypeptides from the 85-locus, and one, with intermediate electrophoretic mobility, from one polypeptide from the 100-locus and one polypeptide from the 85-locus. This locus is called Es-d and therefore the parental cross can be diagrammed as

$$\text{Es-d}^{85}/\text{Es-d}^{100} \times \text{Es-d}^{120}/\text{Es-d}^{80}$$

The progeny phenotypes are given at the right of the diagram. A single butterfly provides enough protein to allow identification of its esterase phenotype. In this way, the genetics of a natural population can be worked out to provide data on the alleles present in it. This work was done by two American researchers, J. M. Burns and F. M. Johnson, on the Texas populations of the butterfly *Hemiargus isola*.

Table 4–3. The occurrence of loci showing more than one allele, i.e., polymorphic loci, in wild populations of *Drosophila pseudoobscura*.

Population	Number of polymorphic loci	Proportion of polymorphic loci
Strawberry Canyon	6	0.33
Wildrose	5	0.28
Cimarron	5	0.28
Mather	6	0.33
Flagstaff	5	0.28

From Lewontin and Hubby, 1966.

species (Table 4–5). Table 4–5 must be looked at with certain reservations, however. The number of loci studied in these species varies from about 15 in wasps to over 70 in humans. Such numbers represent about one or two in every thousand loci present in the genomes studied. And we must state again that the study relies on the detection of soluble proteins; therefore, genes coding for other products have gone unanalyzed. We do not know whether we have a typical or an atypical sample of the total genome, but the general conclusion that significant allelic variation occurs seems to be well founded.

The fundamental theorem of natural selection

Sir Ronald Fisher, a British statistician and a pioneer in population genetics (Box 5–1), showed mathematically that *the rate of evolution in a Mendelian population can be limited by the genetic variability of that population.* The more variability, the faster the possible evolution; the less variability the slower the evolution. This he called the *Fundamental Theorem of Natural Selection.* We will not go into mathematical details here, but in a common-sense way we can grasp what Fisher is saying. His central idea is apparent from Darwin's theory of natural selection. Darwin's fourth point was that heritable variations occur. If they did *not* occur, and if all genotypes were alike and remained alike, then there would be no differences between organisms and no differential survival or differential reproduction. In short, there would be no natural selection. But since variations do arise, different phenotypes occur, and in the context of a struggle for existence, natural selection must also occur. And so the more frequent the variations, the more phenotypes there will be for natural selection to act on, and according to Fisher, evolutionary change will also be more rapid. We will return to this theorem and its mathematical

Table 4–4. Gene loci in *Drosophila willistoni*, coding for 36 different enzymes.

Gene locus	Enzyme coded	Frequency of polymorphic populations	Frequency of heterozygous individuals
Acph-1	acid phosphatase	0.56	0.107
Acph-2	acid phosphatase	0.83	0.148
Adh	alcohol dehydrogenase	0.28	0.075
Adk-1	adenylate kinase	1.00	0.572
Adk-2	adenylate kinase	0.55	0.094
Ald-1	aldolase	0.65	0.157
Ald-2	aldolase	0.63	0.132
Ao-1	aldehyde oxidase	1.00	0.334
Ao-2	aldehyde oxidase	1.00	0.482
Aph	alkaline phosphatase	0.85	0.243
Est-2	esterase	0.67	0.112
Est-3	esterase	0.75	0.107
Est-4	esterase	0.35	0.152
Est-5	esterase	0.24	0.089
Est-6	esterase	1.00	0.285
Est-7	esterase	1.00	0.601
Fum	fumarase	0.50	0.092
Got	glutamate-oxaloacetate transaminase	0.00	0.038
aGpd	alpha-glycerophosphate dehydrogenase	0.02	0.019
G3pd	glyceraldehyde-3-phosphate dehydrogenase	0.65	0.123
G6pd	glucose-6-phosphate dehydrogenase	0.75	0.201
Hbdh	hydroxybutyrate dehydrogenase	0.00	0.015
Hk-1	hexokinase	0.36	0.054
Hk-2	hexokinase	0.78	0.183
Hk-3	hexokinase	0.02	0.042
Idh	isocitrate dehydrogenase	0.06	0.030
Lap-5	leucine aminopeptidase	1.00	0.537
Mdh-2	malate dehydrogenase	0.24	0.073
Me-1	malic enzyme	0.03	0.056
Me-2	malic enzyme	0.85	0.321
Odh-1	octanol dehydrogenase	0.47	0.132
Odh-2	octanol dehydrogenase	0.16	0.106
Pgm-1	phosphoglucomutase	0.62	0.143
To	tetrazolium oxidase	0.15	0.072
Tpi-2	triose phosphate isomerase	0.06	0.023
Xdh	xanthine dehydrogenase	1.00	0.532
	average frequency of polymorphic populations per locus	0.530 ± 0.059	
	average frequency of heterozygous individuals per locus	0.180 ± 0.028	

Here 36 gene loci from over 100 different natural populations were studied through the examination of over 5,000 individual flies. The third column records the frequency of the populations containing more than one allele at the locus studied. These are polymorphic populations. The fourth column records the frequency of individuals heterozygous for the locus studied. (From Dobzhansky et al., Table 2–8, 1977.)

treatment when we deal with the concept of fitness in quantitative terms (Chapters 5 and 6). Here we need to consider the question of genetic variation a bit further in general terms and

Table 4–5. Allelic variation in animal and plant species.

Group	Number of species or forms	Mean number of loci per species	Proportion of polymorphic loci	Proportion at heterozygous loci per organism
Insects				
Drosophila	28	24	0.529 (.030)	0.150 (.010)
others	4	18	0.531	0.151
haplodiploid wasps	6	15	0.243 (0.039)	0.062 (.007)
marine invertebrates	9	26	0.587 (0.084)	0.147 (.019)
Snails				
land	5	18	0.437	0.150
marine	5	17	0.175	0.083
total invertebrates	57	mean = 21.8	mean = 0.469	mean = 0.135
fish	14	21	0.306 (.047)	0.078 (.012)
amphibians	11	22	0.336 (.034)	0.082 (.008)
reptiles	9	21	0.231 (.032)	0.047 (.008)
birds	4	19	0.145	0.042
rodents	26	26	0.202 (.015)	0.043 (.005)
large mammals	4	40	0.233	0.037
total vertebrates	68	mean = 24.1	mean = 0.247	mean = 0.061
annual plants	36/33	11	0.446 (.296)	0.142 (.107)
biennial plants	13/11	18	0.220 (.250)	0.079 (.081)
herbaceous perennials	17/14	12	0.250 (.312)	0.116 (.118)
woody trees and shrubs	10/7	12	0.652 (.314)	0.359 (.102)
total plants	76/65	mean = 12.6	mean = 0.391	mean = 0.149

Entries in the third column are estimates of the number of loci showing more than one allele. These are averages for the species under consideration. In the fourth column the proportion of the loci that are heterozygous, on the average, per organism are estimated. This varies from 3.69 to 35.9%. Basically, values in the third column are obtained by direct observation; values in the fourth column are calculated from the observed frequency of alleles, given as $f_1, f_2, f_3, \ldots f_n$. The probability that an allele is homozygous is the probability of that allele occurring simultaneously in one individual, i.e. $f_1 \times f_1$ or f_1^2. The expected frequency of all homozygotes is, therefore, $f_1^2 + f_2^2 + f_3^2 \ldots f_n^2$. And the frequency of heterozygotes H is one minus the sum of the frequency of homozygotes, since anything not homozygous must be heterozygous. (From Hamrick 1979.)

to introduce an aspect that has been ignored thus far, namely, the environment.

ENVIRONMENTAL VARIABILITY

It takes only a little thought to realize how variable environments can be. The amount of light, and consequently temperature, can vary widely over a 24-hour period as well as seasonally. Moisture and humidity are other variables.

Seasonal variations show a wide range, also. In latitudes close to the poles, these can be quite severe. Winters are periods of sustained cold and summers are short interludes of much warmer weather. During the latter season snow and ice disappear, plants grow, insects reproduce rapidly, and a period of extraordinary proliferation and rapid growth occurs.

In the most stable environments, those deep in caves or in the depths of the ocean, there is no light to fluctuate, and the temperature and the humidity are quite stable. What variations there are depend on the flow of air and water through the cave and the movement of water currents and the settling of debris from surface waters to the ocean depths.

Much more will be said about the effects of the environment on organisms because it is the interactions of organisms with their environment that determines the intensity and direction of natural selection. It follows, therefore, that in a stable environment, once organisms are well adapted to it, their rates of evolutionary change will slow down. But there is always the chance things will change again—Lyell taught very clearly that the earth's surface has always been changing (Box 2–3), so that species, to survive, must have genetic variability (Box 4–6). But how much they have and how fast they can adapt to changes, when they arise, will vary from species to species depending on their past history and local circumstances, including as Fisher made clear, their amount of genetic variability.

Other sources of genetic variation

We will now see what happens to genic variation in the context of a Mendelian population, namely, a sexually reproducing community, but keeping in mind that in addition to genic changes there are also chromosomal and genomic ones (reviewed in Chapter 3). These do not involve qualitative changes in genes (except as chromosomal breaks and rejoining *within* a gene may occur), but rather they are rearrangements of existing genes (additions, deletions, inversions, translocations) within a chromosome or additions of whole sets of chromosomes (aneuploidy or polyploidy). We will have occasion to consider such changes further in subsequent chapters.

The most important point here is to consider the role that sexual reproduction plays in producing new combinations of genetic material. We can make this most explicit by first considering the conservative aspects of asexual reproduction before turning to the inventive capacities of sexuality.

EVOLUTION IN ASEXUALLY REPRODUCING SPECIES

H.S. Jennings, a professor at John Hopkins University, claimed that he had seen evolution occurring before his

Box 4–6. Genetic variability and environmental variation in natural populations.

Because the oceans are thought to be among our most stable environments their inhabitants have attracted a certain amount of attention from geneticists. In the giant clam *Tridacna maxima,* an analysis of soluble proteins revealed that these animals were polymorphic in 20 of the 30 gene loci associated with these proteins. Ayala, Hedgecock, Zumwalt, and Valentine (1973) concluded from their study that "An individual [clam] is heterozygous, on the average at 20.2% of the loci." This study does not, however, make clear the function of the proteins studied—one of the limitations in studying soluble proteins. On the one hand, we do not know if we should expect little or no variability among them relative to the environment. The study is inconclusive in that sense. On the other hand, there is no doubting the observed genetic variability, whatever its cause.

Another study directed at deep-sea forms, whose environment is known to be especially stable, turned up loci determining formation of esterases that were as variable as comparable loci of organisms from terrestrial habitats. The possible reasons for this finding were not discussed by J.L. Gooch and T.J.M. Schopf (1972), who did this work. Below is shown a deep-sea habitat.

This kind of problem can also be approached by looking for genetic variation in that "living fossil," the horseshoe crab (*Limulus polyphemus*). This animal has been around for over 200 million years with little change in phenotype. It might, therefore, be thought to contain a stable, largely uniform genotype, evolution having apparently settled on one obviously successful phenotype in terms of long-term survival. Not so. The study of 24 different proteins formed by 25 different loci, by a team composed of Selander, Yang, Lewontin, and Johnson (1970), revealed that in a sample of 64 crabs from four different localities, the loci studied were "not less genetically variable than some other animals belonging to horotelic [i.e., rapidly evolving] lines."

Finally, there is an elaborate statistical study by Bryant (1974), who looked for correlations between enzyme polymorphisms and weather changes in various terrestrial animals. His conclusion was that about 70% of the variations occurring in the measured heterozygosity of the populations he analyzed "could be accounted for by these measures of temporal variation, i.e., variations over time in the environment." The term "accounted for" is misleading here because it simply means "correlated with" and that makes no necessary causal connection between polymorphism and the environment.

So we are left with the fact that detectable genetic variability is high even when the organisms have survived in a stable environment and that even where genetic and environmental variability occur together we cannot yet show a cause and effect relationship.

eyes—a claim that would have made Darwin envious. In his work, done in the early part of this century, Jennings used certain species of protozoa. Looking at *Paramecium caudatum*, he observed that the unicellular organisms were not all the same size. By selecting individuals of different sizes and allowing them to reproduce asexually over long periods of time, he was able to get what he called "families," which tended to be of a certain average length (Fig. 4–5). When he stopped selecting for different average lengths and still found that the cells produced progeny of those same lengths, he concluded that length was determined genetically.

He made a similar analysis with the shelled ameba *Difflugia corona*. Here he selected for small and large shells, for long and short spines on the shells, and for few or many spines on each shell (Fig. 4–6). Over a period of time, he obtained different phenotypes showing different combinations of the characters studied. And, when he stopped selecting for shell

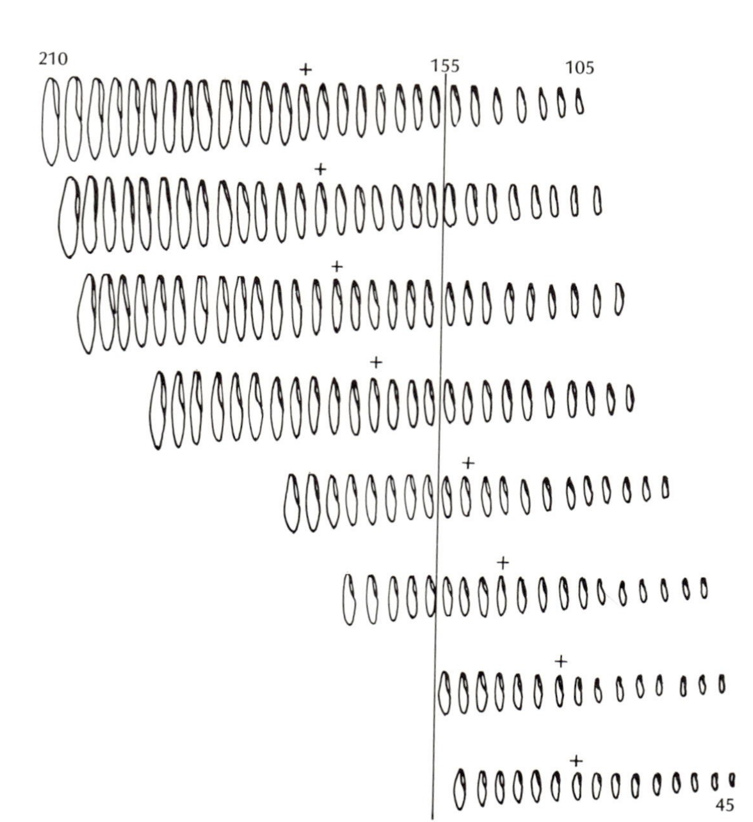

Fig. 4-5: The "families" of *Paramecium caudatum* produced by Jennings through selection in the laboratory. Jennings established that variation within each family (horizontal rows) was due to environmental effects, but that the differences in average size was hereditary. Sizes are given in micrometers.

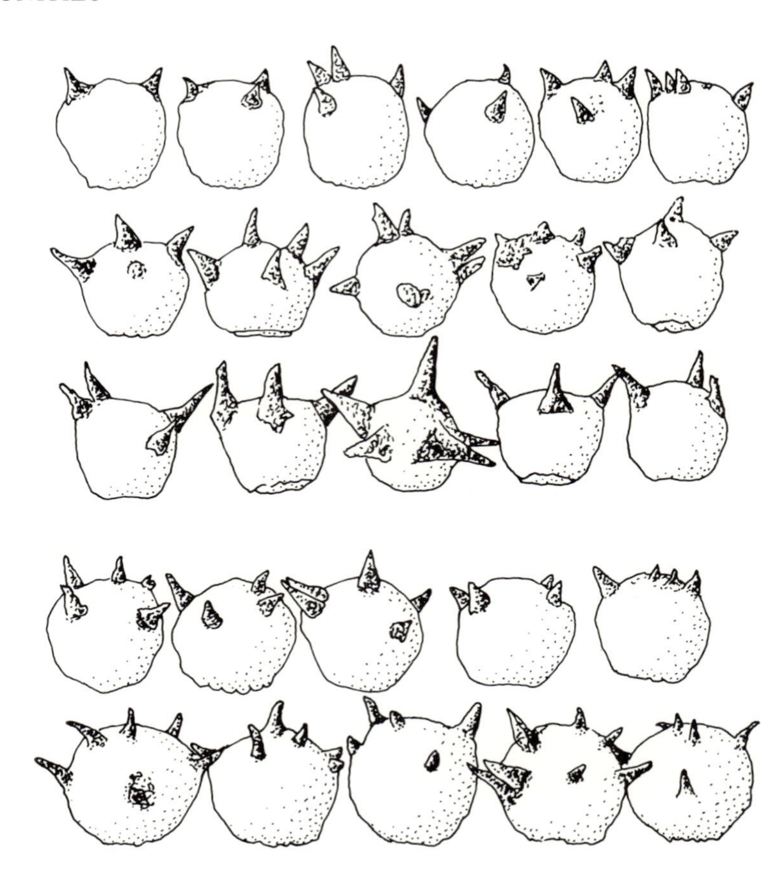

Fig. 4-6: Biotypes, or races, of *Difflugia corona* produced by Jennings through selection in the laboratory. Note the differences in shell size and in the number and size of the spines.

size and spine size and frequency, the phenotypes were inherited asexually, which showed that they were genetically determined.

In 1920, Jennings summarized these and other findings in his book *Life and Death, Heredity and Evolution in Unicellular Organisms*. In that book he wrote, "But when we take a single race and devote all our attention to that alone for years [by selecting certain phenotypes], then we find that real changes do occur; that the race differentiates into many races in the way I have described; that evolution visibly does occur."

Would Darwin have agreed?

Yes and no. Yes, because Jennings quite evidently did produce new, geneticaly controlled phenotypes through selection. No, because Jennings did not produce new species. His paramecia "families" could still conjugate with other members of

P. caudatum. The amebas could not be tested for membership in a reproductive community because they only reproduce asexually. But because Darwin believed variations within a species are the same as those between species, he probably would have been delighted by Jenning's results. Today we are more skeptical because we know that the emergence of new species—new reproductively isolated populations—does not occur the way Darwin thought it occurred. Nonetheless Jennings deserves major credit for showing that selection does produce hereditarily diverse phenotypes. His particular work on paramecia was first published in 1909, long before any comparable work had been done. In modern terms, we say that Jennings selected out alleles for the various phenotypic characters he studied. This means that from a variable genetic pool he experimentally selected out "races" in which one or another allele became predominant. Most probably no new genes, except for point mutations, and no new combinations of genes appeared. This illustrates the limitations of asexual reproduction. But even so, there was significant variation on which selection could act.

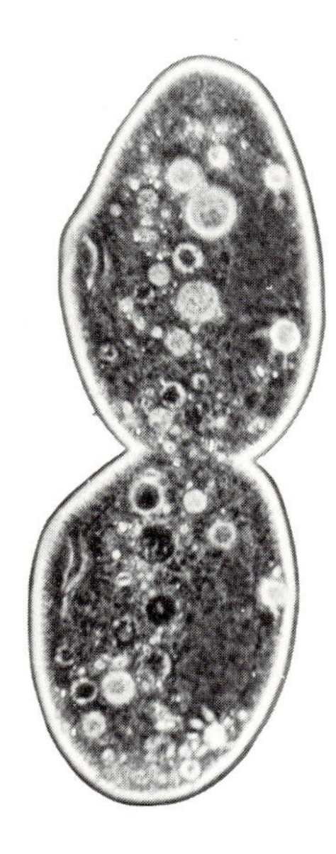

Fig. 4-7: Fission in the ciliated protozoan *Paramecium aurelia.* A single cell constricts at right angles to its long axis to produce two essentially identical daughter cells, about once every six hours in this species, when well-fed cells are grown at room temperature. Fission takes only about twenty minutes. Prior to fission, the parent cell develops all the necessary structures for two complete cells. This includes cytoplasmic structures as well as a macronucleus (polyploid) and micronuclei (diploid). In anticipation of fission, the macronucleus pinches in two, dividing the genetic material fairly equally, and the micronuclei undergo mitosis, so that each daughter cell will receive two of the four new micronuclei.

REPRODUCTION AND SEXUALITY

The unicellular animal *Paramecium aurelia* illustrates precisely the different biological roles of reproduction and sexuality that are important here. Paramecia only reproduce asexually (Fig. 4–7). But because laboratory conditions can ensure survival of all progeny cells and because new cells are produced about every six hours, one cell can produce over 1,000 cells within a three-day period. And in another three days, if optimal growth conditions are maintained, there would be over one million cells. Thus the potential of fission for rapidly increasing the numbers of this species is tremendous, but, except for rare mutations, there is no change in genotypes.

Paramecia also undergo sexual processes, conjugation and autogamy, but here we need only examine the former. What happens during conjugation in *P. aurelia?* First, two cells of complementary mating types must be sexually reactive and able to contact each other. Complementary mating types refer to the fact that only certain paramecia are able to mate, or conjugate: they must always belong to a certain pair of different mating types—there is no conjugation between cells of the same mating type—and they must be physiologically ready, or sexually reactive. When two such cells come in contact, their cilia stick together, the cells align, and the

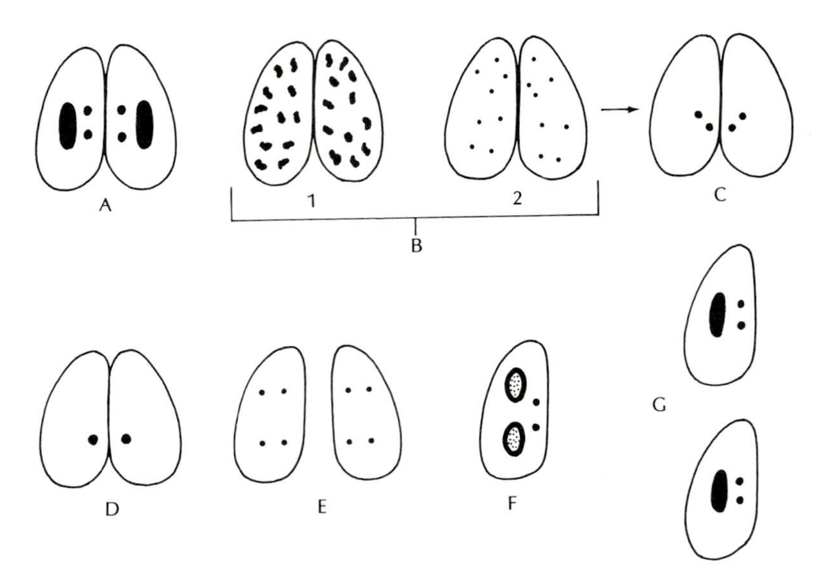

Fig. 4-8: The chief nuclear events of conjugation in *Paramecium aurelia.* In each cell (A), the macronucleus breaks down (B1 shows only the macronuclear fragments). The micronuclei undergo mitosis, and since each cell starts with two diploid nuclei, a total of eight haploid nuclei are produced (B2 shows only the haploid nuclei). Seven of these disappear from each conjugant; the remaining one divides mitotically to produce two gamate nuclei (C), one migratory (male-like) nucleus and one stationary (female-like) nucleus. The migratory nucleus from each cell passes through intervening cell membranes and fuses with its partner cell's stationary gamate nucleus. In this way, both cells are fertilized and the diploid condition is restored (D). The conjugants now separate, and the diploid nucleus restores the characteristic macronucleus and the two micronuclei in each cell (E through G).

remaining events of conjugation then occur as shown in Fig. 4–8.

In that both cells produce migratory, or sperm-like, nuclei and stationary, or egg-like, nuclei they are hermaphroditic cells. That is, the same organism forms both the male and female components needed for fertilization. Therefore, the mating types are not really comparable to male and female sexes, as is sometimes mistakenly stated. Following fertilization the conjugating cells separate. Conjugation starts with two cells and ends with two cells; no reproduction occurs.

How does the sexual process occur in conjugation? Fertilization identifies a sexual process. But what is so important about fertilization? Why are there such elaborate mechanisms to bring nuclear material together from two different individuals? The answer is genetic variability, which produces new phenotypes. And it is on those phenotypes that natural selection acts.

At this point two important definitions are possible.

1. *Reproduction* is a process that increases the number of organisms beyond that represented by the parent or parents.
2. *Sexuality* is the potential for generating new genotypes through the union of nuclei.

Both definitions deserve further discussion.

During asexual reproduction the parent organism often disappears and two new organisms result (Fig. 4–9). But a parent plant, such as a strawberry plant, can send out runners and a new plant can arise where the runner reaches the ground. This type of asexual reproduction occurs in many plants. Even some animals reproduce by budding (Fig. 4–9). Here the parent organism persists, and in addition, new organisms result.

In sexual reproduction there must be both an increase in number beyond that represented by the parents *and* the possibility of new genotypes. Conjugation between two paramecia achieves only the possibility of new genotypes. Hence it represents sexuality—it is a sexual process—but not reproduction.

In the green algae of the genus *Spirogyra* we see that the sexual process decreases the number of individuals, as compared to the two parents. These algae exist as filaments—namely, chains of cells. Each cell is haploid and capable of generating a new filament. In that sense, each one is an individual organism. The filament could also be called a colony, a collection of individual algal cells. At the time of sexual pairing, two filaments align in parallel (Fig. 4–10). Bridges form between adjacent cells of each filament and the contents of one cell merges with the contents of the other cell. The nuclei then fuse in fertilization. In these fused cells, two individuals become one; the number of cells *has decreased*. The fertilized cell, or zygote, then undergoes meiosis to produce four haploid cells, which can then divide mitotically in asexual or uniparental reproduction.

Enough has been said thus far so that we can now ask the following questions and give reasonable answers to them. What are the advantages and limitations of asexual reproduction? Why does an organism reproduce asexually, but retain sexual processes? What are the advantages and limitations of sexual reproduction?

Advantages and limitations of asexual reproduction. The chief advantage is that only one parent is needed to produce more progeny. When that parent finds sufficient food and when other conditions are right, then progeny are produced. And they often have a good chance of persisting and reproducing in their turn. This can be a relatively rapid way—depending on the organism—of increasing the numbers of individuals in a species.

The chief disadvantage is limited genetic variability. When an organism produces more cells by mitosis—for example, a

A

B

Fig. 4-9: Asexual reproduction in multicellular organisms: A. The runner of a strawberry plant gives rise to a new plant. B. In a budding *Hydra,* the new animal extends to the right from the parent.

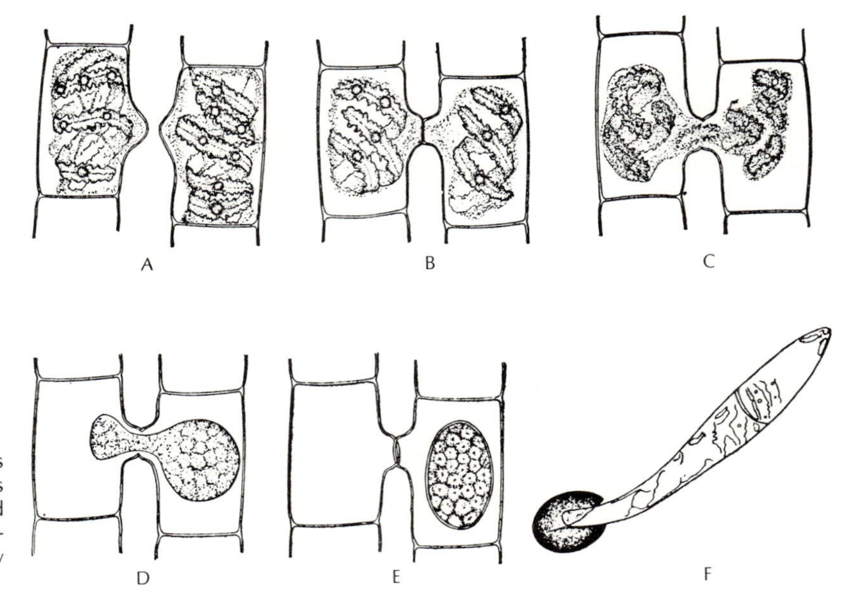

Fig. 4-10: Sexuality in *Spirogyra:* Two cells from different filaments fuse and the contents of one cell migrates into the other (A, B, and C). The cell contents fuse (D); and the zygospore forms, with the emergence of a new filament (E and F).

paramecium dividing, a strawberry plant producing a runner, a hydra budding—genetic material can vary only through mutation. This spontaneous process is a relatively rare event in terms of a given locus, although in terms of all possible mutable loci, mutation can be a significant source of variation, depending on the size of the population. But overall, asexual reproduction is distinguished by genetic stability.

Alternation of asexual and sexual processes. The advantages of alternation now become apparent. An organism such as *Paramecium aurelia* can reproduce rapidly and conservatively by fission, and generate new genotypes by a sexual process (conjugation). These new genotypes are then propagated asexually. But to reproduce sexually, a paramecium mating type must be sexually reactive and locate another paramecium of the complementary mating type that is also sexually reactive. It is quite likely that conjugation in nature does not occur except under rather special conditions, which might be described as follows. There is a sudden growth of bacteria in a freshwater pond as the result, perhaps, of warm fall weather that brings on a more rapid decomposition of fallen leaves by bacteria. These bacteria, in turn, provide food for ciliated protozoans, such as *P. aurelia,* which respond with a burst of growth. After a few days of rapid uniparental reproduction, which, incidentally, inhibits sexual reactivity, the food supply is exhausted by the large number of protozoans, and this induces sexuality in the ciliates. (This phenomenon has been explained by "the hunger theory of sex" and is not unique to ciliates.) Now there is a good chance for appropriate partners to meet. But this can occur only at certain times in certain ponds. Sexuality thus is the exception and not the rule. It seems that such organisms spend most of their time behaving conservatively in terms of fission, and genetically innovative behavior only arises from the rarer sexual processes.

Advantages and limitations of sexual reproduction. In sexual reproduction, all progeny represent new combinations of genes. Let us look more closely at the potential for genetic recombination as a result of sexual reproduction.

Table 4–6 is a simple compilation of the number of recombinants in Mendelian inheritance. As we progress down the table we see a formula emerge for the genetic variability that arises from a given set of chromosome pairs. Note, finally, that the 23 pairs of chromosomes in humans can generate more genotypes than there are humans! [Our human population passed four billion (4×10^9) the summer of 1976.]

Another way of expressing this tremendous potential for recombination is in an example given by the distinguished evolutionist Theodosius Dobzhansky. Assume, for convenience, that a complex organism contains 1,000 gene loci. Assume also that there are ten alleles at each locus. Now assume that through the events associated with sexual reproduction, namely, meiosis and fertilization, these loci and their alleles occur in all possible combinations. (This can happen given enough time and large numbers of organisms.) In this theoret-

Table 4–6. The generation of recombinants through simple Mendelian inheritance.

Chromosomes	Alleles	Genotypic combinations
1 pair	2	3 (e.g., *AA, Aa,* aa)
2 pairs	4	9 (i.e., dihybrid cross)
3 pairs	6	27
n pairs	2*n*	3^n
23 pairs	46	3^{23} *

* This is 94,143,178,827 or approximately 9.4×10^{11}.
It is assumed here, to simplify the discussion, that each chromosome has one locus and at that locus there is only one pair of alleles.

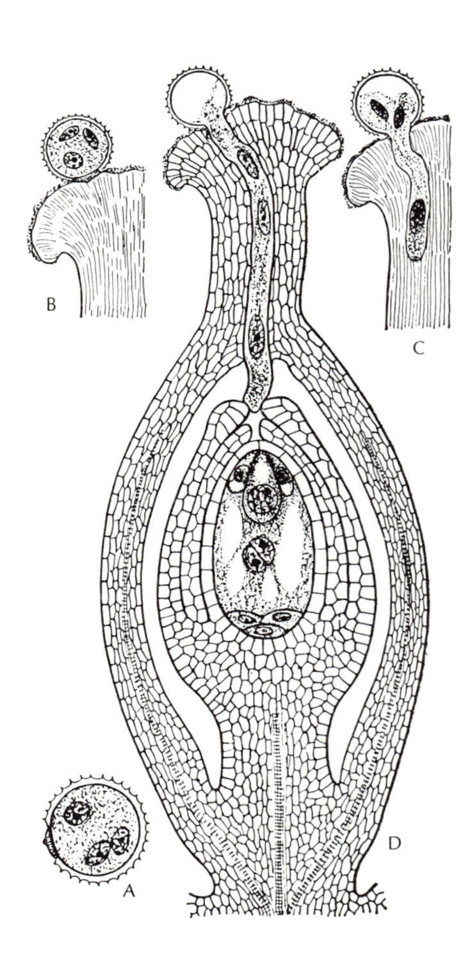

Fig. 4-11: Fertilization in a flower: A mature pollen grain (A) adheres to the stigma (B). The pollen grain extends a growing pollen tube down into the style (C), which contacts the tissues surrounding the ovoid ovule (D).

ical example, $10^{1,000}$ genetic recombinations are possible. To realize the enormity of that figure, one must remember that physicists estimate the total of all subatomic particles in the known universe to be no more than 10^{100}.

Thus the genetic system in organisms with sexuality has an extraordinary potential for generating new genotypes; that is the chief advantage of sexuality.

Are there any disadvantages? Yes. As we saw in our earlier discussion, mating must take place. The two parents must be of complementary types—when not hermaphroditic they will be male and female—and physiologically able to mate. These requirements are often complex. But many plants and animals have evolved ways of life that ensure the meetings of complementary reactive individuals, or, at least, a meeting of their gametes.

Fertilization is likely to be successful only within a species. The fact is that different genotypes can be highly incompatible with one another. That incompatibility is best understood by quickly reviewing certain features of development.

In animals, fertilization consists, essentially, of penetration of the egg membranes by the sperm, entrance of the sperm head into the egg cytoplasm, and fusion of the egg and the sperm nuclei. In plants, somewhat comparable events occur. In flowering plants, for example, the pollen cells grow a pollen tube, which carry two sperm nuclei. This tube extends down from the stigma, through the style, and eventually reaches the egg. Here one sperm nucleus joins the egg nucleus; the other joins the diploid fusion nucleus to form the triploid nucleus of the endosperm cells (Fig. 4–11.) When fertilization occurs, there follow the events that transform the zygote into a new individual. Three processes are involved: namely, cell division, cell differentiation, and morphogenesis. None is completely understood.

Following fertilization there is a period of rapid cell division by the zygote. Although many new cells are produced, there is no increase in dry weight of the cellular components. [These components are essentially proteins, lipids, carbohydrates, and nucleic acids (Box 1–1).] A real increase in weight occurs when the developing system or embryo takes materials from its environment and uses them as building blocks for new cells. The flowering plant embryo does so by absorbing the endosperm; the young tadpole by feeding; and the human embryo by absorbing nutrients provided through its mother's circulatory system. This growth is also accompanied by further cell division. We know a great deal about the synthesis of such macromolecules as proteins, nucleic acids, lipids, and carbohydrates, but we still know very little about how their synthesis is regulated in development.

Differentiation is one of the greatest puzzles of all the developmental processes. The central puzzle is the appearance of phenotypically different cells all of which have the same genotype. Mitosis ensures the latter; somehow development generates different cellular phenotypes. It now seems clear that different cells have genes acting differently in them. This differential action of genes is surely a key aspect of cellular differentiation. But what shuts off certain genes and turns on other genes in the cells of one tissue and, in the cells of an adjacent tissue, shuts off and turns on different genes? We don't know yet. How differential gene action occurs and how it results in the orderly sequence of changes that characterizes normal development in plants and animals must still be worked out.

Morphogeneis is the appearance of adult form during development. This includes the proper placement of developing tissues, organs and organ systems, as well as the molding of body form into recognizable adult structures. In both animals and plants, morphogenesis involves different rates of cellular division and of cellular death. The former selectively increases the masses of certain tissues and the latter does the reverse—it selectively eliminates cells. Also certain cells migrate within the embryo and thus set up new internal relations. (This may well be one mode of regulating cellular differentiation.) Cells are able to recognize and stay in contact with certain other cells; cellular recognition of this nature sorts out cells, and this is also part of morphogenesis. Though these sorts of events are known to occur, how they occur in an orderly sequence in the embryo is still largely not known.

Quite obviously, then, an intricate, yet orderly and reliable series of events transforms the zygote into an adult. Genes must interact effectively with the constituents of the egg cytoplasm. Cell division, biosynthesis of macromolecules, precise timing of gene action, further cellular division, cellular death, migration, and cell recognition must be controlled.

It is hard to overestimate the importance of development in evolutionary studies because it is development that brings genes to expression as phenotypes. And, as we have noted, it is on phenotypes that natural selection acts.

A species as a reproductive community is endowed with a

source of new genes (point mutations), a means of recombining them (sexual processes), and a means of storing variability that is still obscured by the mystery of development. Somehow, all possible phenotypes are not usually expressed; a limited array of phenotypes is seen instead. In some cases, this is the result of diploidy, with dominant alleles suppressing recessive ones. But more important, as the balance theory of population clearly implies, there is a great deal of variation within the relatively uniform adaptive norm selection has apparently forced on a species.

Now let us see if we understand how that comes about.

References

Ayala, F. J., D. Hedgecock, G. S. Zumwalt, and J. W. Valentine, 1973. Genetic variation in *Tridacna maxima,* an ecological analog of some unsuccessful evolutionary lineages. *Evolution 27:* 177–191.

Bradley, Jr., T. B., R. C. Wohl, and R. F. Rieder, 1967. Hemoglobin Gun Hill: Deletion of five amino acid residues and impaired heme-globin binding. *Science 157:* 1581–1583.

Bryant, E. H., 1974. On the adaptive significance of enzyme polymorphisms in relation to environmental variability. *American Naturalist 108:* 1–19.

Burns, J. M., and F. M. Johnson, 1971. Esterase polymorphism in the butterfly *Hemiargus isola. Proceedings of the National Academy of Science, U.S.A., 68:* 34–37.

Dobzhansky, T., 1970. *Genetics of the Evolutionary Process.* Columbia University Press, New York.

Dobzhansky, T., F. J. Ayala, G. L. Stebbins, and J. W. Valentine, 1977. *Evolution.* W. N. Freeman, San Francisco.

Ebert, J. D., and I. M. Sussex, 1970. *Interacting Systems in Development.* 2nd ed. Holt, Rinehart and Winston, New York.

Fisher, R. A., 1929. *The Genetical Theory of Natural Selection.* Clarendon Press, Oxford. (This classic was reprinted by Dover Publications, New York, in 1958.)

Gooch, J. L., and T.J.M. Schopf, 1972. Genetic variability in the deep sea: Relation to environmental variability. *Evolution 26:* 545–552.

Hamrick, J. L., 1979. Genetic variation and longevity. In *Topics in Plant Population Biology,* O. T. Solbrig et al. (eds.). Columbia University Press, New York.

Ingram, V. M., 1963. *The Hemoglobins in Genetics and Evolution.* Columbia University Press, New York.

Jennings, H. S., 1920. *Life and Death, Heredity and Evolution in Unicellular Organisms.* Gorham Press, Boston.

Kimura, M., 1979. The neutral theory of molecular evolution. *Scientific American 241:* 98–126.

Lewontin, R. C., 1974. *The Genetic Basis of Evolutionary Change.* Columbia University Press, New York.

Lewontin, R. C., and J. L. Hubby, 1966. A molecular approach to the study of genic heterozygostity in natural populations. II. Amount of variation and degree of heterozygosity in natural populations of *Drosophila pseudoobscura. Genetics 54:* 595–609.

Markert, C. L. (ed.), 1975. *Isozymes. IV. Genetics and Evolution.* Academic Press, New York.

Mukai, T. 1964. The genetic structure of natural populations of *Drosophila melanogaster.* I. Spontaneous mutation rate of polygenes controlling viability. *Genetics 50:* 1–19.

Nevo, E., 1978. Genetic variation in natural populations: Patterns and theory. *Theoretical Population Biology 13:* 121–177.

Selander, R. K., S. Y. Yang, R. C. Lewontin, and W. E. Johnson, 1970. Genetic variation in the horseshoe crab (*Limulus polyphemus*), a phylogenetic "relic." *Evolution 24:* 402–414.

FIVE

Gene Frequencies: Results of Selection

SOON AFTER THE TURN OF THE CENTURY, an interesting consequence of Mendelian inheritance was noted by scientists in several areas. They pointed out that a population of sexually reproducing organisms of known genotype, under certain ideal conditions, will perpetuate indefinitely its initial genotypes and at the same frequencies. Neither qualitative nor quantitative changes would occur in this population. This view of a Mendelian population clearly implies that evolution does not occur. Those making this point included the American geneticist Castle (1903), the British statistician Pearson (1904), the British mathematician Hardy (1908), and the German physician Weinberg (1908 and later papers). The latter two are associated with the most complete statements and hence their names have become associated with what we now call the Hardy-Weinberg law.

However, we should look carefully at that somewhat mysterious phrase "under certain ideal conditions," used above. It turns out that the ideal conditions are the absence of selection and of mutation, that no organisms migrate into or out of the population in question, that all possible matings occur at random, and therefore, that the population is large enough to ensure random mating. Our careful look emphasizes two points:

(1) These ideal conditions do not exist anywhere. (2) The real conditions must be what cause evolution, since their absence assures no evolution.

It is right here that we begin to see that evolution is a good deal more complex than even Darwin thought. True, natural selection plays an important role, as does mutation—Darwin's heritable variation. But also to be reckoned with are migration, breeding patterns, and population size and structure. Then we must consider the interaction of all these factors. This complexity also makes clearer why evolutionary studies are not the rigorous predictive science its practitioners would like it to be. But this array of factors is helpful in that it identifies precisely the areas that we must analyze to advance our understanding of evolutionary change.

In fact, we shall examine the operation of heredity in reproductive communities of organisms much in the spirit of which we have just described. First, we need to see how the genetics of a Mendelian population works when there are no complications beyond the rules governing the transmission of genes from one generation to the next. Second, we will want to see how selection, migration, mutation, breeding habits, and population size and structure affect genotypes under real conditions.

The Hardy-Weinberg law

The Hardy-Weinberg law states that initial genotypes will be transmitted indefinitely and in their original frequencies under certain ideal conditions of Mendelian inheritance. We have already listed those special conditions, so now let us demonstrate the validity of what is claimed to occur while they prevail. Most of what we will now discuss comes from the work of the early population geneticists (Box 5–1).

The easiest way to demonstrate the Hardy-Weinberg law is to start with a population with just two alleles in it, e.g., *A* and *a*. Let us further assume that there are equal numbers of both alleles and that they can exist as *AA*, *Aa*, or *aa* genotypes in the individuals of this population. Such a population is shown diagrammatically in Table 5–1. The left half of the diagram describes the males the right half the females. The equal frequencies of the two alleles results in a 1:2:1 ratio among the three possible genotypes, *AA*, *Aa*, and *aa*, respectively. This ratio holds for the population as a whole as well as for the two subpopulations represented by males and females.

Box 5–1. The early population geneticists.

Sir Ronald Fisher

Many distinguished scientists have corrected and extended Darwin's views, and some of them are mentioned later in this book. But in a very real sense, a new chapter was opened on the study of natural selection by population genetics.

The Russian S. S. Chetverikov (1880–1959) wrote, in 1926, a remarkable essay on evolution from the genetic standpoint. He recognized the variability present in Mendelian populations and pointed to ways of determining the frequencies of different genes in a species, which he viewed as a reproductive community. His work only became available in an English translation in 1961 and his own career in the Soviet Union was disrupted by politics. He, however, did start a tradition of population genetics in his country, one that has produced much distinguished work.

Sir Ronald Fisher (1890–1962) published *The Genetical Theory of Natural Selection* in 1929. This pioneering work focused on the role of Mendelian heredity in generating variation in natural population and on natural selection acting on that variation. Fisher was very explicit on treating natural selection as the cause of evolution. Additionally, Fisher was one of the leading statisticians of our time and developed several important quantitative research tools. Along with J.B.S. Haldane (1892–1964), a fellow Briton who subsequently left England and continued his statistical work in India, he made outstanding contributions to biometrical genetics.

Sewall Wright (1889–), an American, is another mathematically gifted geneticist. Starting in the early 1930's, he produced a most distinguished series of papers and monographs, explaining in mathematical terms the evolutionary

J.B.S. Haldane

Sewall Wright

behavior of populations. In particular, he clarified the role of population size and the phenomenon of "genetic drift," wherein factors other than selection play a prominent role in evolution.

The work of these early population geneticists laid the foundation for the work of others in this and related areas of biological research.

Table 5–1. Diagram of an ideal population containing two alleles, *A* and *a*, in equal numbers or frequencies.

Males		Females	
Aa 50%		*Aa* 50%	
AA 25%	*Aa* 25%	*AA* 25%	*aa* 25%

Now let us look at the F_1 generation of organisms. Here another aspect of the ideal conditions we are working under must be made clear. Every mating is as fertile as any other and equal numbers of male and female progeny are produced. This can be achieved if, for example, every mating produced two offspring, one male, the other female. Or, the same effect would be achieved if, *on the average*, every mating produced the same number of progeny (whatever that would be) and equal numbers of females and males. These stipulations regarding number and the sex of the progeny are implied by the absence of selection, namely, no one mating has an advantage over any other in terms of numbers of progeny produced and neither sex has a survival advantage.

In Table 5–2 we list all possible matings, the frequency at which they will occur, and the genotypes and frequencies of the progeny that will be produced. All of this is found in an ideal population.

Possible matings are determined by allowing each of the three different male genotypes to be crossed to each of the

Table 5–2. The results of mating under the ideal conditions specified by the Hardy-Weinberg law.

Matings Male Female	Frequency of mating	Progeny *AA*	*Aa*	*aa*
AA × *AA*	¼ × ¼ = 1/16	1/16*AA*		
AA × *Aa*	¼ × ½ = 1/8	1/16*AA*	1/16*Aa*	
AA × *aa*	¼ × ¼ = 1/16		1/16*Aa*	
Aa × *AA*	½ × ¼ = 1/8	1/16*AA*	1/16*Aa*	
Aa × *Aa*	½ × ½ = 1/4	1/16*AA*	1/8*Aa*	1/16*aa*
Aa × *aa*	½ × ¼ = 1/8		1/16*Aa*	1/16*aa*
aa × *AA*	¼ × ¼ = 1/16		1/16*Aa*	
aa × *Aa*	¼ × ½ = 1/8		1/16*Aa*	1/16*aa*
aa × *aa*	¼ × ¼ = 1/16			1/16*aa*
	16/16 = 1	1/4*AA*	1/2*Aa*	1/4*aa*

three different female ones. A total of nine different crosses covers all possibilities.

These possibilities are not all equally probable. There are different frequencies of occurrence. For example, only one-sixteenth of all the matings will be *AA* × *AA*, whereas one-quarter of the matings will be *Aa* × *Aa*. These frequencies are determined as indicated in the table. For example, the frequency one-sixteenth for the mating *AA* × *AA* is determined by the fact that each genotype occurs with the frequency of one-quarter (Table 5–1). When two individuals with these genotypes mate it means, in effect, that both individuals must be present simultaneously. The probability of that occurring is the product of the separate probabilities of each individual genotype, namely $\frac{1}{4} \times \frac{1}{4}$ or $\frac{1}{16}$. In other words, the frequency of the occurrence of a given mating is determined by the frequency of the occurrence of the two partners in the populations; it is the product of those frequencies.

Let us look now at the progeny produced by the matings. The progeny are listed under the three possible genotypes. Note, for example, that the only progeny that can be produced by *AA* parents is *AA* progeny. And since $\frac{1}{16}$ of the matings involve these parents, and every mating provides its proportion of progeny to the next generation, one-sixteenth of the progeny will be *AA*. Looking at the cross between *AA* and *Aa* parents, two genotypes can be produced, *AA* and *Aa*. According to Mendel, those genotypes will occur in equal numbers. Since one-eighth of the matings are *AA* × *Aa*, one-eighth of the progeny will come from that mating and be divided equally between *AA* and *Aa* genotypes. That is, one-sixteenth of all the progeny will be *AA* and another one-sixteenth will be *Aa*.

And so on, down the list of matings, their frequencies, and their contributions to the next generation. To check these calculations, look at the all important bottom line. The sum total of all frequencies of matings adds up to sixteen-sixteenths or one. The total of *AA* progeny is four-sixteenths or one-quarter. And one-half the progeny are *Aa* and the remaining one-quarter are *aa*. The frequency of the progeny genotypes is, therefore, 1 *AA*:2*Aa*:1*aa*. This is the same as the frequency of those genotypes in the parent population (Table 5–1). And if the F_1 produced the F_2 under these ideal conditions, there would again appear the same three genotypes and in the same frequency as before.

GENE FREQUENCIES

Another way to demonstrate the Hardy-Weinberg law is to use mathematical notations that allow for a more general formulation of the law and its implications. Again let us use our hypothetical population of individuals carrying the genotypes *AA*, *Aa*, and *aa*, but we will not use the concurrent frequency of the two alleles *A* and *a*. Instead let p represent the frequency of allele *A* and q represent the frequency of *a*, whatever those values. Note that $p + q = 1$. That is, with p and q, we describe the frequencies of all the alleles at the locus where *A* and *a* occur.

There is a further useful consequence of this notation: The frequency of the genotype *AA* can be expressed as p^2, since it depends on the simultaneous occurrence of two *A* genes. And that, in turn, is given by the product of the frequency of each *A* gene. Similarly, q^2 is the frequency of *aa* individuals. The heterozygote *Aa* is given by $2\,pq$, since there are two ways to generate the heterozygote—one chromosome can carry *A* and the homologous one can carry *a*, in another heterozygous individual the alleles can be switched relative to the chromosomes in question.

This means that the sum of these three frequencies describes the genotypic frequencies of the whole population. Namely,

$$p^2 + 2\,pq + q^2 = 1$$

[This can be checked by substituting actual values for p and q. In our first situation we said that *A* and *a* occurred at equal frequencies (0.5). In that case, $p^2 = 0.25$, $q^2 = 0.25$, and $2\,pq = 0.5$. Compare those values with the frequencies of the three genotypes given in Table 5–1.]

Using p and q to describe the frequency of *A* and *a*, respectively, in our idealized population, let us again look at a demonstration of the Hardy-Weinberg law (Table 5–3).

First, the listing of possible matings has decreased in number. The nine listings of Table 5–2 are here, but they only take up six lines. That is because, for example, male *AA* × female *Aa* and female *AA* × male *Aa*, are both *AA* × *Aa* matings. The same is true for *AA* × *aa* and *Aa* × *aa* matings. Thus it is necessary to double the frequencies of those matings. In the table, the 2 before the parenthesis indicates that fact.

Table 5–3. The operation of the Hardy-Weinberg law described in terms of gene frequencies.

Possible matings		Frequency of Progeny		
Type	**Frequency**	***AA***	***Aa***	***aa***
$AA \times AA$	$p^2 \cdot p^2 = p^4$	p^4		
$AA \times Aa$	$2(p^2 \cdot 2\,pq) = 4p^3q$	$2\,p^3q$	$2\,p^3q$	
$AA \times aa$	$2(p^2 \cdot q^2) = 2p^2q^2$		$2\,p^2q^2$	
$Aa \times Aa$	$2pq \cdot 2\,pq = 4p^2q^2$	p^2q^2	p^2q^2	p^2q^2
$Aa \times aa$	$2(2\,pq \cdot q^2) = 4\,pq^3$		$2\,pq^3$	$2\,pq^3$
$aa \times aa$	$q^2 \cdot q^2 = q^4$			q^4
		p^2	$2\,pq$	q^2

Second, note the way frequency of matings is now expressed. Following the model of Table 5–2 and the logic of mathematical probabilities, we can express these frequencies algebraically, using the symbols p and q. This terminology, as we said before, can now apply to any value for the frequencies of alleles A and a. Substituting our earlier value of 0.5 for p and q, it can be seen that the sum of the frequencies of all possible matings is 1.0.

Third, we can project the frequency of the progeny. And again Table 5–2 provides the model for distributing the algebraic terms under the genotypic headings. To illustrate, look at the matings between the heterozygotic parents $Aa \times Aa$. That type of mating will occur with a frequency of $4\,p^2q^2$, which is also the total of the progeny from such matings. However, the progeny can be of three genotypes and will occur in the frequency of 1 AA: 2 Aa: 1 aa. Hence, the total frequency of $4\,p^2q^2$ is distributed p^2q^2 to the AA category, $2\,p^2q^2$ to Aa, and p^2q^2 to aa.

And fourth, we can check on the validity of these calculations algebraically. The sum of the contributions of the various matings to the AA progeny is

$$p^4 + 2\,p^3q + p^2q^2$$

This can be simplified to

$$p^2\,(p^2 + 2\,pq + q^2)$$

but note that

$$p^2 + 2\,pq + q^2 = 1$$

Hence the sum of the frequency of the AA progeny is p^2, as given in the bottom line of column AA. Similarly the bottom line for the *Aa* column is reducible to 2 pq and for the *aa* column to q^2.

In other words, under the conditions of the Hardy-Weinberg law, a population whose initial gene frequencies are given as

$$p^2 + 2\,pq + q^2 = 1$$

will produce those same frequencies, generation after generation. We now have a general mathematical expression for the Hardy-Weinberg law.

CERTAIN PRACTICAL APPLICATIONS

Thus far we have worked from genes and their frequencies to genotypes and phenotypes and their frequencies. Let us reverse that process, starting from what we can observe directly—phenotypes and their frequency—and see if we can determine the frequency of their genes, which cannot be directly observed.

Blood groups represent easily determined phenotypes, the inheritance of which is well understood. Table 5–4 shows data on the occurrence of M, MN, and N phenotypes in a sampling of the population of Great Britain. These phenotypes are genetically controlled as follows: M occurs when the genes L^mL^m are present, MN when L^mL^n are present, and N when L^nL^n are present. (This situation is known as codominance, i.e., both alleles are expressed in the heterozygote.)

We can now calculate the frequencies of the L^m and L^n alleles in our sample of 1,279 people. Let p represent the frequency of L^m and q that of L^n. The frequency of M or L^mL^m individuals is given by p^2; the frequency N or L^nL^n individuals by q^2. And, of course, 2 pq is the frequency of the heterozygous (L^mL^n) individuals.

Table 5–4. The occurrence of M, MN, and N phenotypes in the British population.

Phenotype	M	MN	N	Total
numbers observed	363	634	282	1,279
frequency in terms of Hardy-Weinberg expectations	0.284	0.496	0.222	1.000
numbers in terms of Hardy-Weinberg expectations	362	637	280	1,279

The value of p can be determined by calculating how many of the L alleles are L^m. In this case, 1,279 diploid individuals means that 2,558 L alleles are present. Of these, 1,360 are L^m alleles and 1,198 are L^n alleles. These last two figures are obtained by adding up all the L^m and L^n alleles in this population. For L^m, each person of the M phenotype has two L^m genes for a total of 726 (2×363) plus 634 in the MN people. This gives a final total of 1,360. Similarily for L^n we find that 2×282 plus 634 adds up to a total of 1,198.

Frequencies are directly calculated, as follows:

$$1360/2558 = 0.532 = \text{frequency of } L^m$$
$$1198/2558 = 0.468 = \text{frequency of } L^n$$

In terms of the Hardy-Weinberg law, we can now state that

$$p^2 + 2\,pq + q^2 = 1.0$$

or

$$(0.532)^2 + 2(0.532)(0.468) + (0.468)^2 = 1.0$$

Solving for p^2, $2\,pq$, and q^2, the phenotypic frequencies, we see that the values obtained are very close to the observed phenotypic frequencies (Table 5–4). This allows us to say that, as regards the M, MN, and N phenotypes, the people studied are in a Hardy-Weinberg equilibrium.

Now let us look at the calculations for a phenotype for which full dominance occurs. The ability of humans to taste phenylthiocarbamide (PTC) is such a phenotype. Individuals of the genotype *tt* cannot taste PTC. Those who are *TT* or *Tt* experience a somewhat unpleasant bitter taste when PTC is placed on their tongues. A population sample, described in terms of these phenotypes, is shown in Table 5–5. Here, because of dominance we cannot directly observe the two different alleles. A different method of calculation is necessary. But if we know the value of p^2, or q^2, it is easy to determine p, or q, by taking its square root. In the case of the tasters, p^2 is

Table 5–5. The occurrence of phenotypes (tasters and non-tasters) in a population.

	Taster	Non-taster	Total
Genotype	*TT* or *Tt*	*tt*	
Number observed	1,680	320	2,000
Frequency	0.84	0.16	1.0

not known, since we cannot distinguish *TT* from *Tt* people. The non-tasters *tt*, however, can be distinguished. Hence, the frequency of *t* is

$$\sqrt{q^2} = \sqrt{0.16} = 0.4$$

and the frequency of *T* is, therefore,

$$p = 1 - q = 1 - 0.4 = 0.6$$

Among the tasters, the frequency of *TT* individuals is $(0.6)^2$ or 0.36, and the frequency of *Tt* individuals is 2 pq or 0.48. Together they make up 84% of the population. (Check against Table 5–5.)

These two examples of blood type and taster phenotypes show how it is possible to go from observations to concepts, in this case, to gene frequencies. When the Hardy-Weinberg law holds, that is, when calculated frequencies are close to observed frequencies of phenotypes, we say that the population is in equilibrium. This means two things: (1) Under these conditions, the phenotypes in question and their frequencies will be perpetuated indefinitely. (2) Selection, mutation, migration, breeding patterns, and population size are having little effect on this population in terms of the alleles being studied.

Now look at Table 5–6 and solve the problem posed there, which is to determine which of the five populations is *not* in Hardy-Weinberg equilibrium. This can be done by looking at the first two populations as examples. Population A provides data that reflect a Hardy-Weinberg equilibrium. Here

$$p = \sqrt{0.4225} = 0.65$$

and

$$q = \sqrt{0.1225} = 0.35$$

Table 5–6. Examples of populations that are or are not in Hardy-Weinberg equilibrium.

	AA	*Aa*	*aa*
Population A	0.4225	0.4550	0.1225
Population B	0.430	0.481	0.089
Population C	0.0025	0.1970	0.8005
Population D	0.0081	0.0828	0.9091
Population E	0.49	0.42	0.09

Find those that are in equilibrium.
(*N.B.* Populations A and E are in equilibrium; the others are not.)

and

$$2\,pq = 2\,(0.65)(0.35) = 0.455$$

Furthermore,

$$p + q = 1$$

By contrast, in population B,

$$p = \sqrt{0.430} = 0.656$$

and

$$q = \sqrt{0.089} = 0.295$$

Here $p + q$ does not equal 1.0, but only 0.95, and the calculated value for 2 pq is 0.387, which differs significantly from the observed value of 0.481. Going further in this population, because of codominance we can add up the occurrence of the alleles directly and calculate their frequencies. (Assume there are 1,000 individuals in population B, hence there are 2,000 loci.) This provides the following values

$$p = 1341/2000 = 0.671$$
$$q = 659/2000 = 0.329$$

These values differ from those calculated previously from the square roots of p^2 and q^2.

This lack of agreement between the frequencies expected with the Hardy-Weinberg law and the observed frequencies tells us that population B is not in equilibrium. What is causing the disequilibrium is not clear from the data. But a researcher seeing the disequilibrium could pursue that problem. The answer is predictably the result of one of the factors already discussed, i.e., selection, mutation, migration, breeding habits, size and structure of the population, or some combination of these factors. We turn next to see how these factors might cause such a disequilibrium.

Selection

In light of the Hardy-Weinberg law we suspect that selection is occurring when a genotype is present in a frequency higher or lower than that expected. Selection is favoring a genotype showing an unexpectedly high frequency and is acting against one that is lower than expected. This is another way of referring to differential reproduction or natural selection.

To be perfectly clear about selection and its action on gene frequencies let us digress from Mendelian populations that reproduce sexually for a moment and look at selection acting on bacterial populations, a much simpler case. Here we are looking at asexual, haploid cells. The bacterial species *Diplococcus pneumoniae* causes pneumonia in human beings. One method of treatment is to administer an antibiotic. This treatment stops the growth of bacterial cells, which are then destroyed by the defense mechanisms of the host organism, notably the white blood cells.

One antibiotic that is effective against *D. pneumoniae* is penicillin. We can see this in a situation in which large numbers of cells grow in the absence of penicillin, but not in

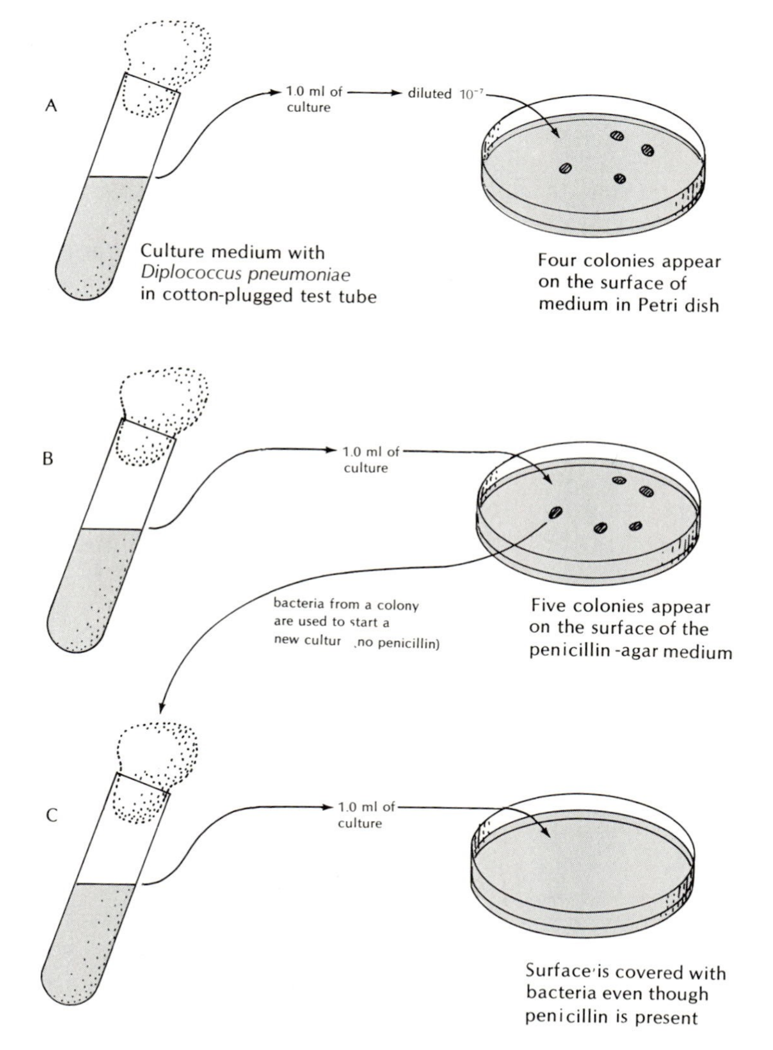

Fig. 5-1: Selection acting on a bacterial population. The initial population is grown in a fluid culture medium, in a test tube.

its presence (Fig. 5–1). This bacterium can be grown in the appropriate culture medium in a test tube. When 1.0 ml of this thriving culture is diluted 10 million times and then a 1.0 ml inoculum is placed on a semi-solid culture medium in a Petri dish (Fig. 5–1A), a small number of colonies, let us say four, of bacteria are found growing on this agar medium. Since each colony represents a clone (the asexual progeny of one original cell), it means that at least four cells were in the 1.0 ml sample that was spread on the surface of the agar. Since that many cells are there after dilution, it means forty million cells were present in the 1.0 ml taken from the test tube.

If penicillin were added to the agar before the bacteria were plated, then, no colonies would grow. Even at low dilutions and with thousands of inoculated bacterial cells, the usual story is no colonies. However, if undiluted culture medium is spread on the penicillin-agar plate a few colonies are found. And if these are cultured in a test tube and a sample plated on the penicillin-agar medium of a Petri dish, the surface of the agar would be covered with colonies, unless the sample had been diluted.

What is happening? The tested and accepted answer is that exposure to penicillin selects those cells resistant to it. With mutations occurring at a frequency of around one in a million cells (Table 4–2), a sample of 10^7 cells could be expected to have a few mutants in it. These would be detected by the growth of the corresponding number of colonies when a sample of around 10^7 cells is placed on the penicillin-agar medium. And all members of each colony would be penicillin resistant and would transmit that trait to their progeny. This is what was described above. Now let us describe those events in terms of the effect of selection on gene frequencies.

Assume as a plausible starting condition that only one cell in a million cells is resistant to penicillin. Such resistant cells are haploid and contain the gene P^r. Sensitive cells carry the gene P^s. If p is the frequency of P^s and q of P^r, their frequencies in the population are 0.999999 and 0.000001, respectively. When a population with these gene frequencies is exposed to penicillin, there is a dramatic change. All the survivors now carry gene P^r. When these cells are cultured in a test tube without penicillin the cells will all be P^r except for rare mutations of P^r to P^s. (This probably occurs at a rate of around one in a million cells or 10^{-6}.) Hence p is now around 0.000001 and q around 0.999999, a reversal of their previous values. In other words, because of selection a population that was originally close to 100% P^s is close to 100% P^r. This result of selection is called *allele substitution.*

At this point it must be made clear that the primary and direct action of selection is on the phenotype, not the genotype. What an organism *does* is what determines its survival. Therefore the *functional phenotype,* the capability of an organism, is what is crucial in natural selection.

To be sure, function depends on structure and structure depends on gene products. Therefore, phenotype and genotype are connected, and population geneticists and evolutionists will thus speak of selection acting on certain genes. It must always be understood that what is meant is selection acting against (or in favor of) certain functional capabilities that are presumed to be genetically determined. Therefore the statement that selection is acting for or against a certain gene is simply shorthand for the statement that selection affects genes through the phenotypes expressed.

Keeping this in mind, we can return to diploid organisms. There are four types of problems here: (1) selection against a dominant gene, (2) selection against a recessive gene, (3) selection against both homozygotes of a pair of alleles, and (4) selection against the heterozygote. These can be represented in terms of the alleles *A* and *a* as (1) selection against *AA* and *Aa,* (2) selection against *aa,* (3) selection against *AA* and *aa,* and (4) selection against *Aa.*

SELECTION AGAINST A DOMINANT GENE

Dominant genes are of various kinds, considered from the point of view of selection. Some bestow a selective advantage, others a selective disadvantage. The most extreme disadvantage is, of course, lethality. Genetic lethality, however, can be expressed in various ways. It always refers to death occurring from genetic causes, but that death can appear at different times in an individual's life. It can occur early in development, as an embryonic lethal, or it can occur sometime after birth, but before sexual maturity. From an evolutionary viewpoint, if lethality occurs before an individual can reproduce, that organism is genetically dead because its genes are not transmitted. On the other hand, a gene that causes early death after the organism has reproduced is not considered to be a lethal gene. It may reduce the number of progeny produced, but the gene can be perpetuated through those progeny. Just now we are concerned with genes that cause genetic death.

If we consider selection as the only factor acting against a dominant lethal gene—mutations and other factors are ignored for the present—the consequences are very straightforward. The gene is never established in a population because it is eliminated whenever it appears.

If by chance environmental conditions changed suddenly and a certain dominant gene and its phenotypes (*AA* and *Aa*), which were not previously lethal now conferred lethality, then these phenotypes would quickly disappear from the population. In fact, the next generation would lack the *A* gene. Selection against a dominant lethal gene is the most severe kind of natural selection.

This can be contrasted with a lethal gene that causes death later in life, such as that for Huntington's chorea in humans. It is a dominant gene and is usually expressed in people who are in their late 30's or 40's as a progressive motor incoordination, mental deterioration, and, finally, death. As far as is now known, all cases of Huntington's chorea in the United States can be traced back to two individuals who brought the dominant gene from England to what was then called Boston Bay, in 1630. From early colonial times to now the gene has persisted in the white population of this country. And it has done so because carriers of the gene can reproduce before they die. In conventional terms the dominant gene causing Huntington's chorea is lethal—it kills people—but only after they have reached the age of reproduction. Therefore it has not been eliminated by natural selection.

A different example of selection against a dominant gene is seen in cases where dark-colored insects, normally at a selective advantage in areas darkened by industrial pollution, stray into unpolluted areas. This has been most intensively studied and is best understood in the case of the peppered moth *Biston betularia*. The work of H.B.D. Kettlewell and his associates at Oxford University has demonstrated that birds prey on these moths. During the day the moths often rest on tree trunks and their predators search for them there. In an unpolluted area the tree trunks are a mottled grey. The normal phenotype makes the peppered moth very inconspicuous on this background. Its dark-colored, or melanic form, is readily seen (Fig. 5–2). This latter phenotype is the result of a dominant gene, either in the homozygous or heterozygous condition; the normal phenotype is controlled by the homozygous recessive allele. In unpolluted areas then, the dominant gene is at a definite disadvantage, and it is selected against.

Naturally, in polluted areas, the recessive gene and its

Fig. 5-2: *Biston betularia,* normal and dark forms resting on backgrounds, one darkened by industrial pollution, the other undarkened.

Fig. 5-3: Distribution of phenotypes of *Biston betularia* in the British Isles.

phenotype is selected against, as can be seen in the map in Fig. 5–3. Downwind from industrial centers there is, of course, heavy pollution. (The Industrial Revolution started in England in the middle of the last century and coal was the primary energy source. The unburned carbon or soot is the major industrial pollution affecting the lives of peppered moths.) In areas of heavy pollution the frequency of dark forms is very high and the light-colored normal forms are correspondingly less frequent. The reverse is true in unpolluted areas. This example of what is called industrial melanism illustrates selection against either the dominant or recessive allele, depending upon the environment. More precisely, selection, in the form of predation by birds, is acting on the phenotypes of these moths; this in turn has an effect on the frequencies of the alleles responsible for the phenotypes.

Incidentally, since the passage of the anti-pollution laws after World War II, pollution has been decreasing in Great Britain. It is reported that the frequencies of the light phenotype is increasing since the time of the first studies. Sample data for one area show that the frequency of occurrence of light moths went from 5.2% in 1961 to 8.9% in 1964 and reached 10.5% in 1974. Industrial melanism is not limited to moths, but is found in various insects. Furthermore it occurs in many heavily industrial areas, e.g., the Ruhr valley in West Germany and around Pittsburgh in the United States.

SELECTION AGAINST A RECESSIVE GENE

It is worth considering both rigorous selection against a recessive gene and weak selection against it. The consequences are somewhat different in the two cases.

Recessive lethal genes. As can be guessed from what was just said about dominant lethal genes, selection is most rigorous against recessive alleles when they are lethal. That is, in a homozygous condition the recessive allele generates a phenotypic effect that ensures death for that individual. And here, as with the dominant gene, death can occur anywhere in the sequence of events leading from the zygote to the mature adult, but in any case death occurs before sexual maturity.

One of the clearest examples of such selection would occur if sterilization laws were invoked in humans regarding some trait that was agreed upon as undesirable (Box 5–2). Let us consider a hypothetical situation. Society has decided a certain recessive gene is undesirable because, when it is expressed, the resulting phenotype is undesirable. Such a recessive is

Box 5–2. Sterilization and human values.

At this point, examples involving humans have been brought into our discussion. The purpose is to show that humans, too, are subject to selection. However, there is special concern when humans themselves are the agents of selection, as in the case when sterilization is invoked, for here human values are involved, along with the biological consequences.

In 1951, 27 states in the United States had sterilization laws and their purpose was relatively clear. It was to stop perpetuation of traits considered, by at least certain members of society, to be undesirable. With time it has become quite clear that the designation of traits as undesirable was too often racist or elitist and that sterilization laws often reflected a minority rather than a genuinely democratic opinion. On the other hand, sterilization has been used to prevent mentally retarded persons from having children whom they cannot take care of or may accidently injure and who even may die through neglect. Here the intent is clearly humanitarian.

Furthermore, it is clear that certain traits thought by some people to be undesirable are not heritable traits, for example, promiscuity among welfare recipients. Those urging sterilization did not want to see more children produced who would be dependent on welfare and might in turn be promiscuous and produce more children and increase the welfare burden. There is no evidence that such a situation is hereditary. It is most probably due to a complex of socioeconomic factors. In fact, socioeconomic factors lead to *voluntary* acceptance of sterilization by certain women, as a certain means of preventing unwanted pregnancies. Important as these issues are, we will have to leave them as digressions from the main purpose of our discussion.

Except for the mistaken notions of heredity and "social diseases" mentioned above, our society does not permit sterilization as a means of ridding itself of unwanted genetic diseases. One very important reason for the difficulties in accepting sterilization as a treatment for genetic diseases is that sterilization brings about the genetic death not only of the unwanted gene, but of *all* other genes in the sterilized individual. That could very well be too high a price to pay. Another reason is the inefficiency of sterilization as shown in the hypothetical example discussed in the text.

This is not to be confused with another area where our society is allowing its members—but not without protest from dissenters—to rid itself of unwanted genetic disease. We refer to voluntary abortions of fetuses in whom undesirable traits, such as the lethal Tay-Sachs disease, are detected early in the pregnancy by the technique of amniocentesis.

The issues touched on here are complex, but the fact remains that humans do indulge in a variety of behaviors that affect the transmission of genes from one generation to another. Sterilization and celibacy, of course, terminate all transmission. Abortion terminates transmission to one individual. Birth-control devices and family planning aim to control the rate at which progeny are born and, also, to contribute to an improved home environment for raising children. Genetic counselling is available to help people protect the quality of the biological heritage passed on to children. Directly or indirectly, consciously or unconsiously, these practices can affect the nature of the gene pool our society is perpetuating.

Those interested in pursuing these topics further are urged to read the appropriate parts of works by Cavalli-Sforza and Bodmer and by L. J. Kamin, C. Stern and the volume on *Genetic Screening* cited at the end of this chapter.

only detected when homozygous, i.e., *aa;* hence, only *aa* individuals are sterilized. Since they then cannot reproduce, from a genetic and evolutionary point of view *aa* will represent a lethal condition. Suppose, further, that the frequency of *a* in the population is low; it is 0.02. This is reasonable in that *a*, being undesirable, is associated with poor health or disability in some way. That is already a kind of selection against the gene; hence, it is already being acted against as evidenced by it relatively low frequency. (The frequency of its allele or alleles is 1-q or 0.98.) We are now assuming that society has decided to speed up the process of elimination by means of sterilization. How long will it take, if all *aa* individuals are sterilized, to reduce the frequency of q by one-half? That is, to reduce it from 0.02 to 0.01? (Note that if $q = 0.02$, q^2 is 0.0004 and so there are four *aa* individuals in every 10,000 members of the population. When $q = 0.01$, the frequency of *aa* is one in 10,000.)

The frequency of q after various generations is

$$q_n = \frac{q_0}{1 + nq_0}$$

where q_n is q after n generations of selection and q_0 is the initial frequency of q. (The derivation of this formula is given in Box 5–3.) What we are asking is how many generations will it take for q_0 to be reduced by one-half? That is, for $q_n = q_0/2$. We note that $nq_0 = 1$ when n is the reciprocal of q. Therefore if $q_0 = 0.02 = 1/50$, then $n = 50$, or $50/50 = 1$. Hence for

$$\frac{q_0}{1 + nq_0} = \frac{q_0}{2}$$

$n = 50$. It will, therefore, take 50 generations of sterilization to achieve the desired reduction of q. In humans a generation is normally considered to be, on the average, between 20 to 30 years. Some couples have children when they are under 20, some when they are over 30. But the average, it appears, is between 20 to 30 years of age. This means 50 generations of selection will take 1,000 to 1,500 years to reduce the frequency of q by one-half.

In the light of this most people feel such a selection program to improve our society is of no immediate value. Eugenics is the name applied to such endeavors aimed at improving the quality of our lives. The foregoing analysis represents one sound argument against such programs. But in na-

Box 5–3. Derivation of the formula for determining q after generations of selection.

After one generation of selection, selection removes all *aa* individuals,

$$q_1 = \frac{pq}{p^2 + 2\,pq}$$

The heterozygotes *Aa* are the only source of new *a* genes. Their frequency is 2 *pq*, but only one-half their genes are *a*. Hence the frequency of that allele is *pq*, as given in the numerator. The denominator $p^2 + 2\,pq$ gives the surviving alleles after selection has removed *aa* individuals. Thus

$$p^2 + 2\,pq + q^2 = 1$$
$$p^2 + 2\,pq = 1 - q^2$$

The new value q_1 is determined by the frequency of the remaining *a* genes divided by the frequency of the remaining alleles at the *A* locus. Therefore,

$$q_1 = \frac{pq}{p^2 + 2\,pq} = \frac{(1 - q)\,q}{(1 - q)\,(1 + q)} = \frac{q}{1 + q}$$

After two generations

$$q^2 = \frac{q}{1 + 2\,q}$$

and after *n* generations

$$q_n = \frac{q}{1 + nq}$$

where *q* on the right of the equation is the initial value of *q* or q_0.

ture where time spans of hundreds of years are commonplace, rigorous selection against a recessive gene will have its effect.

A general description of the change in *q* over time when *aa* is either a lethal or is selected out is given in Fig. 5–4.

Low selection against the recessive gene. In low selection against *aa* individuals, only a small proportion of these individuals show the effect of selection. At this point the concept of fitness must be introduced along with the selection coefficient.

Fitness is the ratio of the progeny actually produced to the progeny expected from Mendelian inheritance. Fitness is, therefore, always relative. The *selection coefficient* (*s*) measures selective disadvantage (or advantage). Fitness and the selection coefficient are discussed in more detail in Box 5–4.

In the case of low selection against the recessive phenotype, let us see what happens when $s = 0.001$. That is, when only one in a thousand individuals who show the recessive phenotype will be acted on by selection. Haldane, one of the pioneering population geneticists (Box 5–1), did the calcula-

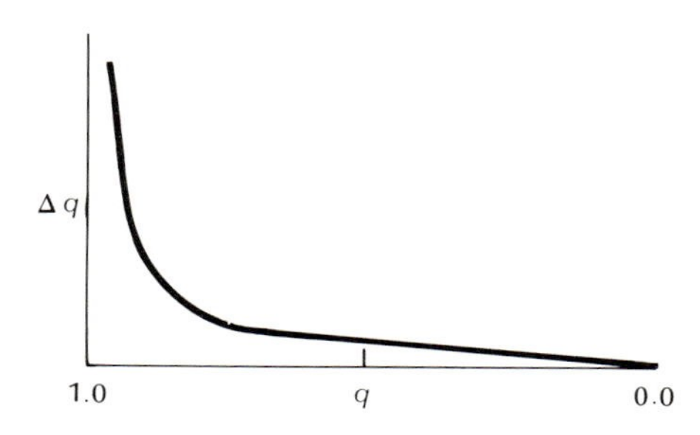

Fig. 5-4: Changes in q (Δq) when *aa* individuals show a lethal phenotype or are selected against, for example, by sterilization, as if they were lethal. At high frequencies of q, change is rapid because there are many *aa* individuals in the population against which selection can act. At lower frequencies, change is progressively slower because there are fewer *aa* individuals.

Box 5–4. Fitness and the selection coefficient.

Fitness and the *selection coefficient* are complementary terms that can be expressed quantitatively; they are important in the mathematical development of population genetics.

Starting from a familiar cross, such as that between two heterozygous parents, we can easily predict the genotypes of the progeny and the relative frequence of their occurrence.

Parents	$Aa \times Aa$
Progeny genotypes	$AA : Aa : aa$
Progeny frequencies	1 : 2 : 1

If the progeny represent a large number, such as 400 (as with insects), the expected numbers of progeny are

$$100\ AA : 200\ Aa : 100\ aa$$

If, however, only 85 *aa* individuals are actually born or survive to sexual maturity, selection against the recessive gene could be occurring. The fitness of the recessive phenotype (and genotype) is then expressed by the ratio of actual to expected progeny, i.e., 85/100 or 0.85.

From this value we can also obtain the value of the selection coefficient s. Since

$$s = 1.0 - \text{fitness}$$

in this instance

$$s = 1.0 - 0.85 = 0.15$$

Or a selection coefficient of 0.15 means that 15% of the expected progeny are removed by selection.

Selection can act against any genotype. As was seen earlier, the dominant gene can be selectively removed from a population. For a dominant lethal gene, the selection coefficient is 1.0 because its fitness is 0.0. Or, in words, 100% of the progeny are removed by inability to survive.

When there is no dominance and all three genotypes are expressed as phenotypes, selection can act against both homozygous genotypes or only against the heterozygous genotype. This is discussed later in the chapter.

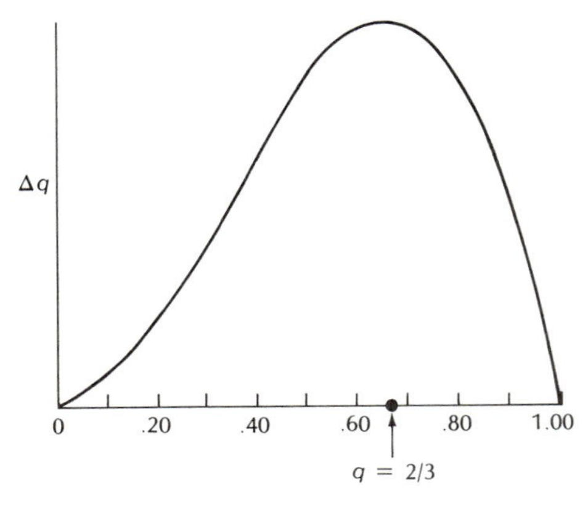

Fig. 5-5: A graph of Δq at low selection pressures. Here $\Delta q = -spq^2/(1 - sq^2)$ (see Box 5-5).

tions shown in Table 5–7. It is of interest that the rate of change of q is most rapid at intermediate values (Table 5–7). But when q is either low or high, changes are quite slow, and they are clearly much slower at the lower values. This can be shown graphically in Fig. 5–5 and its mathematical derivation in Box 5–5. Compare Fig. 5–4, which applies to cases where selection is high against *aa* individuals, to Fig. 5–5. This comparison makes clear why high and low selection against the recessive gene are best treated as separate cases.

SELECTION AGAINST BOTH HOMOZYGOTES

Selection against both homozygous genotypes can occur only

Table 5–7. Generations needed to achieve certain changes in q when s is 0.001.

Reduction in q	Change in q	Generations required
0.99999–0.99	0.00999	12,563
0.99–0.5	0.49	5,595
0.5 –0.1	0.4	102,595
0.1 –0.00001	0.09999	99,896,918

when they are distinguishable from the heterozygous condition. This means either that dominance does not occur or that both alleles are equally dominant. The latter is called codominance and is best indicated by alleles designated A and A'. The genotypes in a Mendelian population are, therefore, AA, AA', and $A'A'$. We will now consider the consequences of selection against AA and $A'A'$. In other words, selection favors the heterozygote AA'. This can be represented as follows:

Parental genotypes	AA	AA'	$A'A'$
Expected numbers of progeny	100	200	100
Observed numbers of progeny	85	200	65

The fitness of the AA individuals is 0.85, the selection coefficient s_A, 0.15. Similarly, the fitness of the other homozygote, $A'A'$, is 0.65, s_A', 0.35.

What is the long-term consequence of selection against both homozygotes? In contrast to what we saw happening to selection against dominant and recessive genes, i.e., a tendency to loss at rates depending on the value of s, here both alleles are preserved. This is essentially so because the heterozygote, containing both alleles, is at an advantage. In fact, there will be an equilibrium point at which both alleles will be preserved at frequencies that can be predicted from s_A and $s_{A'}$. First, let us see how the equilibrium point is determined and then we can better understand a graphic representation of what is happening.

From Box 5–5 we see that the frequency of A', or q, changes according to the following mathematical expression

$$\Delta q = \frac{pq(s_A p - s_{A'} q)}{1 - s_A p^2 - s_{A'} q^2}$$

Then at equilibrium, by definition, $\Delta q = 0$, which means the numerator on the right is 0. (This is so because 0 divided by whatever number one chooses is always 0.)

Box 5–5. Changes in q due to selection.

The means for calculating the rates of change in gene frequencies as a result of selection can be seen from what follows here.

1. Changes in q when selection is low against the homozygous recessive genotype.

	AA	*Aa*	*aa*	Total	Frequency of *a*
Initial frequency	p^2	$2\ pq$	q^2	1	q
Fitness	1	1	$1-s$		
Frequency after selection	p^2	$2\ pq$	$q^2(1-s)$	$1-sq^2$ *	$\frac{pq+q^2(1-s)}{1-sq^2}$ **
					$=\frac{pq+q^2-sq^2}{1-sq^2}$
					$=\frac{q(p+q-sq)}{1-sq^2}$
					$=\frac{q(1-sq)}{1-sq^2}$

* Derived from

$$p^2+2\ pq+q^2(1-s)$$
$$=p^2+2\ pq+q^2-sq^2$$
$$=1-sq^2$$

** The frequency of those genotypes containing the *a* gene divided by total frequency after selection. Note that pq is one-half $2\ pq$, since only one-half of the heterozygote genotype contains the *a* gene.

$$\Delta q=\frac{q(1-sq)}{1-sq^2}-q=\frac{q(1-sq)}{1-sq^2}-\frac{q(1-sq^2)}{1-sq^2}$$
$$=\frac{q-sq^2-q+sq^3}{1-sq^2}=\frac{-sq^2+sq^3}{1-sq^2}$$
$$=\frac{sq^2(1-q)}{1-sq^2}=\frac{spq^2}{1-sq^2}$$

$$0=pq(s_Ap-s_{A'}q)$$
$$0=pq(s_Ap)-pq(s_{A'}q)$$
$$pq(s_{A'}q)=pq(s_Ap)$$
$$s_{A'}q=s_Ap$$

Or the change in the frequency of A' is equal to the change in A, and these changes are given by the product of the selection coefficient against an allele multiplied by the frequency of that allele.

One further step is needed before we can calculate the value of p or q at equilibrium. Since, at equilibrium, $s_{A'}q=s_Ap$,

The change in q is expressed as Δq. It is given by the frequency of q after selection minus the initial frequency. When that value is negative it tells us that q is decreasing.

2. Change in q (Δq) when selection goes against both homozygotes.

	AA	*AA'*	*A'A'*	Total	Frequency of q
Initial frequency	p^2	$2pq$	q^2	1	q
Fitness	$(1-s_A)$	1	$(1-s_{A'})$		
Frequency after selection	$p^2(1-s_A)$	$2pq$	$q^2(1-s_{A'})$	$1-p^2s_A-q^2s_{A'}$*	$\dfrac{q(1-qs_{A'})^{**}}{1-p^2s_A-q^2s_{A'}}$

* Derived from $p^2 - p^2s_A + 2pq + q^2 - q^2t$
$= (p^2 + 2pq + q^2) - p^2s_A - q^2s_{A'}$
$= 1 - p^2s_A - q^2s_{A'}$

** Derived from

$$\frac{pq + q^2(1-s_{A'})}{1 - p^2s_A - q^2s_{A'}} = \frac{pq + q^2 - q^2s_{A'}}{1 - p^2s_A - q^2s_{A'}} = \frac{q(p+q) - qs_{A'}}{1 - p^2s_A - q^2s_{A'}} = \frac{q(1-qs_{A'})}{1 - p^2s_A - q^22s_{A'}}$$

$$\Delta q = \frac{q(1-qs_{A'})}{1-p^2s_A-q^2s_{A'}} - q = \frac{q(1-qs_A) - q(1-p^2s_A-q^2s_{A'})}{1-p^2s_A-q^2s_{A'}}$$

$$= \frac{q - q^2s_{A'} - q + qp^2s_A + q^3s_{A'}}{1-p^2s_A-q^2s_{A'}} = \frac{qp^2s_A + q^3s_{A'} - q^2s_{A'}}{1-p^2s_A-q^2s_{A'}}$$

$$= \frac{qp^2s_A + q^2s_{A'}(1-p) - q^2s_{A'}}{1-p^2s_A-q^2s_{A'}} = \frac{qp^2s_{A'} + q^2s_{A'} - pq^2s_{A'} - q^2s_{A'}}{1-ps_A-q^2s_{A'}}$$

$$= \frac{qp^2s_A - pq^2s_{A'}}{1-p^2s_A-q^2s_{A'}} = \frac{pq(ps_A - qs_{A'})}{1-p^2s_A-q^2s_{A'}}$$

then

$$0 = s_Ap - s_{A'}q$$
$$0 = s_Ap - s_{A'}(1\text{-}p)$$
$$0 = s_Ap - s_{A'} + s_{A'}p$$
$$s_{A'} = s_Ap + s_{A'}p$$
$$\frac{s_{A'}}{p} = s_A + s_{A'}$$
$$p = \frac{s_{A'}}{s_A + s_{A'}}$$

Substituting 0.35 for $s_{A'}$ and 0.15 for s_A, we find that p at equilibrium is 0.7. The comparable value for q is 0.3.

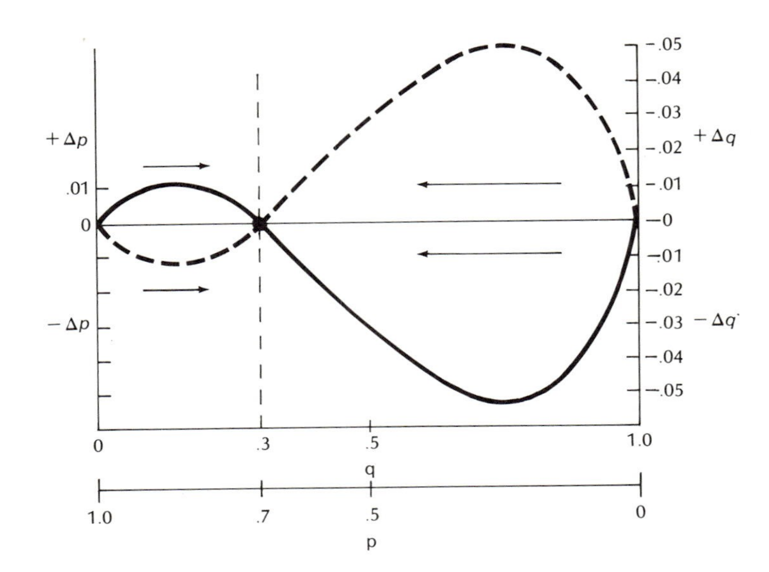

Fig. 5-6: The equilibrium point is at 0.3 when selection is against both homozygotes, and $S_A = 0.15$ and $S_{A'} = 0.35$ (q, solid line; p, dotted line).

This case of selection against both homozygotes is given in Fig. 5–6. In particular, note what happens when the values of p and q do not lie at the equilibrium values. If p is high, for example, 0.9, then q must be low, in this example 0.1. This says that there are proportionately many more A genes in the population than there are A' genes. Necessarily, then, there will be proportionately more selection against A than against A'. Hence more A will be lost and p will decrease and q increase. The reverse will occur if q is high and p is low. Selection in both cases drives p and q to more intermediate values. And, depending on the actual values of s_A and $s_{A'}$, equilibrium will be reached at those intermediate values where the change in p is the same as that in q.

The result in terms of the phenotypes in the population is called *balanced polymorphism*. *Polymorphism* refers to the three phenotypes, namely, AA, AA', and $A'A'$, that are retained by selection for the heterozygote and *balanced* refers to the equilibrium, the tendency to preserve a constant ratio among the phenotypes.

A classic demonstration of balanced polymorphism in human populations involves the gene for sickle-cell anemia. Normally, as was described earlier (Fig. 4–2), the beta hemoglobin molecule has glutamic acid in position six, but in the presence of the sickling gene, glutamic acid is replaced by valine. The genotypes possible when the normal allele Hb^A and the sickling allele Hb^s are present are: Hb^AHb^A, Hb^AHb^s, and

Hb^sHb^s. The resulting phenotypes are normal (no anemia), very mild or no anemia, and severe anemia, respectively. Those individuals with severe anemia rarely reproduce, since the Hb^s allele acts very much like a lethal. It is, therefore, a surprise to see that the frequency of this gene is so high in many parts of the world (Fig. 5–7A). One would, without much further thought, expect a gene as harmful as this one to occur at a very low frequency because of selection against it.

A look at Fig. 5–7B provides the information needed to explain the high frequency of Hb^s. Malaria is an important disease in tropical areas. Only since the European colonization of the New World has the disease been important in tropical America. In the Old World, malaria has been a scourge for millennia. There are various forms of the disease depending on the species of *Plasmodium* causing the infection. Here we are considering only P. *falciparum,* which causes the most severe type of malaria. Death occurs in about 10% of the infections in humans. In *falciparum* malaria, the human red blood cells are infested with the parasite. It turns out that the presence of the sickling hemoglobin in red blood cells significantly reduces that infestation. This is so important in areas where malaria occurs that individuals with the Hb^AHb^s genotype have an advantage over Hb^AHb^A individuals, who are much more seriously affected by this type of malaria.

Overall then, in areas where malaria occurs, selection favors the heterozygote and acts against the homozygotes. The Hb^sHb^s individuals are weakened by severe anemia, and the Hb^AHb^A individuals are more seriously ill when they contract

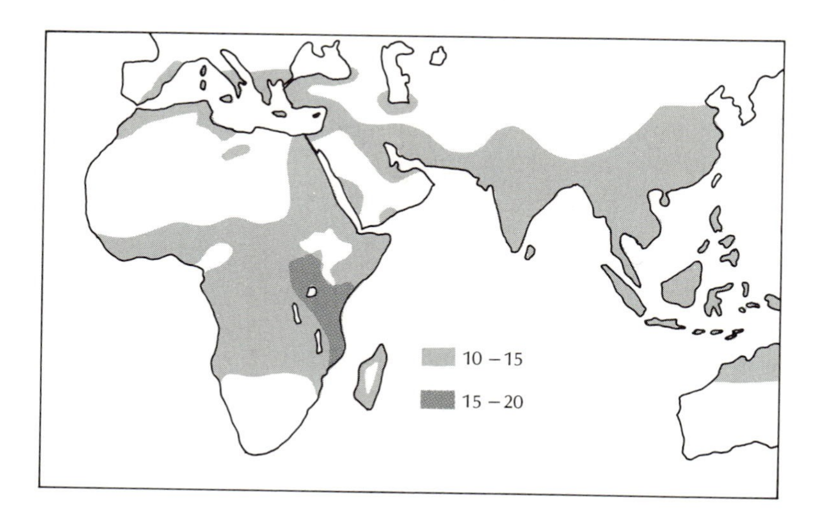

Fig. 5-7: The distribution of Hb^s allele in Africa and parts of Asia (dark shading). The distribution of malaria caused by the protozoan parasite *Plasmodium falciparum* (light shading).

malaria. The result is a balanced polymorphism and this accounts for the otherwise unexpectedly high frequency of the sickling gene.

Preservation of a lethal gene. It is instructive to look at another case of balanced polymorphism wherein the heterozygote has a slight selective advantage and one homozygote acts as a lethal genotype. The situation can be summarized in this way:

genotype	AA	AA'	$A'A'$
fitness	0.99	1.0	0
selection coefficient	0.01	0	1.0

Solving for q at equilibrium, i.e., $\hat{q} = s_A/(s_A + s_{A'})$, we see that the frequency of the A' gene, which is lethal when homozygous, is 0.0099 or approximately 1%. At this frequency, $A'A'$ individuals would occur once in about 10,000 members of the population. This is a low frequency, but it is a significant one and, therefore, we can conclude that only a slight advantage to the heterozygote will keep even a lethal gene in the population.

This raises the very important question as to how many apparently recessive genes—as is the case here—are kept in a population by balanced polymorphism. To determine the slight selective advantage of the heterozygote is a difficult task, for it requires that accurate data on the fitness of the AA and AA′ phenotypes be collected. Such data can only be collected under laboratory conditions. And those conditions may eliminate selective factors operating in nature or may introduce new factors; either way, the laboratory results will not reflect the natural conditions and will, therefore, confuse the study.

Finally, we can remark that circumstances that preserve a lethal gene may be one explanation of the occurrence of lethals in natural populations. We found this in Chapter 4, when variations in natural populations were being discussed. We could not, at that time, explain how such lethals were preserved. As we will see, however, another explanation lies in the fact that a gene that is lethal under certain conditions, may, under other conditions, be of some advantage to an organism.

SELECTION AGAINST THE HETEROZYGOTE

The fourth and final category of selection involves a disadvantage for AA' individuals. It is only necessary to examine two cases, since they will make it clear why, under this kind of selection, only one allele will persist.

Box 5–6. The effect of selection against a lethal heterozygote when p and q are not equal.

genotypes	AA	AA'	$A'A'$
fitness	1.0	0	1.0
selection coefficient	0	1.0	0
Frequencies:			
initial	0.36	0.48 lethal	0.16
first generation	0.476	0.428 lethal	0.096
second generation	0.689	0.282 lethal	0.029

The frequency of the AA and $A'A'$ genotypes can be easily calculated. The frequency of AA individuals in the first generation is 0.476. This is arrived at by assuming a sample of 100 individuals in the initial generation. Of these 36 are AA, 48 AA', and 16 $A'A'$. All the AA' die leaving only AA and $A'A'$ organisms. Among these the frequency of the A gene p is

$$\frac{\text{all } A \text{ alleles}}{\text{all alleles at this locus}} = \frac{72}{72 + 32} = 0.69$$

Therefore,

$$p^2 = (0.69)^2 = 0.476$$

Or, more simply,

$$p = \frac{0.36}{0.36 + 0.16} = 0.69$$

$$q = 1.0 - 0.69 = 0.31$$

and

$$q^2 = 0.096$$

Similarly, for the second generation,

$$p = \frac{0.476}{0.476 + 0.096} = \frac{0.476}{0.572} = 0.83$$

$$p^2 = 0.689$$

The values of q and q^2 for the second generation can also be calculated.

A transient or unstable equilibrium will occur if $p = q = 0.5$. In this case selection against the heterozygote will affect the frequencies of A and A' equally. This, however, is the only value of p and q where this holds; variation away from a value of 0.5 will be followed by further changes.

This can be seen by looking at a second case in which, for example, $p = 0.6$ and $q = 0.4$. In Box 5–6 the results of two

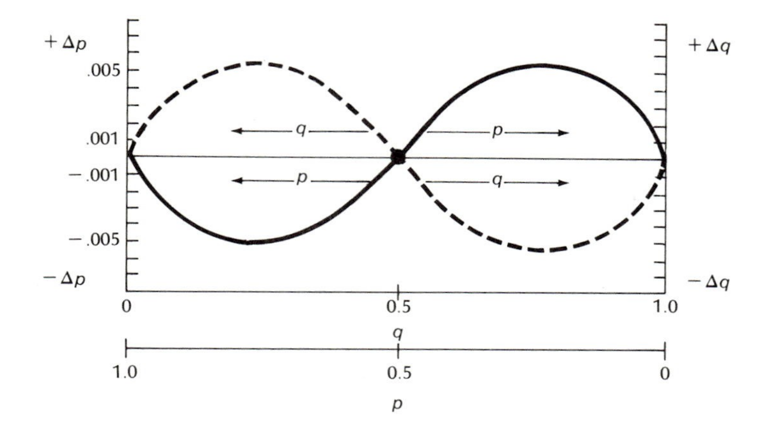

Fig. 5-8: Selection against the heterozygote: Any deviation from the transient equilibrium ($p = q = 0.5$) will send one allele to fixation, the other to extinction (p, solid line; q, dotted line).

generations of reproduction are summarized where AA' is acting as a lethal condition. Note how p^2 is increasing rapidly and how q^2 is thus decreasing. This means that p and q are also changing; in the second generation, $p = 0.83$ and $q = 0.17$, whereas the initial values were 0.6 and 0.4, respectively.

This rapid change occurs because, when the heterozygotes die out, there is, *proportionately,* a greater loss of A' genes than of A genes. Take, for example, 100 individuals in the initial population. In this sample there are 120 A genes (72 in AA and 48 in AA' individuals) and 80 A′ genes. When the heterozygotes die they remove 48 genes from each type of allele. Proportionately the loss is greater for the A' allele. And this type of loss continues in each generation. The result is, as we see in Box 5–6, a trend that will drive A' to extinction, i.e., $q = 0$, and p to fixation, i.e., $p = 1.0$. This is shown graphically in Fig. 5–8.

THE GENERAL EFFECTS OF SELECTION

Here we discuss three types of selection and, from the foregoing discussions, we can project their effects on a population. These are *stabilizing selection, directional selection,* and *disruptive* or *diversifying selection,* and their effects are shown in Fig. 5–9.

Stabilizing selection. Also called normalizing selection, this type of selection acts to preserve a certain array of phenotypes because of their selective advantage. Because of mutation and the various kinds of recombination possible as a result of sexual reproduction, any Mendelian population can generate an enormous array of genotypes. From among that variety, selection preserves those of selective advantage. This is the basis for Fisher's remark that ". . . natural selection is a mechanism for generating an exceedingly high degree of improbability." That is, out of all the possible genotypes and their phenotypes (refer again to Table 4–6), only a few are preserved by natural selection. Or, to make the point another way, try to imagine the appearance of a Mendelian population if there were no selection. There would be little resemblance among the various individuals; lethal genes would be rampant in the population; individual genomes might or might not be compatible with one another; there would be biological chaos. Selection imposes the order of adaptive phenotypes on a natural population and that, in a universe tending to disorder or entropy, is what Fisher means by a "high degree of improbability."

In Fig. 5–9A, normalizing selection, over periods of time, tends to reduce further the array of adaptive phenotypes expressed in a population. The bell-shaped curve that emerges, also termed an *adaptive norm,* describes the phenotypes around which the population is stabilizing due to selection.

Directional selection. When new selection pressures appear a population will respond or not. Response is shown in Fig. 5–9 B by the appearance of a new adaptive norm. The new norm is represented by phenotypes and their corre-

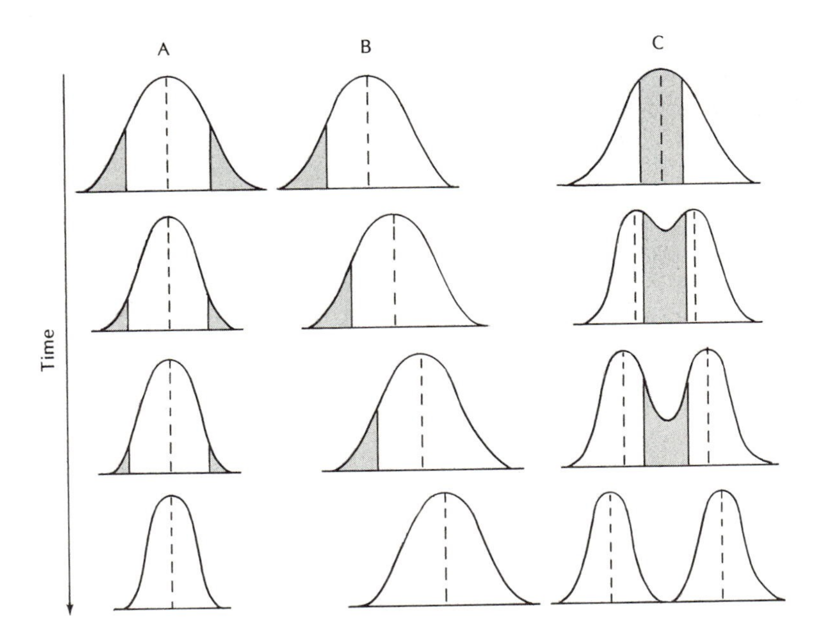

Fig. 5-9: Three kinds of selection and their effects. Each curve has the same coordinates: horizontal, phenotypes available in a population; vertical, fitness or frequency, which are interchangeable here, since high fitness is necessary for high frequency and high frequency implies a high degree of fitness. Time progresses as one moves down the vertical rows of diagrams, from top to bottom, under A, B, and C, respectively. A. *Stabilizing selection.* Certain phenotypes are preserved and the shape of the curve tends to narrow through elimination of less desirable phenotypes and their genes (shaded area). B. *Directional selection.* Changes in selection pressures generate a new norm over time. Phenotypes, originally occurring at low frequencies, are now highly advantageous and appear at higher frequencies. C. *Disruptive* or *diversifying selection.* Selection generated two norms, depending on environmental heterogeneity and, as discussed in Chapter 6, on population size. Balanced polymorphism can result in two phenotypes and the superficial appearance of disruptive selection. More importantly, the two adaptive norms represent different adaptive tendencies, due to different selection pressure, on one Mendelian population.

sponding genotypes and genes that previously occurred at low frequencies. New environmental conditions and the new selection pressures they generate demand a new adaptive norm, and the population must respond or it does not survive.

Disruptive or diversifying selection. If we considered only one pair of alleles in drawing Fig. 5–9C, then the two

Box 5–7. Fisher's Fundamental Theorem of Natural Selection.

We first discussed Fisher's Theorem in general terms, saying, in essence, that if heritable variations do *not* occur, then natural selection has no genetically determined phenotypic differences on which to act and therefore evolutionary change by natural selection will not occur. Conversely, we can say that if there is significant genetic variation on which selection can act there will be significant evolution. In fact, as a general rule we claim that the greater the variation, the greater the evolutionary change.

Fisher was, of course, much more precise on this issue. In his great book, *The Genetical Theory of Natural Selection,* first published in 1929, he wrote, "The rate of increase in fitness of any organism at any time is equal to its genetic variance in fitness at that time." His argument is therefore developed not in terms of general evolutionary change, but in terms of fitness. We realize, however, that an increase in fitness means an increase in the frequency of the more fit genes, which, in turn, is an evolutionary change.

A demonstration of these changes in fitness in mathematical terms can be done several ways—in terms of a single locus or in terms of the effects of linkage and epistasis or dominance of one locus over other loci. We will consider here only the simplest case, that of alleles at a single locus with random mating. [This treatment is adapted from Solbrig and Solbrig (1979). For another treatment of this and more complex cases see Crow and Kimura (1970).]

Although Fisher states his theorem in terms of "the rate of increase of fitness in any organism at any time . . . ," it is better to think in terms of any randomly mating Mendelian population at any time. We assume such a population now and consider two codominant alleles, A_1 and A_2. The frequency of A_1 is p, that of $A_2 q$. The fitnesses of the three genotypes A_1A_1, A_1A_2, and A_2A_2, are ω_{11}, ω_{12}, and ω_{22}, respectively. The average fitness for the whole population is $\bar{\omega}$, which is equal to $\omega_{11}p^2 + 2\ \omega_{12}pq + \omega_{22}q^2$, and change in that fitness is expressed as $\Delta\omega$. Where the value for $\Delta\omega$ is positive there is an increase in fitness. Our problem now is to show that an increase in fitness is equal to what Fischer calls "the genetic variance in fitness." Variance, a statistical term for deviation from some mean or average, is the square of the calculated deviations. Hence *genetic variance in fitness* refers to the square of the deviations from the genetically determined average fitness. Fisher is saying, in effect: If there is no deviation, and the population is therefore genetically uniform, then no change in fitness will occur; but if there is genetic variation then changes will occur in proportion to the size of the variation (or, more precisely, the variance as a measure of that variation). Here is the argument.

After one generation of selection, the frequency of A_1 is p_1, which is determined as the occurrence of A_1 in A_1A_1 and A_1A_2 genotypes divided by the occurrence of all genotpyes. In mathematical terms,

$$p_1 = \frac{\omega_{11}p_0^2 + \omega_{12}p_0q_0}{\omega_{11}p_0^2 + 2\ \omega_{12}p_0q_0 + \omega_{22}q_0^2} = \frac{\omega_{11}p_0^2 + \omega_{12}p_0q_0}{\bar{\omega}}$$

peaks shown in the figure would represent two surviving phenotypes, as in the case of sickle-cell anemia, although the peaks would be drawn much more sharply in that case. However, what we are considering here, and in all the graphs of the figure, is the totality of the adaptive phenotypes generated by a given gene pool. The adaptive norm is not restricted to the

---

(The initial values of p and q are p_0 and q_0, respectively.)

The change in $\bar{\omega}$ after one generation of selection is

$$\begin{aligned}\Delta\bar{\omega} &= \bar{\omega}_1 - \bar{\omega}_0 \\ &= (\omega_{11}p_1^2 + 2\,\omega_{12}p_1q_1 + \omega_{22}q_1^2) - (\omega_{11}p_0^2 + 2\,\omega_{12}p_0q_0 + \omega_{22}q_0^2)\end{aligned}$$

Omitting here the details of the algebraic manipulations needed, we can simplify the foregoing equation to

$$\begin{aligned}\Delta\bar{\omega} &= \omega_{11}[(p_1 - p_0)(p_1 + p_0)] + 2\,\omega_{12}[(p_1 - p_0)(1 - p_0 - p_1)] + \\ &\quad \omega_{22}[(p_1 - p_0)(p_1 + p_0 - 2)] \\ &= (p_1 - p_0)[\omega_{11}(p_1 + p_0) + 2\,\omega_{12}(1 - p_0 - p_1) + \omega_{22}(p_1 + p_0 - 2)]\end{aligned}$$

Then recalling that

$$p_1 - p_0 = \Delta p$$

and

$$p_1 = \frac{\omega_{11}p_0^2 + \omega_{12}p_0q_0}{\bar{\omega}}$$

we can again simplify to get

$$\begin{aligned}\Delta\bar{\omega} &= (\Delta p^2)(\omega_{11} - 2\,\omega_{12} + \omega_{22} + \frac{2\,\bar{\omega}}{p_0q_0}) \\ &= 2\,p_0q_0[\omega_{11}p_0 + \omega_{12}(1 - 2\,p_0) - \omega_{22}q_0]^2 \\ &= 2\,p_0q_0(\omega_{11}p_0 + \omega_{12}q_0 - \omega_{12}p_0 - \omega_{22}q_0)^2\end{aligned}$$

Now the deviation from the average fitness $\bar{\omega}$ of the genotypes containing the A_1 allele, which we shall call E_1, is

$$E_1 = \omega_{11}p_0 + \omega_{12}q_0 - \bar{\omega}$$

And for the A_2 allele it is

$$E_2 = \omega_{12}p_0 + \omega_{22}q_0 - \bar{\omega}$$

Therefore

$$\begin{aligned}E_1 - E_2 &= (\omega_{11}p_0 + \omega_{12}q_0 - \bar{\omega}) - (\omega_{12}p_0 + \omega_{22}q_0 - \bar{\omega}) \\ &= \omega_{11}p_0 + \omega_{12}q_0 - \omega_{12}p_0 - \omega_{22}q_0\end{aligned}$$

and so

$$\Delta\bar{\omega} = 2\,p_0q_0(E_1 - E_2)^2$$

Squaring the deviations E_1 and E_2 gives us a variance. Thus the change in fitness after one generation of selection is equal to twice the product of the frequencies of each allele multiplied by the deviations in fitness of the two alleles. This is a precise statement of Fisher's theorem.

Note that there will be no change in $\bar{\omega}$ if there is no genetic variability, i.e., either $p = 0$ or 1 and $q = 1$ or 0. Only when both alleles are present can there be a non-zero value for $\bar{\omega}$. That is, genetic variability is necessary for evolutionary change.

---

phenotypes generated by one locus and its alleles, it is all the adaptive phenotypes generated by all the loci in a given population. The appearance of two adaptive norms indicates that the population is diverse enough to be adapting differently in two different areas of the region it occupies. The two peaks are two subpopulations within the larger interbreeding population. This also indicates that the environment occupied by this larger population is heterogeneous. It contains areas of differing selection pressures due to that heterogeneity.

And, of course, an environment is not limited to two different regions. It may have several or many different subareas, in which case there might be equally many adaptive norms. A bimodal graph, like that of Figure 5–9C, is not the only indicator of diversifying selection. It could be trimodal or show some greater degree of polymodality.

An extreme example of such a polymodal or polytypic species is found in the pocket gopher *Thomomys bottae* (Fig. 4–4). Here local races or demes are apparently adapted to a great variety of adaptive norms.

In summary, we can do no better than refer again to two aspects of Fisher's work. We can now take a more rigorous look at his Fundamental Theorem of Natural Selection. One of its mathematical formulations is given in Box 5–7. We can also consider his comment that selection produces a high degree of improbability. That is so, for the reasons already made clear. But there are further implications. The resources of this earth are finite and therefore the Darwinian struggle for survival is inevitable. Given heritable variations and the struggle for survival, natural selection, too, is inevitable. Hence, the high degree of improbability or order generated by natural selection is also inevitable.

Adapted organisms, which make up populations that can exploit their environments and maintain themselves long enough to reproduce themselves and at least some of their variations, are the products of natural selection.

Problems for Chapter 5

(Answers in Appendix II)

1. In a population in which the Hardy-Weinberg law applies, calculate the frequency of gene A when its allele A' is homozygous in the following populations and in the frequencies indicated:

Population	Frequency of $A'A'$
I	0.64
II	0.09
III	0.36
IV	0.16

2. In a group of Americans tested in New Mexico, it was found that 305 were of blood type M, 52 MN, and four N. Recalling the genetics of this situation as described in this chapter, answer the following questions.

(a) What is the frequency of the L^m and L^n genes in this population?

(b) Is the population in Hardy-Weinberg equilibrium?

(c) If the population size remains constant, what proportion of the progeny are derived from mothers that are L^mL^n?

(d) What proportion of the children from L^mL^n fathers will be L^mL^n?

3. To help you understand the Hardy-Weinberg law more clearly, consider the following problems.

(a) What would be the fate of a gene pool that fulfilled the conditions of the Hardy-Weinberg equilibrium for thousands of generations? Does it matter if the environment changed during this period?

(b) Can the Hardy-Weinberg equilibrium apply to one locus and not another? Does linkage have any bearing on your answer?

(c) Consider a population in Hardy-Weinberg equilibrium and suppose the frequencies of the genes were changed by a sudden influx of new genes, for example by immigration, which then ceased. How long will it take, i.e., how many generations, to establish a new equilibrium?

4. Recall that those who taste phenylthiocarbamide (PTC) have at least one T gene and that people homozygous for the t gene cannot taste PTC. In a sample of 1,138 people it was found that 884 could taste PTC.

(a) If there is no choice of marriage partners due to taster or non-taster phenotypes (i.e., "random mating" regarding this trait), what will be the frequncy of marriages between two tasters?

(b) What is the ratio of taster to non-taster children in marriages between parents both of whom are tasters?

(c) What is the ratio of taster to non-taster children in marriages between a taster and a non-taster?

5. The occurrence of phenotypes determined by genes A and A' was studied in two populations. The following results were obtained:

	AA	AA'	$A'A'$
Population I	4	32	64
Population II	81	18	1

Assume that somehow these populations unite to give a third population,

	AA	AA'	$A'A'$
Population III	85	50	65

(a) Are these three populations in Hardy-Weinberg equilibrium?
(b) If any population is not in equilibrium, determine whether there are more or fewer heterozygotes than expected.
(c) In how many generations will the population not in equilibrium reach equilibrium? What will be the frequencies then of the three phenotypes?

References

Allison, A. C., 1964. Polymorphism and natural selection in human populations. *Cold Spring Harbor Symposia on Quantitative Biology 29:* 137–149.

Bodmer, W. F., and L. L. Cavalli-Sforza, 1976. *Genetics, Evolution, and Man.* W. H. Freeman, San Fancisco.

Crow, J. F., 1962. Population genetics: Selection. In *Methodology in Human Genetics.* W. J. Burdette (ed.). Holden-Day, San Francisco.

Crow, J. F., and M. Kimura, 1970. *An Introduction to Population Genetics Theory.* Harper & Row, New York.

Kamin, L. J., 1975. *The Science and Politics of IQ.* Wiley, New York.

Kettlewell, H.B.D., 1973. *The Evolution of Melanism.* Clarendon Press, Oxford.

Li, C. C., 1976. *First Course in Population Genetics.* Boxwood Press, Pacific Grove, Calif.

National Research Council, 1975. *Genetic Screening. Programs, Principles and Research.* National Academy of Sciences, Washington, D.C.

Solbrig, O. T., and D. J. Solbrig, 1979. *An Introduction to Population Biology and Evolution.* Addison-Wesley, Reading, Mass.

Stern, C., 1973. *Principles of Human Genetics.* 3rd ed. W. H. Freeman, San Francisco.

Wright, S. S., 1931. Evolution in Mendelian populations. *Genetics 16:* 97–159.

Wright, S. S., 1968–1978. *Evolution and the Genetics of Populations.* 1968. *Vol. I. Genetics and Biometric Foundations.* 1969. *Vol. II. The Theory of Gene Frequencies.* University of Chicago Press, Chicago. (For Volumes III and IV see references for Chapter 6.)

SIX

Gene Frequencies: Mutation, Migration, Breeding Patterns, and Population Size

IN ADDITION TO SELECTION, there are four more factors that must be considered to round out our understanding of how reproductive communities become well-adapted populations. These are mutation, migration, breeding patterns, and the size and structure of the population. We will treat each factor in turn and summarize them by an examination of Sewall Wright's peak-valley or adaptive landscape model. That model is a useful way to bring the interaction of selection and the other factors into clearer view. It allows us to see what determines whether a population dies out or survives and continues to evolve.

Mutation

We will deal here with mutations as rare, random, but stable changes in a gene. Chromosomal and genomic changes will be examined later on, especially in Chapter 9.

The role of mutation in a Mendelian population can be better understood if we treat the subject in four interrelated steps: (1) the fate of a single mutation; (2) the effect of recurring mutations; (3) the role of reverse mutations, and (4) the interaction of mutation and selection.

THE FATE OF A SINGLE MUTATION

Again we start with a model system so as to isolate the issue under study. Let us assume we have a population composed solely of *AA* individuals, i.e., the frequency of *A* is $p = 1.0$ and the frequency of *a* is $q = 0$. Now assume that one *A* allele mutates to *a*. Hence, there is one *Aa* individual among all the others who are *AA*.

What will happen to this new gene? (Remember, no factors other than the normal operation of Mendelian genetics are in force here.)

The *Aa* individual can only mate with an *AA*. If they produce only one offspring, there is a 50% probability that the new allele *a* will be lost because the chances of having a child whose genotype is *AA* is 50%.

If the *Aa* and *AA* parents have two children, then the chance of losing the *a* allele is 25%. This comes from the realization that the probability of the first child being *AA* is 0.5 and the probability of the second child being *AA* is also 0.5; thus, the chance of two *AA* progeny is the product of the probability of each one being born, namely, $0.5 \times 0.5 = 0.25$ or 25%.

Conversely, the probability of increasing the one parental *a* allele to two *a* alleles in the progeny is 25% because the probability of both offspring being *Aa* is 0.25.

What these simple calculations tell us is that chance alone can play an important role in whether the new gene is lost or preserved. And even if selection were operating, there might be no effect. (We are excluding the situation in which *a* would act as a dominant lethal.) If a selection coefficient of 0.1 were in effect, chance alone would still be much more important than selection in determining the fate of the new mutation.

Thus far we have considered the fate of the mutant over two generations. What about many generations? Here we find that Sir Ronald Fisher (Box 5–1) has provided some answers. His conclusions are shown in Table 6–1. The following assumptions have been made: (1) The *average* number of progeny is two per one set of parents to assure a constant population size. (2) Because of this averaging we realize that some parents have no progeny, some one, and others two, three, or four. A few may even have more than four. The mathematical treatment of this situation involves what is called a Poisson series, which is used when averages are so low that a significant number of samples are zero due to random variation. (Here it refers to parents with no progeny.) Such a situation

Table 6–1. The probability of loss of a new, mutant gene.

Generation	No selective advantage	1% selective advantage
1	0.3679	0.3642
2	0.5315	0.5262
3	0.659	0.6197
4	0.6879	0.6811
5	0.7319	0.7246
6	0.7649	0.7572
7	0.7905	0.7825
15	0.8873	0.8783
31	0.9411	0.9313
63	0.9698	0.9591
127	0.9897	0.9729

results in an asymmetrical distribution of numbers around the mean—the numbers are cut off by zero on the low side and extend indefinitely on the high side, although the further they are from the mean, the fewer there are of them. Because of this the Poisson series gives the results in the first generation when there is no selection as 0.3679 rather than the 0.25 we calculated previously. (3) We note that the probability of survival or of extinction is the sum of the successive probabilities for succeeding generations. The probability of survival or extinction is, therefore, cumulative. (4) Finally, we note that Fisher gives us his results as the probabilities of loss or extinction.

From the middle column of Table 6–1, we see that chance alone will in all probability eliminate the mutant allele given one hundred or more generations. The right-hand column is even more intriguing. It indicates that even with a selective advantage—a low one, to be sure—extinction will occur in all probability.

The conclusions are obvious. Most mutant alleles are lost by chance alone and within a few generations of their origin. This is true even when they have a slight selective advantage. The question then becomes, How does a new gene become established in a population? The answer has two parts. First, there must be a significant selective advantage for that gene. Second, a mutation does not occur once only; rather, mutations are recurrent. We avoided the latter point by using a model system in which we followed the fate of a single new mutation. Now we will look at recurring gene mutations.

THE EFFECT OF RECURRING MUTATIONS

Although a mutational event at a given locus is rare, it can recur. That mutations are unavoidable is described as *mutation pressure*. Thus we are now really inquiring into the effects of mutation pressure on a Mendelian population.

The rate of mutation for a given gene can be measured (Table 4–2). It seems to have a fairly constant value, but it varies from gene to gene. If we consider the mutation of A to a at a rate μ, i.e.,

$$A \xrightarrow{\mu} a$$

and if q is the frequency of a, then the change in q, Δq, is given by

$$\Delta q = \mu p$$

That is, the additions to a from A are what change the frequency of a. For example, if p is 0.2 and μ is 10^{-5} per generation, then

$$\Delta q = 0.2 \times 10^{-5} = 2 \times 10^{-4} = 0.0002$$

Under these conditions Δp would be proportional to Δq and negative, namely, p is decreasing while q is increasing.

The long-term result of this situation is straightforward. Given only the mutation of A to a, the whole population will eventually become a and $q = 1$, while $p = 0$. Recurrent mutation or mutation pressure will eventually drive one allele to fixation and the other to extinction.

THE ROLE OF REVERSE MUTATIONS

Obviously, the foregoing discussion does not reflect the whole situation because it ignores mutation in the reverse direction, that is, a to A. That can occur at a rate designated as ν. A more accurate description of mutational events can be written as

$$A \underset{\nu}{\overset{\mu}{\rightleftarrows}} a$$

This then means we can better define Δq as

$$\Delta q = \mu p - \nu q$$

This reflects the gain in q, i.e., μp, and the loss in q, i.e., νq.

The first term comes from A mutating to a and the second from a mutating to A.

If we look at the results of this reverse mutation, we can see what is happening by referring to the situation we described in the preceding section, in which one-way mutation pressure had changed a population into all aa individuals such that $p=0$ and $q=1.0$. In that population, the only mutations possible are a to A, or $\Delta p=\nu q$. The frequency of A will increase due to reverse mutation pressure, and in time the number of A will be so frequent that they will mutate in significant numbers to a. Intuitively one can expect an equilibrium situation. Such an equilibrium will occur when the rate of change in q is equal to that of p, i.e., when $\Delta q=\Delta p$.

Note that

$$\Delta q=\mu p-\nu q$$

and

$$\Delta p=\nu q-\mu p$$

so when

$$\Delta q=\Delta p$$

then

$$\mu p-\nu q=\nu q-\mu p$$

or

$$\mu p=\nu q$$

From this formula we can see that at equilibrium, i.e., at $\hat{p}$ and $\hat{q}$, $\hat{p}=\nu/(\nu+\mu)$ and $\hat{q}=\mu/(\mu+\nu)$. The derivation of these formulae is given in Box 6–1.

To illustrate its application we can assume mutation rates of $\mu=3\times10^{-5}$ and $\nu=2\times10^{-5}$.

$$\hat{q}=\frac{3\times10^{-5}}{(3\times10^{-5})+(2\times10^{-5})}=\frac{3}{5}=0.6$$

and $p=0.4$. At these values of p and q the changes in each will be equal and the population will be stabilized at those frequencies of A and a.

THE INTERACTION OF MUTATION AND SELECTION

Two general situations should be considered—one in which mutation and selection are complementary and the other in which they are antagonistic.

Box 6–1. Derivation at equilibrium of $\hat{p}$ and $\hat{q}$ when only mutation rates are under consideration.

$$\mu p = \nu q$$
$$0 = \nu q - \mu p$$
$$= \nu(1-p) - \mu p$$
$$= \nu - \nu p - \mu p$$
$$\nu = \nu p + \mu p$$
$$\frac{\nu}{p} = \nu + \mu$$
$$p = \frac{\nu}{\nu + \mu}$$

And

$$p = \hat{p}$$

Similarly

$$\hat{q} = \frac{\mu}{\mu + \nu}$$

Note the parallel here to the equilibrium generated by selection against different alleles when the heterozygote is favored. The derivation of

$$\hat{p} = \frac{s_{A'}}{s_{A'} + s_A}$$

from $s_{A}p = s_{A'}q$ was given earlier. The derivation given there is slightly different from that given above. But clearly, the two are interchangeable.

We saw that a low selection pressure (Table 6–1) could reduce by a small amount the extinction probability of a new gene. This tells us that recurrent mutation and a significant selective advantage are needed to establish a new allele in a population. Once these conditions are met then a rapid change in gene frequencies can favor the new gene. Its persistence in the population is assured.

When there is antagonism between mutation pressure and selection pressure, the situation is more complicated. First, if the two antagonistic forces are equal in strength, an equilibrium is reached. That is, as a first approximation, if $s = \mu$, where s is the selection against gene a, then selection removes a genes at a rate equal to their appearance as mutants from A. If $\mu = 3 \times 10^{-5}$, then s is correspondingly low, indicating a very low selection pressure.

A more detailed treatment of this problem depends on the calculation of Δq in terms of both selection and mutation.

$$\Delta q = \mu p - \nu q$$

and

$$\Delta q = s_A p - s_a q$$

If the increase in q from mutation is equal to the decrease from selection, then an equilibrium can be reached. The equilibrium point will depend, of course, on the mutation rates and the selection coefficients.

When there is low selection pressure against a gene, for example, $s = 0.01$ to 0.1, the frequency of that gene will be reduced, but the gene will still persist in the population. This is because the frequency of occurrence of its allele is now so high that mutation pressure keeps the disadvantageous gene in the population. Its frequency will be close to that of the mutation rate, since this rate favors its occurrence.

VESTIGIAL ORGANS

At this point it is worth looking at mutation and vestigial organs. The appendix is sometimes cited as such an organ, although the case is not entirely clear that we do not use our appendices. Anyhow, it is clear that vigorous, happy lives are possible without an appendix. The explanation for the occurrence of such vestigial organs in any organism is that at an earlier time selection favored the organ, but later that selection was greatly relaxed. At the earlier time, then, genes that did not contribute to the maintenance and function of the organ were selected against, but when the organism no longer needed the organ, mutations that no longer maintained the organ would accumulate because they were unopposed by selection, and the organ would begin to regress. The process of regression would be slow, since there would be little or no selection either for or against the mutants. But in time, when mutation pressure is unopposed by selection pressure, the former will render a previously improbable entity more probable—remember Fisher's comment on the role of selection—and in the process dilute and change the former character of that entity.

DOMINANCE AND CODOMINANCE

When there is strong selection against a gene, as, for example, a dominant gene that allows individuals to be born, but causes death before they reproduce, then the frequency of that gene is determined by its frequency of origin by mutation. If the mutation rate is 10^{-6}, then $q = 10^{-6}$, which is also the frequency of the phenotype of the dominant lethal in the population.

Let us, however, consider a recessive lethal. Indeed, look at the case studied earlier (Chapter 5) in which selection against the $A'A'$ genotype is complete, i.e., $s_{A}' = 1.0$, but the heterozygote has a 1.0% advantage over AA individuals. Here, the frequency of A' was 0.0099 or close to 1.0%, and the frequency of $A'A'$ individuals or q^2 would be $(0.01)^2$ or 0.0001. This is one in 10,000 members of the population, or 10^{-4}, and it is significantly higher than the frequency due to mutation alone.

If we next look at a recessive lethal gene that is kept in the population simply by mutation pressure, the value of q^2 would be very low. Selection would be driving the gene out of the population, but more and more slowly as q decreased. If, however, the mutation rate of 10^{-6} was all that kept the allele in the population, then q^2 would be 10^{-12}.

HETEROZYGOTES AS CARRIERS

The importance of heterozygotes as carriers of the recessive lethals can be illustrated using data from studies on phenylketonuria (PKU) in humans. This genetic disease can result in mental deficiency due to the accumulation of phenylalanine and phenylpyruvic acid in the blood of affected individuals. The genetic cause of this is the absence of the enzyme phenylalanine hydroxylase, which normally breaks down phenylalanine in the body. The enzyme is formed when the dominant allele p^+ is present. When individuals are homozygous for p^0, then the signs of PKU may be expressed. It is worth noting that a special, low-phenylalanine diet can significantly reduce the extreme effects of the PKU genotype.

The frequency of PKU among white persons in the United States is about one in 10,000; $q^2 = 0.0001$ and $q = 0.01$. This is a frequency much higher than would be expected from mutation pressure alone. The cause of this high frequency is not known, but given this frequency, the occurrence of heterozygotes or carriers of p^0 is $2pq$ or 0.0198. Now note how these carriers contribute to the p^0p^0 individuals found in the next generation (Table 6–2). The p^0p^0 progeny from parents, both of whom are heterozygotes, will make up about one in every 10,000 of the next generation. The same progeny genotypes from the remaining two types of parent matings provide only about two in 100,000 and one in 10^8, respectively, who

Table 6–2. The frequency of appearance of individuals who can show the PKU phenotype when the parental population has a frequency of $p^0 = 0.01$.

Parental genotypes	Frequency of occurrence	Frequency of $p\,p$ progeny
$p^+p^0 \times p^+p^0$	$(2\,pq)^2$	$p^2q^2 = 0.00009801$
$p^+p^0 \times p^0p^0$	$2\,(2\,pq)(q^2)$	$2\,pq^3 = 0.000019602$
$p^0p^0 \times p^0p^0$	$(q^2)^2$	$q^4 = 0.00000001$

are potentially PKU individuals. Clearly, the great majority of children with a PKU genotype come from normal, carrier parents. Whatever keeps the p^0 gene in the population above a frequency determined by mutation pressure retains the largest numbers of those genes in heterozygous individuals.

Migration

Migration can be summarized as *gene flow,* the movement of genes within a Mendelian population. Three factors within a population determine the rate and extent of that flow. They are the mobility of individuals, the mating habits of individuals, and the nature of the environment. Let us look briefly at each of these factors, but especially at the last one.

FACTORS AFFECTING GENE FLOW

Animals are, in general, much more mobile than plants. And among animals, the ones that can fly are more mobile than the ones that burrow, for example. However, generalities of this sort are not always useful in discussing gene flow. For example, although the arctic tern (*Sterna paradiseae*) migrates annually from one polar region to the other and back again—a round trip of around 24,000 miles—it breeds in a restricted locality within the Arctic Circle. So significant is this restriction that the tern returns to the same, very limited nesting area year after year, as do its progeny. Hence, when we come to gene flow we find that mates are chosen from a restricted population in one or another nesting colony situated on rocky arctic islands. This means that mobility, in the literal sense, is not what is important here. Of special concern is whether mates are chosen by closely related individuals or by unrelated ones. This point will be developed more fully when we turn to breeding habits. We can now put mobility in perspective by asking how far an individual will move from its birthplace to find a mate. The general answer to that question is that mobility varies from species to species. In self-pollinating plants the movement is essentially zero. In wind-pollinated plants, the windborne pollen can travel for many miles before it finally lands on the stigma of a receptive flower of the same species and fertilizes it.

Further, after genes migrate, what ensures a successful fertilization? In the case of wind pollination, the determining factor is the chance event given above. Among animals, the male in search of a mate must find a receptive female, in species that copulate. When gametes are released into the environment, as among such marine invertebrates as the sea urchin and other echinoderms, the release of sperm and eggs must be coordinated to the extent that both are viable and are present in the same area at the same time. Release of sperm without eggs, or vice versa, would be an enormous waste of reproductive energy and material.

Fig. 6-1: An attempted mating between a toad and a frog.

Synchrony between males and females of the same species is essential for successful matings. Occasionally members of different species try to mate (Fig. 6–1), but biologically this is wasted effort and selection will work against such behavior because either no progeny will arise or they will be sterile or weakened; in brief, they will be at a selective disadvantage.

The environment plays a major role in determining gene flow. Quite obviously, rivers can be barriers to movement across them by terrestrial animals. Similarly, island populations are isolated from each other depending in degree on the mobility of their members. Mountain ranges and deserts interrupt forests and grasslands; cold and warm currents are thermal barriers in the ocean; and so on. In the final analysis we must realize that the area occupied by a species is usually heterogeneous to varying degrees and limited by geographic factors.

Only a small population occupies a really homogeneous living space. But the very fact that it is small means gene flow is restricted to that limited population. An alternative would be to think of a Mendelian population as being composed of many small subpopulations or demes (Fig. 4–4). If individuals could migrate from one subpopulation to the next, then a potential for gene flow exists. As regards small populations, a word of caution is needed relative to the phrase "a really homogeneous living space." Living space refers, in this context, to what organisms do for a living in their allotted space. It is really a matter of ecological niche rather than geographic habitat. A niche, it has been said, refers to an organism's profession, whereas its habitat is just its address.

A niche is composed of many factors. They include such nonliving components of the environment as light, temperature, oxygen supply, and availability of food and water, home sites, and shelter, as well as various living components. These latter include competitors, predators, and so on. Quite obviously then a niche can be very complex. If, however, these components of a niche vary little or not at all, especially as one encounters them in different parts of the habitat, then the environment can be said to be homogeneous. If, on the other hand, each, or at least many, of the components vary greatly, the niche is heterogeneous.

GENE FLOW

When there is a heavy gene flow throughout a population, it will appear that there is one adaptive norm, regardless of the heterogeneity of the environment or niche. Heavy gene flow will counteract any regional differences that might have been significant if the flow were weaker. In such a case, we could expect a good deal of variability within the adaptive norm. This polymorphism would occur because selection would favor the variety of genes that would be of adaptive value in the somewhat different environments occupied by this species.

The effect of gene flow is comparable to that of mutation. In both instances, new genes are added to a population or old genes are lost. However, there are some important differences. Whereas in mutation the loss of one allele always means the addition of another, in migration a gene is either simply lost from (emigration) or added to (immigration) the population. In a large population that difference between mutation and migration is negligible. The second difference can be more important. It is that migration rates can be much higher than mutation rates.

The migration coefficient m refers to the population of newly immigrating or emigrating genes in each generation. If genes are added (immigration) m is positive, if they are lost (emigration), m is negative. It is also useful here to distinguish between the frequency of an allele in the whole species and its frequency in a subpopulation. The average value of p or q is given as $\bar{p}$ or $\bar{q}$, and the value for a local population is given by the traditional p or q. The terms p_0 and q_0 are used if we refer to the frequencies at some original (starting) point. When we continue to watch the effects of migration for a certain number of generations (n), the new local frequencies would then be given as p_n or q_n.

If population size remains constant, the result of immigration from the larger population into the subpopulation will be seen as some genes of frequency p replaced by genes of frequency $\bar{p}$. Since the rate of replacement is m, the loss from the subpopulation is mp and the gain is $m\bar{p}$. The new gene frequency after one generation is

$$\begin{aligned} p_1 &= p_0 - mp_0 + m\bar{p} \\ &= p_0(1-m) + m\bar{p} \end{aligned}$$

For the next generation, it is

$$p_2 = (p_0 - mp_0 + m\bar{p})(1-m) + m\bar{p}$$

In each subsequent generation the new value of p, i.e.,p_1, p_2, etc., is obtained by multiplying p of the previous generation by $(1-m)$ and adding $m\bar{p}$. After n generations the value of p_n is given by

$$p_n = p_{n-1}(1-m) + m\bar{p}$$

The difference between p_n and $\bar{p}$ after one generation is

$$p_1 - \bar{p} = p_0 - mp_0 + mp - p = (1-m)(p_0 - \bar{p})$$

After a second generation it is

$$\begin{aligned} p_2 - \bar{p} &= (p_0 - mp_0 + m\bar{p})(1-m) + m\bar{p} - \bar{p} \\ &= p_0 - mp_0 + m^2p_0 + mp - m^2\bar{p} - \bar{p} \\ &= (1-m)^2(p_0 - \bar{p}) \end{aligned}$$

This can be generalized to n generations as

$$p_n - \bar{p} = (1-m)^n(p_0 - \bar{p})$$

Here there are five factors (p_n, $\bar{p}$, p_0, m, and n); thus, if any four are known, the fifth can be calculated.

An interesting application of this is the following example taken from a study by Bentley Glass and C. C. Li (1953). In white and black populations in the United States, the frequency of one of the alleles involved in a blood characteristic called the Rh factor was determined. This allele, R^0, showed a frequency of 0.446 for blacks and 0.028 for whites. Glass and Li wanted to see if these values could be used to estimate the amount of gene exchange between these two groups. Genes from the black population certainly entered the white population and vice versa. However, with whites forming a relatively much larger population than blacks, it is possible to view the situation in this way. The frequency of R^0 in whites is taken as $\bar{p}$ and that in blacks as p. Furthermore, the blacks have been isolated from their parent population in Africe for over two centuries and, therefore, p_0 would be the frequency of R^0 in East Africa, the area from which most blacks were taken as slaves. In East Africa the frequency of R^0 is 0.630. Glass and Li estimate that the blacks have remained in the United States for a period equal to ten successive generations. This means we now have values for four of the five factors used in our last equation. We can therefore solve for m, the rate of gene immigration from the white into the black subpopulation.

The previous equation can be rewritten as

$$(1-m)^n = \frac{p_n - \bar{p}}{p_0 - \bar{p}}$$

and substituting the appropriate values we have

$$(1-m)^{10} = \frac{0.446 - 0.28}{0.630 - 0.28} = 0.694$$
$$1 - m = \sqrt[10]{0.694} = 0.964$$
$$m = 0.036$$

Thus, since slaves were introduced to this country they have kept close to 70% of their African genes. The loss of African genes is due to gene migration from the white population at a rate of about 3.6% per generation. Other genes, such as the gene for sickle-cell anemia, have also been studied in terms of the effect of gene immigration into the black population and the results agree with this study for the R^0 gene.

GENE FLOW INTERACTING WITH OTHER FACTORS

Just as we saw how selection and mutation could interact in either a complementary or an antagonistic fashion, the same considerations apply to gene flow. The general conclusions can be stated quite simply. When selection favors a certain gene and it is large for that gene, the values of p and q can change rapidly. When selection and migration pressure work together, mutation pressure will have little effect. It will keep the other alleles in the population at frequencies close to their mutation rates.

If selection is strong and migration pressure is weak, there is a chance that the population will differentiate into various subpopulations or demes. An equilibrium, determined by the competing effects of selection eliminating a certain allele and migration continuously reintroducing it, may occur.

If migration pressure is stronger than selection pressures, then local differences will be counteracted by gene flow, and local values of p and q will not differ significantly from $\bar{p}$ and $\bar{q}$. This will result in a rather homogeneous population in which selection effects on the alleles in question will be minimal.

In this regard, Sewall Wright (Box 5–1) has summarized some complex mathematical calculations by pointing out that an m as high as 5% has an effect close to random mating. That tells us that, in local subpopulations, the values of p and q will differ from $\bar{p}$ and $\bar{q}$ when gene flow is below 0.05. In other words, a certain degree of reproductive isolation is necessary for local selection pressures to be effective in shaping the adaptive norm of subpopulations or demes.

This brings us again to a consideration of population size. When a population is small, then mutational events have little effect, selection may or may not be effective, and migration could be very effective if it is significantly large. However, in a small population there is a real possibility of near relatives mating with each other (inbreeding) on the basis of chance alone, and there are increased chances for non-random matings (assortative rather than random). Both situations may be antagonistic to the effects of selection—even strong selection—as we shall see shortly. Hence, migration may be the only pressure that keeps certain desirable genes in a population. If that is so, it is a precarious way to ensure survival, since many environmental factors can act to reduce the immigration of desired alleles, and the small population would be at the mercy of chance events rather than shaped by selective forces into a well-adapted population.

Larger populations have greater mutational resources than small ones, which renders less significant those factors that might undermine the effect of selection, namely, inbreeding and certain non-random breeding behaviors. Here, gene flow could keep the population fairly homogeneous, but because selection would be effective, the population would not depend on gene flow for survival. A low value of m could allow the differentiation of local populations and so lead to more genetic variation and gene flow would assure continual sharing of those genetic resources.

There have been various studies on the topic of gene flow

among natural populations, of both plants and animals. One that is outstanding because of the length of time devoted to it, the clarity of its results to date, and the general implications drawn from it is that of Paul Ehrlich (1978), of Stanford University, California, and his various coworkers on the checkerspot butterfly *Euphydryas edithi*. This work has focused on three neighboring populations living in the hills of California. Work started in 1960 and the populations have been under continuous surveillance since then in terms of population size and migration between the populations; various other ecological and genetic factors were also studied. The populations exist in grassy areas stretching along about two miles of a locality known as Jasper Ridge. The life cycle and feeding habits of the butterflies have been extensively studied, and data on population size and movement are available.

The most interesting results from these studies, which have also been greatly expanded to include other populations of this butterfly throughout California and in Oregon, Nevada, and Colorado, show that the species is not an ecological unit or even a genetic one. Since Ehrlich and his coworkers found little or no evidence of movement from one population to another, they concluded that this lack of gene flow must mean that the various populations are reproductively isolated even though interbreeding could apparently occur between contiguous populations if the opportunity arose. The utilization of different food plants and of habitats as different as deserts and upland meadows show strikingly different adaptations among these butterflies. These workers were puzzled as to why these insects continue to look alike and to share common genes (as determined by protein electrophoresis) when they live in such different environments. They concluded that among these forms and others with a similar population structure, the individual populations or demes are the units of evolution rather than the morphologically defined species.

This raises important theoretical questions as to what factors control the phenotypes of the checkerspot butterflies and whether there is one or more species with that phenotype. At present we can conclude that the reproductive community is still the unit of evolution, but that such communities may be coextensive with one large gene pool, or that they may be demes, as in these butterflies, showing little or no gene flow even though they share a common phenotype.

Now let us examine just population size and breeding patterns and learn how they affect Mendelian populations. We treat them together because they are necessarily interconnected.

Population size and breeding patterns

First, how do we determine the size of a population? The obvious answer is to take a census—count everybody in the population. This can be done and is referred to as the *actual size*. From an evolutionary point of view, however, our interest rests primarily on those members of a population that produce—actually or potentially—the next generation. This is called the *breeding size* of a natural Mendelian population. It excludes from the census all those organisms too old to reproduce or those that are permanently sterile for reasons of bad health or injuries. It does count immature forms that are potential parents even though they are incapable of reproduction when the census is taken. Finally, there is the *effective size* of a population. Here all individuals are excluded who cannot reproduce at the time the census is made. That means the following are not counted: old individuals, diseased and damaged individuals, immature juveniles, and even pregnant females. This census only counts those organisms capable of providing progeny to the next generation. (Offspring already conceived are considered part of the present generation.)

Quite obviously then the actual size provides a larger count than the breeding size and both are larger than the effective size. Which measure is most useful? As a result of studies done largely by Sewall Wright we find that effective size is the measure most often used.

But note some of the problems associated with that measure. If in a population of 1,000 individuals there are 200 females capable of mating and producing progeny and there are only five males to mate with, what is the effective size? Because the males can probably mate with and inseminate many females, the effective size will be more than five, but considerably less than 200. Wright has proposed a mathematical expression such as the following

$$N_e = \frac{4\, N_f N_m}{N_f + N_m}$$

Here N_e is the effective population, N_f is the number of females ready to become parents, and N_m the comparable number of males. In this case N_e is just about 20, a figure much lower than the actual number of 1,000 individuals. (See Box 6–2 for derivation of this formula.)

INBREEDING

When a population is small the probability of inbreeding is high. The most intense sort of inbreeding is self-fertilization, in which the breeding population is one individual. But even here the situation must be looked at carefully. The key question is how the gametes that unite to form a new individual are formed.

Self-fertilization. For example, if in a hermaphroditic flatworm the testes produce a number of sperm and the ovaries produce a number of eggs, then fertilization can be viewed as resulting from a random association of one egg and one sperm from the two populations of gametes. In this case, although the oogonia and spermatogonia probably have iden-

Box 6–2. Derivation of Wright's formula for the effective size of a population when numbers of males and females are unequal.

The total individuals in the population N are N_m males plus N_f females. Even though $N_m \neq N_f$, the two sexes necessarily make the same contribution to the next generation through one egg and one sperm for each new individual.

The probability that two genes in different progeny individuals are both of male origin is $0.5 \times 0.5 = 0.25$ (There is for each gene a probability of 0.5 that the gene came from a male and 0.5 that it came from a female. That they came from the *same* male is $0.25\ N_m$.) The probability of both genes coming from the same female is $0.25\ N_f$. Finally, the probability that they come from the same individual, regardless of sex, is

$$\frac{1}{4\,N_m} + \frac{1}{4\,N_f} = \frac{1}{N_e}$$

where N_e is the effective number of individuals in the population. Notice that if $N_m = N_f = N/2$, then N_e and N are the same. When the sexes are not equally represented, then the effective number and actual number are different.

We can solve the foregoing equation for N_e.

$$1 = \frac{N_e}{4\,N_m} + \frac{N_e}{4\,N_f}$$

$$4 = \frac{N_e}{N_m} + \frac{N_e}{N_f}$$

$$4\,N_m = N_e + \frac{N_e N_m}{N_f}$$

$$4\,N_m N_f = N_e N_f + N_e N_m$$

$$= N_e(N_f + N_m)$$

$$N_e = \frac{4\,N_m N_f}{N_f + N_m}$$

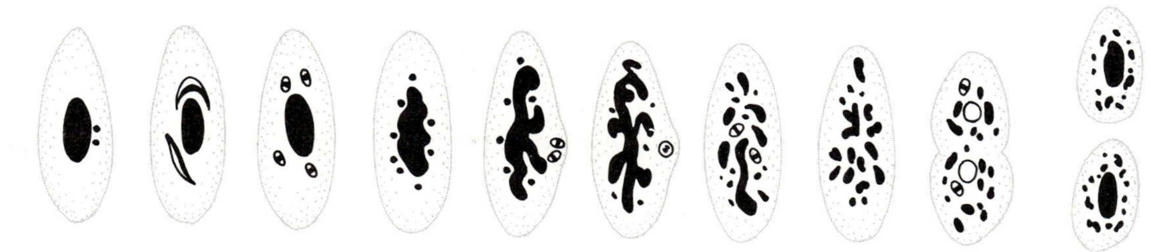

Fig. 6-2: Autogamy in *Paramecium aurelia:* Self-fertilization in a population of gametes.

tical genotypes, the resulting egg and sperm can have a wide range of differing genotypes. The events of meiosis assure that because of a random assortment of chromosomes and exchanges, such as crossing-over or translocations, between chromosomes.

Now consider a different example, one that represents the ultimate in inbreeding because the two gametes that unite are as close to identical as biologically possible. This is the case of autogamy or self-fertilization in the ciliated protozoan *Paramecium aurelia*. When we discussed sexuality in these paramecia earlier (see Fig. 4–9), we dealt only with conjugation. In that process two ciliates came together and exchanged male nuclei with the result that each partner fertilized the other. There was mutual fertilization. Autogamy represents a sexual process wherin the male nucleus, rather than migrating to a partner cell, remains within the parent cell, since there is no partner. In this case (Fig. 6–2), note how the male and female nuclei arise. They are the mitotic products of a single haploid nucleus. The only chance of a difference between these nuclei is that a rare mutation might have occurred during the mitotic replication of DNA that preceded their formation. This union of a single male and a single female gamete nucleus is not a random union between populations of sperm and eggs. (Population geneticists have said, in their own charming language, that "self-fertilization can be regarded as random mating within a population of one.") Rather, in the case of autogamy, the union of male and female nuclei is a necessary fusion of the only two gametes available. The resulting homozygosity is unavoidable.

It is here that we see most clearly the genetic consequence of inbreeding. It results in homozygosity, which entails a loss of genetic variability. Even if a population were initially heterozygous, the consequence of inbreeding of the sort seen in autogamous paramecia is homozygosity in one generation.

If we look at the flatworm type of inbreeding the process takes somewhat longer. Such a situation is the more common kind of inbreeding; it is often encountered in self-pollinating plants. In Table 6–3 we see the results. In four generations, heterozygous individuals are one-sixteenth as frequent as they were initially, and the frequencies of homozygotes have increased proportionally. In time—how long a time depends on the size of the population—the heterozygotes would be reduced to one individual and, by chance alone, the progeny might all be homozygotes and the heterozygotes would have disappeared.

This can be expressed mathematically by pointing out that, eventually, the frequency of each homozygote genotype will equal its initial frequency plus one-half the frequency of the heterozygotes. That is, the final frequency of *AA* is p^2+pq. The same is true for the frequency of *aa,* which can be expressed as q^2+pq. Note also that

$$q^2+pq=p(p+q)$$
$$q^2=p^2$$

This indicates that the final frequencies of each homozygote will equal the frequency of their alleles in the population. (The frequencies of *AA* and *aa* do *not* equal p^2 and q^2, respectively; since there is no random mating, there is no Hardy-Weinberg equilibrium.)

Thus extreme inbreeding can achieve homozygosity in one sexual generation; less intense inbreeding takes longer to achieve homozygosity. The next step is to consider even less intensive inbreeding—brother and sister matings. As a general rule, such matings also increase homozygosity and decrease heterozygosity, only it takes longer than if self-fertilization were occurring.

Table 6–3. The result of inbreeding.

Generation	Frequency of genotype *AA*	*Aa*	*aa*
0	d	h	r
1	$d+\frac{1}{4}h$	$\frac{1}{2}h$	$r+\frac{1}{4}h$
2	$d+\frac{1}{4}h+\frac{1}{8}h$	$\frac{1}{4}h$	$r+\frac{1}{4}h+\frac{1}{8}h$
3	$d+\frac{7}{16}h$	$\frac{1}{8}h$	$r+\frac{7}{16}h$
4	$d+\frac{15}{32}h$	$\frac{1}{16}h$	$r+\frac{15}{32}h$
limit	$d+\frac{1}{2}h$	0	$r+\frac{1}{2}h$

Table 6–4. The result of brother-sister mating.

Generation	Relative heterozygosity $\frac{H_n}{H_0} = P$	Decrease in heterozygosity $\frac{(H_0 - H_n)}{H_0} = f = 1 - P$
0	1 or 1/1	0
1	1 or 2/2	0
2	3/4	1/4
3	5/8	3/8
4	8/16	8/16
5	13/32	19/32
6	21/64	43/64
limit	0	1

PANMICTIC INDEX AND THE INBREEDING COEFFICIENT

At this point we can usefully introduce symbols that will help us draw parallels with the other factors we have discussed in considering adaptation in Mendelian populations. The panmictic index P is a measure of heterozygosity in a population as determined by breeding patterns. If there is perfect random mating, $P = 1.0$. When there is less than perfect random mating, then a decrease in heterozygosity necessarily follows. That decrease is given by f, the inbreeding coefficient. When $P = 1.0$, $f = 0$. Therefore $f = 1 - P$.

The panmictic index can also directly measure the relative heterozygosity in a population, namely, $P = H_n/H_o$, where H_n is the frequency of heterozygotes at generation n and H_o is their original frequency. This allows us to express f in terms of heterozygote frequencies.

$$\begin{aligned} H_n &= PH_0 = H_0(1 - f) \\ &= H_0 - H_0 f \\ H_0 f &= H_0 - H_n \\ f &= \frac{H_0 - H_n}{H_0} \end{aligned}$$

Using these terms, the result of brother-sister matings is given in Table 6–4. The progressive changes in relative heterozygosity can be calculated in two ways. The more laborious technique is to work out the result of all possible brother-sister matings, generation by generation. A quicker way is to follow a rule of thumb, so to speak, that gives essentially the same results: (1) obtain the denominator by doubling the previous denominator and (2) obtain the numerator by generating what is called a Fibonacci series. (Add the previous two numbers in the series. It is for that reason that the first two values of P in Table 6–4 are given as 1 or 1/1 and 1 or 2/2.) Given the value of P, the figures in the third column are given by $1-P$. By the sixth generation more than one-half of the heterozygosity initially present has been lost. In time, all of it will be lost.

The inbreeding coefficient can also be defined another way, which allows another view of the consequences of inbreeding. Before presenting this new definition we need to understand the terms autozygous and allozygous. *Autozygous* refers to the fact that the two alleles at a given locus in a diploid individual both come from one ancestral allele. In an *allozygous* individual the alleles are of independent origin (Table 6–5). When selfing continues, as in autogamous paramecia, the parent haploid nucleus is the source of identical alleles. In brother-sister matings, as shown in Table 6–5, the grandparent is the source of identical alleles.

With this in mind we can now define f in either of two ways: f is the probability that an individual is autozygous, namely, that a pair of alleles in two gametes that form an individual are derived from the same parent allele. Therefore, an individual with an inbreeding coefficient of f has that probability of being autozygous. Conversely, that individual has a probability of $1\text{-}f$ of being allozygous.

In other words, an allozygous individual has alleles of independent origin and is, therefore, not an inbreeder. The frequency of the genotypes of such individuals are given by a binomial formulation where p_1 is the frequency of allele A_1 and p_2 is the frequency of A_2. Here, then, $p_1+p_2=1$. On the other hand, autozygous individuals will have frequencies that equal the respective frequencies of each allele, i.e., p_1 and p_2. This is shown in Table 6–6, where the frequency of the three possible

Table 6–5. The autozygous and allozygous origin of alleles found in one individual.

	Origin of an autozygous individual	Origin of an allozygous individual
P	$A_1A_2 \times aa$	$A_1A_2 \times aa$
F_1	$A_1a \times A_1a$	$A_1a \times A_2a$
F_2	A_1A_1	A_1A_2

Here we consider only one locus and only the A allele at that locus. (The different A alleles are designated A_1 and A_2.)

Table 6–6. Genotypic frequencies of alleles A_1 and A_2 with frequencies of p_1 and p_2, respectively.

Genotype	Allozygous		Autozygous
A_1A_1	$p_1^2(1-f)$	+	p_1f
A_1A_2	$2\,p_1p_2(1-f)$		
A_2A_2	$p_2^2(1-f)$	+	p_2f

genotypes is given by the sum of their allozygous and autozygous origins. Note that when $f=0$, i.e., when mating is random, then the formulae with non-zero values total $p_1^2+2p_1p_2+p_2^2$, which is the Hardy-Weinberg binominal formulation. When $f=1$, i.e., when there is intensive inbreeding or self-fertilization, the non-zero values are p_1+p_2, and the population is completely homozygous. The latter result is the same as that derived above for the situation given in Table 6–3.

SOME CONSEQUENCES OF POPULATION SIZE

It will be useful now to discuss what are simply called small, intermediate-sized, and large populations. Small populations are those with under 100 members; large populations, over 100,000 members; and intermediate-sized populations, a few thousand members. There are mathematical reasons for such figures, largely having to do with whether or not mating is random, but details are not needed here.

Genetic drift. One of the most important consequences of small size in a population is that chance events can be more important than factors that lead to adaptive changes. The result will be that maladaptive or even non-adaptive changes will become established in such a population. Such changes are often called *genetic drift*. The term "drift" refers to the vagaries of chance as determining the fate of a population. For example, successive generations of inbreeding might—by chance alone—render a maladaptive gene homozygous and thus drive out a more adaptive allele.

Another way to show the importance of chance relative to population size is given by the following numbers. (These were used initially by Sewall Wright.) Let us look at three populations containing 50, 5,000, and 500,000 individuals, respectively. If we assume that population size will remain constant, the next generations will be formed from 10^2, 10^4, and 10^6 gametes (eggs and sperm), respectively. Now, let us further assume that we are looking at the effect of chance in the dis-

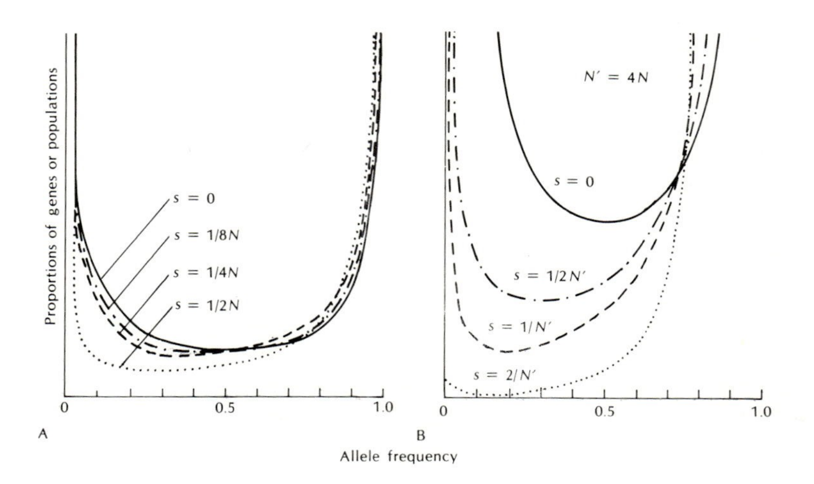

Fig. 6-3: Gene frequencies in small populations. The horizontal axis is allele frequency; the vertical axis can be read at least three ways: (1) frequency of gene occurrence in multiple samples from one population, (2) number of subpopulations in which a certain gene occurs, and (3) different genes of a population. The first two may be quite similar. In graphs A and B, the third interpretation is probably the most useful. A. Extinction or fixation of gene frequencies. The solid line describes events as $s = 0$ (no selection). The other curves describe the effects of differing amounts of selection, i.e., $s = 1/8\ N$, $1/4\ N$, and $1/2\ N$, where N is population size. When $s = 1/2\ N$, selection is strong enough to skew the shape of the curve significantly. But there is still a strong tendency to extinction or fixation. B. In a population four times larger than that in A, the effect of selection can be quite important. Again, when $s = 0$, gene frequencies tend to go to extinction or fixation, but with the highest degree of selection shown here ($s = 2/N$), extinction is uncommon and the favored gene is fixed.

tribution of alleles *A* and *a* in the next generation. The frequency of *A* is the same as that of *a*, i.e., 0.5.

If we take random samples of gametes from these populations, the relative numbers of *A* and *a* will vary with the sample size. A small sample will have more variation (or *variance,* as it is called statistically) than a large one. This can be illustrated by considering the relative occurrences or frequencies of heads and tails when a coin is flipped a few times (small sample) or a few hundred times (larger sample). Common sense tells us that about half of the time heads will occur and the other half, tails. But in a small sample the results may vary from our expectation—they *deviate from expectations,* to use another statistical term. And in a large sample, the deviations from expectations are much reduced. The statistical treatment

Table 6–7. The effect of chance on the distribution of alleleic frequencies in populations of differing sizes

A. Frequencies of alleles in the next generation	
Large population (500,000)	$p = q = 0.5000 \pm 0.0005$
Intermediate population (5,000)	$p = q = 0.500 \pm 0.005$
Small population (50)	$p = q = 0.50 \pm 0.05$
B. Range of variation in frequencies of alleles when 95 of 100 generations are considered (or 95 of 100 possible samples are considered)	
Large population (500,000)	0.4990–0.5010
Intermediate population (5,000)	0.490–0.510
Small population (50)	0.40–0.60

We assume the frequency of $A = p =$ the frequency of $a = q = 0.5$ and a Mendelian population of constant size.

is that the usual or standard deviation is given as the square root of the variance, Hence,

$$\text{Standard deviation} = \sqrt{\text{variance}} = \sqrt{pq/n}$$

where p and q are the frequencies of the two alleles and n is the number of individuals in the population. We can calculate these frequencies, and their range of variation, as in Table 6–7. In the upper part (A) of the table, compare the size of the three standard deviations for the three different populations. The standard deviation for the population of 50 individuals is ± 0.05, which means that in one generation, due to chance alone, we could expect $p = 0.55$ and $q = 0.45$, or vice versa. Deviations of that range are the more common ones in a small population, i.e., they will occur about 67% of the time. If we ask what range of deviations we can expect 95% of the time (or in 95% of the generations studied), the deviation is twice as big. This is shown in the lower part (B) of Table 6–7.

Thus, however the genetics of the situation are described, we find that in small populations events having nothing to do with selection pressures or adaptive responses, in general, can play a significant and, even on occasion, a decisive role in the fate of such a population because of genetic drift. The summary shown in Fig. 6–3A makes this apparent.

Figure 6–3B shows the fate of gene frequencies in a somewhat larger population. Here Sewall Wright has chosen a population four times larger than the previous one. Though not an intermediate-sized population by our definitions, the differences between it and a smaller population are striking. Here, a strong selection pressure, i.e., $s = 2/N$, can result in a situation in which extinction is not common and frequencies higher than 0.6 predominate.

Another analysis by Wright is summarized in Fig. 6–4. The population, in this case, is of intermediate size. (However, note carefully the qualifications.) Here, when the selection coefficient s is large enough, an equilibrium is reached between mutation and selection. Selection keeps the favored gene in the population at a rather narrowly defined frequency.

Finally, let us look at Fig. 6–5. Here the effects of migration is summarized. In this graph the abscissa is best considered as a plot of separate subpopulations or demes. Hence, we are examining the number of demes at a given frequency of the gene in question as a result of different rates of migration (m). When migration is low, the demes are isolated and, in effect, become a series of small, separate populations. Hence,

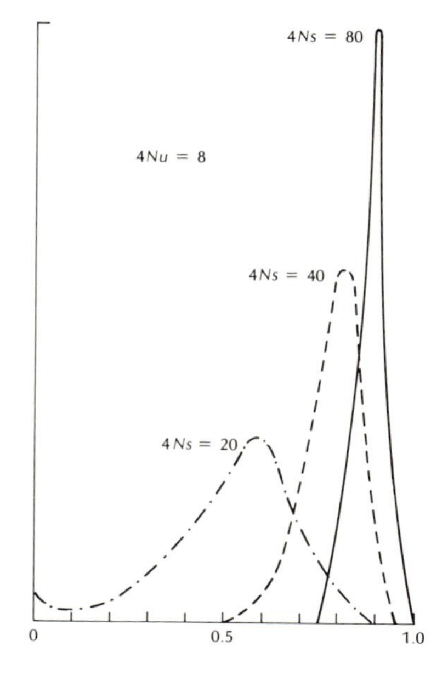

Fig. 6-4: Interaction of mutation and selection in a population of intermediate size. (*N.B.* Wright indicates population size when he tells us, in this case, that $4N\mu = 8$, where μ is the mutation rate removing the allele selection is trying to preserve. If $N = 5{,}000$, $\mu = 4 \times 10^{-4}$, a rather high value. If $\mu = 10^{-5}$, a more common value, then $N = 200{,}000$, a large value.) Although the assumptions implicit here are not entirely clear, the effect of different selection pressures is clear. When $4Ns = 80$, or $s = 20/N$, equilibrium between selection and mutation pressures is reached at about a frequency of 0.9.

the frequency of the gene in question goes to extinction or fixation in the separate demes. As m increases to $m = 4/N$, gene flow produces less variance, and the distribution of the gene frequency tends more and more to whatever the mean value is for all the demes (0.5, in this case). Gene flow, then, is an obvious factor in determining the effective size of a population, and it takes very little migration to make a significant difference. Note that when $m = 1/4\ N$, there is one migratory individual in every four generations. (The population is assumed to be constant in size and hence 4 N means four generations of population of size N.) When one individual is migratory per generation ($m = 1/N$), the change in the resulting distribution of the gene frequency is dramatically different from that when $m = 1/4\ N$ (Fig. 6–5).

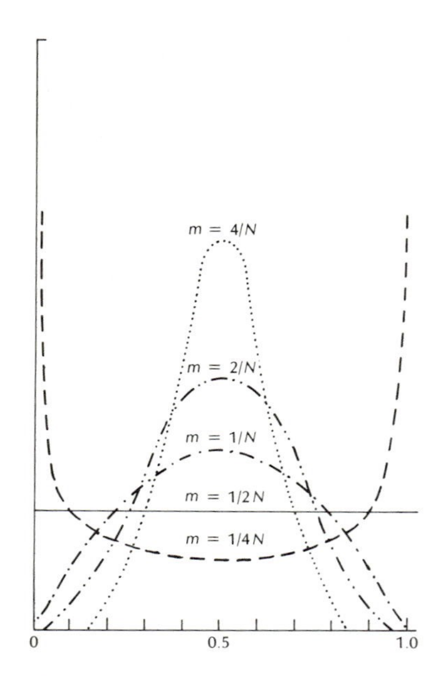

Fig. 6-5: Effect of migration on gene frequencies. When low migration effectively isolates subpopulations or demes, gene frequency tends to 0.0 or 1.0. Higher values of migration (m) change that situation markedly.

THE PEAK-VALLEY MODEL

In 1932, Sewall Wright summarized much of his thinking in what is now called Wright's peak-valley or adaptive peak model. Since then, the original ideas have been reworked and refined, and in 1977 and 1978 Wright restated them in a revised model. This model of Wright's allows us to visualize how close a population comes to occupying the adaptive peak that assures long-term survival. It includes ideas on what sort of population size and structure is best suited to occupation of an adaptive peak and how populations can move from one position to another on the adaptive landscape; in particular, how a population reaches an adaptive peak. The latter point is included in Wright's shifting balance theory. We here review the peak-valley model and what it implies and then look at the shifting balance theory.

Basic assumptions. The adaptive peak model is a diagrammatic summary of the interaction of selection (s), mutation (μ and ν), migration (m), breeding patterns (f), and population size (N). To treat the interaction of five variables mathematically is extraordinarily complex, hence the usefulness of adaptive peak diagrams.

If we start with a specific ecological situation, it will be easier to envision the basic concepts contained in the peak-valley model. Therefore, let us start with a piece of landscape containing two freshwater ponds connected by a stream (Fig. 6–6A). Further, let us assume we are discussing a species of freshwater animal, for example a fish or a water-beetle. We will be concerned to see how close a Mendelian population of this species comes to being well-adapted to the niche it occupies.

The two ponds can be represented in terms of adaptation by Fig. 6–6B. Here the landscape of the previous figure is replaced by two adaptive norms. They are located at the site of the two ponds and indicate the ideal situation represented by the norms. That is, we specify the genes present (horizontal axis) and their fitness (vertical axis) to describe an optimally adapted population. The vertical axis could also refer to gene frequencies if we assume that the fitter genes will occur in higher frequencies. This is a two-dimensional graph, but it can be made three dimensional by considering the horizontal axis as a two-dimensional plane rather than a one-dimensional line. Each point on that plane becomes a gene available in the gene pool of the species we are studying; the vertical dimension is again the frequency of a given gene. Because of these three dimensions (two in the horizontal plane and one in the vertical plane), the curve of the adaptive norm can now be visualized as a three-dimensional adaptive peak or an adaptive landscape.

Our problem now becomes one of seeing how close an actual population comes to occupying its adaptive peak. In other words, does it have the genes and in the needed frequencies to be at the very top? If it does sit there on the peak, it is, by definition, well adapted. This problem is best discussed by returning again to small, intermediate-sized, and large populations.

Small populations. These populations often contain fewer than 100 individuals. Such populations rarely, if ever, occupy an adaptive peak. Wright argues that, typically, they are located in a fixed position on the lower slopes of the adaptive peak (Fig. 6–7), for the following reasons. Because of the small size, inbreeding is bound to occur (f is high), and therefore genes will go to extinction or fixation regardless of their selective value. Chance events are more important than selection pressures. In a small population, mutation has little effect in increasing genetic variation and what variation is present is being lost through inbreeding. Migration from one population to the other may be low or non-existent if animals highly adapted to the quiet waters of the pond cannot survive in the stream between the ponds. Because of these combined effects there is little likelihood of well-adapted genes being preserved and whatever genes are in the population will occur at a high frequency (fixation). Hence, there is little variation for selection to act on, even if it could act.

In summary, small populations are probably doomed to extinction. They are not only poorly adapted to the niche in which they are trying to live, but they have such little genetic variation in their gene pool that if the environment changed—

Fig. 6-6: A freshwater pond habitat represented in terms of adaptive norms or adaptive peaks. A. A landscape, with two ponds, described by contour lines. Two streams flow into the upper pond, one stream connects the two ponds, and there is an outflow from the lower pond (lower left). B. The same landscape, but with adaptive norms (genes) replacing the ponds. The genes and their frequencies define populations well adapted to a freshwater pond niche (horizontal line of graph, different genes available to the gene pool of the populations in question; vertical line, gene fitness or gene frequency, if it is assumed that the more fit genes are more frequent). The highest part of the curve, the adaptive peak, represents those genes conferring adaptive advantage.

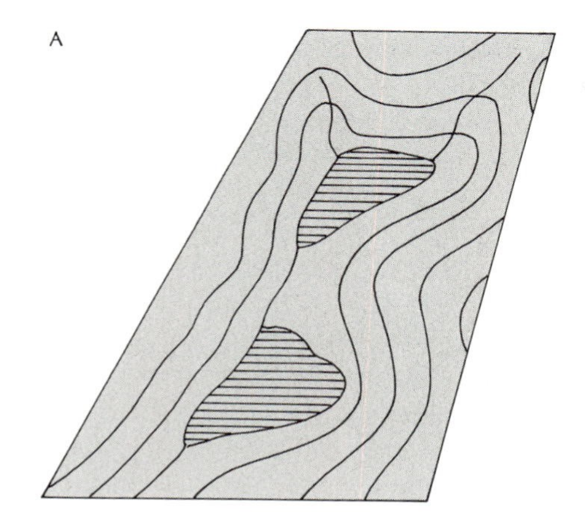

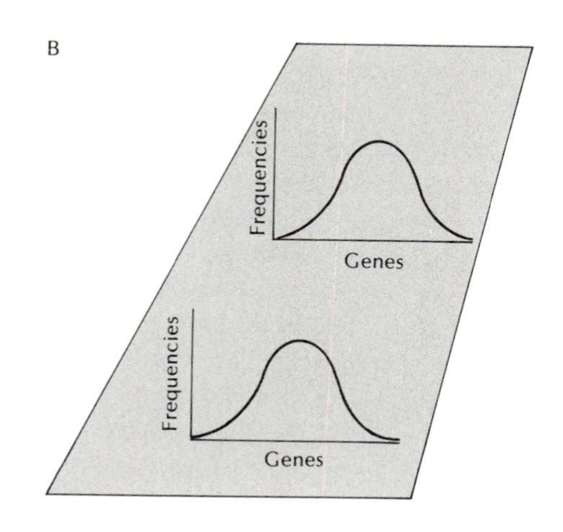

Fig. 6-7: Locations of small and intermediate-sized populations on their adaptive peaks. A. Small populations are fixed in positions somewhere on the lower slopes of their adaptive peak. The two different positions shown here indicate that, depending on chance events, they will rarely be the same. But in all probability, the positions will not be at the top of the peak. The solid boundaries indicate that fixation of genes allows little variability in a population and hence little chance of changing its genotypes. B. Intermediate-sized populations occupy areas closer to the top of the adaptive peak. Because they can vary significantly, their location is not fixed (note the dotted boundaries).

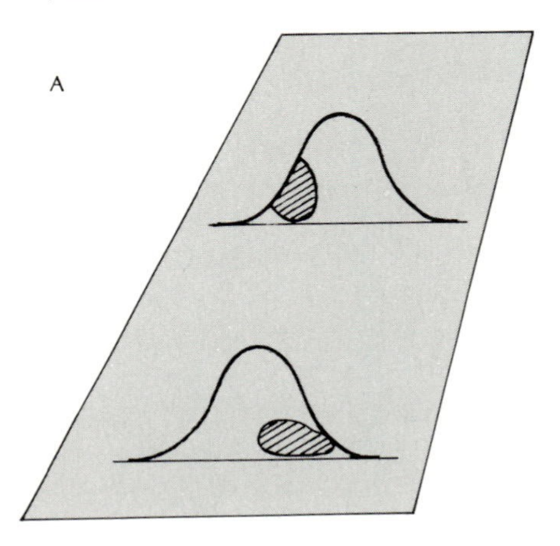

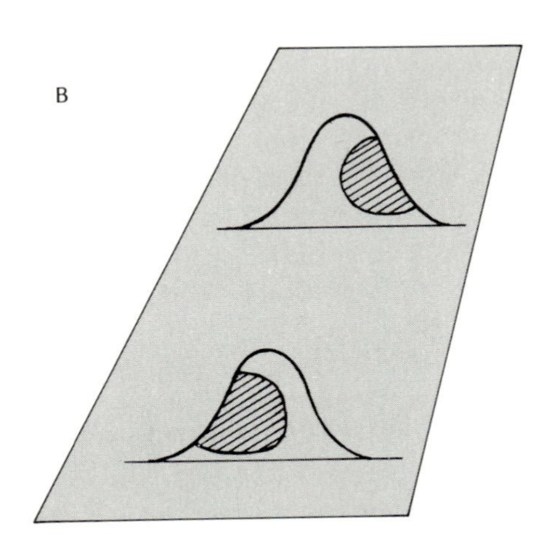

a drought might change the pond to a marshy area—they could not adapt.

Intermediate-sized population. Populations containing a few thousand individuals fall into the category of intermediate sized. Predictably, they are closer to the top of their adaptive peaks because selection can play a more significant role when chance events are not so important. Enhancing this role of selection is the fact that there is less inbreeding and mutation has more of a chance to contribute genetic variability. Migration may or may not be important. These factors will produce populations in which many members are well adapted to their niche.

In summary, intermediate-sized populations will move around the upper slopes of their adaptive peaks. Their polymorphic variability ensures that they will not be fixed in one position and selection keeps pushing them up toward the top (Fig. 6–7).

Large populations. In these populations, there are typically 100,000 individuals or more. Whether our freshwater ponds could support a population of fish that large is questionable, but a water-beetle population of that size is plausible. Or, looked at another way, the freshwater habitat we are discussing may represent only part of a larger watershed that might contain millions of individuals of the species studied.

Wright depicts the fate of a large population in Fig. 6–8A. Note that the population completely covers each peak and extends throughout the whole freshwater system. This means that there are well-adapted forms present and others that, because of genetic variability, can survive in areas away from the top of the peak. In fact, Wright's diagram indicates local areas where gene frequencies differ from those on top of the peak. The large population has areas of adaptation or local differentiation into subpopulations or demes.

That being so, we can refer to the large population as polytypic, and therefore, it is better represented by another diagram (Fig. 6–8B). The various adaptive peaks shown in this figure tell us that some demes are adapted to shallow water and some to deeper water, some to faster running water (the streams) and some to marshy areas, and so forth.

This differentiation of a large species into locally adaptive subpopulations occurs because significant variation is present and natural selection thus is effective. Migration throughout the whole area reduces inbreeding as does the large number of possible mates. Of course, if migration rates were so high as to guarantee effects equivalent to random mating, then the single adaptive peak (Fig. 6–8A) is the more appropriate description. But if there were low migration rates and random mating within each subpopulation or deme of 10,000 or so members, then the multiple peak-valley model (Fig. 6–8B) better describes what is going on. This is a polytypic gene pool and each deme is predictably polymorphic.

Mutation is now a significant source of variants. And the elimination of chance as an important factor means that, in the long run, selection is most important in determining the fate of the population.

In summary, large polytypic populations with locally adapted demes or subpopulations appear to adapt the best. Within such a population are certain individuals containing the genotypes appropriate to the adaptive peaks or norms. Furthermore, because of the wide range of adaptations possible, sudden changes in the environment can be met successfully. In drought, some members of the population will be already well adapted to stagnant, marshy waters. A pond that has been mostly filled up by a landslide will still be exploited by forms able to live in shallow waters. The large, polytypic population has the best chances for long-term evolutionary survival because of its present and future adaptedness.

Shifting balance theory. This great contribution of Wright to the evolution of gene pools provides insights as to how well-adapted demes can move across the adaptive landscape to reach adaptive peaks and how their genotypes can spread throughout the gene pool.

We start with the conclusions just arrived at in the preceding discussion: Among the various kinds of population size and structure that are possible, polytypic species are the most likely candidates for long-term survival and evolution. This is why many of the populations are, of course, examples of the balance theory of population structure. Wright's concern here is to show how the polymorphic balance of different loci and their alleles can shift depending on the factors determining the evolution of the deme or demes in question. This is his shifting balance explanation for the way populations of one size and structure can evolve into another and, presumably, better-adapted one.

There are three steps in this evolution toward better adaptation—toward reaching an adaptive peak. We start with a polytypic species whose demes or subpopulations show little gene flow among them: they are almost reproductively iso-

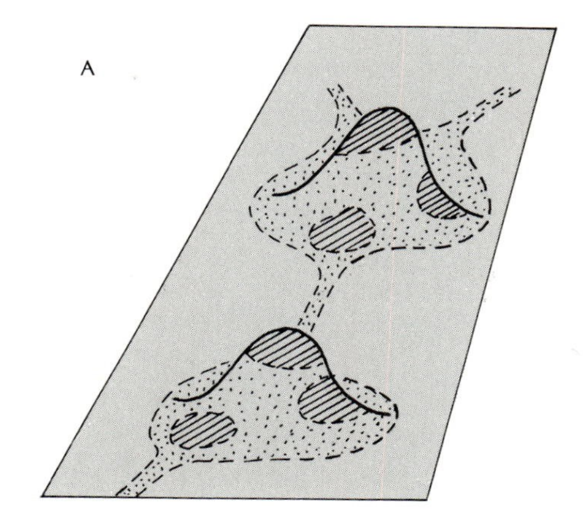

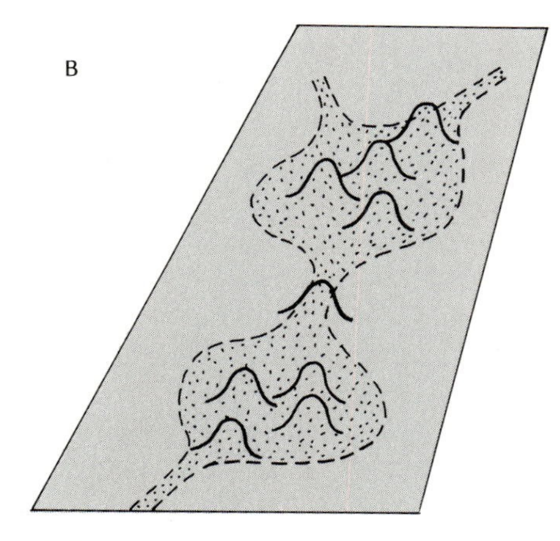

Fig. 6-8: Location of a large population on its adaptive peak or peaks. A. Large populations are more widely spread, but more importantly, because genetic variability is high and selection is effective, the population covers its adaptive peak. Local areas, where gene frequencies differ from those of the population as a whole, represent subpopulations or polytypic populations. B. A polytypic population, with a number of adaptive peaks, shows that a Mendelian population carries significant adaptive differences within it, with a potential for further adaptive evolution.

lated. Depending on the size of the demes, various fates are possible. Let us take the most problematic case, which is the small population. Its most likely fate is to have its gene frequencies determined by genetic drift. By chance it may locate on the side of a peak where it is reasonably well adapted. It may then increase in size and evolve into an intermediate-sized, polymorphic deme.

Now a second phase can occur, wherein natural selection rather than drift, is important and this subpopulation moves higher up toward the adaptive peak. Then the third phase takes over, and it is interdeme selection that now becomes important. Well-adapted demes continue to grow and expand, which leads to increased gene flow, and their genes are exchanged with other demes through migration; thus favorable genes are widely dispersed. Those new combinations of genes of the greatest fitness are the ones selected for and they determine the continuing location of demes on their adaptive peaks through a continual shifting balance of the polymorphism that keeps each deme well adapted.

This provides us with a model as to how reproductive communities maintain the genetic variability that is basic to their ability to evolve (recall Fisher's fundamental theorem) and how they can evolve into the highly adapted subpopulations of a polytypic species. In this way we can understand how the factors affecting the structure of reproductive communities can assure survival of the well-adapted ones. In brief, we see that reproductive communities are the units of adaptive change and, therefore, of evolution.

Problems for Chapter 6

(Answers in Appendix II)

1. Assume that there is a population of 10 million individuals and of constant size. If a gene in this population mutates at a rate of 10^{-5} per generation, what will be the number of such mutant genes, on the average, in the next generation?

2. Are assortative mating and inbreeding always different kinds of departures from random breeding?

3. The following data are from a census taken in south India of their local antelope, the blackbuck (*Antilope cervicapra*).

Adult males	Immature males	Adult females	Fawns	Total
52	14	405	22	493

What is the actual size of this population? What is its breeding size? What is its effective size?

4. In humans, infantile amaurotic idiocy is determined by a recessive lethal gene. Assume that this gene arises with a frequency of 3×10^{-6} per generation from its normal allele. This means selection is removing the gene when it is homozygous and mutation is restoring the gene to the population. An equilibrium can arise here. Calculate the frequency of the lethal gene at equilibrium. (Let $\hat{q}$ be this frequency.)

Recall that selection against a recessive is given in Fig. 5–16A as the change in q after one generation of selection or

$$\Delta q = spq^2/1 - sq^2$$

The rate of introduction of the lethal gene by mutation is μp, where μ is rate of mutation from the normal to the lethal and p is the frequency of the normal gene. At equilibrium the loss of the gene will equal its introduction by mutation. Namely,

$$spq^2/(1 - sq^2) = \mu p$$

Solve for q in terms of μ to determine their relation at equilibrium.

5. In humans, retinoblastema is a dominant lethal. Assume that this gene arises with a frequency of 3×10^{-6} per generation from its normal allele. Here again there is equilibrium, with mutation and selection pressures opposing each other. How would you calculate the frequency of the lethal gene at equilibrium? (Let $\hat{p}$ be this frequency.)

From calculations not included in the text, it can be shown that loss of the dominant lethal can be expressed as follows

$$\Delta p = sp(1 - p)^2$$

The introduction of the lethal is given by νq. Hence, at equilibrium loss and introduction will be balanced

$$sp(1 - p)^2 = \nu q$$

Solve for p in terms of ν.

6. What will be the frequency of the homozygotes of lethal genes described in Problem 4? In problem 5?

References

Crow, J. F., and M. Kimura, 1970. *An Introduction to Population Genetics Theory*. Harper & Row, New York.

Cullenward, M. J., P. R. Ehrlich, R. R. White, and C. E. Holdren, 1979. The ecology and population genetics of an alpine checkerspot butterfly, *Euphydryas anicia*. *Oecologia 38:* 1–12.

Ehrlich, P. R., 1978. The butterflies of Jasper Ridge. *The Sciences 18* (November, No. 9): 10–15.

Ehrlich, P. R., and P. H. Raven, 1969. Differentiation of populations. *Science 165:* 1228–1232.

Glass, B., and C. C. Li, 1953. The dynamics of racial intermixture—an analysis based on the American Negro. *American Journal of Human Genetics 5:* 1–20.

Kimura, M., and J. F. Crow, 1963. The measurement of effective population number. *Evolution 17:* 1279–1288.

Levin, D. A., and H. W. Kerster, 1974. Gene flow in seed plants. *Evolutionary Biology 7:* 139–220.

Smith, J. M., and R. Holliday, 1979. The evolution of adaptation by natural selection. *Proceedings of the Royal Society, London. B. Biological Sciences 205:* 433–608.

Spiess, E., 1977. *Genes in Populations*. Wiley, New York.

Wright, S. S., 1932. The roles of mutation, inbreeding, cross-breeding and selection in evolution. *Proceedings VI International Congress of Genetics 1:* 356–366.

Wright, S. S., 1938. Size of population and breeding structure in relation to evolution. *Science 87:* 430–431.

Wright, S. S., 1968–1978. 1977. *Vol. III. Experimental Results and Evolutionary Deduction*. 1978. *Vol. IV. Variability Within and Among Natural Populations*. University of Chicago Press, Chicago. (For references to Volumes I and II see Chapter 5.)

3

Speciation

1. *Geospiza magnirostria*
2. *Geospiza fortis*
3. *Geospiza parvula*
4. *Certhidea olivasea*

Darwin's finches

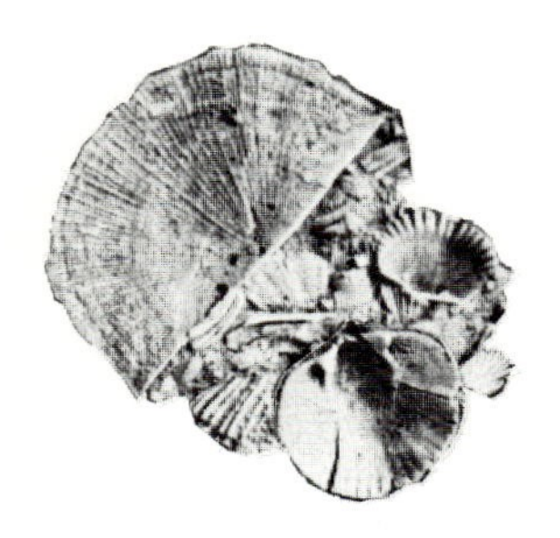

SPECIATION IS THE EMERGENCE of new species, or, to use Darwin's phrase, the origin of species. However, the subject is approached today from a perspective not available to Darwin. The Darwinian view of a species, derived from Linnaeus, was that of a static morphological type to which actual organisms were compared and then the decision was made whether or not they belonged to the species represented by that type. Deviations from the ideal represented by the typical form were ignored. The new perspective offered by population genetics is that of a reproductive community in which variation is integral to the community, since it leads to adaptive fitness. Ernst Mayr, a leading contemporary evolutionist, has said "The replacement of typological thinking with population thinking is perhaps the greatest conceptual revolution that has taken place in biology." [From *Animal Species and Evolution* (1965), Belknap Press, Cambridge, Mass., p. 5.]

With that in mind, let us examine the critical question of the origin of species. We are now informed in considerable detail as to what species do as reproductive communities. But how do they arise in the first place and then establish themselves? That is the problem we face in this section. Chapter 7 examines how new reproductive communities first emerge. Chapter 8 looks at the problems of their establishment and survival as new species. And Chapter 9 introduces us to the consequences of continuing speciation in terms of an awareness of the continuing history of life.

SEVEN

The Origin of New Species

THE NAME OF DARWIN'S FIRST WORK ON EVOLUTION, *The Origin of Species*, might suggest that there is one path to the formation of new species. Conceptually, Darwin argued that to be true and his viewpoint is summarized in his phrase "Varieties are incipient species." By this he meant, of course, that the appearance of variations within a species is the source of the differences that will finally generate new species. That is the concept; but he also knew that the reality was more complex since he admitted to deep ignorance regarding the source and nature of heritable variations and he knew that selective pressures can differ for different species.

Where are we today in the analysis of the origins of species? Our viewpoint differs from Darwin's of over 100 years ago in that (1) we now know a great deal about the genetics of variation; (2) we now view species as reproductively isolated communities; and (3) we have a clearer view of the action of selection and other factors that shape the adaptations of natural populations. But, as we shall see, we are still very poorly informed as to just what occurs at that critical time when a new gene pool emerges.

Speciation

To put things into a clearer perspective we must first stand back and see species formation or speciation in its broadest outlines. From that vantage point we will proceed to narrow the problems down to key issues, foremost among which will be the appearance of reproductive isolation. Probably the first person to state the central importance of reproductive isolation was the German morphologist Naef. He said, in 1919,

> Nature presents us first of all with individuals (we speak only of multicellular forms) which occur in all degrees of similarity or difference. But among them there belong together natural groups of the specially similar ones that form a reproductive community, which is to say, that some of them can stand physically related to each other as ancestors and descendents and some as mates. Such a reproductive community, insofar as it is not produced by artificial means and it persists, that is, is viable through many generations, is called a species. There is no other possible basis for this concept, above all no morphological one and we can abandon at this point any further attempts in that direction. (*Idealistische Morphologie und Phylogenetik,* pp. 44–45.)

There are two general modes of speciation: transformation and divergence.

SPECIES TRANSFORMATION

By transformation is meant the cumulative changes over time whereby one gene pool changes into another that is presumably reproductively isolated from the first. We must say "presumably reproductively isolated" for there is no way to test for the fact of reproductive isolation: the first population no longer exists, having been transformed into the later one. The reason for using the criterion of reproductive isolation comes, of course, from its universal importance in defining the basic biological discontinuity among natural populations. Its practical application is, therefore, a problem in this case.

Transformation is apparent from the fossil record (Fig. 7–1). Fossils provide many examples of changes accumulating over many years, wherein forms were replaced by their significantly changed descendants. If we compare such changes to similar, but reproductively isolated forms living today, we see a way around the problem of fossils as forms that cannot be tested for reproductive isolation. Living forms can provide a

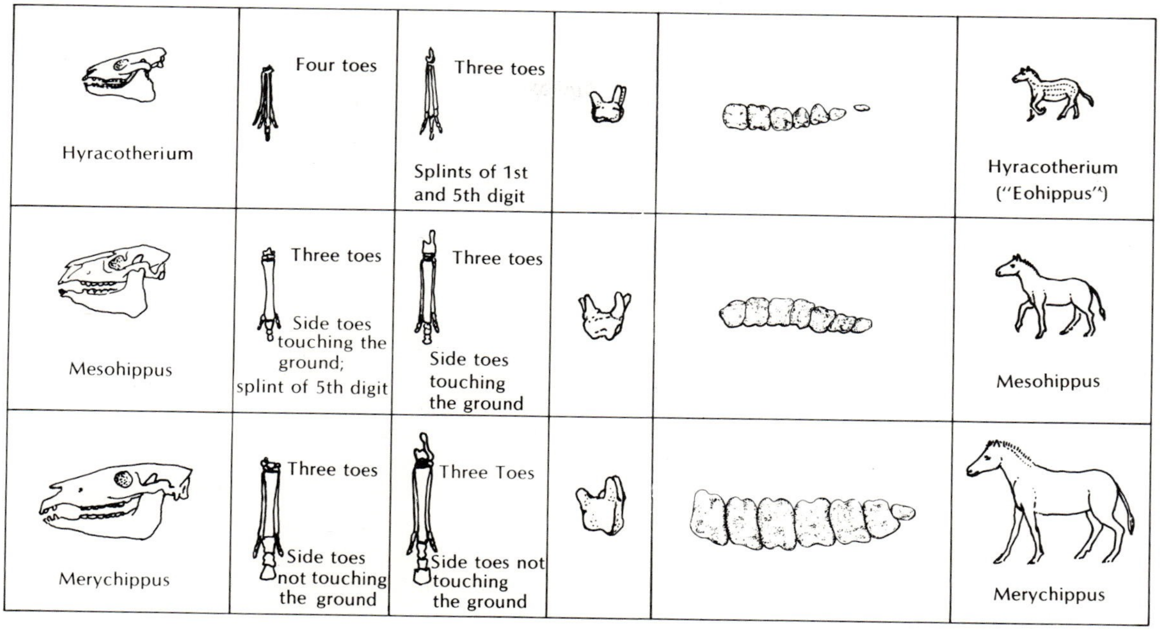

Hyracotherium	Four toes	Three toes Splints of 1st and 5th digit			Hyracotherium ("Eohippus")
Mesohippus	Three toes Side toes touching the ground; splint of 5th digit	Three toes Side toes touching the ground			Mesohippus
Merychippus	Three toes Side toes not touching the ground	Three Toes Side toes not touching the ground			Merychippus

Fig. 7-1: Reconstruction of early horses from the fossil record. Skulls, forefeet and hindfeet, and teeth are important in deciding what the browser *Hyracotherium,* as well as its descendents *Mesohippus* and *Merychippus,* looked like. The latter is a grazer.

standard of how highly morphological difference can be correlated with reproductive isolation. This standard is not a precise one because morphological differences are essentially absent between sibling species—such species differ physiologically and behaviorally rather than morphologically. But, if it is used conservatively, an experienced worker can propose two fossil forms as being sufficiently different as to very probably represent two different gene pools. In other words, by arguing carefully from known species, one can fairly safely identify different fossil species, even in the absence of any breeding studies.

Change over time involves at least two processes. Either the environment is changing or the species is becoming more specialized in the way or ways it exploits the environment. If more than one way of survival emerges, we either find a polytypic species or the emergence of two species where originally there was one. This is species diversification.

SPECIES DIVERSIFICATION

Species diversification is the appearance and establishment of more than one reproductive community from one parental community. In this process, the parent community may be changed (actually, transformed) as it diversifies into two or more new species or it may persist along with the one or more new species that have arisen from it. Here, then, breeding studies, both in nature and in the laboratory, can be attempted to determine the degree of reproductive isolation among the coexisting communities.

Some biologists have tended to make speciation synonymous with diversification. Indeed, Darwin emphasized diversification in *The Origin of Species*. The reason for his emphasis is easy to see. Diversification deals with living forms and is, therefore, easier to analyze. Also, it gets to the heart of the conceptual problem in organismic diversity, which is the increase in number of species and not just the transformation of one into another. Nonetheless, both transformation and diversification make up speciation.

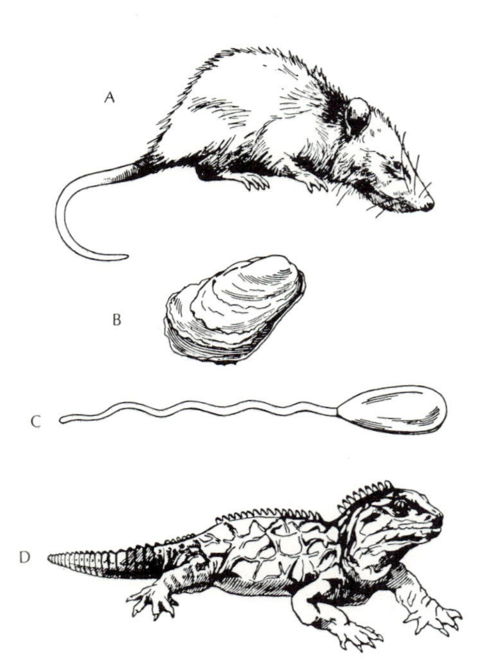

Fig. 7-2: Animals that have shown little structural change over long periods of time: opossum (A); oyster (B); *Lingula* (C); *Sphenodon* (D).

A few more comparisons of these two modes of obtaining new species will be useful at this point. In both modes, natural selection, acting to preserve useful, chance mutations, is always present. If the environment being exploited by a species population is very stable then little or no speciation will occur. The result will be a species existing essentially unchanged for millennia (Fig. 7–2). When environmental changes occur, transformation or diversification will occur, depending on the breeding patterns of the population. If migration and breeding habits assure something close to random breeding throughout the population, than the whole population will change—transform—in response to selection and other factors. On the other hand, if breeding is not random and the population differentiates into subpopulations and demes, the changes that occur can include transformation *and* diversification.

Transformation and diversification are not mutually exclusive processes; rather they are often complementary. The fate of a given Mendelian population, or gene pool, depends, then, on how its biological factors—size and structure, breeding habits, mutability, migration, adaptations—interact with its environment.

Finally, we must realize that speciation does not occur automatically as a result of population-environment interactions. As we have said, in a stable environment there may be little or no speciation even though natural selection continues to operate. Furthermore, environmental changes can be cyclical, e.g., droughts come and go as do excessively hot summers and cold winters. Populations can transform in response to these changes, but they can also revert to original gene frequencies when conditions return to normal. Thus a deme can become geographically isolated from the parent species, but subsequently reestablish contact with it. In such a case, the deme can be resorbed into the parent species. On the other hand, the isolated deme could also suffer the fate expected of most small populations, i.e., extinction.

In general terms, then, speciation depends on survival in the face of change. Therefore the changes must be adaptive. They can take place within a species, which is transformation, or they can generate a separate species, which is diversification. And at this point, we emphasize again the importance of reproductive isolation: It excludes maladaptive genes from the adapted and coadapted genes already in a gene pool. Adaptation is the embodiment of that slogan that praises specialization, i.e., "Jack of all trades and master of none." In a well-adapted reproductive community there are phenotypes and their genotypes that have been preserved by natural selection. This assures that the organisms in that community will be "masters," to some degree, at surviving in their environment. And reproductive isolation protects that mastery from dilution by an influx of maladaptive genes from other species. Fisher's comment on the high degree of improbability generated by natural selection again comes to mind as a description of a well-adapted reproductive community. And reproductive isolation maintains that improbability by preventing the inflow of other genes.

Let us now consider speciation as two processes: (1) the emergence of a new gene pool and (2) the establishment of this gene pool as a new species. Both processes are inherent in transformation and diversification, but are much more conspicuous in the latter because both isolation and adaptive success in what is initially, in most cases, a small population are required. This requirement, as we have seen, is a most difficult one, biologically speaking. Hence, if we can successfully understand what goes on during diversification, we are probably solving the most difficult part of the problem of how a new species originates.

Here we will concentrate on how new populations that have the potential of becoming new species *emerge*. In Chapter 8 we will examine what we know of how such populations can go on to *establish* themselves as new species.

Initial questions

The discussion that follows will be more theoretical than experimental and observational, since, as R. C. Lewontin emphasizes in his book, *The Genetic Basis of Evolutionary Change,* "we know virtually nothing about the genetic changes that occur in species formation" (p. 159). This, at first glance, seems to be a gross exaggeration. But we see that it is not when we look at the questions we must ask and then at the possible answers. To do that let us follow Lewontin's point of view a bit further. He sees three steps in species formation or what we have called speciation through diversification. These are

1. Following on divergence in the ecological niches being exploited and geographical isolation of two populations, genetic differences appear that severely restrict gene flow between the populations when they again contact each other.
2. Reproductive barriers to mating become stronger between the two populations.
3. The two populations continue to evolve along their own lines as separate reproductive communities.

We are concerned with the first point here—the appearance of one or more new populations with the potential of becoming new species. (The second and third points will be examined in the next chapter.) Regarding this first point, Lewontin raises questions of this sort:

1. How much of the gene pool is involved in the early divergence of one population from another?
2. How much of this divergence affects reproductive isolation?
3. What are the genetic bases of the phenotypic differences between diverging populations?
4. How are the genotypic changes and the functional phenotype on which selection acts related?

These questions interconnect and overlap. They are all fundamentally one basic question: What are the first genetic changes accompanying population divergence? Related to that basic question are two more that are necessary to answering it, namely, What phenotypes arise from those first genotypic changes? and Why are just those changes (genotypic and phenotypic) correlated with divergence?

It is to questions such as these that we have little—"virtually nothing" in Lewontin's words—in the way of real answers. Thus, work in this area is necessarily that of following the usual tactics of scientific inquiry at such a juncture. It is the constructing and testing of hypotheses that may elucidate the problems at hand. Therefore, our discussion of species origins must be more speculative than factual.

What, then, are the theories and hypotheses that are now under scrutiny?

The two general categories of explanations for divergence are *rapid diversification* and *gradual diversification.* Since most workers today favor gradual diversification, let us look at that category first.

GRADUAL DIVERSIFICATION

In this category various explanations have been suggested as to what goes on in speciation. The three that we shall emphasize can be shown diagrammatically (Fig. 7–3), to clarify their differences and similarities.

Breaking a chain. The German term *Rassenkreis,* circle of races, refers to a polytypic species that has differentiated to the point where certain of its subpopulations or demes behave as if they were real species. They are, in short, reproductively isolated at the point of contact, but not otherwise. A well-known example is that of certain gulls. In Great Britain two phenotypically differentiated gulls are called the lesser black-backed gull (*Larus fuscus*) and the herring gull (*Larus argentatus*). However, as one studies the distribution of these two apparent species one finds they are members of one gene pool.

This gull story is summarized in Fig. 7–4. Experts recognize various phenotypically discrete populations designed principally as *L. argentatus, L. fuscus, L. cachinnans,* and *L. glaucoides;* each has several subpopulations.

It seems that groups of individuals of a widespread gull population, originally like *L. argentatus,* were isolated by the ice age of about 2½ million years ago—the Pleistocene epoch. One of these groups, today called *L. cachinnans,* stayed near the Aral and Caspian seas in what is now the Soviet Union. It gave rise to *L. fuscus,* in northern Europe, and expanded into the Mediterranean region. The original *L. argentatus* seems to have survived on the eastern coast of Siberia or in Alaska and subsequently spread into the northern half of North America and then across the Atlantic to Great Britain where it met with *L. fuscus.* This North American form also met up with another relict group, which has now become established as a separate

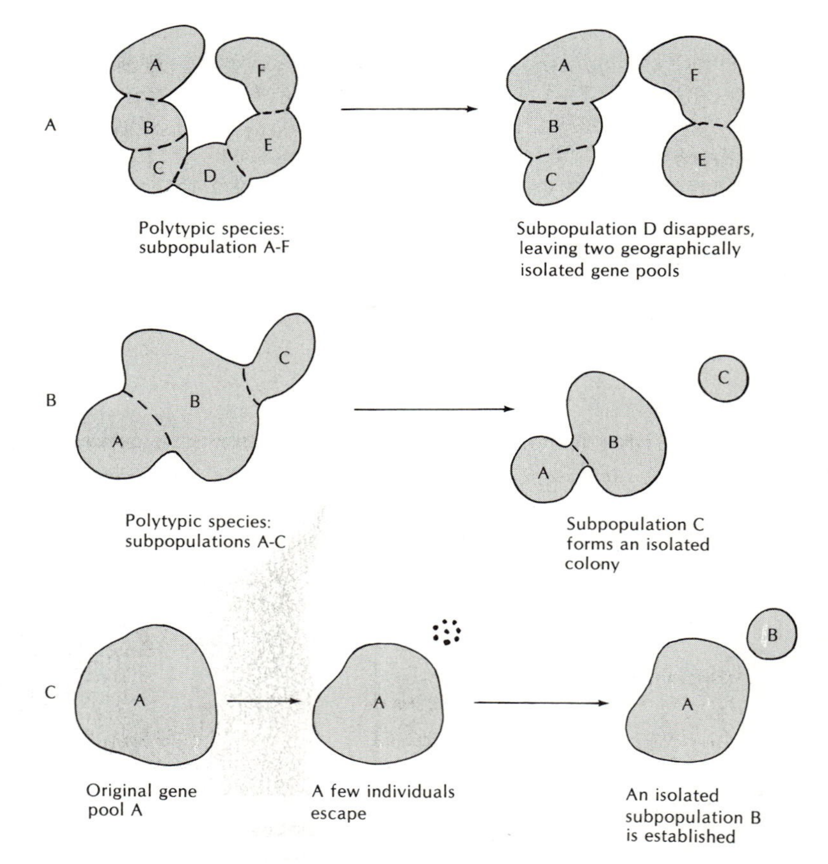

Fig. 7-3: Three modes of gradual diversification. Breaking a chain of polytypic subpopulations (A); colony isolation, I (B); colony isolation, II (the founder principle) (C).

species, *L. glaucoides*, in the Arctic Ocean near Baffin Island and Greenland. The differing degrees of interbreeding between the various *L. argentatus* and *L. fuscus* subpopulations indicates differing degrees of isolation. The Pleistocene isolation was apparently long enough to allow local differences to become established, but not long enough to allow full reproductive isolation, except subsequently in Great Britain between *L. fuscus* and *L. argentatus*, and in the case of *L. glaucoides*. Among the rest there is sufficient interbreeding so that we can argue that they all belong to the same gene pool, which is what also binds together *L. fuscus* and *L. argentatus* as the terminal links in a chain of circumpolar populations.

As another example of the *Rassenkreis* situation we can look at certain California salamanders (Fig. 7–5). These amphibians inhabit the streams of the hilly areas of California. They are absent from the great central valley of that state. Apparently these animals first evolved in the hills toward the northern end of the central valley and then spread southward along both sides of the valley. When the ends of this distribution meet at the southern ends of their range in the San Joaquin valley, they do not interbreed. They behave as if they are members of two different species, although they are members of the same polytypic gene pool.

Many other examples of such polytypic species could be given. Our interest in them is to see how they could give rise to new species (Fig. 7–3A). The mechanism would be (1) a break in the chain of interbreeding subpopulations so that, in effect, one interbreeding gene pool would be separated into two and (2) the first gene pool had already tended toward reproductive isolation. The original polytypic condition took many thousands of years to establish. The advent of some geographic change, which then wipes out the intermediate forms or otherwise isolated the two extremes of the chain of demes, would then promote the emergence and establishment of two separately adapted gene pools.

We must emphasize that no one has actually seen this sort

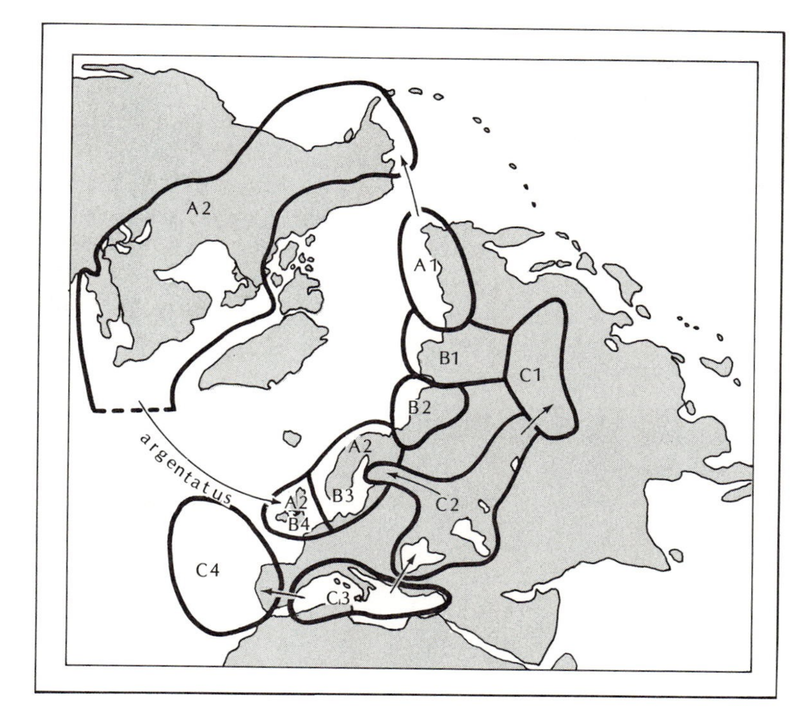

Fig. 7-4: The circumpolar distribution of the herring gull (*Larus argentatus*) and the lesser black-backed gull (*Larus fuscus*). The gene pool is thought to have originated in northeastern Siberia; the *argentatus* group (subpopulations A1 and A2) spread eastward to Alaska, through North America, and eventually reached the Western British Isles. The *fuscus* group (subpopulations B1, B2, B3, and B4) spread westward to reach the British Isles. Subpopulations C1, C2, C3, and C4 are another subspecies of *Larus*. And D1, D2, and D3 are *Larus glaucoides*, reproductively isolated from the other gulls as a good species.

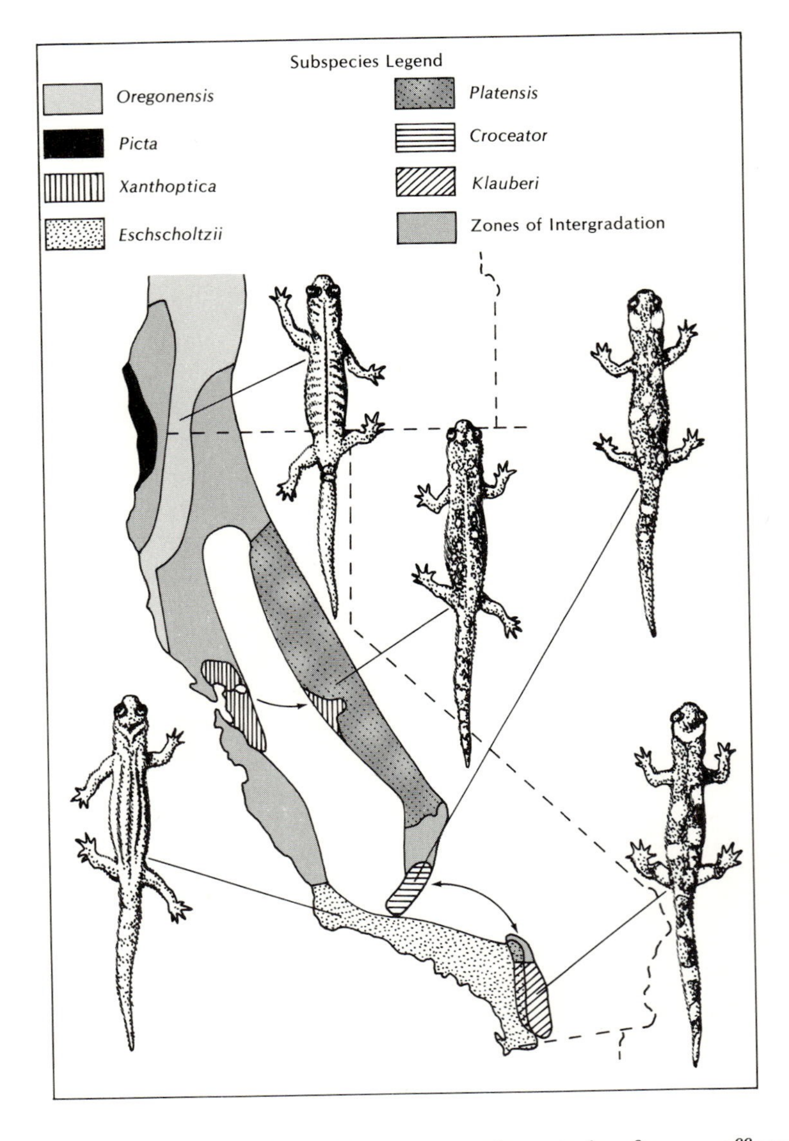

Fig. 7-5: The circle of subpopulations (*Rassenkreis*) formed by California salamanders (*Ensatina eschscholtzii*). Because no intermediate forms have been found between *E. eschscholtzii* and variety *E. e. klauberi,* they may be reproductive isolates. To the north, *E. e. eschscholtzii* and *E. e. xanthopicta* apparently interbreed, as do *E. e. klauberi* and *E. e. croceator* to the northwest. There is also evidence of interbreeding (intermediate forms) among other adjacent subpopulations or subspecies, as recognized by taxonomists. But electrophoretic and immunological studies do not agree with species classification based on morphology, as the work of Larson and Highton shows; hence, salamander studies, as demonstrations of ongoing speciation, must be cautiously interpreted.

of speciation take place. However, such a circle of races offers perhaps our clearest view of a single species on the verge of forming two species. It does show us that *a single species can differentiate to the point where certain of its members behave toward each other as if they were members of different species.*

Colony isolation I. The terms *geographical theory* or *conventional model of species formation* have been applied to this mode of diversification. Neither term is especially helpful, since geographic isolation is a very common requisite in most, but not all types of diversification, and the term conventional is both ambiguous and uninformative.

What is thought to happen is this. A subpopulation at the periphery of its gene pool becomes isolated from the rest of the species (Fig. 7–3B). This could occur in a variety of ways: environmental changes could shrink the original population leaving behind locally adapted demes, or geographic changes might isolate the subpopulation, and so on. In any case, the isolated subpopulation is of substantial size (what we earlier called an intermediate-sized population) with a gene pool well adapted to its environment.

Further evolution would include (1) further adaptation to the local environment and (2) no gene flow from the parent population. These factors are interrelated. The absence of gene flow means that local selection pressures will determine the gene frequencies of this colony to a great extent. It can, thereby, improve its adaptive response to its environment. But it will no longer have the genetic resources of the larger population to draw on in the event further environmental change occurs since it is now on its own. If no catastrophic changes occur, then the colony, starting out well adapted, is able to evolve without interruptions. Gene flow from the rest of the population would represent interruptions, since there would be maladaptive genes coming from other parts of the pool. (Such genes could, of course, be adaptive in other subpopulations.)

Colony isolation II. This mode of diversification occurs when the new colony is smaller than the colony discussed under Colony isolation I. A small number of individuals somehow leave or migrate from the parent population and survive in geographic isolation. Theoretically, the minimal size of such an emigrant population could be one individual, providing that individual were one of three kinds: (1) an individual who could reproduce asexually, with progeny capable of sexual reproduction; (2) an individual capable of self-fertilization; or (3) a pregnant female, carrying progeny of both sexes. Any of these three could reestablish a Mendelian population.

In any case, the new colony or population would face all the problems of a small population. Since it starts with a small sample of the original gene pool and since it is only partially preadapted to the area it is colonizing, the new colony would probably have to generate a quite new gene pool to survive. The size of the sample (i.e., few genotypes) and the continuing selection to generate a new adaptive norm would combine to produce a significantly different gene pool. And it might occur relatively rapidly if the initially small population swiftly grows into a larger, better-adapted one. Mayr has called that change, potentially drastic, between the old parent gene pool and the pool of the new colony, genetic revolution. This growth and differentiation of a new gene pool subsequent to geographic isolation of the founding colony is receiving considerable attention by researchers today. The answers are not yet in as to the size of the revolution postulated by Mayr.

The emergence of a new species in this way has been designated by Mayr as the *founder principle*. Others refer to it as the *quantum model of species formation* or *quantum speciation*. Whatever the name, the important features of this mode are clear. There is a small initial population that somehow achieves geographic isolation. If it succeeds in surviving—presumably because it has some preadaptations for its new habitat—and grows to a size at which selection can maintain and refine its adaptations, then it can establish a gene pool that can vary remarkably from its parent pool. However, the bottleneck of small size probably means that many more evolutionary attempts, according to the founder principle, fail than succeed.

Comparative comments. The foregoing three modes of species diversification are compared in Fig. 7–3; they are compared in more detail in Table 7–1. Even though they show certain real differences, it should be remembered that they represent three points on a spectrum of differences. The three modes can intergrade into each other in terms of the size of the populations and their mobility. Size and mobility are intimately connected in an inverse way, since large populations are not as mobile as small ones. Or, more precisely, large populations cover a large area and although their mobility within that area may be great, their migration to new areas may not be so obvious as with a small population of one or a few individuals.

Furthermore, beyond the size and mobility question, all three models depend on geographic isolation. The parent population is separated into two parts or somehow a part of it is isolated. Some researchers have, in fact, argued that geographic isolation is a prerequisite for species diversification. Certain others have claimed that *sympatric speciation*—species formation within a common geographic region—can occur. We will examine those claims after we finish our discussion of Table 7–1.

The differences between the three modes of gradual species diversification come down to differences in size. In break-

Table 7–1. A comparison of selected features of the three modes of species diversification.

	Breaking the chain	Colony isolation I	Colony isolation II
Initial gene pool	*Rassenkreis* (polytypic)	usually polytypic	homogeneous, polymorphic, or polytypic
Means of geographic isolation	disruption of subpopulation chain	emigration of a large colony or relict formation through contraction of the parent population	usually emigration of a small population
Size and structure of the new population	large and polytypic, but smaller than parent population	intermediate-sized or smaller; polymorphic or homogeneous	small and homogeneous
Subsequent fate	survival highly probable	survival possible	survival unlikely
Nature of new gene pool	much like situation preceding disruption	probably differs somewhat from parent population	if it survives, probably very different from parent population
Rate of change (depends, in all cases, on differences between old and new habitats)	slow change, if any	can be fairly rapid change	if it survives, rapid change

ing the chain of polytypic subpopulations, the surviving populations are large and already well adapted to their environments. Their survival would be relatively assured. The chief changes in their gene pool would be moves higher up their particular adaptive peaks. They could so move as a consequence of being freed from the effects of immigration of new genes from the rest of the pool.

In colony isolation of the first sort discussed here, the new colony would be intermediate to small in size. It might well remain in an area to which it is already adapted or it might move further away from the parent habitat as it continued to adapt. In either case, it is significantly preadapted to the environment within which it is located. If of intermediate size, it has a good chance of moving well up on its adaptive peak and continuing its evolution.

In colony formation involving the founder principle, the gene frequency changes that occur in the small population involve important limiting factors that make survival improbable. The chief of these are limited preadaptations, a limited gene pool, and limited effectivenes of selection, unless the population rapidly increases in size. But in the long run, even if the rate of survival is low, from repeated emigration of a potential founder there would accumulate a significant number of successes and this mode of speciation is a valid mode of species diversification.

One especially important result of the founder principle,

which has been borne out by experimental work, is the variety of its successful outcomes. Starting with a small gene pool in which chance plays a large role in the initial development of the population, we would predict a variety of genetic outcomes in the final population. This was tested by Dobzhansky and Pavlovsky. They started 10 populations from 20 founder individuals of *Drosophila pseudoobscura* and grew them in parallel under controlled conditions. Another 10 populations, each started with 4,000 founder fruit flies, were grown, as far as possible, under the same conditions. The results, after 17 months, are given in Fig. 7–6. The chief character studied here was the frequency of occurrence of a given chromosome. All experimental populations started with this chromosome at a frequency of 50%. The large populations all ended up with frequencies between 20 to 34%; the small populations between 16 to 48%, a much higher variance.

Sympatric diversification. As noted above, the types of speciation already discussed all depended upon geographic isolation. The argument is that such isolation is necessary to allow two gene pools to diverge significantly. In the case of colony isolation, the word *escape* is sometimes used to describe the breaking away of the colony. The connotations of that term are intriguing; we will return to them shortly. But first we

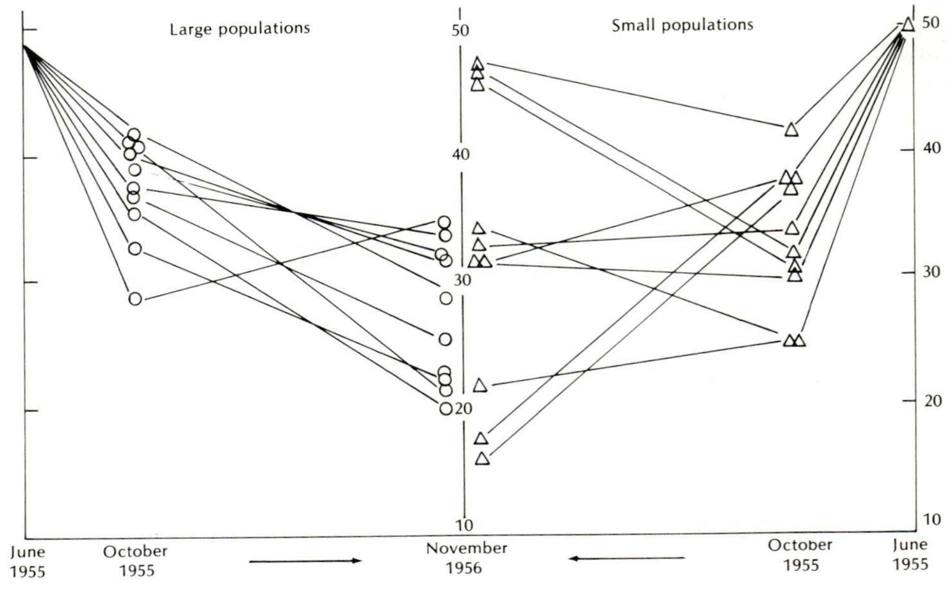

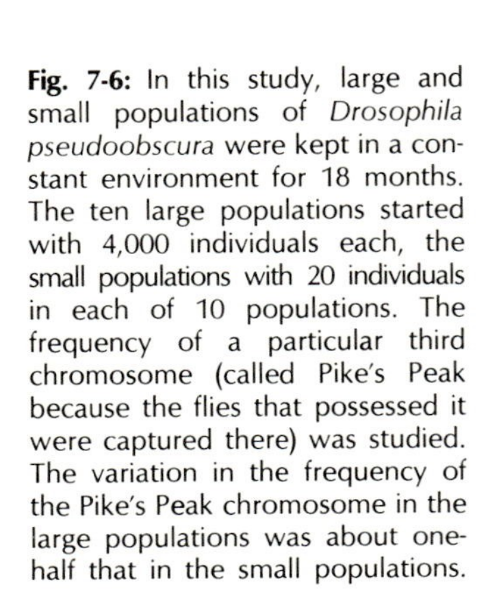
Fig. 7-6: In this study, large and small populations of *Drosophila pseudoobscura* were kept in a constant environment for 18 months. The ten large populations started with 4,000 individuals each, the small populations with 20 individuals in each of 10 populations. The frequency of a particular third chromosome (called Pike's Peak because the flies that possessed it were captured there) was studied. The variation in the frequency of the Pike's Peak chromosome in the large populations was about one-half that in the small populations.

need to explore further the importance of geographic isolation. Most students of evolution argue that it or something like it is necessary to achieve species diversification.

The reason isolation is necessary, according to this argument, is that it allows differences to build up in the different gene pools in response to local selection pressures. Such differences presumably accumulate such genotypic differences that hybrids, if they are formed between two such gene pools, will be at a selective disadvantage. That being so, mating *within* the gene pools will be selected for and mating *between* the gene pools will be selected against. The final outcome is now determined; selection pressure will inevitably separate the gene pools further into separate reproductive communities by favoring mating within each pool.

The crucial point in species diversification is, therefore, selection against hybrids. This leads to the question of how genetic differences underlying hybrid disadvantage might arise between two subpopulations of one original population. Is geographic isolation necessary or are we really arguing for *any* kind of isolation? The latter is the case, but it then means we must discover other kinds of isolation. Therefore, our attention to sympatric speciation does not reject geographic or allopatric speciation, but, rather, searches for alternate modes of effective isolation leading to species diversification.

What are the possibilities? Three examples can be given to illustrate them. In the first one, *allochronic speciation,* the isolation of breeding individuals is time dependent, or seasonal. The northern spring field cricket (*Gryllus veletis*) breeds in the spring (May to July) and is sympatric with the northern fall field cricket (*G. pennsylvanicus*), which breeds in the fall (July to October). The mating seasons are thus temporally fairly well isolated. Furthermore, the fall cricket survives the winter as fertilized eggs, which then develop into immature forms in the following spring and early summer and finally molt into adults in the late summer, when mating occurs. The spring cricket has a different cycle. Its eggs, laid in the summer, promptly hatch into juveniles. They go through the winter in the late juvenile stage and molt into adults in the spring, when they mate. Alexander and Bigelow, who have studied these forms carefully, think this situation arose sympatrically, since both species occupy the same geographic area.

In Alexander and Bigelow's view, the key point is that early juveniles are not cold resistant. Hence, if the original gene pool had crickets mating throughout the summer (and perhaps giving rise to two generations each year) then a decrease in temperature, such as occurred in the Pleistocene ice ages, could separate the two gene pools by eliminating forms breeding in the middle of the summer. Such crickets would have had their cold-sensitive forms killed by fall frosts. Only the early breeding forms, whose juveniles were in a cold-resistant late period, or late breeding forms, whose zygotes were cold resistant, would survive. This effectively divided the single original gene pool into two pools, by seasonal or allochronic isolation. All of this would have occurred sympatrically, according to Alexander and Bigelow.

However, Mayr points out that this is not the only interpretation possible. He suggests that geographic isolates in refuges produced by the ice age were separated into spring- and fall-adapted subpopulations. When the ice age ended, these populations expanded and became sympatric, but because of their adaptations they were effectively isolated into two different species. Thus, this one case of possible sympatric speciation could also be explained by allopatric factors or geographic isolation. It is still ambiguous.

The second possible case of sympatric formation of species is put forward by Bush; it involves parasitic flies of the family *Tephritidae*. These flies depend on hosts (either plants or animals, but the most convincing case involves flies with plant hosts), which have undergone demic differentiation within a given habitat. The parasites adapt quite specifically to their hosts. It is as if the host's adaptation to local environmental differences magnify the parasite's adaptations to the host. This produces differences among the parasites that have led to speciation in a quite restricted geographic area. Although the flies surely move through a shared geographic space—they are sympatric—they do not share the same breeding areas because of their precise adaptations to their hosts.

In criticizing this analysis by Bush, we can point out that though the flies appear to differ phenotypically, as would be expected of species, there is no hard evidence that they cannot interbreed. And, further, this work raises the question as to what constitutes geographic isolation. The breeding activities of the flies appear to be geographically isolated, i.e., allopatric, among the different subpopulations of their hosts. When they are not breeding and flying about they are sympatric. So even if we agree that genuinely isolated reproductive communities have arisen through specialization on host differences, is this really an example of sympatric diversification? The problem has now become one of how sympatry and allopatry are defined.

The third and final case involves isolation by disruptive selection. In 1966, the British evolutionist J. M. Smith argued on theoretical grounds that disruptive selection could lead to isolation of gene pools. An example from nature that seems to illustrate Smith's hypothesis is the work of the Taubers, from Cornell University, in their study on lacewing insects. They found that two sibling populations, *Chrysopa dounesi* and *C. carnea,* showed no evidence of interbreeding because of differences in habitat preferences (different species of trees in the same geographic area) and temporally different reproductive times. These sibling species, however, were fully interfertile, producing fully viable F_1 and F_2 progeny. This made possible a genetic analysis that allowed this husband and wife team to conclude that alleles at only three loci determine habitat preference and time of reproduction. They argue that disruptive selection acts on the variability of this ostensibly sympatric population to separate it into two isolates.

Again some further comments are called for. First, the Taubers point out that the two sibling species differ genetically by more than the three loci identified, but that the three studied can account for the observed isolation in nature. Second, the designation of *C. dounesi* and *C. carnea* as species can be challenged if we demand that reproductive isolation is the necessary criterion for a species, since the two groups are reciprocally fertile. The challenge is met by the fact that, in nature, no hybrids are found and, because of habitat and seasonal isolation, cannot occur. Practically speaking, these forms are isolated. Third, it appears that the Taubers may well have found an extremely early stage in speciation in which habitat and seasonal isolation *preceded* genuine reproductive isolation. Presumably, the latter will emerge later. And fourth, the Taubers suggest that isolation through disruptive selection might be "not uncommon among animals with habitat differences." From an evolutionary point of view, the central point that *some kind of isolation is requisite to achieve new gene pools* remains. At the least, in the great majority of cases straightforward geographic isolation fulfills that requirement.

The origin of reproductive isolation. There are many ways to achieve reproductive isolation. Biological factors include the appearance of hybrids that are at a selective disadvantage, the inability to form normal embryos, and various mating problems ranging from inactivation of gametes after copulation to the inability to exchange gametes and to lack of recognition of mating partners. The physical factors are geographic and temporal isolation, which may precede or follow the biological factors. In any case, the biological factors ultimately guarantee genuine isolation. Let us briefly consider them now; they will be discussed further in the next chapter when we examine isolating mechanisms in greater detail. At this point, we are particularly interested in the emergence of the initial differences that lead to reproductive isolation.

The first difference must be some adaptive response to an environmental difference between members of one gene pool. In these terms we are close to Darwin's view that varieties (or demes) are incipient species. But we differ from Darwin when we apply the founder principle. Here there may be a homogeneous parent population—one without varieties or other recognizable subpopulations—from which founders escape. The essential point is, then, that there is some sort of environmental difference that results in different selection pressures for different members of a gene pool.

Quite obviously environmental differences can cover a wide range, from physical ones, such as differences in humidity and temperature, to biological ones, such as type of food, availability of mates, and avoidance of predators. Adaptations to these factors must be genetically determined, to some degree, to be of evolutionary importance. So quite obviously we must first talk about genetic changes. How might such changes lead to reproductive isolation?

The answer is frankly disappointing. We really do not know. We are back at Lewontin's point regarding our virtual ignorance in such matters. And so again we speculate as to what might be happening in the hope that our speculations will lead us to meaningful research.

The central problem at this juncture lies with developmental biology. We still are a very long way from relating genotype to ecologically meaningful phenotypes. Thus the unitary relation between a factor and a character, first elucidated by Mendel, is now recognized to be the exception rather than the rule above the level of polypeptide sequences (see Chapter 3). A cell organelle, such as a cilium or a mitochondrion, is formed from the products of many genes. A cell is an even more complex product of genic interactions. And the complexity increases as we go to multicellular organisms. In ecological terms, the whole organism survives by interacting with its environment, an environment that includes the physical and biological factors that determine selection pressures. Therefore how genetic factors relate to a given ecological niche is both subtle and complex. The role of genetic factors in reproductive isolation is no less difficult to analyze. Or, briefly and emphatically, we realize that most characters are polygenic (i.e., controlled by many genes) and that any one gene is pleiotropic (i.e., it affects more than one character). And, last, we know that the environment can play an important role in the expression of a gene. All of these act to obscure any simple relationship between genotype and phenotype.

So where are we? Two directions for further work can be established: (1) continue developmental studies that advance our understanding of the relation between genotype and the functional or ecological phenotype and (2) focus on natural or laboratory conditions that represent or illustrate, respectively, the emergence of reproductive isolation.

The genotype-phenotype relationship. Little would be gained in reviewing the present state of developmental genetics, but what is helpful, however, is to view developmental genetics from the perspective of species diversification. Again, we consider Professor Mayr's point of view. He uses such phrases as "harmoniously integrated gene complex," "the unity of the genotype," a "well-buffered epigenetic system," and "genetic cohesion of gene pools." To this we add Dobzhansky's concept of *coadapted genes* and single out the embryologist's view of development as *epigenesis.* Then the needed perspective emerges.

We begin with epigenesis, in which development is considered to be a sequential process; it is an unfolding of developmental events that brings the genotype to expression in an orderly way. In classical descriptive terms, this refers to a succession of events as shown in Fig. 7–7. In terms of the mechanics of development, it is clear, for example, that germ layers cannot form (in the gastrula) until the blastula has invaginated into the gastrula. And the neurula similarly depends on the gastrula, which precedes it. Development is thus a complex interweaving of successive effects of gene action, cell proliferation, cellular movement, and differentiation of cellular organs and organ systems.

To think of an organism as the simple sum of its gene products is hopelessly naive. Closer to reality is the epigenetic view of subtly modulated interactions of genes and their products that achieve the balanced effect that culminates in a functional living system. The genotype of an organism is really an epigenotype, a storage of information that comes to expres-

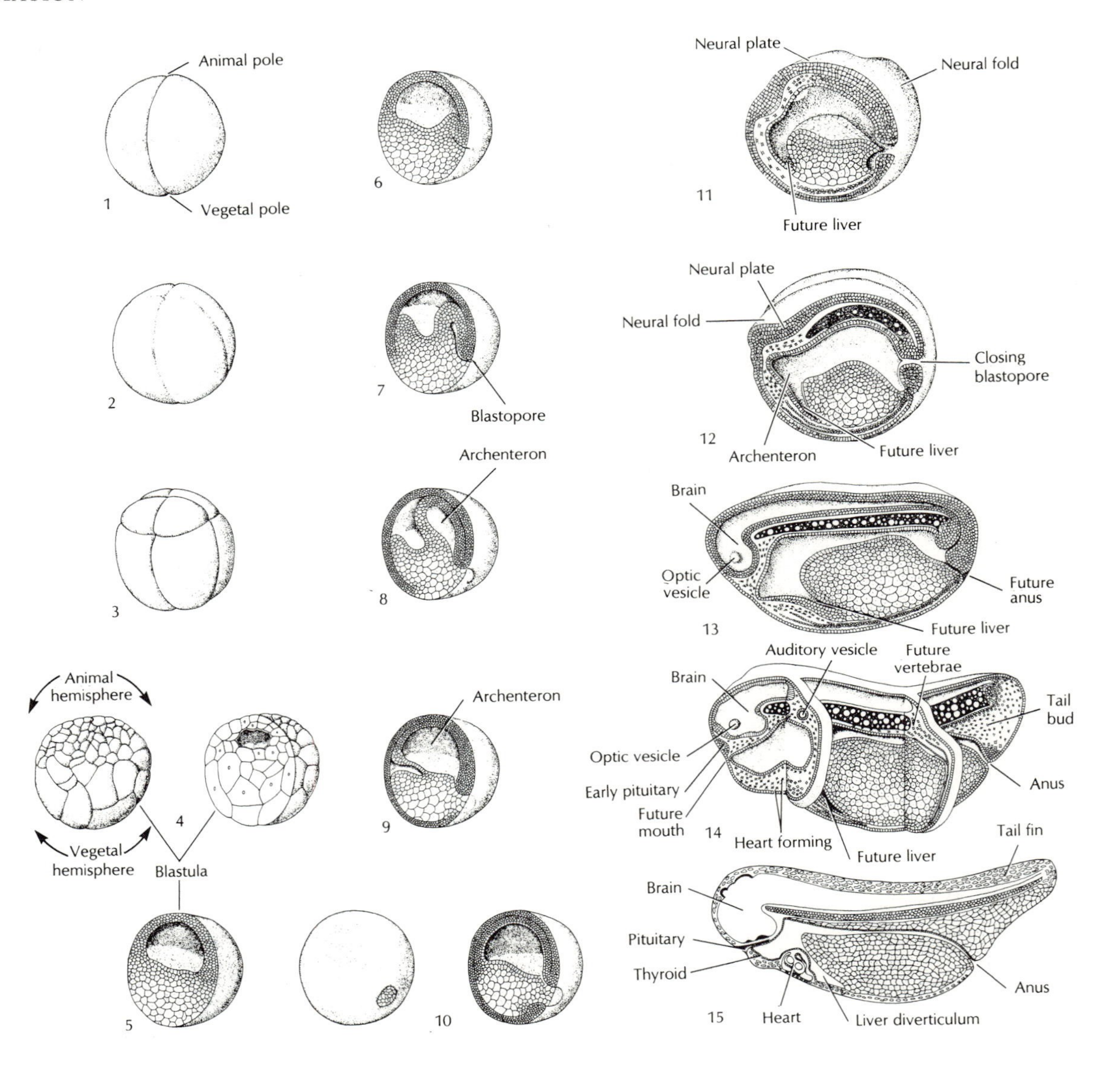

Fig. 7-7: Early stages in frog development. Stages 1-3: Early cleavage. Stages 4-6: Blastula. Stages 7-10: Gastrula. Stages 11-12: Neurala. Stages 13-15: Further development of body layers and differentiation of major organ systems.

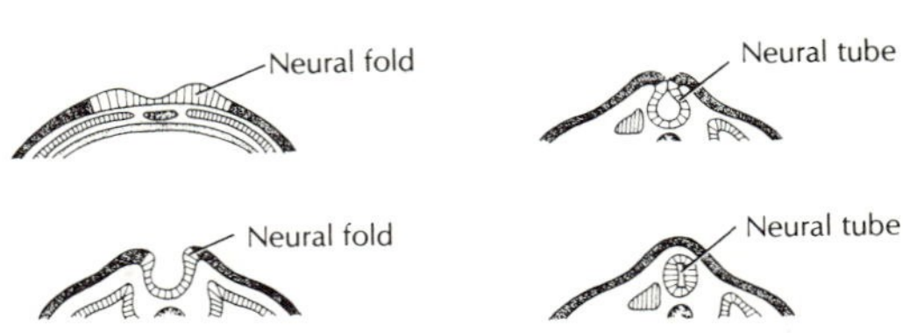

(Above) Four successive cross-sectional views of the development of the neural tube between stages 11–13

sion through the progressive unfolding and balanced interaction of its many products.

Now genic adaptation takes on meaning. A change in one gene can affect the whole phenotype through its epigenotypic effects. Hence, from the evolutionary point of view, the epigenotype of a gene pool represent a collection of genes that act harmoniously to meet adaptive norms. And that leads us to the genetic cohesion of the gene pool that Mayr discusses so eloquently:

> The phenotype is the product of the harmonious interaction of all genes. The genotype is a "physiological team" in which a gene can make a maximum contribution to fitness by elaborating its chemical "gene product" in the needed quantity and at the time when it is needed in development. There is extensive interaction not only among the alleles of a locus, but also between loci. The main locale of these epistatic interactions is the developmental pathway. Natural selection will tend to bring together those genes that constitute a balanced system. The process by which genes are accumulated in the gene pool that collaborate harmoniously is called "integration" or "coadaptation." The result of this selection has been referred to as "internal balance." Each gene will favor the selection of that genetic background on which it can make its maximum contribution to fitness. The fitness of a gene thus depends on and is controlled by the totality of its genetic background. [From *Animal Species and Evolution,* Balknap Press of Harvard University Press (1963) p. 295.]

From this perspective we can see how genotypes interact to achieve phenotypes appropriate to the adaptive norm. We also see why the cohesiveness of the old gene pool must be dissolved to achieve a new gene pool. If gene flow persists, the new pool can never respond independently to its own adaptive norm. Some kind of isolation is necessary. Escape from the parent gene pool is a prerequisite to the formation of a new reproductive community. And the resulting genetic changes establish new epigenotypes, which show their own coadaptations, integrations, and internal balance.

We now have an idea of why hybrids between two highly integrated, coadapted gene pools are often at a selective disadvantage; why embryos may not develop normally; why gametes fail to meet; and why individual organisms fail to mate. At best, conceptually, we begin to see how reproductive isolation emerges. Epigenotypes from two different gene pools, since they are not on the same developmental team, cannot collaborate harmoniously to produce progeny.

The genetics of reproductive isolation. Another way to state the above is that we can observe the phenotypic effects of reproductive isolation. They include everything from maladaptive hybrids to the absence of any attempt to mate. This, again, is a spectrum of effects. Presumably there are fewer gene differences between the epigenotypes of two parents that can produce viable hybrid progeny than between those whose zygotes develop only to the point of a malformed embryo. And in the latter, the parental epigenotypes are less different than those of two individuals who make only an abortive attempt to mate. (Exceptions to these generalizations, especially the latter one, may well exist. It may be that a few key behavioral differences will suppress effective mating that otherwise might have achieved viable hybrids.) However, differences between epigenotypes can become greater as the respective gene pools continue to diverge.

Experimentally, the point of attack on the emergence of the genetic basis for reproductive isolation is the point where the progeny first show evidence of a selective disadvantage. And that point, as we have said earlier, is hybrid disadvantage. That individuals of two different epigenotypes can reproduce at all is a sign that their genes have the potential for significant coadaptation or integration. If the progeny they produce showed a fitness equal to or better than those from matings within each gene pool, then there is no evidence for significant divergence. But if the hybrids show any kind of decreased fitness, then selection will work against the matings that produce them. Other matings—those within rather than between gene pools—will produce more fit progeny and, therefore, will be favored by selection.

Have there been any experiments that select against hybrids and, if so, what were the results? Not unexpectedly, a good deal of work has been done. We can cite one well-known study here. In 1962, Thoday and Gibson, working with *Drosophila melanogaster,* found that by selecting against flies with an intermediate number of body bristles they could produce two discrete populations, one with a high and the other with a low number of bristles. This occurred after 12 generations of selection (Box 7–1). This disruptive selection produced a bimodal population, which, when members from each mode were interbred, showed clear evidence that genes for high and low numbers of bristles had been selected for. [In some ways this work is like that of Jennings (Fig. 4–6), who selected for long and short spines on the ameba *Difflugia corona.* However, Jennings worked with asexual forms, which may well account for the greater period of selection needed before stable phenotypes were obtained.] Thoday and Gibson were careful to point out, however, that their results did not demonstrate that this was the basis for speciation. They and others are reluctant to see such simple selection as being operative in nature. On the other hand, they are impressed with the rapidity of the effects of disruptive selection in the same way that J. M. Smith is and with its apparent confirmation in the work of the Taubers on lacewings.

Box 7–1. The effect of selection on the number of body bristles in *Drosophila melanogaster.*

The light histograms show the distribution of bristle number in the progeny of females selected for low number of bristles, the dark histograms the progeny of females selected for high bristle number. The experiment started with four non-virgin females captured in the wild (a garbage can on the outskirts of Cambridge, England). Their progeny were combined. Two generations later, twenty virgin females and twenty males were examined for bristle number. From each sex, eight flies with the highest and eight with the lowest number of bristles were selected and all thirty-two flies were allowed to mate at random for 24 hours. The males were discarded and the females were again separated into high and low categories of eight flies each. Then each of these was separated again at random into groups of four. There were now four cultures and the progeny from these were collected after mating had occurred and again eight high females and eight low females were selected and subdivided into groups of four each. This was repeated for twelve generations giving the results seen in the figure. The results in the tenth generation arose where "circumstances were unusual," according to the researchers Thoday and Gibson, i.e., there were "electric power cuts which affected the culture room temperature and . . . special arrangements, which involved a 24-hr. delay in mating, were made to accommodate the Christmas holiday." [Thoday and Gibson (1962) *Nature* p. 1165.] The results in the twelfth generation are what are striking. There is a complete separation of the high and low distribution of bristle numbers. There was no sexual isolation between the high and low phenotypes, however; perfectly viable hybrids and hybrid progeny were found.

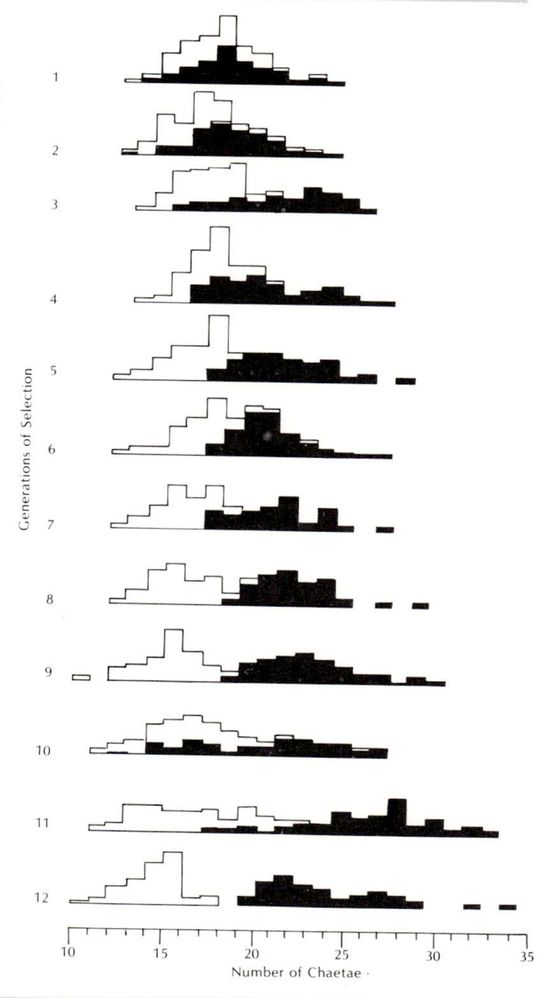

A final comment on this work: Because of its intrinsic importance to gene pool divergence, others have tried to repeat it, without notable success, but Lewontin's conclusions on this type of work are worth noting. He says,

> Attempts to produce sexually isolated populations within a species by selecting against hybrids between them have succeeded, but the isolation that appeared during the course of the selection disappears if the populations are again allowed to interbreed. . . . If there is any element of the theory of speciation [namely, species diversification] that is likely to be generally true, it is that geographical isolation and the severe restriction of genetic exchange between populations is the first, necessary step in speciation. [From *The Genetic Basis of Evolutionary Change*. Columbia University Press (1974) p. 101.]

We can draw two conclusions in the light of the foregoing discussion.

1. Isolation of some sort between populations, and in the vast majority of cases it is geographic, initiates species divergence.
2. The epigenotypes of the two populations then become so different as to be incompatible in terms of producing fit progeny, i.e., they are less fit than those progeny formed within each population.

Or briefly, isolation and hybrid disadvantage are the two first steps in the emergence of gradual species diversification.

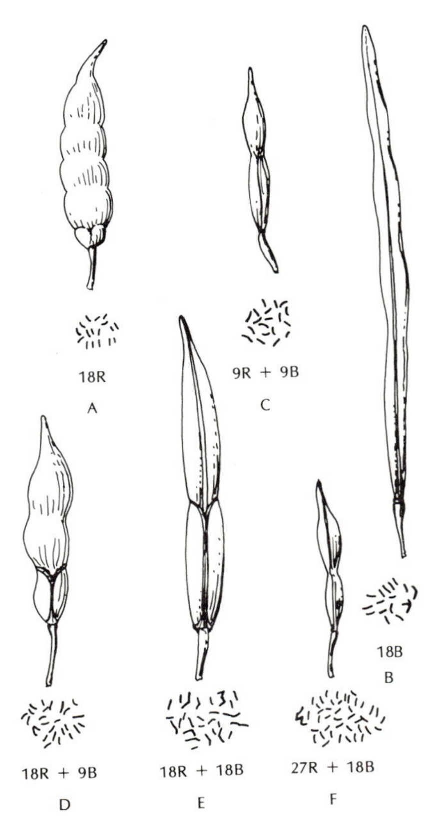

Fig. 7-8: Seed pods and chromosomes from *Raphanus sativa* (A) and *Brassica oleracea* (B). Shown are the diploid hybrid (C), the triploid hybrid (D), the allotetrapoloid *Raphanobrassica* (E), and a pentaploid condition (F).

RAPID DIVERSIFICATION

At one time certain biologists were of the opinion that although the gradual accumulation of gene mutations could generate differences among demes or races or subspecies, a different kind of mutation caused species to differentiate. Such mutations had major effects on the phenotype and were called macromutations, as compared to micromutations in which their effects were limited to differences below the species level. A few such mutations could rapidly change the individuals containing them into members of a species, or so it was thought. Suffice it to say that speciation by macromutation has no evidence to support it. It is only of historical interest now.

But there is another kind of major change in the genetic constitution of individuals that can bring about differences that are important to species differentiation. It is polyploidy, an increase in the whole genome. Evolutionary change by polyploidy is, in fact, well known in plants. We must now look at its role in species diversification.

Let us look at the story of *Raphanobrassica* (Fig. 7–8). In this study, done by the Russian geneticist Karpechenko in 1928, members of two different genera were crossed—the radish (*Raphanus sativa*) with the cabbage (*Brassica oleracea*). Low fertility was the rule in the hybrids except in certain plants. Typically the sterile hybrids have the same chromosome number as the parents, i.e., 18 chromosomes with pollen and ovule carrying the haploid genome of nine chromosomes. However, the gametes of the hybrid progeny that were fertile contained 18 chromosomes. Here, because of an abortive meiotic first division there was no reduction of chromosome number; the gametes therefore were diploid (containing two haploid sets, each from a different species). The general name for this is *allopolyploidy*, defined as the condition in which two or more different genomes from two or more species occur in the same individual.

The importance of this for speciation is that homologous chromosomes would pair. But in this case there was little homology between these two haploid genomes. That was what led, initially, to abnormal meiosis and sterility followed because the gametes could not be formed properly. Allopolypoloidy changed all of that. Now each plant was, in effect, a tetraploid. It had a diploid set each of *Raphanus* and of *Brassica* chromosomes. Meiosis could proceed normally, since each *Raphanus* chromosome could synapse with a genuine homolog as could each *Brassica* chromosome. The allopolyploid hybrids formed functional gametes with 18 chromosomes each.

When two allopolyploids were crossed, fertile progeny were formed whose diploid number was, of course, 36 chromosomes. When these same plants were crossed back to plants like either parent plant, the progeny were sterile because, although they were viable, meiosis in a triploid condition is highly abnormal. There would be two normal homologous chromosomes trying to pair, but disturbed by the attempted pairing of yet a third chromosome. Abnormal meiosis again led to abnormal gametes and sterility. The picture was, then, of the allopolyploid (a tetraploid compared to the parent plants) producing fit progeny with other similar tetraploids but unable to do so with diploid plants (here, the parents). The allotetraploid, *Raphanobrassica*, even had a phenotype different from either *R. sativa* or *B. oleracea*. But most importantly, as the breeding results showed, *Raphanobrassica* was reproductively isolated. One could say that Karpechencko had produced, through allopolyploidy, a new plant species in one generation.

Studies in the intervening years has shown that many species of flowering plants use polyploidy in their speciation. Both allo- and autopolyploidy seem clearly to have occurred. In the latter, chromosome doubling occurs in the diploid genome of one species. Apparently, in the successful cases, having four homologous chromosomes does not seem to inhibit normal synapsis into two sets of paired homologs.

It is worth noting that autopolyploidy can be readily induced by treating developing pollen and ovules with colchicine. This compound destroys the meiotic spindle and so aborts meiosis. The result is a doubling of the chromosome number through chromosomal replication, but because there is no cell division the gametes are polyploids. Plant breeders used this technique widely for large-scale production of polyploid strains in the hope of developing new strains of commercial use. Results have been disappointing. Stebbins, a leading researcher on plant evolution, reports only a few cases in which an autopolyploid seems to have been successful in establishing itself in nature and only one case of commercial success with triploid sugar beets. (The beets are reproduced not by seeds, but asexually by planting parts of the sugar beet root that grows into a new plant. This gets around the problems of a triploid meiosis.)

The role of polyploidy in rapid evolution, and polyploidy is the only known mechanism here, raises further questions: (1) What are the evolutionary advantages of polyploidy? (2) What is the subsequent fate of a species originating by alloploidy? (3) What is the role of polyploidy in animals? We can answer these questions most effectively by reversing their order.

The role of polyploidy is minimal in animals. The reasons, given in Chapter 3, can be briefly restated here. Primarily because sex determination is genotypic in animals, polyploidy is not a workable genetic innovation. It upsets the balance between the sex determining chromosomes and thus greatly reduces reproductive capabilities. Where animal polyploidy occurs it is usually limited to forms reproducing parthenogenetically. One special exception are tetraploid, hexaploid, and even octoploid frogs that reproduce sexually—but they apparently have no sex chromosomes.

The long-term fate of polyploid plants is a topic best deferred to the next chapter in which the establishment, rather than emergence, of new species is discussed. The essence of the problem can be mentioned now, and it seems to come down to finding a balanced interaction of the new genotypes. In particular, the interacting of haploid genomes can have a disrupting effect, whereas the doubling of each genome has a stabilizing or conservative effect.

The evolutionary advantages of polyploidy are indicated by the answer to the preceding question. Wherever heterozygotes are at an adaptive advantage, there the allopolyploid forms will be favored. The proportion of homozygotes to heterozygotes is 1 : 7, hence if the latter are favored there are fewer of the former to eliminate. When we deal with diploids the homozygote to heterozygote ratio is 1 : 1. But another point to be remembered is that, with comparable selection pressures, tetraploids respond one-half as rapidly as diploids. Polyploidy emerges as a mechanism for conserving the effects of genic variation or heterozygosity. This is especially true when allopolyploidy occurs between quite diverse forms.

From polyploidy we can go on to one of its important consequences, *hybrid introgression*. G. Ledyard Stebbins, of the University of California at Davis, says of this phenomenon that "In actively evolving plant species, it is one of the commonest sources of new variability" (*Processes of Organic Evolution* 2nd ed., p. 130). What he is referring to is the sequence of hybridizing, then back-crossing to a plant of one of the parental species, and stabilizing the backcross progeny through selection. If the hybrid is an allopolyploid the backcross will provide a complete set of chromosomes of one species plus chromosomal segments of the other. The latter are the source of new genetic materials and new genic variations. The origin of corn, *Zea mays*, includes evidence of introgressive hybridization in its history. In another, special kind of introgression, a species *AA* hybridizes to form tetraploids with two other species, i.e., *AABB* and *AACC*. These latter, Stebbins points out, can often hybridize and exchange genes because the *A* genomes in each species act to buffer the different *B* and *C* genomes and allow them to interact more effectively. These effects of introgression of genes from different species, aided by polyploidy, provide a quite important route in plants for producing some otherwise impossible combinations of epigenotypes.

Overview of species emergence

The emergence of species is summarized by Fig. 7–9.

Once more it is useful to start with Darwin. Such an approach provides a perspective on how our knowledge has grown since his day. Also, it allows us to see most clearly the limitations that were mentioned earlier regarding Darwin's remarks on "hybridism" as a test of his theory of natural selection (Chapter 2). It may be recalled that it was just this point

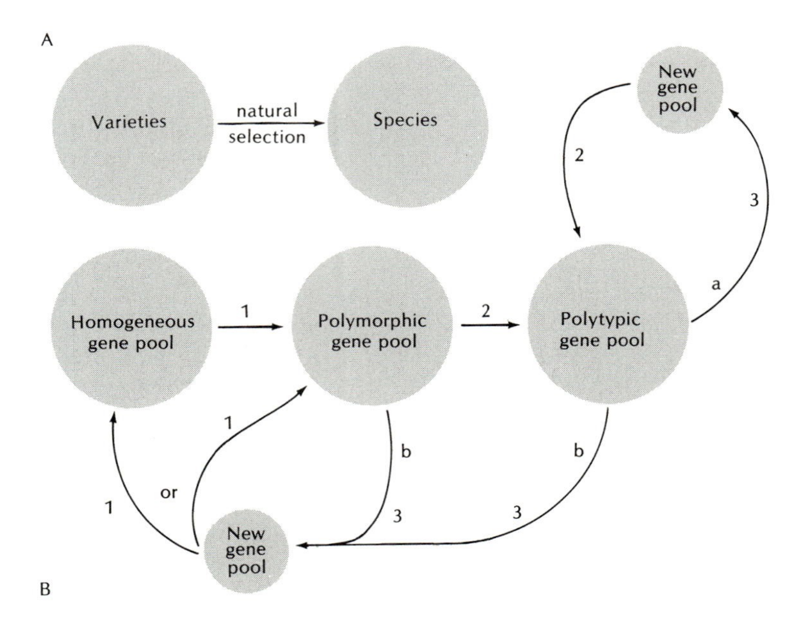

Fig. 7-9: A summary of species diversification.

that William North Rice attacked most vigorously as evidence against, rather than for, "Mr. Darwin's theory." Part A of Fig. 7–9 encapsules Darwin's thinking as represented in his view that varieties are incipient species; part B is a critique of that view as well as a statement of where we are today when considering the origin of species in the light of species diversification.

Quite simply, Darwin was aware, despite his Linnaean and, therefore, typological view of species, that significant variation was a normal occurrence within species. And such variation could even be expressed as the presence of varieties or races within a species, with this latter giving rise to new species. Because Darwin had no conception of species as reproductive communities, he saw no difference in the way selection acted within a phenotypically homogeneous species, between varieties, or between species. It was all selection for beneficial, heritable variations. Beneficial, it is understood, meant gaining in the struggle for existence by achieving better adapted organisms. To Darwin, the problem of hybrid vigor and hybrid sterility was puzzling and deeply intriguing, but it did not really help in the question of the origin of species. In Darwin's view, when species interbred, their hybrids showed everything from hybrid vigor to hybrid sterility. He interpreted that range of effects as indicative of progressive divergence among species.

From our perspective today, a key point is that what we call one polytypic species Darwin would have called many different species. We call it one species because there is gene flow among the different subpopulations. Darwin would have called these subpopulations different species because they are phenotypically distinguishable. Hence, he did indeed find hybrids ranging from vigorous to sterile when crosses were made between them. Our concept of reproductive isolation totally changes that picture.

Looking at part B of Fig. 7–9, we can start by observing the horizontal listing of different kinds of gene pools (reproductive communities or species). The listing reflects the phenotypic variation that can be observed. A homogeneous gene pool shows very little. A polymorphic pool shows at least the variation that comes from balanced polymorphism, namely, variation as a result of selection for different alleles. This may involve polymorphism at various loci; it is not necessarily restricted to one locus. The homogeneous and polymorphic gene pools are characterized by one adaptive norm; the polytypic gene pool by more than one adaptive norm (see Fig. 6–9). And, finally, there is the newly emergent gene pool, reproductively isolated from other ones.

Next, we need to examine the selective forces that act to change one type of pool into another one. These are indicated in Fig. 7–9B by encircled numbers. Selection indicated by the number 1 is selection within a gene pool that is usually either stabilizing or directional selection. In general, it acts to move populations to the top of their adaptive peaks. Selection represented by the number 2 is disruptive selection. Here there is an advantage in having more than one adaptive peak between which there is gene flow. In all three types of gene pools, selection acts to remove maladaptive genes as populations shift to or near their respective adaptive peaks. The third type of selection (number 3) concerns hybrid disadvantage and gives rise to a new gene pool. Here the action of selection differs from the preceding cases in terms of its effect. (It still proceeds, of course, in the Darwinian terms of preserving the more fit individuals.) The effect is to generate reproductively isolated gene pools rather than to preserve gene flow within one pool. It most notably removes whole genotypes, i.e., those of hybrids that are less fit than non-hybrids, rather than focusing on maladaptive genes. In these two regards the result of selection is quite different when a new gene pool emerges than when an already existing pool is preserved. Finally, we need only mention a fourth aspect of selection, the establishment of a new gene pool. It will be examined fully in Chapter 8. Its results are reinforcement of isolation and further refinement of adaptive norms. There is minimal selection against hybrids now simply because they will be rare. This is, again, stabilizing or directional selection.

Last, let us look at the speciation process as shown in Fig. 7–9. The letter [a] refers to the breakage of a chain of subpopulations in a polytypic species. The new gene pools may retain significant polytypic variation as a result. (See arrow connecting the *new gene pool* with the *polytypic gene pool* and labeled [a].) The letter [b] refers to colony isolation, either through cutting off or emigration of an intermediate-sized population or of a small founder population. These can arise from any of the gene pools we have been discussing. The new gene pool so formed can, in its turn, be homogeneous or polymorphic, depending largely on its size. It will have to evolve further in establishing itself.

The emergence of a new species by diversification leads us to complications not understood by Darwin. But the very act of conceptualization is what leads us to the observational work in field and laboratory that will test our concepts. No area of evolutionary study is a more exciting mixture of theoretical and experimental work than this one.

References

Alexander, R. D., and R. S. Bigelow, 1960. Allochronic speciation in field crickets, and a new species *Acheta veletis*. *Evolution 14:*333–346.

Bogart, J. P., and A. O. Wasserman, 1972. Diploid-polyploid cryptic species pairs: a possible clue to evolution by polyploidization in anuran amphibians. *Cytogenetics 11:*7–24.

Bush, G. R., 1974. The mechanism of sympatric host race formation in the true fruit flies (Tephritidae). In *Genetic Mechanisms of Speciation in Insects,* M.J.D. White (ed.), pp. 3–23. Reidel Publishing Co., Boston.

Dobzhansky, T., and O. Pavlovsky, 1957. An experimental study of interaction between genetic drift and natural selection. *Evolution 11:*311–319.

Grant, V. 1971. *Plant Speciation*. Columbia University Press, New York.

Larson, A., and R. Highton, 1978. Geographic protein variation and divergence in the salamanders of the *Plethodon welleri* group (Amphibia, plethodontidae). *Systematic Zoology* 27*:*431–448.

Lewontin, R. C., 1974. *The Genetic Basis of Evolutionary Change*. Columbia University Press, New York.

Mayr, E., 1963. *Animal Species and Evolution*. Belknap Press of Harvard University Press, Cambridge, Mass.

Mayr, E., 1970. *Populations, Species and Evolution*. Belknap Press of Harvard University Press, Cambridge, Mass.

Naef, A., 1919. *Idealistiche Morphologie und Phylogenetik*. Fisher Verlag, Jena.

Smith, J. M., 1966. Sympatric speciation. *American Naturalist 100:*637–650.

Stebbins, G. L., 1971. *Chromosomal Evolution in Higher Plants*. E. Arnold, London.

Stebbins, G. L., 1977. *Processes of Organic Evolution*. 3rd ed. Prentice-Hall, Englewood Cliffs, N.J.

Stebbins, R. C., 1949. Speciation in salamanders of the plethodontid genus *Ensatina*. *University of California Publications in Zoology 48:*377–525.

Tauber, C. A., and M. J. Tauber, 1977. Sympatric speciation based on allelic changes at three loci: Evidence from natural populations in two habitats. *Science 197:*1298–1299.

Thoday, J. M., and J. B. Gibson, 1962. Isolation by disruptive selection. *Nature 193:*1164–1166.

Thoday, J. M., and J. B. Gibson, 1970. The probability of isolation by disruptive selection. *American Naturalist 104:*219–230.

White, M.J.D., 1977. *Modes of speciation*. W. H. Freeman, San Francisco.

EIGHT

The Establishment of New Species

THE POINT OF NO RETURN in species diversification is apparent when hybrids between two populations are less fit than other progeny, for selection then continues to separate the two populations.

Various biological mechanisms aid in establishing this growing divergence. Collectively, they are called isolating mechanisms because they contribute to the establishment of reproductive isolation. These mechanisms have been studied intensively in an effort to clarify our understanding of the origin of species and, in particular, species divergence.

At this point, it is worth making explicit a perspective on isolating mechanisms. Biologists are familiar with the observational data obtained from dividing cells, such as onion root tips, which establish the stages of mitosis. We realize, as we look at that material, that what we see are different stages in one continuous process. Even as we identify a telophase cell here, a prophase there, and then another, earlier prophase, and then anaphase and metaphase stages, and so on, we realize that these are all examples of one unbroken series of events. It is that flow of events that is most important, since it informs us about chromosomal behavior during cell division. Similarly, speciation is a continuous series of events. Breaking it up into emergence and subsequently establishment of species differences is arbitrary, although it is useful for analytic purposes. And breaking up establishment of species differences into different kinds of isolating mechanisms is that same arbitrary process taken further. But it is necessary in our larger effort of trying to understand the origin of species because species divergence in most cases is a slow process. We cannot expect to watch it occur in one lifetime or even one thousand lifetimes. Our only hope of grasping it observationally is to find the process in progress at its different stages. Putting these stages together, we can sense the whole process, just as seeing cells at different stages of nuclear division gives us a good sense of mitosis as a whole.

Isolating mechanisms

Let us try to reconstruct the steps that might occur as diversification becomes established. In Table 8–1 are listed the various mechanisms that are recognized today as operative in promoting reproductive isolation. Actually two lists are given in the table. On the left two major categories of prezygotic and postzygotic mechanisms are listed, along with their subcategories. On the right is the list we will follow here. Note the subcategories in both lists, those in one list are essentially the reverse of those in the other because postzygotic or hybrid

Table 8–1. Isolating mechanisms.
The list on the right is a sequence going from situations with less to situations with more genetic difference between individuals considered as parents or potential parents. Recall, however,, the discussion in Chapter 7, in which disruptive selection could apparently result in seasonal or temporal isolation, but with fully viable and fertile progeny. That is an exception to the sequence being suggested here. The list on the left is another, more conventional breakdown of the same isolating mechanisms. It is taken from Dobzhansky (1970).

A. Premating or Prezygotic Isolation	A. Hybrid Disadvantage
1. Ecological or habitat isolation	1. Hybrid breakdown
2. Seasonal or temporal isolation	2. Hybrid sterility
3. Sexual or ethological isolation	3. Hybrid nonviability
4. Mechanical isolation	B. Gametic Isolation
5. Gametic isolation	4. Physiological or chemical isolation
B. Postmating or Zygotic Isolation	5. Mechanical isolation
6. Hybrid inviability	C. Mate Isolation
7. Hybrid sterility	6. Ethological or behavioral barriers
8. Hybrid breakdown	7. Seasonal or temporal barriers
	8. Ecological barriers

disadvantage mechanisms are usually the first to appear in gradual species diversification. (The exception is isolation by disruptive selection discussed in Chapter 7.) Hence, we will start our discussion there and progress down the right-hand list. This progression is consistent with our effort to present findings in a way that allows reconstruction of the process of species divergence. These then will be the steps to be placed in sequence as we try to establish, observationally, what is actually happening. In Chapter 7 we established, at least theoretically, the central importance of selection against hybrids. In more detail now, just what does that entail?

SELECTION AGAINST HYBRIDS

There are three subcategories here: hybrid breakdown, hybrid sterility, and hybrid nonviability.

Hybrid breakdown. The term hybrid breakdown refers to hybrids who can mate, but whose progeny are weak and sickly or whose numbers are much reduced.

Examples of this are found in the results of crossing certain populations of fruit flies that phenotypically look like *Drosophila pseudoobscura*. In one particularly informative study, done by Prakash and reported in 1972, it was found that flies collected from the area of Bogota, Colombia, in South America, showed the very beginnings of reproductive isolation. When male flies from Bogota are crossed with females taken from populations of *D. pseudoobscura* collected from other localities, the progeny are fit in terms of vigor and fertility. But that does not hold when the reciprocal cross is made between females from Bogota and males from other populations. In such crosses, though the females are normal, the males are all sterile. This represents a case of what is now known as Haldane's law, discussed first in Chapter 3. This law states that weakness or sterility is more to be expected in the heterogametic (XY, XO, ZW, or ZO) sex than in the homogametic one (XX or ZZ). The law can be explained as follows: Let us consider the XX and XY situation. The heterogametic sex is called hemizygous for the reason that genes on the X chromosome do not have alleles on the Y chromosome, and vice versa. Hence, recessives can come to expression under those conditions, whereas in XX females they coexist with an allele that may be dominant. When flies from different populations are mated it is not unlikely that disharmonious genes will be present; in the male they can be expressed, whereas in the female they might well be suppressed. For these reasons Haldane argued that decreased viability and sterility will be first expressed in the heterogametic sex.

Returning to Prakash's study, it is not entirely clear why male progeny are sterile in crosses only with Bogota females, but not in the reciprocal crosses. Nevertheless, his results are as reported; male sterility occurs, with a clear-cut breakdown in the ability of the hybrids to produce male progeny.

The same phenomenon is seen in the progeny of crosses between different populations of *Drosophila paulistorum*. The gene flow is so limited between these populations that they can be designated as a semispecies. That places them somewhere between good species and races. Crosses between them result in viable hybrids, both male and female. The females are fertile; the males are sterile. In the laboratory, members of one semispecies show a real preference for mating with flies of that same semispecies. There are behavioral characters during courtship that strengthen the preferences. Such behavior is a kind of isolating mechanism, which will be discussed below.

A last example of hybrid breakdown is that reported from Sonneborn's studies on the 16 sibling species of what were formerly called one species, *Paramecium aurelia*. In some of the few crosses that are possible between such species, the hybrids are vigorous. But when they mate or undergo autogamy to produce another sexual generation, that second generation rarely produces viable progeny.

Hybrid sterility. In this case, no progeny are produced by the hybrids. They are totally sterile. The familiar example of the mule is a case in point. A mule is produced from a mating between a horse and a donkey. Almost without exception, mules do not reproduce. The reason is chromosomal; the donkey and horse chromosomes are so different that they do not pair or synapse properly in meiosis and that upsets meiotic cell divisions, which, in turn, result in abnormal gametes and sterility. For all practical purposes mules are sterile, since matings between mules or between a mule and a horse or a mule and a donkey are not successful except for a few very rare exceptions. The vigor of a good mule is legendary and is a fine example of hybrid vigor.

Cases such as this remind us that reproductive isolation as a criterion for a species does not always mean absence of hybrids. Reproductive isolation really means no gene flow.

Hybrid nonviability. This refers to instances where hybrids are produced, but their fitness is reduced. Many cases of this sort are known. The cause of this is commonly developmental, namely two different haploid epigenotypes do not cooperate harmoniously in producing a new organism. Such disharmony could result not just in developmental sterility, but in weaknesses of various parts of the organism. Quite obviously the overall result would be reduced fitness.

An excellent example of this is seen in what were originally recognized as races or subspecies of the frog *Rana pipiens*. Such authorities as Dobzhansky now feel that this complex of populations is better understood as various sibling species. Much of the original work was done by Moore, who showed a certain pattern of latitudinal or north-south differences in the ability to produce variable hybrids. If gametes from frogs collected in Vermont fuse with gametes from Florida frogs, development of the zygote is very abnormal. The nature of the abnormalities differ, depending on the donor of the egg and the donor of the sperm. Apparently the egg cytoplasm plays a critical role in supplying a background in which the epigenotypes express themselves.

If the eggs and sperm are taken from frogs less distantly separated, in a north-south direction, development is more normal. And when gametes come from frogs as close together as New Jersey and Vermont, the offspring are only slightly abnormal or are normal. The reason for emphasizing a north-south distribution is that east-west differences are minor, even if the distances are great. Gametes from Vermont and Wisconsin frogs produce normal frogs.

The north-south differences correlate with temperature adaptations. In the north, the breeding season is May to June; in Florida, breeding occurs the year round. Also, Moore found that the temperatures at which development is normal are 5 to 28 °C in the north and 11 to 35 °C in the south. The probable conclusion is that the genetically controlled developmental adaptations to different latitudes make the formation of fit frogs impossible.

GAMETIC ISOLATION

We can now turn to another group of isolating mechanisms wherein mating occurs, but the gametes never unite. This gametic isolation (absence of fertilization) can be divided into two important subcategories: *physiological* or *chemical isolation* and *mechanical isolation*.

Physiological or chemical isolation. Fusion of gametes culminates a complex series of events. It includes both the formation of functional gametes and bringing them together under conditions favorable to fertilization.

In animals, fertilization can be either internal or external, as it can in plants. Many aberrancies can occur in these processes to abort fertilization. Let us look at a few of them.

Some of the best known examples are found, as might be expected, in the genus *Drosophila*. Normally, in fruit flies, mating or copulation results in placement of sperm in the female genital tract. The sperm are contained in the seminal receptacles. As eggs mature and pass down the genital tract, they pick up sperm as they pass the receptacles. Fertilization then occurs and, eventually, the zygote passes out of the female's body and proceeds to develop into a larva and then a pupa, and finally it hatches into a fly. In cases of hybrid nonviability, development of the zygote is affected. In physiological isolation, fertilization never occurs because the sperm become nonfunctional. The cause can differ in different crosses. In some cases the sperm become solidified into a packet in the seminal receptacles and never get out again. In other cases, the sperm are rendered immobile in the receptacles. And in yet other cases, the sperm are expelled from the genital tract of the female before fertilization can occur.

Somewhat parallel cases are known in plants in which normal function of the pollen grain is inhibited (Fig. 4–11). In cases of normal pollination in a flowering plant, the pollen grain alights on the stigma and proceeds to put out a pollen tube that grows down the style to an ovule. The pollen initially contains one tube nucleus and two male nuclei. All of these move down the pollen tube with the tube nucleus in the lead. The pollen tube enters the ovule and then the embryo sac, which contains the egg and other structures. Fertilization occurs when one of the male nuclei and the egg nucleus unite. Each fertilized ovule gives rise to a seed that is, in effect, a plant embryo with a coating to protect it until it germinates and develops into a young plant.

The process can, of course, be affected anywhere; causes may vary, from the lack of adherence of the pollen onto the stigma to poor development in the young plant. In the case of physiological isolation, we are concerned with events involving the growth of the pollen tube and its penetration to the egg in the embryo sac.

As an example, it is worth looking at some work done on the cobwebby gilias (members of the phlox family). This work is similar to that done on *Rana* in that examples of several cat-

Table 8–2. Results of crosses in the cobwebby gilias (*Gilia* populations, races, and species).

Cross	Percentage of successful crosses	Seeds per flower	Hybrids per ten flowers
within a population	100	17.8	22
between races	73	15.2	12
between species of a section	43	3.7	3
between species of different sections	2	0.004	0.038

From Grant (1971).

egories of isolating mechanisms are in operation, besides the one of particular interest. Specifically, the data in Table 8–2 show us that physiological isolation occurs, but also that there is some selection against the hybrids. Recall what we have just said about the seed containing an embryo. Hence, seed formation is one measure of successful fertilization and early development. And seed germination is a further stage in development. The work cited was done in 1963 by Verne Grant, a leader in plant speciation. He was studying the occurrence of isolating mechanisms among different populations of *Gilia*. He attempted 174 crosses and his results are the ones we are looking at (Table 8–2). As we proceed down the left-hand column of the table, we are following the percentage of successful crosses, using as an arbitrary standard of 100% success the values taken from crosses between members of one population. Lack of a successful cross means that fertilization did not occur or that there was gametic isolation. The middle and right-hand column express hybrid disadvantage as seen in seeds per flower and seeds germinating per every ten flowers.

In each column, the values decrease as one goes down the columns, which indicates that geographically more distant populations are more isolated from one another in terms of decreasing ability to form hybrid progeny.

The basis of this physiological isolation is not clear here, but it could involve any one of the following: The pollen simply might not germinate on the stigma. Or it might germinate, but the pollen tube might not grow. Or, even, the ovule might be reached, but the embryo sac might not be penetrated. The most divergent populations would have the least success, e.g., no germination of pollen; and the less divergent ones would come closer to achieving fertilization, e.g., no penetration of the embryo sac.

Looking next at external fertilization in animals, we can mention a number of examples of chemical isolation. Here, of course, we are dealing with gametes released by their parents. The classic studies in this regard go back to the research of Lillie, in 1921, on sea urchins, in which both eggs and sperm are released in the water. Work since his time has elucidated some of the chemical mechanisms underlying species-specific attractions of egg and sperm.

It is now known that the egg of the sea urchin (*Stronglylocentrus purpuratus* has been studied most intensively in this regard) releases a glycoprotein called fertilisin. In water this acts as a chemical attractant for the sperm. The sperm, in turn, release anti-fertilisin. Its nature is less clear. The two together play an important role in the physiology of fertilization. For our purposes the reciprocal and apparently rather specific interaction of fertilisin/anti-fertilisin is of interest. If that interaction is disrupted, subsequent union of gametes could also be disrupted.

In his original studies, Lillie showed that species-specific (homogamic) mixtures of gametes resulted in more frequent fertilizations than did interspecies (heterogamic) mixtures (i.e., gametes from different species). The varying ability of specific attractants to function heterogamically could account for these results, and this could, therefore, provide a chemical explanation for gametic isolation.

Comparable systems for homogamic external fertilization in fishes and amphibians are also known. A little thought on this subject will show why such systems would be of real selective advantage. Sperm and eggs are released into an environment that already contains an abundance of other material. And even though the gametes may be present in large numbers, they can be diluted rapidly, especially the motile sperm, or destroyed by various agents. Anything that promotes their union, and especially the union of gametes with complementary and harmonious epigenotypes, will have a selective advantage.

Mechanical isolation. Mechanical isolation is gametic isolation occurring as a result of physical hindrances to fertilization. Some convincing examples of mechanical isolation come from mechanisms that inhibit cross-pollination between flowers of different plants.

In Fig. 8–1, there is an example from the primrose *Primula officinalis.* Here the arrangement of floral parts aids in cross-pollination between different plants. In effect, this is a means of assuring cross-pollination. These flowers represent two phenotypes. In the one on the left, called *pin,* the anthers are

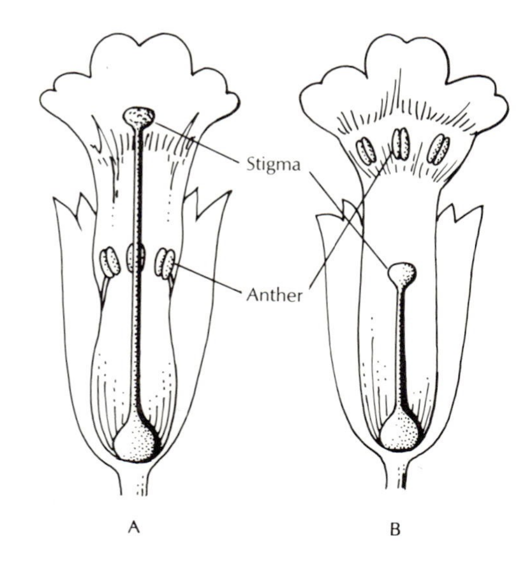

Fig. 8-1: Flowers of the primrose (*Primula officinalis*) showing the structural differences between the pin (A) and the thrum (B) phenotypes. Note especially the position of the anthers and the stigma.

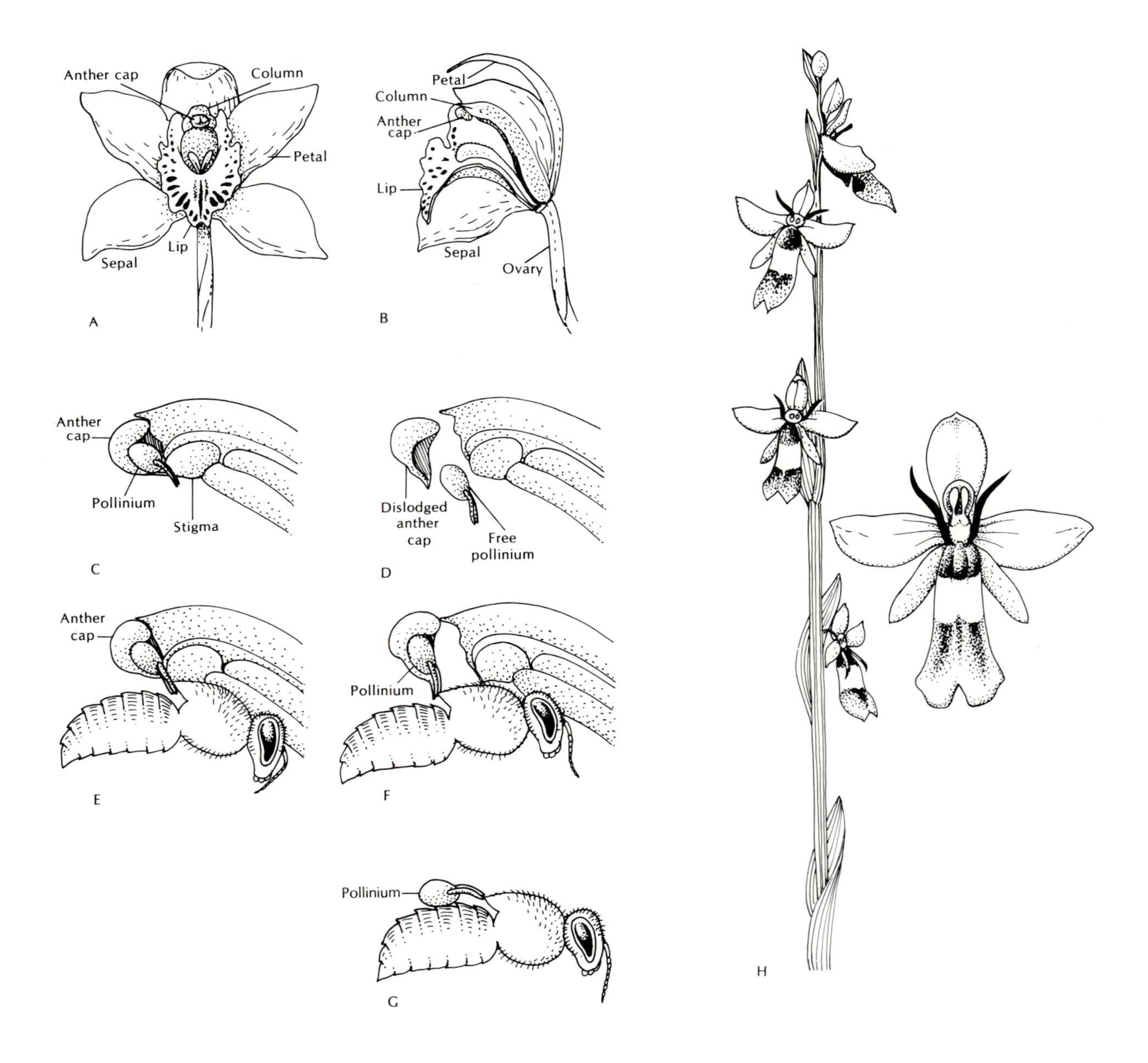
Anther cap
Column
Petal
Lip
Sepal
A
Petal
Column
Anther cap
Lip
Sepal
Ovary
B
Anther cap
Pollinium
Stigma
C
Dislodged anther cap
Free pollinium
D
Anther cap
E
Pollinium
F
Pollinium
G
H

Box 8–1. Pollination in orchids.

Probably nowhere is the precise interaction of animal pollinators and the flowers they pollinate more precisely integrated than among certain insects and the orchids. Here we give only two of many such examples. In the first, the insect is attracted to the flower primarily by its scent, as far as we can tell. A and B show the major features of the flower of an orchid belonging to the genus *Cattleya*. The showy, colorful petals are visible to insects, but odor seems to be of special importance in attracting certain insects of the size and vigor to crawl into the flower over the petal (the lip). That part of the flower called the column is a complex structure containing a special anther with pollen and the stigma (C, D). In E, F, and G an insect, here a bee, enters the flower and forces its way below the column, apparently in search of the source of the odor. When the bee backs out, as it eventually must, it picks up a pollen packet (pollinium) on its back. When it enters the next flower the pollinium becomes lodged in the stigma of that flower, thus leading to cross-fertilization [redrawn from Stephens and North (1974) *Biology* Wiley].

The second example of pollination comes from the small, north European field orchids of the genus *Ophrys*. These orchids are remarkable for the way in which the flower mimics the female of a given species of insect. Attempts by males to copulate with the flowers of different plants (H) lead to transportation of the pollen from one flower to another.

placed below the stigma. On the right is the *thrum* phenotype with anthers above the stigma. There is a single gene difference between these two phenotypes; pin is a recessive homozygote and thrum is a heterozygote. The insect entering a pin flower in pursuit of nectar will pick up pollen on the midportion of its body. If it then goes to a thrum flower, the adhering pollen is in a position to be deposited on the stigma of that flower. At the same time it will get thrum pollen on its posterior end. And that, when a pin flower is visited, is important in placing pollen on the pin stigma.

So far these observations concern the promotion, not the prevention of pollination. But the barriers to pollination can be clearly seen in terms of the insects attracted to a flower. In many cases, the flowers are attractive to particular species of insects. This is nowhere more apparent than in the orchids and their pollinators (Box 8–1). Here the orchid flowers mimic so perfectly a female insect that the males of the species try repeatedly to copulate with the flower. In the process there is a transfer of pollen from one flower to another. It has been found that this relation between the flower and the insect is species specific for both plant and animal. The physical appearance (mechanical structure) of the plant attracts only certain male insects and not others. The attraction is achieved not

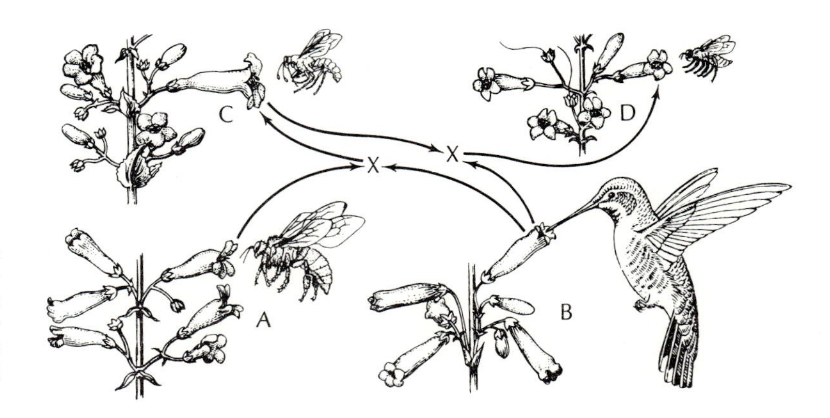

Fig. 8-2: Pollinating mechanisms in different species of the genus *Penstemon* (beard-tongues) of California: mountain penstemon (*P. grinneli*) (A); scarlet bugler (*P. centranthifolius*) (B); showy penstemon (*P. spectabilis*) (C); and a scarlet bugler hybrid (*P. clevelandii*) (D).

only by the shape of the plant, but by its colors and its odor. (The latter are, of course, more chemical than mechanical.)

This adaptation to specific pollinating agents, or vectors as they are also called, is also seen in the genus *Penstemon* or the beard tongues. Here the mountain penstemon (*P. grinelli*) is attractive to carpenter bees (*Xylocopa*) (Fig. 8–2A). The white flower with blue or pink on the wide lower lip of the corolla (fused petals) is especially adapted to this vector. This species of plant occurs in pine forests. The scarlet bugler (*P. centranthifolius*) has a bright red flower that is attractive to hummingbirds (Fig. 8–2B). Its flower, with a tubular corolla, is shaped to accommodate the slender beak and long tongue of these pollinators. These plants occur on drier slopes than the

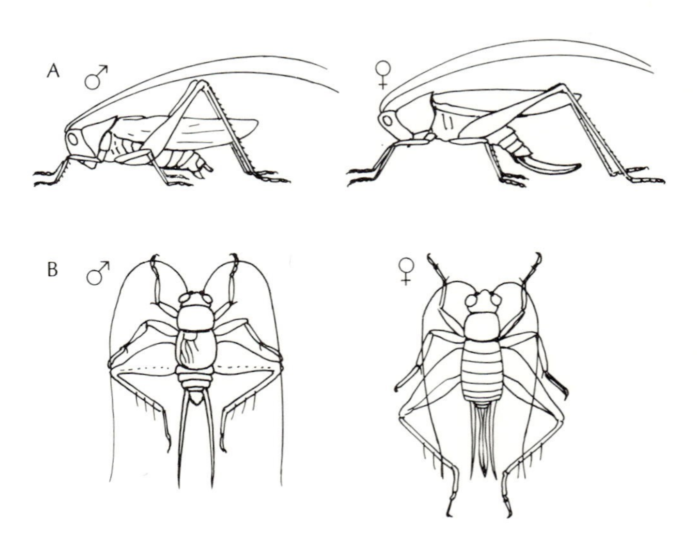

Fig. 8-3: External genitalia in certain grasshoppers: male and female of *Orchelimum vulgare* (A) and male and female of *Hydronemobius alleni* (B).

mountain penstemon. Apparently through hybrids that could adapt to brush-covered hills and were pollinated by wasps, a third species called the showy penstemon (*P. spectabilis*) (Fig. 8–2C), with bright blue or purple flowers, evolved. Further hybridization with the scarlet bugler produced another hybrid (*P. clevelandii*) (Fig. 8–2D), which diverged and adapted to pollination by solitary bees or occasionally by hummingbirds. Here there is clear evidence of the coordinated evolution of flower parts and the pollinating animals. This is an example of *coevolution*, a phenomenon, found widely in nature, in which species interact in cooperative rather than competitive ways.

In animals it was previously thought that the spines on the terminal segments of insect bodies (Fig. 8–3) functioned as mechanical means to isolation. A sort of lock-and-key concept was implied wherein properly adapted males were thought to fit their genitalia into the female genitalia and achieve sperm transfer. Other males could not and so were believed to be blocked from effective copulation. More recent work has not always supported this initially plausible idea.

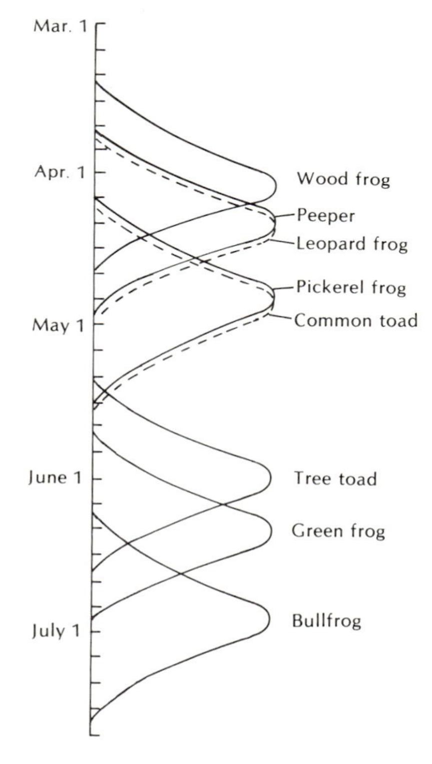

Fig. 8-4: Seasonal peaks of mating activities in frogs and toads near Ithaca, New York. The data were obtained over several years and therefore have a broader spread than data representing mating activities in any single year.

MATE ISOLATION

Here potential mates are prevented from mating. No gametes are transferred from one organism to another if internal fertilization normally takes place, and in cases of external fertilization, the release of gametes is out of synchrony, or some other factors prevent the coming together of gametes. Three subcategories are usefully recognized here: *ethological* or *behavioral barriers, seasonal* or *temporal barriers,* and *ecological isolation.* Though it is easy to find examples for each of these subcategories, it is much more difficult to find examples in which there is only one type of isolation. Populations at this point of isolation in their divergence usually show several kinds of isolating mechanisms.

This can be seen in the following two studies on frogs. Figure 8–4 describes the peaks of breeding activities for various species of frogs and one species of toad in the rolling hills around Ithaca, New York. It is clear that as spring progresses the different species are fairly well separated by different temporal periods of activity, with two exceptions. These are the peeper and the leopard frog, which breed around early April, and the pickerel frog and the common toad, which breed most actively the end of April. In these two cases, behavioral and ecological factors are also important.

In the first case, peeper frogs inhabit woodland ponds and

rather shallow water, whereas leopard frogs are found in marshy swamps. Also the calls of the male frogs are quite different in the two species, and responding females would move toward a male of their own species in order to mate. Here, then, ecological and behavioral differences isolate these two species; there is no temporal isolation.

In the second case, also, ecological and behavioral factors are important in the absence of temporal isolation. The pickerel frog occurs only in upland streams and ponds, but the toad does not appear at all fussy since individuals are described as "using any ditch or puddle as a mating ground." Again mating calls are quite different in the two species.

Another study, done by John Moore, of Columbia University, on seven species of North American frogs, is summarized in Table 8–3. In this study Moore estimated hybrid nonviability, in crosses between the species, and degree of seasonal, ecological, and geographic isolation. Seasonal isolation refers to the peak of the breeding season, ecological isolation to habitat preference. Geographic isolation (which can be treated as part of ecological isolation) refers to the spatial distance between members of the various species. In some crosses hybrid nonviability is complete and so reproductive isolation is solidly established. In other cases, as between *Rana pipiens* and *R. palustris,* there is no hybrid death. But the combination of other factors are very probably sufficient to isolate these two species. These factors cause complete or nearly complete isolation in all cases.

Another informative case is a study of diversification in two populations of toads, initially described as two separate species. When the problem was first attacked in 1941 by Blair, he thought that he was observing the coming together of two species, *Bufo americanus* and *B. fowleri.* These toads are widely distributed in the eastern United States and he was looking at their point of contact in Indiana. It seemed to him that these animals may have been separated into two populations by the Pleistocene glacier and cold weather. The withdrawal of the glacier and recent changes caused by the white inhabitants in the environment, such as farming, clearing of forests, and damming of streams, brought the two populations into contact again. Blair reported, in 1941, that 9.4% of the toads in the wild had a phenotype intermediate between the two species. Furthermore, these forms, which were apparently hybrids between *B. americanus* and *B. fowleri,* showed no evidence of decreased fitness. Blair also noted some ethological

Table 8–3. Incipient divergence in American species of the grass frog *Rana pipiens* and the relative importance of various isolating mechanisms.

Females \ Males	*Rana sylvatica*	*Rana pipiens*	*Rana palustris*	*Rana clamitans*	*Rana catesbeiana*	*Rana septentrionalis*
Rana sylvatica		G 29	G 61	G 59	G 68	G 80
		E 70	E 40	E 30	E 70	E 60
		S 60	S 100	S 100	S 100	S 100
		H 100	H 100	H 100	H 100	H ?
Rana pipiens	G 59		G 74	G 67	G 62	G 88
	E 70		E 70	E 70	E 85	E 80
	S 60		S 40	S 100	S 100	S 100
	H 100		H 0	H 100	H 100	H 100
Rana paulstris	G 13	G 0		G 3	G 23	G 72
	E 40	E 70		E 60	E 70	E 50
	S 95	S 40		S 95	S 100	S 100
	H 100	H 0		H 100	H 100	H ?
Rana clamitans	G 28	G 0	G 24		G 18	G 22
	E 30	E 70	E 60		E 30	E 20
	S 100	S 100	S 95		S 50	S 0
	H 100	H 100	H 100		H 100	H 100
Rana catesbeiana	G 51	G 0	G 47	G 29		G 93
	E 70	E 85	E 70	E 30		E 30
	S 100	S 100	S 100	S 50		S 50
	H 100	H 100	H 100	H 100		H ?
Rana septentrionalis	G 0	G 0	G 37	G 36	G 79	
	E 60	E 80	E 50	E 20	E 30	
	S 100	S 100	S 100	S 0	S 50	
	H ?	H 100	H 100	H 100	H 95	

Key: G, geographic isolation; E, ecological barriers; H, hybrid nonviability. Complete isolation is represented by 100; absence of isolation by 0; varying intermediates by numbers between 0 and 100 (from Moore 1949).

and ecological differences between the two species: their calls were noticeably different, their breeding sites differed somewhat, and *B. americanus* was an early breeder and *B. fowleri* a late breeder. This led him to conclude that isolation had not established differences so marked that diversification would continue. Rather, he thought, interbreeding was blurring the differences and he was observing the reestablishment of a single, polytypic gene pool.

Some 30 years later, in 1972, Jones reported on what had happened to the two toad populations in Bloomington, Indiana. By this time further work had lead to the merging of *B. fowleri* with another phenotypic species called *B. woodhousii*. The new name, logically enough, was *B. woodhousii fowleri* and so Jones was looking at what had happened to this species and *B.*

americanus. The result was not what one would have expected from Blair's work. No hybrids were found. Both species were smaller, having converged on a similar phenotype and, hence, the individual toads captured were not always identifiable as one species or the other. The ethological barriers noted by Blair had persisted in differences of breeding sites, of breeding season, and of mating call. Jones concluded that, at least in the Bloomington area, the two species had remained distinct. Presumably, there was some selective disadvantage to the hybrids.

It is not clear why character displacement affected ethological or behavior characters and not morphological ones. Speculation might suggest that predation by snakes or hawks forced a common, most favorable morphological phenotype on all the toads. Breeding traits were therefore the only area free to show divergence. These traits would have included behavioral (mating calls), temporal (time of breeding), and ecological (breeding sites) isolating mechanisms. Apparently they are effectively separating the Bloomington populations into distinct species.

Genetic differences between species

When species diversification effectively generates two gene pools where formerly there was one, exactly how different are these two pools? More generally speaking, what are the genetic differences between two established, closely related species? As might be expected, we have no solid, single answer, but various answers have been offered. Let us look both at some earlier breeding studies and some recent molecular efforts in this regard.

BREEDING ANALYSIS

We start with one of the earliest studies (1924), which was done in Germany by Baur on snapdragons. In these flowers, the species *Antirrhinum majus* and *A. molle* were crossed. The F_1 hybrids were viable and fertile. Phenotypically they were intermediate between the parent species and showed the same variation as either parent. When, however, the next generation was obtained by crossing the F_1 plants, the observed variability was called "spectacular." All kinds of recombinations of parental traits were found, but there was no recovery of the parental phenotypes. (This would have depended on the highly improbable instance of all the chromosomes from one parental species segregating to one gamete, e.g., the pollen, and then that happening again in the egg, and then, an added improbability, precisely these two gametes uniting in fertilization.) Overall, Baur guessed that there were more than 100 genes different between the two species.

In 1942, Pontecorvo, in Great Britain, analyzed the genetic basis for hybrid sterility in fruit flies from crosses between *Drosophila melanogaster* and *D. simulans*. These two species are quite similar in appearance, yet their hybrids are sterile and the males or the females, depending on which species contributes the egg, are nonviable. As we have seen, *Drosophila* species are often quite similar and various complexes of semispecies and sibling species are known. Pontecorvo estimated that at least nine loci were responsible for the nonviability. This and subsequent work lead to the concept that species differ in blocks of genes or in patterns of genic interaction. Again we are lead to the epigenotype. Sensing this, Harland, in 1936, stated that "the modifiers really constitute the species."

Harland's point is exceedingly important. In more modern terms, what is being said is this: More than one genotype can produce the same adaptive norm. It is the function of modifiers, or buffering genes or regulatory genes—here the terms are essentially interchangeable—to bring different genotypes to the same adaptive phenotype. In this way a species can have an enormous adaptive variability with apparent phenotypic uniformity—hence the phenotypic similarity of semispecies and sibling species.

An excellent example of this harmonious interaction of the epigenotype and its disruption in hybrids comes from the work of Gordon and Rosen (1951). They studied the fate of hybrids between two fish species, the platyfish (*Xiphophorus maculatus*) and the swordtail (*X. helleri*). The action of certain genes are listed in Table 8–4. In the left-hand column are the genes and their phenotypes when the genes are acting normally in their usual epigenotype of *X. maculatus*. When these genes, as a result of crossing, interact with *X. helleri* genes, the result is distinctly detrimental, leading to malignant cell proliferation.

MOLECULAR STUDIES

Turning now to more recent studies, we can use molecular techniques to get more information. (Recall what was said in Chapter 4 about the relative advantages and disadvantages of breeding analyses and molecular techniques in determining

Table 8–4. Gene action in two species of platyfish, i.e., *Xiphophorus maculatus* and *X. helleri* (from Gordon and Rosen 1951).

Phenotype in *X. maculatus*	Gene	Phenotype in *X. helleri*
spotted pattern	*Sp*	cutaneous melanoma
broad black band on flank	*N*	melanotic tumors on black flank band
dark spots on dorsal fin	*Sd*	melanotic tumors on dorsal fin
darkening of ventral spots	*Sb*	melanomas on midventral line
horizontal lines	*Sr*	no tumors, but exaggerated F_1 effects; flank tumors in $F_1 \times$ *X. helleri* progeny

genetic variability.) Table 8–5 gives the results of a study by Hubby and Throckmorton, who compared proteins in different species of fruit flies. Ten species similar to *Drosophila virilis* were examined. In terms of their chromosomal similarities, four species were put into a virilis subgroup and the remaining six species into a montana subgroup. The number of proteins studied ranged from 29 to 42. These researchers determined what percentage of proteins was unique to a species, what percentage was shared within a subgroup virilis or montana, and what percentage was shared by all ten species. If we assume that this technique identified unique proteins, then we are

Table 8–5. Shared and unique proteins in a group of ten species of fruit flies separated into a virilis and a montana subgroup (from Hubby and Throckmorton 1965).

Subgroup and species	Number studied	Percentage unique to species	Percentage common to subgroup	Percentage common to group
virilis subgroup				
D. americana	38	5.3	23.7	71.1
D. texana	42	21.4	16.7	61.9
D. novamexicana	38	7.9	21.1	71.1
D. virilis	38	2.6	21.1	76.3
average	39.0	9.3	20.7	70.1
montana subgroup				
D. littoralis	39	28.2	25.6	46.2
D. ezeana	35	8.6	29.7	65.7
D. montana	37	18.9	37.8	43.2
D. lacicola	29	20.7	20.7	58.6
D. borealis	42	19.0	28.6	52.4
D. flavomontana	29	10.3	37.9	51.7
average	35.2	17.6	29.4	53.0
overall average	36.6	14.3	25.9	59.8

looking at as many loci as there are proteins. So our sample is of 36 loci, on the average, for each species. This is a small sample of the 5,000 or so loci thought to be in a haploid genome of a fruit fly. But given this sample we find these species to have around 60% of their proteins in common. The unique proteins were, on the average, only about 14% of the total, ranging from 2.6 to 28.2%. If we extrapolate from this limited sample to the whole genome, we can argue that if 14% of the proteins are unique, then over 500 loci in a species might have unique functions. (Compare this with Baur's estimate of over 100 different genes between snapdragon species.)

In another, similar study, the same authors found that sibling species shared only about 50% of the proteins studied. This too suggests great genetic differences between phenotypically similar species. Again we must think of Harland's modifiers and the way different epigenotypes can produce very similar adaptive phenotypes.

One of the most exciting stories on the genetic basis of species diversification has come from comparative analyses of frog evolution with the evolution of placental mammals. These two groups are compared in Table 8–6. Note especially that frogs are placed in one order with more than 3,000 species in it, whereas the 4,600 species of placental mammals are distributed among 16 to 20 orders. Furthermore, frogs have been on earth about twice as long as the placental mammals. In these two groups, then, the placental mammals have evolved much faster into many more diverse forms than have the frogs.

When we look at the molecular evolution of these two groups, a different story emerges. The biochemical work from A. C. Wilson's laboratory at the University of California, Berke-

Table 8–6. Rates of evolution in frogs and placental mammals.

Property	Frogs	Placental mammals
number of living species	3,050	4,600
number of orders	1	16–20
age of the group ($\times 10^6$ years)	150	75
rate of organismal evolution	slow	fast
rate of albumin evolution	standard	standard
rate of loss of hybridization potential	slow	fast
rate of change in chromosome number	slow	fast
rate of change in number of chromosomal arms	slow	fast

The number of frog species is that estimated in 1974 and many new species are being described each year (from Wilson, 1976).

ley, rests on two sets of data. The first comes from work on the protein albumin. This macromolecule is a chain of 580 amino acids. From the amount of change in the sequences it is apparent that this molecule has evolved at about the same rate in frogs and mammals. Wilson also expresses it this way: "Species which are similar enough in anatomy and way of life to be included within a single genus of frogs (e.g., *Rana*) can differ as much in their albumins as does a bat from a whale." The results with other proteins, which have been studied less extensively, are the same as those with albumin.

The second area of molecular research involves differences in DNA. When the DNA from two sibling species of frogs belonging to the genus *Xenopus* were studied, very large differences were found. In terms of their morphology, these frogs were often classified as members of the one species *X. laevis*. But since the DNA between the species differs more than the DNA between humans and South American monkeys, two distinct species, *X. laevis* and *X. borealis,* are now recognized.

From the foregoing Wilson proposes a fascinating general conclusion: *Evolutionary change at the organismal level is largely dependent on regulatory genes*. His argument is somewhat indirect. He assumes that genes are either structural and give rise to specific sequences of amino acids or else they are regulatory and determine gene action. Since the changes seen at the organismal level (diversity of species) in frogs and placental mammals does *not* correlate with molecular changes in proteins, structural genes cannot be the cause of organismal change. That leaves only the regulatory genes as the other alternative.

In support of this idea he has several lines of evidence. Since organismal change (species diversification) is relatively slow in the frogs, there should be relatively little change in the regulatory genes, even if the structural genes are changing as fast as they are in mammals. Changes in structural genes can be measured by amino acid sequences or even more rapidly by comparing the immunological responses (antigen-antibody reactions) of proteins. Changes in regulatory genes can be seen developmentally: similar regulation will result in successful hybridization between species, and dissimilar regulation will result in little or no hybridization. Therefore, Wilson predicted that frogs would show successful hybridization (little regulatory change), but significant change in albumin (much structural gene change), whereas mammals would show little successful hybridization and little albumin change. Figure 8–5A shows that his prediction was fulfilled.

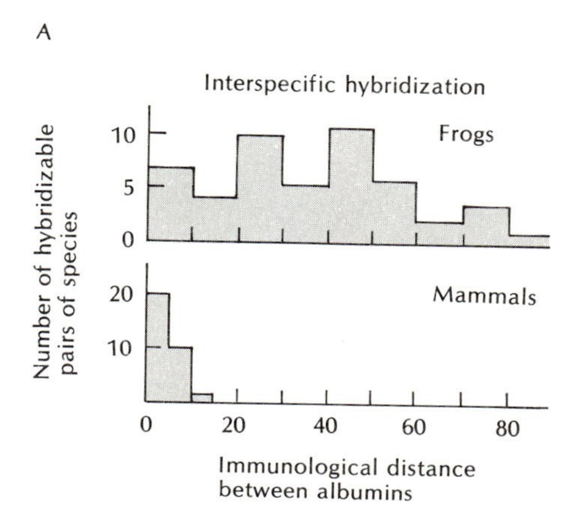

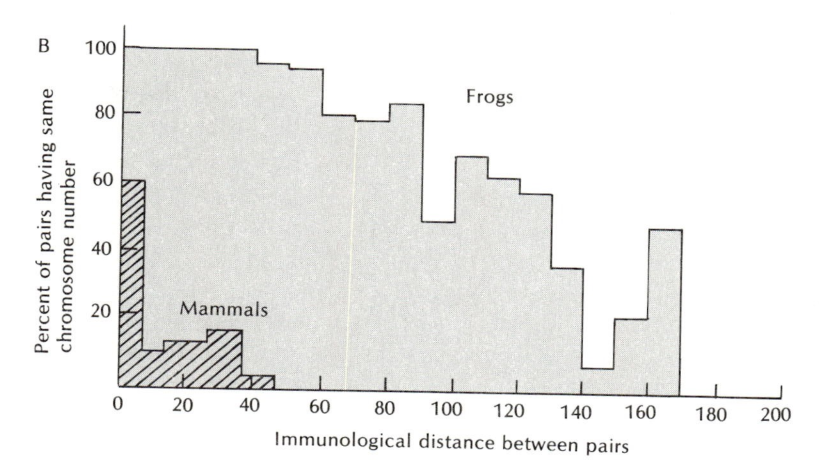

Fig. 8-5: Molecular evolution relative to other measures of evolutionary distance in frogs and in mammals. A. Molecular evolution as determined by differences in albumins (measured immunologically) compared to formation of viable hybrids. Here 50 pairs of frog species and 31 pairs of mammalian species were studied. B. Molecular evolution as determined by differences in albumins between pairs of species compared to the same species pairs having identical chromosome numbers. Here 373 pairs of frog species and 318 pairs of mammalian species were studied.

Chromosomal studies also confirmed Wilson's conclusion. Placental mammals show much more diversity in number of chromosomes and arrangements within chromosomes than do frogs. Such changes are thought to be related to the regulatory functions of genes. If pairs of species having the same chromosome number are plotted against the immunological differences in their albumins (Fig. 8–5B), it is clear that the chromosomal configurations of frogs are much more conservative (show less change) than those of mammals.

Finally, we should comment on the fact that Wilson has found that humans are much more like chimpanzees in molecular terms than in organismal morphology. Humans (*Homo sapiens*) and chimpanzees (*Pan satyrus*) are placed in different families. But species of mice, frogs, and flies can differ more in their DNA than do humans and chimpanzees. And when proteins are studied in humans and chimpanzees, it is found that over 99% of their proteins are identical. These two primates are about as different as semispecies. Clearly something other than structural genes has changed. Again Wilson concludes it must be the regulatory genes. This modern analysis confirms Harland's comment regarding modifiers.

All of this is both helpful and somewhat confusing. The confusion comes largely from the emphasis on regulatory genes, which represent an area of extensive ignorance in present day genetics. But, in one sense, we have already prepared ourselves for this by reminding ourselves of how little we know about genes and development. The epigenotype summarizes both our ignorance and points to an important biological reality. The concept of regulatory genes refines the idea of the epigenotype and in that sense it is helpful.

Looking directly at the question we have been examining in this section, i.e., what are the genetic differences between established species?, we see why no answer is possible today. If Wilson's thinking is correct, the answer largely ties in what we have yet to learn about regulatory genes.

Chromosomal evolution.

This is the last topic in this chapter. We study it here because it represents a transition to trans-specific evolution or macroevolution. Our basic problem in understanding evolution is to explain the diversity of organisms, of discontinuous species hierarchically related to each other. That is, organismic diversity is represented by reproductive communities showing differing degrees of similarities or differences. The origin of species derives from the microevolution of diversity within species and appears as macroevolution when we see how emergent and established species continue to diversify. That they continue to diversify is a story that can be read in their chromosomes. That is why we turn now to chromosomal evolution.

CHROMOSOMES IN THE GENUS *DROSOPHILA*

The polytene chromosomes of insects, in general, and fruit flies, in particular, are a gold mine of genetic and evolutionary information. Their special feature is that we can see differences along the length of a chromosome. From breeding analyses came the theory that genes line up in single rows along the length of a chromosome (see Chapter 3). But, however convincing that theory, it could not be tested directly until the polytene chromosomes were discovered (Fig. 8–6). Here, using only the light microscope, an investigator could see huge chromosomes banded in a unique way along their length. And not only were they banded, but even more important, the banding was a regular feature of a given chromosome and the bands could be rather precisely related to genetic maps.

By careful study it is possible to recognize quickly which chromosome is being studied in a given genome. But in such studies, one must recognize the special characteristics of the polytene chromosomes. First of all, as their name implies, they are composed of many copies (polytene means many threads) of single strands of DNA with their associated proteins and

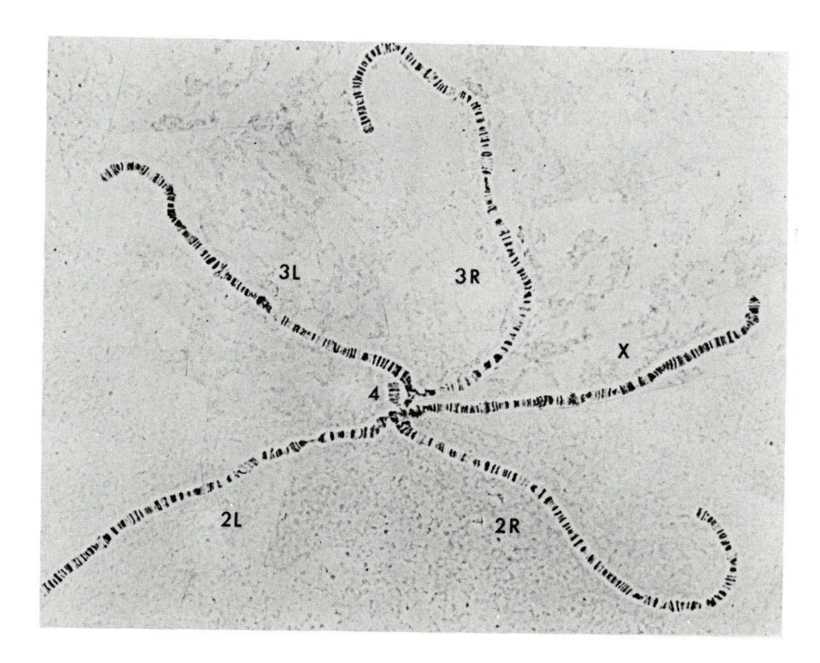

Fig. 8-6: A complete set of salivary gland chromosomes in *Drosophila melanogaster.* Homologous chromosomes are paired, and their kinetochores are joined (compare Fig. 3-1) (X, paired X chromosomes; 2R, 2L and 3R, 3L, right and left arms and the second and third chromosome, respectively; 4, the small fourth chromosome).

RNA. Second, the banding represents local regions of heavy coiling of the DNA, which stain intensely for DNA as compared to other regions. Nonetheless, Watson-Crick double helices of DNA run the length of the chromosome and estimates are that there are about 1,000 per chromosome. Third, homologous chromosomes are paired. This means that in *Drosophila melanogaster* the eight chromosomes are paired into four polytene configurations. And fourth, the centromeres of all the chromosomes lie attached to a chromocenter (Fig. 8–6).

Keeping the above in mind, how do we interpret the details of the banding in terms of genes and their possible rearrangements? In the first place, the hope that the bands were directly equivalent to genes was false. On the larger chromosomes (the X and the second and third chromosomes) there are easily 1,000 bands. But the *Drosophila* genome carries some 5,000 genes, and so there can be many genes per band. However, by means of deletions, when these are large enough, the visual identification of a lost band or bands has, in some cases, been correlated with the loss of a genetic marker. In other words, visual differences within a chromosome can be correlated with the genetic map.

More importantly, from our point of view, inversions can be

located quite precisely. As we will see, it is these changes in a chromosome that allow us to decipher the evolutionary history of related species of fruit flies. An expert can locate the break that occurred at one end of an inversion to within about two bands of the actual break. Given around 1,000 bands on the chromosome, this means that the site of a break can be identified at any of about 500 positions along the chromosome. This is a considerable degree of precision.

Now recall that an inversion necessitates two breaks (Fig. 3–10). Each break can be at one of 500 identifiable spots on a large chromosome and, therefore, the total number of identifiable inversions is about 125,000. This figure is obtained from $500 \times 499/2 = 124,750$; the calculation derives from the concept that a break at point p_1 can be associated with 499 other breaks, at point p_2 with 498 different other breaks, at p_3 with 497 different other breaks, and so on. Hence if the total number of breaks is t, then

$$t = 499 + 498 + 497 \ldots 3 + 2 + 1$$
$$= (499)!$$

A quick way to calculate this is $500 \times 499/2$.) Given one inversion on a chromosome, that one could subsequently have another of the possible 125,000 inversions, and so on, practically indefinitely. If two inversions occur independently on the same chromosome, there is less than one chance in 125,000 that they will be identical, assuming equal probabilities of any break along a chromosome. Hence, if two apparently identical inversions are seen, it can be assumed that they most probably had a common ancestor rather than that they arose twice, independently of each other. That is the first working assumption behind chromosomal evolution in *Drosophila*.

The second assumption is that each observed inversion is the result of only two breaks. The estimated frequency of inversions in *Drosophila*, based largely on genetic evidence, is once in every 100,000 cells or 10^{-5}. This implies a frequency of single breaks of around one per 300 copies of a given chromosome. [The probability of two such breaks occurring simultaneously is $(1/300)^2$ or 1/90,000, which is approximately 10^{-5}.] Therefore, if inversions involved three breaks they would be 300 times less frequent than two-break inversions. Thus, the inversions observed in the vast majority of cases are assumed to be two-break inversions. Each one can be followed by another two-break inversion, and so on.

Reading inversions. We now come to the crux of the

matter: the reading of chromosomal history. The essentials are given in Box 8–2. When the method is applied to a large number of inversions found in many populations of flies, and even in flies of different species, we find that evolutionary history can be reconstructed.

The chromosomal history from three species of *Drosophila* is summarized in Fig. 8–7. Note one form of the third chromosome has not been found yet. It is the one labeled Hypothetical in the figure. Either it exists somewhere in the western United States or, through competition with better adapted forms, it has become extinct. Note also that it is not possible, at present, to say actually where the ancestral form is located. It could be any one of Standard, Hypothetical, Santa Cruz, or Tree Line. It may be that, in terms of the third chromosome and its configurations, we cannot determine the ancestral form. The nature of ancestral forms is, in general, a difficult problem, as we will see in the next chapter. Nevertheless, this figure presents evidence of the continuing evolution of well-established species.

CHROMOSOMES IN THE GENUS *DATURA*

A quite different manner of reading evolutionary history from chromosomes occurs in flowering plants of the genus *Datura*. These also have the common name of Jimson weed, or thorn apple, the latter because of their prickly seed capsules.

We need to start with a quick description of the chromosomal situation or karyotype of these plants. Their seven pairs of chromosomes are V-shaped, or metacentric, because the centromere is located in the middle of the chromosomes. These chromosomes and their arms are labeled as consisting of pairs of numbers, i.e.,

$$1\cdot2 \quad 3\cdot4 \quad 5\cdot6 \quad 7\cdot8 \quad 9\cdot10 \quad 11\cdot12 \quad 13\cdot14$$

The first chromosome carries arms 1 and 2 and the seventh chromosome carries arms 13 and 14. These chromosomes and their arms are not visibly differentiated the way the *Drosophila* chromosomes are. Nonetheless, as we will see, once a given set of chromosomes is taken as a standard, those that differ from it can be identified. The essential difference followed in *Datura* are translocations, the exchange of arms between nonhomologous chromosomes (see Fig. 3–6).

We next follow out the consequences of translocations so as, eventually, to see how the phylogenetic history of these plants has been worked out. Starting with a single trans-

Box 8–2. Reading a sequence of chromosomal inversions.

A. An inversion recurs on the third chromosome of *Drosophila pseudoobscura*. B. Chromosomes with different inversions pair, as in the salivary gland chromosomes. First sequences of inversions such that an inversion in the first chromosome can give rise to either chromosome II or IV occur. Another inversion in chromosome II gives rise to chromosome III and another inversion in chromosome IV gives rise to chromosome V. The lower diagrams show how chromosomes I and II would pair if they occurred in the same cell and how other combinations of the inverted chromosomes would pair. Note especially that if there is a difference of only one inversion between the paired chromosomes, a single loop results (pairings I/II, I/IV, II/III, and IV/V), but if two inversions have occurred the pairings show a figure-8 configuration (I/III and I/IV).

How does one interpret these configurations in evolutionary terms? In this sort of analysis we are working back from observed configurations to the events that caused them. The inversions at the top of the preceding figures gave rise to the pairings seen at the bottom. In practice, that process is reversed: we proceed from observed pairings to a reconstruction of the events that caused them.

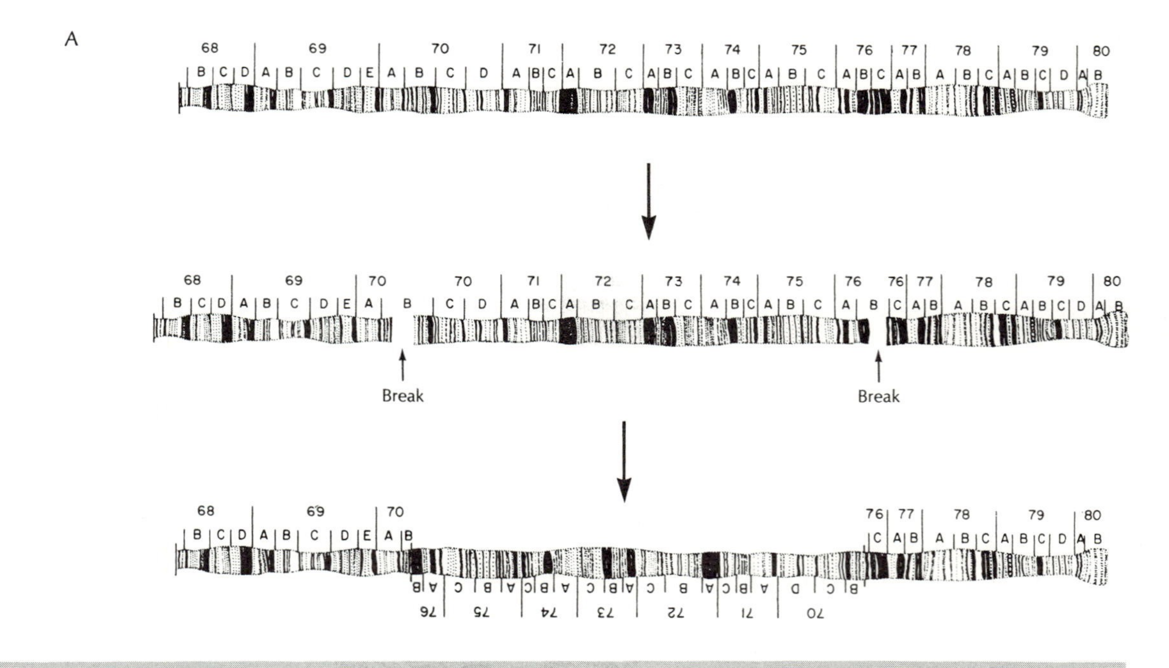

location between the 1·2 and 3·4 chromosome, we get chromosomes 2·3 and 4·1. Thus at meiosis the configuration of chromosomes is characteritic of translocations when a translocated set tries to pair with one that is not translocated (Fig. 8–8). The result is a cross-shaped configuration that

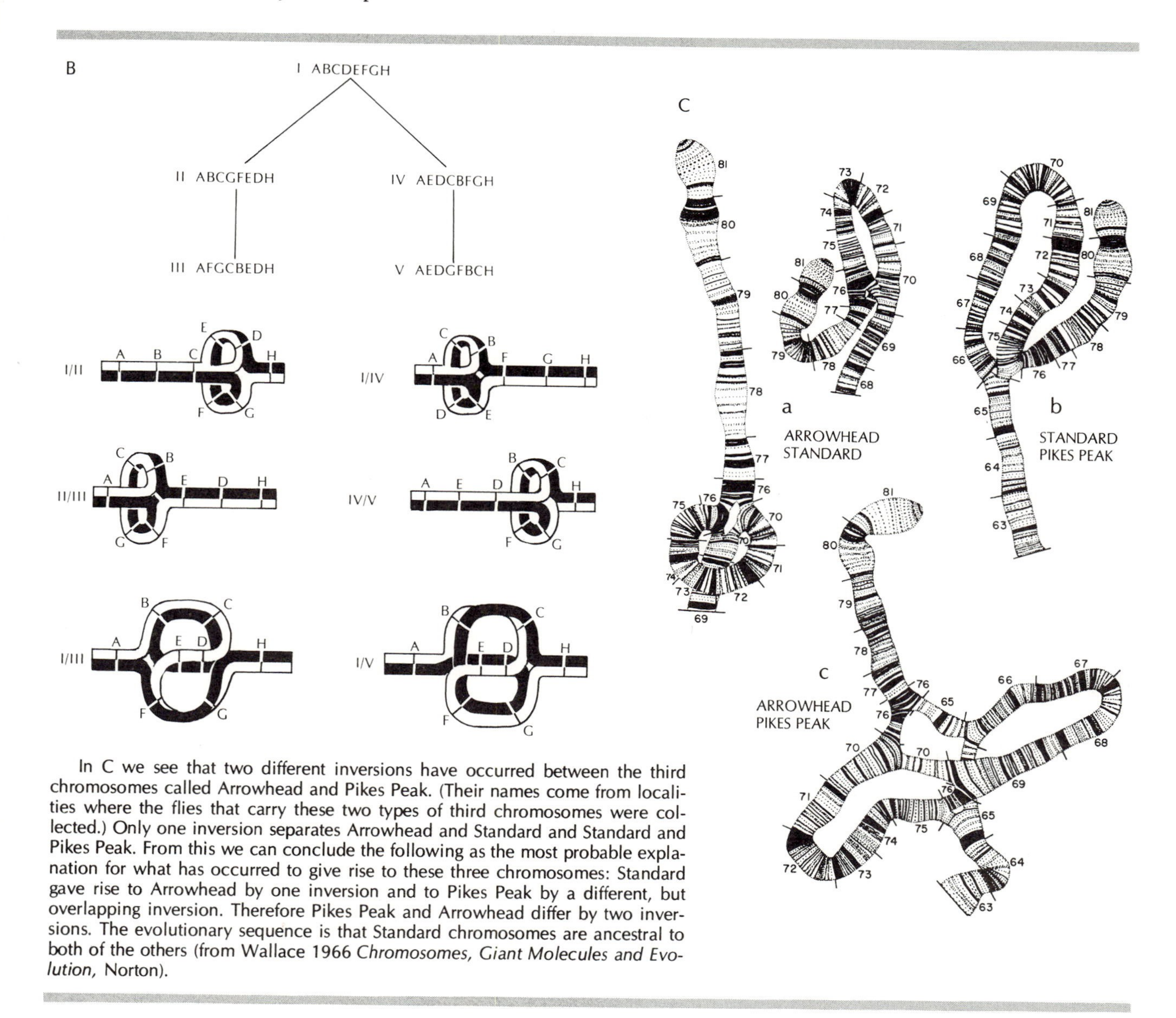

In C we see that two different inversions have occurred between the third chromosomes called Arrowhead and Pikes Peak. (Their names come from localities where the flies that carry these two types of third chromosomes were collected.) Only one inversion separates Arrowhead and Standard and Standard and Pikes Peak. From this we can conclude the following as the most probable explanation for what has occurred to give rise to these three chromosomes: Standard gave rise to Arrowhead by one inversion and to Pikes Peak by a different, but overlapping inversion. Therefore Pikes Peak and Arrowhead differ by two inversions. The evolutionary sequence is that Standard chromosomes are ancestral to both of the others (from Wallace 1966 *Chromosomes, Giant Molecules and Evolution,* Norton).

aligns homologous arms with each other and involves four different chromosomes—1·2 and 3·4, which are not translocated, and 2·3 and 4·1, which are translocated. In this configuration the four centromeres lie close together in the center of the cross, at the point where each chromosome bends. In ana-

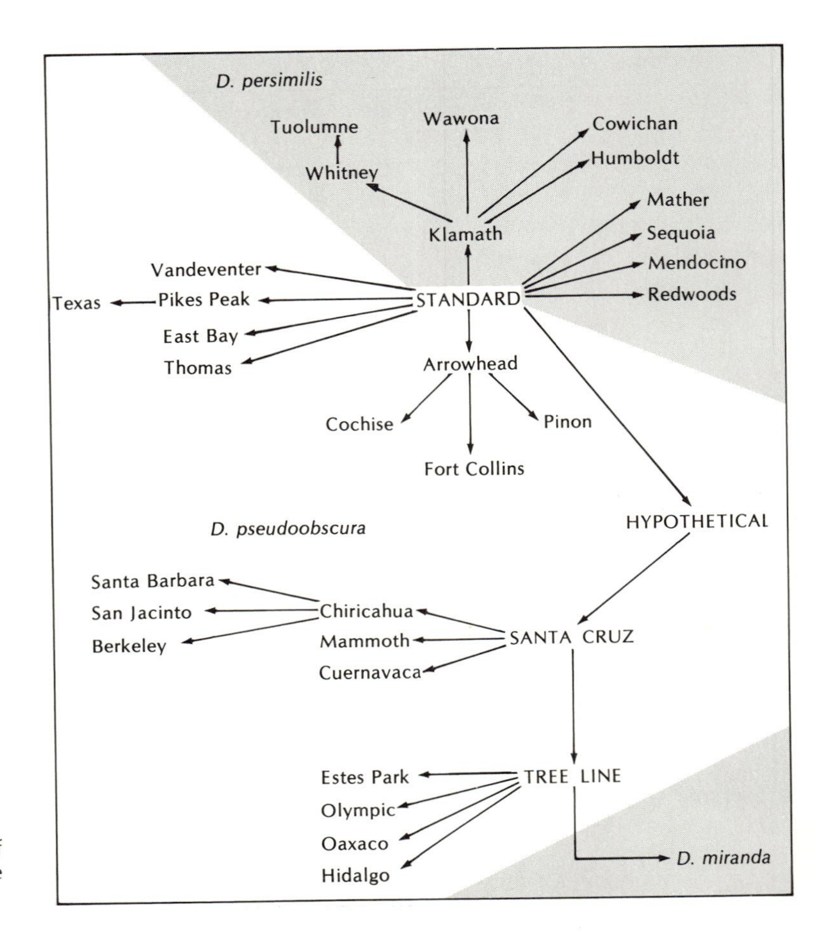

Fig. 8-7: A phylogeny of three species of *Drosophila* based on inversions in the third chromosome.

phase, two centromeres go to one spindle pole and two to the other pole. Seen in side view and allowing for the three-dimensional reality of what is happening, one is aware of a ring of chromosomes in which the centromeres alternate in going to opposite poles (Fig. 8–8).

Now we can see an important consequence of the translocation. The normal chromosomes (1·2 and 3·4) *both* go to one pole, and both translocated chromsomes go to the other pole. There is a perfect, stable segregation of the two types. This will be repeated whenever fertilization results in an egg with the chromosomes 1·2, 3·4, 2·3, and 4·1. And at meiosis, when the chromosomes are segregating, we will again see one ring of chromosomes and five normally paired sets of chromosomes. If, however, a plant were homozygous for the translocations, i.e., if it carried two 2·3 and 4·1 chromosomes, there would be seven pairs of normally paired chromosomes.

Next, we determine the consequences of one translocation being followed by another, much as we followed two consecutive inversions in *Drosophila*. Consider the situation shown in Box 8–3. Here there are four different karyotypes, originating and organized as shown. Now, what are the consequences of crosses among them?

The different sets can be indicated by the letters given be-

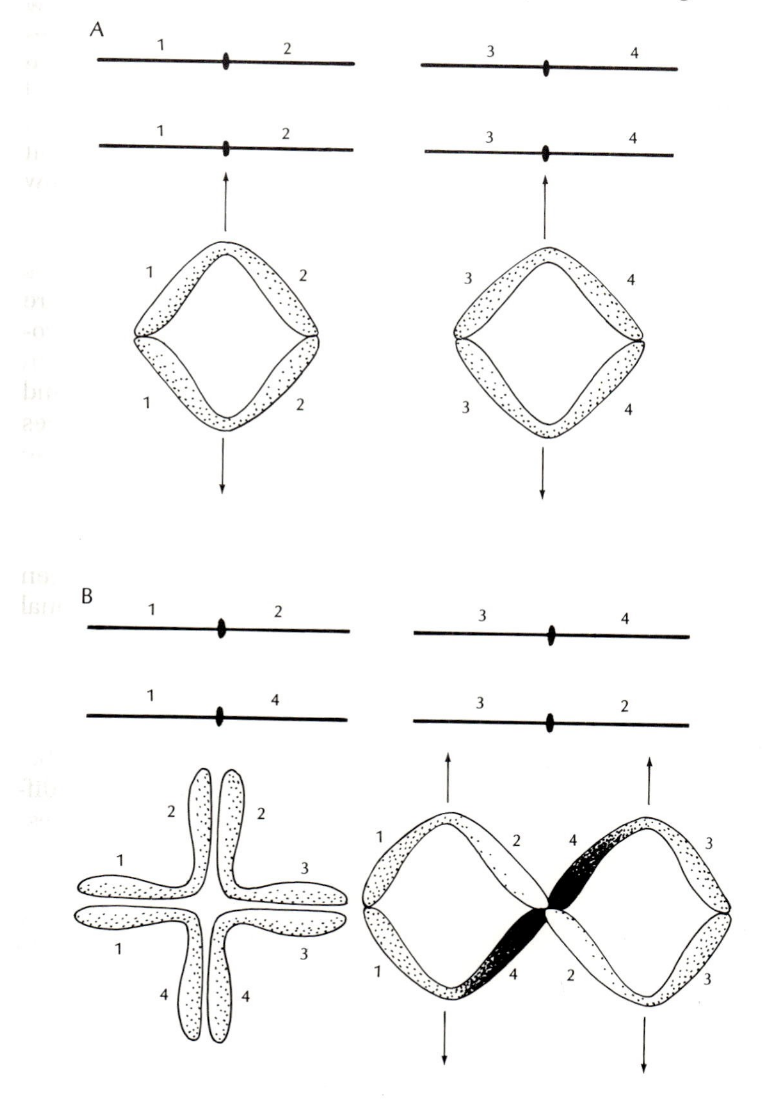

Fig. 8-8: Translocation of chromosomes. A. Untranslocated chromosomes separate at anaphase. The centromeres lie adjacent to the base of each arrow, which shows the direction of chromosomal movement. B. After translocation, the chromosomal arms are joined 1•4 and 3•2. At metaphase there is a cross-shaped configuration; at anaphase (side view) the chromosomes reveal an undulating ring (superficially, a figure eight).

side them. Hence an *A/A* plant is one whose parents are both of the original type. A plant designated B/C has one B parent and one C parent, and so on. Box 8–3 also shows the karyotypes of progeny plants and the visible meiotic configurations they would produce, especially as regards the conspicuous chromosomal rings.

At this point we can begin to unravel what happened historically. Remember that we set up the problem as if we knew what had happened—we went from Fig. 8–8 to Box 8–3. In reality the reverse order is what we have to deal with. What we observe is what is given in the last column of Box 8–3 and from that, plus our knowledge of the crosses we made to get those configurations, we must deduce the prior events. That is, we can reconstruct the karyotypes of the parents and how they originated.

How do we do that?

First, remember that the one circle of four chromosomes and the five pairs of the other chromosomes tell us that there is one translocation difference between the two parental karyotypes (Box 8–3). The plants designated A/B, B/C, and B/D fit that situation. But A/C and C/D give us two circles of four and two pairs. Therefore, there are *two* translocational differences between A and C and between C and D. Thus we can propose the following sequence of translocations

$$A \rightarrow B \rightarrow C$$

This accounts for the single translocational difference between A and B and between B and C, and the double translocational difference between A and C. Also we can propose

$$B \begin{smallmatrix} \nearrow C \\ \searrow D \end{smallmatrix}$$

This accounts for single differences between B and C and between B and D. It also explains the two translocational differences between C and D. The overall picture then becomes

$$A \rightarrow B \begin{smallmatrix} \nearrow C \\ \searrow D \end{smallmatrix}$$

There remains the circle of six and four pairs, in A/D, to be accounted for. From the last diagram we would say two translocations have occurred (A→ B→ D) and they both involve the same chromosomes. This would produce the circle of six rather than two circles of four.

In this way we have worked out the history of transloca-

Box 8–3. Successive translocations in terms of karyotypes.

First, consider the following four sets of chromosomes. Set A gives rise to set B through translocations between chromosomes 3 · 4 and 9 · 10. Then in set C, another translocation occurs independent of the first one, and in set D yet another occurs, which overlaps the first one.

	A. Original karyotype					
1·2	3·4	5·6	7·8	9·10	11·12	13·14
	B. Translocated karyotype					
1·2	3·10	5·6	7·8	9·4	11·12	13·14
	C. Translocated karyotype (independent of B)					
1·13	3·10	5·6	7·8	9·4	11·12	2·14
	D. Translocated karyotype (overlaps B)					
1·2	3·8	5·6	7·10	9·4	11·12	13·14

When plants are bred to bring together the foregoing karyotypes, the following meiotic figures are observed.

Cross	Karyotype	Observed results
A/B	1·2 3·4 9·10 5·6 7·8 11·12 13·14 1·2 4·9 10·3 5·6 7·8 11·12 13·14	one ring of four chromosomes
A/C	1·2 14·13 3·4 9·10 5·6 7·8 11·12 2·14 13·1 4·9 10·3 5·6 7·8 11·12	two rings of four chromosomes
A/D	1·2 3·4 9·10 7·8 5·6 11·12 13·14 1·2 4·9 10·7 8·3 5·6 11·12 13·14	one ring of six chromosomes
B/C	1·2 14·13 3·10 5·6 7·8 9·4 11·12 2·14 13·1 3·10 5·6 7·8 9·4 11·12	one ring of four chromosomes
B/D	1·2 3·10 7·8 5·6 9·4 11·12 13·14 10·7 8·3 5·6 9·4 11·12 13·14	one ring of four chromosomes
C/D	1·13 14·2 3·10 7·8 5·6 9·4 11·12 13·14 2·1 10·7 8·3 5·6 9·4 11·12	two rings of four chromosomes

tions as they have occurred in five of the seven chromosomes present here. Chromosomes 5·6 and 11·12 show no changes in the A, B, C, or D karyotypes. How would translocations in them be detected? If a translocation between these chromosomes had occurred, then a ring of four and five pairs would appear whenever the plant, call it E, containing it is crossed to all the other four. In that way, these five karyotypes (A through E) would have involved all seven chromosomes in one or another type of translocation. Any new strain could be identified as either like or unlike any of the first five and its evolutionary history could be deduced from the results of the crosses.

Using this sort of study, evolutionary botanists have been able to read in significant detail the history of speciation in various species of Jimson weed (*Datura*) (Fig. 8–9).

A very similar story is found in the evening primrose (*Oenothera lamarckiana*). This species shows a ring of 14 chromosomes. Overlapping translocations have occurred throughout the whole karyotype. The result is a regular segregation of one precise half of the genome from the other half (Fig. 8–10). This behavior is the same as if the diploid chromosome number were two—the theoretically lowest possible number. But that is not the end of the intriguing genetic and evolutionary complications in this species.

Balanced lethals are present. (See Chapter 4, especially Table 4–3.) In *O. lamarckiana* these lethals are recessive and lie in each of the segregating genomes or karyotypes or superchromosomes. Thus no homozygous karyotypes survive, because when an egg is fertilized by pollen of the same karyotype, the homozygous lethal is expressed. Hence, only heterozygotes carrying one of each karyotype can survive.

Oenothera plants are found with all the combinations of

Fig. 8-9: Reconstruction of chromosomal phylogeny in *Datura* species. (C, *D. ceratocaula;* D, *D. discolor;* F, *D. ferox;* I, *D. innoxia;* L, *D. leichhardtii;* M, *D. metel;* Md, *D. meteloides;* P, *D. pruinosa;* Q, *D. quercifolia;* S, *D. stranomium;* and Hy, hypothetical form.)

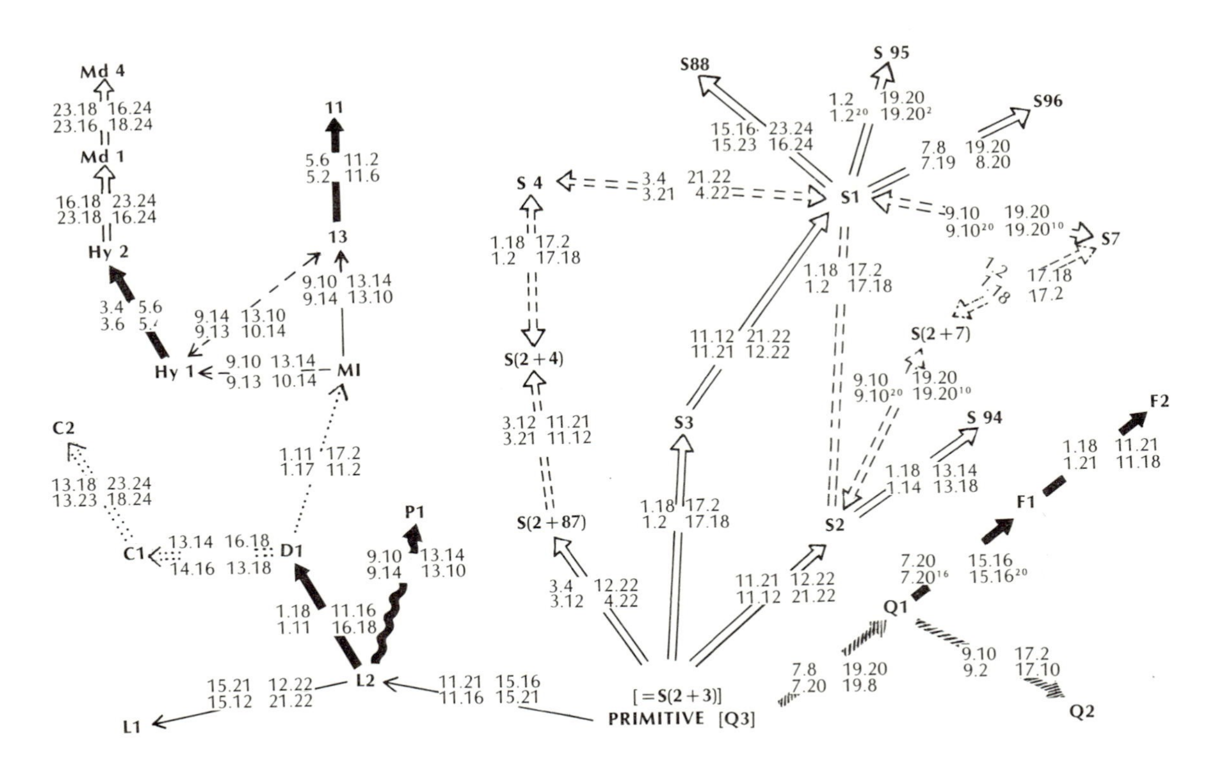

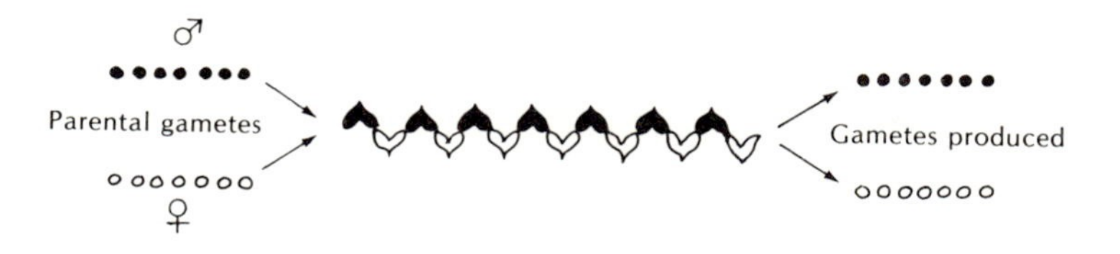

Fig. 8-10: Precise segregation of maternal and paternal genomes in *Oenothera lamarckiana*. Chromosomal behavior: The genome introduced by the pollen is shown solid black; the genome from the egg is white. The haploid number is seven. The alternating row of paternal and maternal chromosomes (middle) actually forms a ring, but for purposes of illustration, the ring is broken and extended.

circles and paired chromosomes possible between a circle of 14 and 7 pairs. The populations, often called "species," with paired chromosomes, usually have large flowers, and cross-pollination is common. Those populations with circles of 14 chromosomes have small flowers and are self-pollinated. Self-pollination is correlated, therefore, with species carrying two super-chromosomes and enforced heterozygosity.

All this adds up to the paradoxical situation that evening primrose species with rings of 14 chromosomes maintain maximum variability through heterozygosity, which is achieved by intensive inbreeding, i.e., self-pollination. This is, of course, precisely the opposite of what we would expect from self-pollination. But it is the outcome of the fascinating system of multiple translocations and balanced lethals that have evolved here. The whole system is selected for because it guarantees maximal heterosis or hybrid vigor in coadapted heterozygous genotypes.

The analysis of these populations with their different distributions of rings and pairs has allowed the same phylogenetic analysis we saw in *Datura* (Fig. 8–9). *Oenothera* is, as Dobzhansky has remarked "a genetic *tour de force* . . . ." Another *tour de force* has been the elucidation of its genetics and evolution. Together, the analysis and the system analyzed, provide us with an extraordinary insight into the chromosomal history of this genus of flowering plants.

References

Baur, E., 1924. (Original in German. For summary and references see Dobzhansky, 1970, below.)

Blair, A. P., 1941. Variation, isolating mechanisms and hybridization in certain toads. *Genetics 26:* 398–417.

Cleland, R. E., 1972. *Oenothera: Cytogenetics and Evolution.* Academic Press, New York.

Dobzhansky, T., 1970. *Genetics of the Evolutionary Process.* Columbia University Press, New York.

Dobzhansky, T., F. J. Ayala, G. L. Stebbins, and J. W. Valentine, 1977. *Evolution.* W. H. Freeman, San Francisco.

Gordon, M. 1948. Effects of five primary genes on the side of melanomas in fishes and the influence of two color genes on their pigmentation. In *The Biology of Melanomas.* Special Publication of the New York Academy of Sciences *4:* 216–268.

Grant, V., 1971. *Plant Speciation.* Columbia University Press, New York.

Grant, K. A., and V. Grant, 1964. Mechanical isolation of *Salvia apiana* and *Salvia mellifera. Evolution 18:* 196–212.

Hubby, J. L., and L. H. Throckmorton, 1965. Protein differences in *Drosophila.* II. Comparative species genetics and evolutionary problems. *Genetics* 52: 203–215.

Jones, J. M., 1975. Effects of thirty years of hybridization on the toads *Bufo americanus* and *Bufo woodhousii-fowleri* at Bloomington, Indiana. *Evolution 27:* 435–448.

Moore, J. A., 1950. Further studies on *Rana pipiens* and racial hybrids. *American Naturalist 84:* 247–254.

Pijl, L. van der, and C. H. Dodson, 1966. *Orchid Flowers: Their Pollination and Evolution.* University of Miami Press and Fairchild Tropical Garden, Coral Gables, Florida.

Pontecorvo, G., 1943. Viability interactions between chromosomes of *Drosophila melanogaster* and *Drosophila simulans. Journal of Genetics 43:* 51–66.

Prakash, S., 1972. Origin of reproductive isolation in the absence of apparent genic differentiation in a geographic isolate of *Drosophilia pseudoobscura. Genetics* 72: 143–155.

Sonneborn, T. M. 1957. Breeding systems, reproductive methods, and species problems in Protozoa. In *The Species Problem,* E. Mayr (ed.), pp. 155–324. AAAS Publication, Washington, D.C.

Srb, A., and B. Wallace, 1964. *Adaptation.* 2nd ed. Prentice-Hall, Englewood Cliffs, N. J.

Wallace, B., 1966. *Chromosomes, Giant Molecules, and Evolution.* Norton, New York.

Wilson, A. C., 1977. Gene regulation in evolution. In *Molecular Evolution,* F. J. Ayala (ed.), pp. 225–234. Sinauer Associates, Sunderland, Mass.

NINE

Speciation and Phylogeny

WE STUDY EVOLUTION to understand the diversity of life. The problem of organismic diversity can be divided into two parts: speciation (the origin of species) and phylogeny (the history of species diversity). Only when we understand the first can we understand the second. The preceding chapters have been devoted to elucidating the origin of species, and we now turn to the second problem phylogeny. Because phylogeny arises from the process of speciation it is appropriate to start its study in Part 3, which is devoted to speciation. In Part 4, we will be almost exclusively concerned with phylogenetic problems.

Speciation and organismic diversity

Before addressing phylogeny and its problems, it is necessary to look at the species concept one more time. This time we are viewing it as a foundation on which to build phylogenetic studies.

THE REALITY OF SPECIES

When we view species as reproductive communities we have objective criteria for defining a species. We realize that a species is a dynamic entity and that any species we look at is at some stage in its evolution. Depending upon that stage, its reproductive isolation will be apparent or measurable in certain terms. It may form no hybrids with other species. Or it may form some, but for various reasons the hybrids may be at a selective disadvantage. Or there may simply be unsuccessful attempts at mating. And so on, depending on the isolating mechanisms in force.

The species is the taxon in which reproductive isolation operates. No other taxon is so clearly definable in biological terms. All the others, from semispecies to subspecies and deme, to variety, clinal variety, or other differentiated group within the species, are more arbitrarily defined. Since they can interbreed, their differences must be determined phenotypically. If measurements are used, their differences are often expressed as tendencies toward certain phenotypic norms. Also in the taxa extending from genera to the kingdoms, the differences are somewhat arbitrary—they reflect the experience of those who have found useful ways of distinguishing taxa. Here phenotypic differences may be qualitative rather than quantitative, but they differ from the concept of reproductive isolation.

This special status of a species is reinforced when we consider the function of a species. Mayr says that a species has three functions. They are (1) adaptation to an environment, (2) successful competition with other species, and (3) maintenance of reproductive isolation. Above the species level, no comparable functions can be attributed to any taxa—to genera, families, orders, classes, phyla, or kingdoms. Below the species level, certain adaptations and competitive features are meaningful, as in the local adaptive norms of polytypic species. But because of gene flow, the local phenotypes exist within the gene pool and can be shared. Therefore these differences between demes are best described not as isolated differences, but as differences in gene frequency when the local frequencies of alleles differ from those of the whole pool.

Thus a species is a functional evolutionary unit. It is an objectively defined unit of coadaptations maintained as an isolated gene pool. That same unit, therefore, must be the basis of phylogenetic studies.

Non-Mendelian populations as species. Even though species as Mendelian populations make up the great majority of species, there are some interesting exceptions. We refer to groups of phenotypically similar organisms in which sexuality is minimal or absent. Many years ago, Fisher suggested that they might be called *ecospecies* and Mayr, commenting on them, suggested that the members of such an ecospecies probably occupied one adaptive peak. What is the basis for these interpretations of asexual populations?

The first point is that a non-Mendelian population would appear to be a perfectly good species to Linnaeus. That is, in morphological terms, Mendelian and non-Mendelian populations are equally good species. Their members tend to look more like each other than they look like any other population. In fact, one of the species first described by Linnaeus was the giant ameba, which, presumably because of its unorganized and ever-changing appearance, he called *Chaos chaos* (Fig. 9–1). We can still recognize a member of this species when we find it, and no one has ever discovered any sexual processes going on among its members. It reproduces exclusively by asexual means. All the blue-green algae are also apparently devoid of sexuality and so are the euglenoid algae. In fact, asexual reproduction by fission is the predominant mode of reproduction throughout the Monera and Protista. Also, asexual reproduction is a common mode of proliferation among the Metaphyta and Fungi. It is even known among the Metazoa. However, in the present discussion our emphasis will be on organisms that have a recognizable species phenotype but as yet show no sexuality whatever.

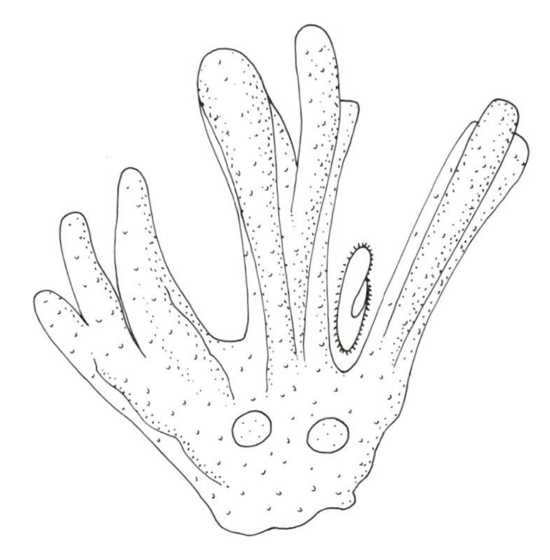

Fig. 9-1: A drawing of the ameba *Chaos chaos*, named by Carolus Linnaeus and first described in the eighteenth century. Subsequent work resulted in other names, such as *Pelomyxa carolinensis*. Overlapping descriptions of the same species occur when modern workers, using better microscopes, are able to see more detail. They assume that they are looking at a different species, to which they assign a different name, but in time, a single taxonomic name is finally assigned.

There then arise questions as regards the nature of the gene pool in asexual ecospecies. Do they even have one? No, they do not have one pool; they have many. Each line of asexual descent is a clone. Each clone is reproductively isolated from every other one. What then keeps the different organisms looking like each other? Where is the cohesiveness of the gene pool we have talked about so much? Where are the coadapted epigenotypes in the absence of gene flow? They are absent. Since there is no gene flow, there is no interbreeding between clones; there is no sharing of genes. There is only descent from one asexual generation to the next, in parallel but completely isolated lines of clonal continuity in which genetic variation is derived only from gene mutations.

We must expand Fisher's ecospecies concept here. He felt that such populations were adapted to a certain ecological niche and selection kept them relatively uniform. Selection was pushing them to the top of their adaptive peak. Any organism on that peak survived, and it did so because it had the needed functional or ecological phenotype. Such a concept also suggests why sexuality is not found in these species. In sexuality there is continual experimentation in the production of new genotypes. But sexuality seems unnecessary when the needed phenotype has already evolved. In fact, it simply tends to threaten the organisms' position on their adaptive peak. They have the needed genes and so natural selection apparently put an end to genetic experimentation by means of sex. The asexual organism now functions well with what is already working well.

If all of this is true, it brings us to the final conclusion that the absence of sexuality in ecospecies is not a primitive condition, but an evolved one. Sexuality was lost, because in these forms losing it was an advantage. The implications of this line of thinking have still not been fully explored. It suggests the possibility that the number of species we have on this earth represent a certain number of ecological niches that provide the matter and energy for the maintenance and reproduction of life. Evolution by natural selection is the process that helps living things locate these niches. Such niches are predictably variable in their stability: some are constantly available, whereas others are changeable. (They could be changeable in several ways—durability, size, etc.) Asexual species or ecospecies represent adaptation to the few very stable niches. Sexually reproducing outbreeders represent adaptations to highly unstable niches. And then in between, across the spectrum from stable to unstable niches, are species with varying degrees of reliance on sexuality as a source of new genotypes, ranging from asexual ones to the inbreeders and then to the outbreeders. Much of this range of breeding patterns are found among the sibling species of the *Paramecium aurelia* group of ciliates, as shown by the American geneticist Tracy Sonneborn in a pioneering analysis of species structure in these protistans.

Such a conclusion again emphasizes our view of the species as the basic unit of evolution. And we also see again the never-ending struggle for existence as a process wherein the survivors are simply those that reach and stay on top of an adaptive peak. Surprisingly, the lowly ameba may be one of the most successful of all organisms, if success is to be measured in terms of remaining securely on such a peak without any recourse to sexually generated genetic variations.

TRANSSPECIFIC EVOLUTION

The distinguished German biologist Bernhard Rensch used transspecific evolution as a term to cover all evolutionary change beyond the establishment of a good species. From the perspective of a species and natural selection, there is only further adaptation as a species, since all organisms exist only as members of a species (with or without sexuality). From the perspective of continuing divergence among species, we see what Rensch and others mean. Transspecific evolution is meaningful only as a comparative study, in which we see how descendants of an ancestral species actually continue to diverge from one another.

Transspecific evolution, and what we earlier referred to as macroevolution, are the same thing. But in microevolution, as in species adaptation, we examine what goes on within a single species. The transition from micro- to macroevolution, from speciation to transspecific evolution, is the origin of species, either by transformation or by diversification.

This means, then, that the major process underlying transspecific evolution is still natural selection. There was a time when certain biologists included in their thinking the possibility of macromutations. Such mutational events were believed to have consequences that generated the differences that not only demarcated species from one another, but they also allowed differentiation among genera, families, or other higher taxa. Thus, the origin of the higher taxa was not continuation of the origin of species, but depended upon special kinds of genetic changes. As we said before, when we touched on this subject, no evidence has been presented to support this view of evolution above the species level; it has been critically reviewed by Sewall Wright in 1978. We will proceed on the basis that transspecific evolution is the result of the processes of natural selection, which are also seen at work at the species level.

How much evolutionary history can we hope to read? If life is an unbroken series of successive generations, it represents an extraordinary continuum going back, unbroken, to its origins. How much of that history can we decipher? And, of what we decipher, how much detail will it contain? How do we read the history of life on this earth? That is the ultimate problem of phylogenetic studies. It can be broken down into two questions: Can we reconstruct the history of life? Can we understand the reconstruction? Both questions are essential because it is one thing to describe the sequence of historical events and it is another thing to understand that sequence so as to explain why some things happened and others did not. This historical analogy is a useful one to keep in mind when we consider the fundamentals of phylogenetic analysis.

The problem area of phylogeny. We have now opened up

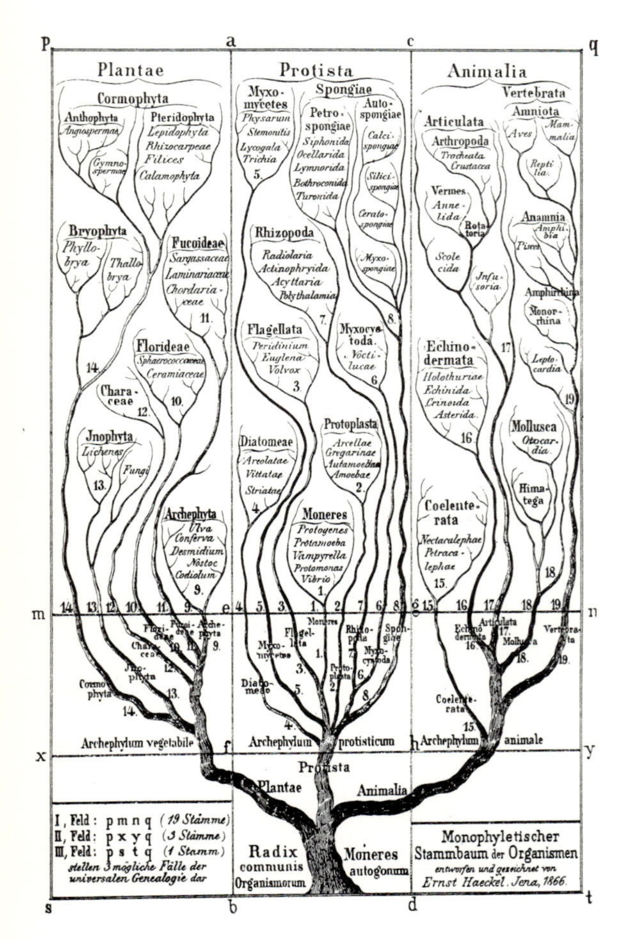

Fig. 9-2: Haeckel's phylogeny of living things. Note the attempt to depict a tree with a stout trunk and progressively thinner branches and twigs. Haeckel's three kingdoms—Plantae, Protista, and Animalia—are placed where the first branches come from a common ancestral "trunk," *Radix communis organismorum* or *Moneres autogonum.* This latter genus is the source of the name Monera for the prokaryotes. (The term, or concept, of prokaryote was unknown to Haeckel.) Just below Protista, at the top of the middle column, is placed Spongiae. Note Haeckel's placement of Fungi as a small group just below the middle of Plantae.

the possibility of including, under phylogeny, the whole area of organismic diversity, both as a present-day problem and as a historical one. Thus we are going to need information ranging from the origin of life to the present and ranging from species to kingdoms and across all the kingdoms. Because it is a herculean task we can only try to cover its essentials in these pages. And what are the essentials?

First, we need a scientifically reliable analytical method. In earlier chapters we have seen how work on phylogeny was based on chromosome studies. We have also seen how molecular studies of the proteins of different species help reconstruct the history of species. Systematics, too, can reflect hierarchical relations among species by using degrees of difference or similarity among them. All of these clearly suggest that we can read evolutionary history. But what common methodology, if there is one, underlies them? How can we tell if we are indeed reading the history of life? And if we are, how can it be applied to pursue phylogenetic studies? And most importantly, do we have a method by which we can recognize the reproductive community as the fundamental evolutionary unit and natural selection as the predominant agent for evolutionary change?

There is probably no area of biology today in which there is sharper disagreement than in phylogeny. Ernest Haeckel, in the late nineteenth century, was one of the first biologists to apply Darwin's ideas on natural selection and evolution to rather detailed phylogenetic speculations. He summarized his thinking in what is now called a dendrogram or a phylogenetic tree (Fig. 9–2), but it is hard to reconcile that dendrogram with some of the ideas we hold today. For example, Haeckel recognized three kingdoms, Plantae, Protista, and Animalia. In the protists he included the bacteria and the sponges (see Appendix I and Chapter 1), in the belief that these groups originated from one ancestral stock.

But the differences are not just those between workers of 100 years ago and better informed, modern researchers. Among the latter there are many differences of opinion. There is still no good agreement on how eukaryotic cells originated from prokaryotic ones; how land plants evolved; and how the flowering plants developed. The origin of multicellular animals and of the chordates are also areas of considerable controversy. In brief, there are many problems in reading the full history of life. But we must come back to our original question: What method is being used here? There is agreement, one hopes, on how to go about the job, even though there is real disagreement on the conclusions emerging from all this work.

HOMOLOGY: THE PREREQUISITE FOR PHYLOGENY

Although it has been in use as long as the idea of natural selection itself, there is still considerable confusion regarding the concept of homology. Some evolutionists insist that the concept is central to the study of phylogeny, others that it is worthless. We shall take the former position because, as we will show, homologies define the action of natural selection. For that reason phylogeny rests on the study of homology.

Let us define the concept of homology by looking first at its history. Darwin's contemporary, the British anatomist Richard Owen, was the first to use the idea of homology. He recognized serial homology and special homology. By serial homology he meant similar structures as one proceeded along the length of an animal's body. The legs of a lobster (Fig. 1–7) are similar in structure, especially the larger walking legs. Our ribs and our vertebrae are further examples. But serial homology is less important than special homology, which Owen defined as similarities between structures in different organisms. He argued that our arms were homologous to the fore-flipper of a whale, the wing of a bat or a bird, and the foreleg of a crocodile or a frog (Fig. 9–3). As an anatomist Owen was simply describing general patterns in what he observed. But all this was thrown into a new light by Darwin's idea of evolution by natural selection. Natural selection could explain Owen's descriptions with the idea of descent from a common ancestor. Since natural selection preserves adaptive charac-

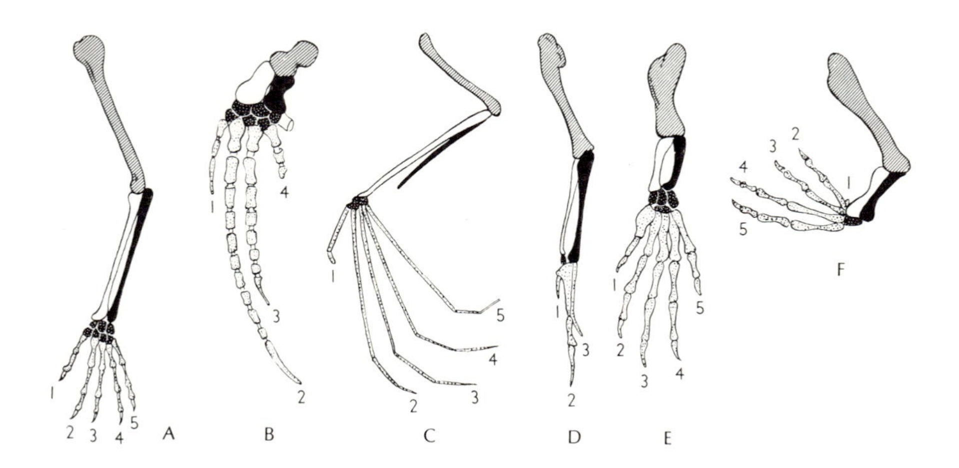

Fig. 9-3: Homologies in the anterior appendage of selected vertebrates: forearm of a human (A); flipper of a whale (B); wing of a bat (C); wing of a bird (D); foreleg of a crocodile (E); and foreleg of a frog (F). Homologous bones are identified by cross-hatching (humerus), solid black (ulna), white (radius), dark stippling (metacarpals), and light stippling (phalanges). The phalanges are numbered as are the human fingers.

ters, species with a common ancestor would continue to be similar. In modern times, the descendants still shared certain features of their epigenotypes, and this appeared as phenotypic similarities.

Because of this Darwinian explanation, the concept of homology changed from a statement describing similar structures to the idea of similarity due to descent from a common ancestor, and then, further, of similarity due to certain common genes or certain common developmental capacities (presumably determined by genes).

This approach lead to certain difficulties of interpretation, which were brought to a focus by the distinguished German biologist Adolf Remane. He pointed out, in a classic discussion of homology published in 1956, that people were confusing identification of homologies with explanations of homologies. The operational problem is to look at two structures in different organisms and state whether or not they are homologous. If they are, then we can say something about descent from a common ancestor. That is why phylogeny depends on homology.

It must be emphasized that Remane did not deny the genetic-developmental explanation of homologies, but he did point out that such an explanation was often useless in identifying homologies. For example, how do we determine that the same gene (or genes) determines the formation of the human arm and the whale flipper? Certainly, breeding analyses cannot be done. Further, recall how little we know of the complex steps that lie between genes and the development of a major phenotypic character such as an arm or a flipper. Even a comparison of limb formation in the whale and human embryos might not be convincing. In a discussion of the problems associated with homology, the British developmental biologist de Beer points out that structures long considered homologous, such as the digestive system of vertebrates, can develop in different ways in embryos of different species.

Remane's fundamental contribution provides us with the key criteria for identifying homologies. And, these criteria are based on the assumption that complex, point-to-point similarities can only exist in species with a common ancestor. The explanation of homologies is implicit in the criteria for identifying them.

Defining homology. Remane proposed three criteria for identifying homologies—the *positional,* the *compositional,* and the *serial.* The positional criterion emphasizes that homologous

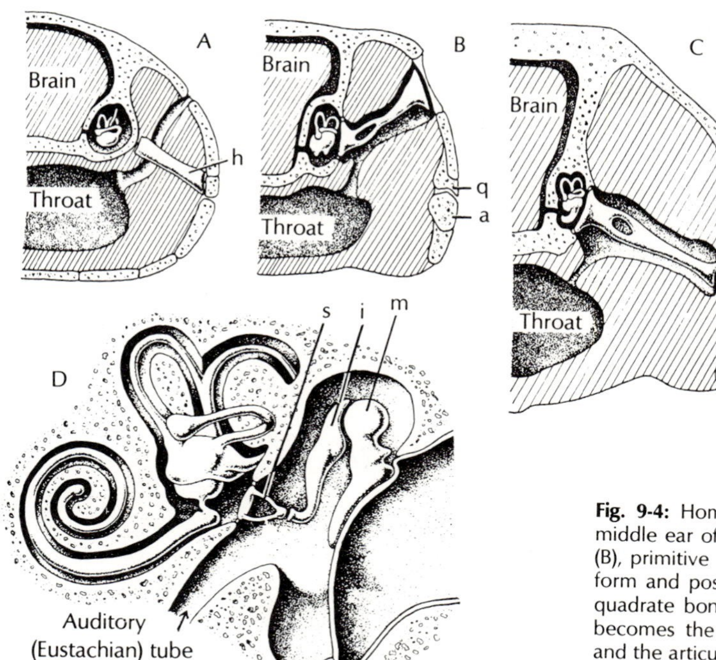

Fig. 9-4: Homologies in the cranial bones and bones of the middle ear of certain vertebrates: fish (A), primitive amphibian (B), primitive reptile (C), and mammal (D). Note the changes in form and position of the articular (a), hyomandibular (h), and quadrate bones (q), and especially the hyomandibular, which becomes the stapes (s). The quadrate becomes the incus (i) and the articular the malleus (m).

structures must lie in similar positions in the organisms where they are found. (Note the similar positions of arms, foreflippers, wings, and forelegs in Figure 9–3.) The compositional criterion says that homologous structures have similar parts—that is, similar in number, similar in the way the parts of the structure are attached, and similar in the substances of which they are made up. (Again look at the skeletal structures in Fig. 9–3.) The serial criterion, as we shall see, is the most important one for phylogeny. It says that if otherwise dissimilar structures can be related through intermediate structures, which are homologous by the positional and compositional criteria, then the dissimilar structures are also homologous. A classic example of this is shown in Fig. 9–4. Here are shown bones that are on the surface of the skull in fish and bones that lie inside the skull in mammals. Although they are quite dissimilar in shape and in the way they fit next to each other in these extreme forms, these bones are homologous, since there are intermediate forms that form a series in which neighboring forms fulfill the positional and compositional criteria and thus relate the extremes of the series.

Now let us step back from these three criteria to see more clearly just what Remane's methodology means. The answer is straightforward. It means that the identification of homologous forms depends on point-to-point comparisons. This is seen in greater detail in Fig. 9–5. It further means that homology is a comparative study that expresses a *relationship* between two or more entities. Fundamentally, the relationship is that of

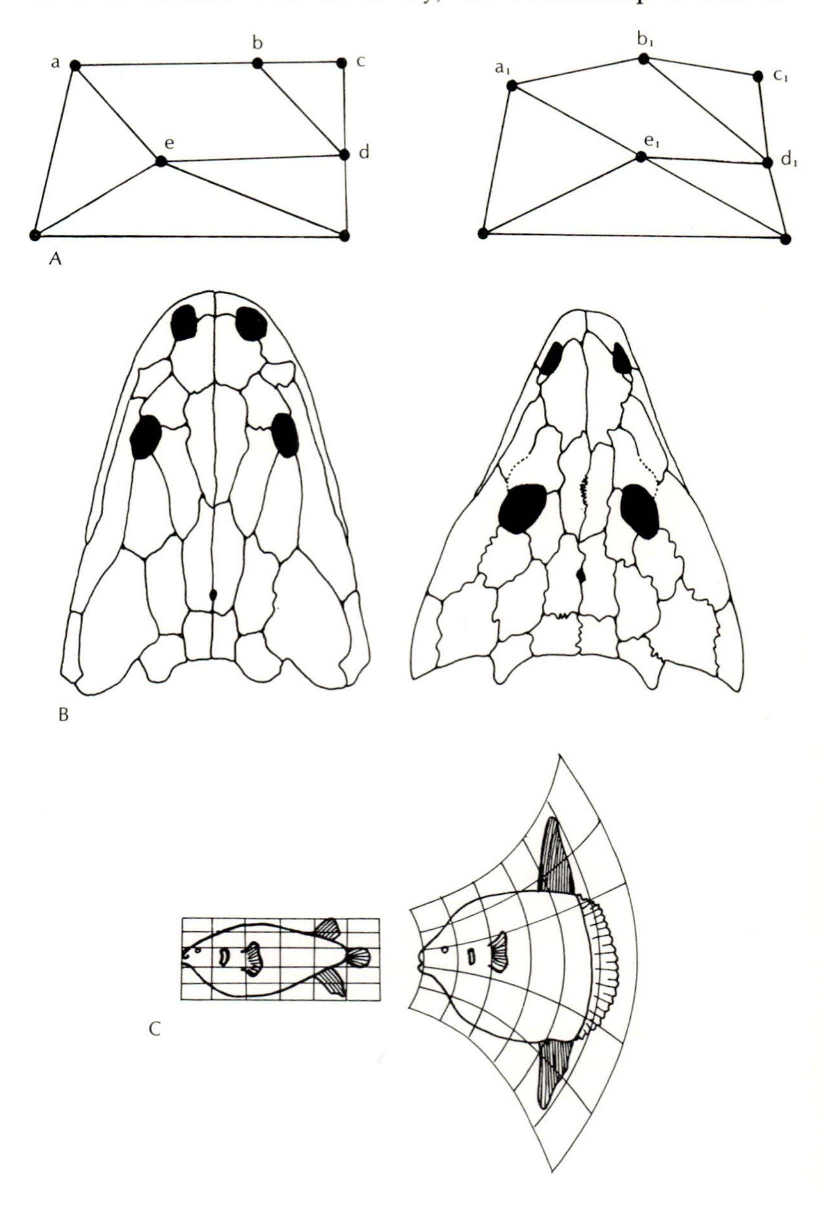

Fig. 9-5: Point-to-point comparisons as the basis for identifying homologous relations between structures. A. Two geometrical figures show obvious similarities among identifiable labeled points, even though the points are somewhat displaced in the two figures. B. Skulls of two different fossil reptiles. A close study reveals that the bones in each skull are precisely homologous, despite some differences in shape and size. C. Coordinates show how the body of the fish *Diodon* can be homologized to that of *Orthagoriscus.* All three examples are aposemic homologies.

similarity, but it can be extended, by serial relationships or the serial criterion, to dissimilar entities. Owen would have had a difficult time accepting that latter point. Remane also pointed out that complex behavior, when it is inherited, as it is for example, in the food gathering and courtship habits of various animals, can be homologized.

All of this means that the idea of homology can now be redefined in recognition of its application to similar and dissimilar entities, which are both structures and functions. Here we will discuss first the revised definition and then its applications.

Homology is a direct or a derived similarity of structures and functions of different organisms. The direct similarities are determined by positional and compositional relationships and the derived ones by serial relationships. In all cases point-to-point comparisons are used to determine similarities or their absence.

Homologous structures. The application of the foregoing definition to structures is quite obvious, at least in terms of the morphology and anatomy of readily visible organisms and their major parts. But biological structures include tissues, cells, and molecules. The application of homology to these microscopic and submicroscopic structures needs some further discussion.

As we saw in Chapter 1, there are only two major types of organisms—those with prokaryotic cells and those with eukaryotic cells. Immediately two questions come to mind: Are all prokaryotic cells homologous? And, are all eukaryotic cells homologous? The best answer, just now, is probably yes. Figure 1–2 described the major features of these cells; in compositional terms they appear homologous. The positional relationship is meaningless when we try to compare cells in unicellular organisms, since the cell is the organism. What is more meaningful is to consider parts of the cell, which are, of course, parts of the organism, and at that level rigorous application of the positional and compositional relationships is possible. In those terms, it appears convincing that all prokaryotic cells are homologous, as are all eukaryotic ones.

We say appears convincing because an important warning regarding homologies at the cellular level has been issued by some biologists. The British biologist C.F.A. Pantin has characterized the problem as "limitations of organic design." By this he means that just as human engineers are limited by their materials—steel, wood, concrete, etc.—so cellular systems are limited by their materials—macromolecules, such as proteins, lipids and sugars, and the ways they can interact. Therefore, if there is a selection pressure in favor of locomotion, it may be that something like a cilium is the best answer and may be invented or evolved independently several times over. A cilium is remarkably similar in composition in all the ciliated eukaryotic cells. But, Pantin is asking, Does that similaritiy mean descent from a common ancestor or does it mean more than one independent attempt at invention has come up with the same answer? The former is homology and the latter is convergence. Homology is essential to correct phylogenetic interpretation and convergence often confuses those interpretations.

As we will see, the compositional and serial relationships are the keys to uncovering and identifying convergences. We will discuss that shortly. Here we must complete our consideration of structural homologies at the cellular and molecular levels.

The essentials of Pantin's concern for the limitations of organic design leave us very wary of homologizing all the organelles and other fine structures of cells. The tendency today is to consider the relative uniformity of such cell organelles as cilia and flagella, microtubules, plasma membranes, nuclei and chromosomes, mitochondria, the Golgi apparatus, endoplasmic reticulum, and ribosomes (Fig. 1–2) as unresolved in terms of homology. However, eukaryotic cells show many specializations (Fig. 9–6), and it is among these that the search for homologies is more meaningful.

It is more meaningful because such specializations are often limited to certain taxa, and so they set these taxa apart from others. It is just that situation which is of interest to the phylogenist. It allows him or her to use the serial criteria to describe what group was ancestral to the divergent one (or ones) and to use natural selection to explain the cause of divergence.

In summary then, the two distinct types of cells are seen as two major adaptive responses resulting in a certain set of common features in each group. What evolution has done in particular cases with those two engineering solutions in designing a functional cell is what primarily engages our interests in phylogeny.

As for the molecular level, three kinds of studies are of phylogenetic interest. These are (1) sequence studies of amino acids in proteins or nucleotides in nucleic acids; (.2) physical characteristics of macromolecules, such as electrophoretic be-

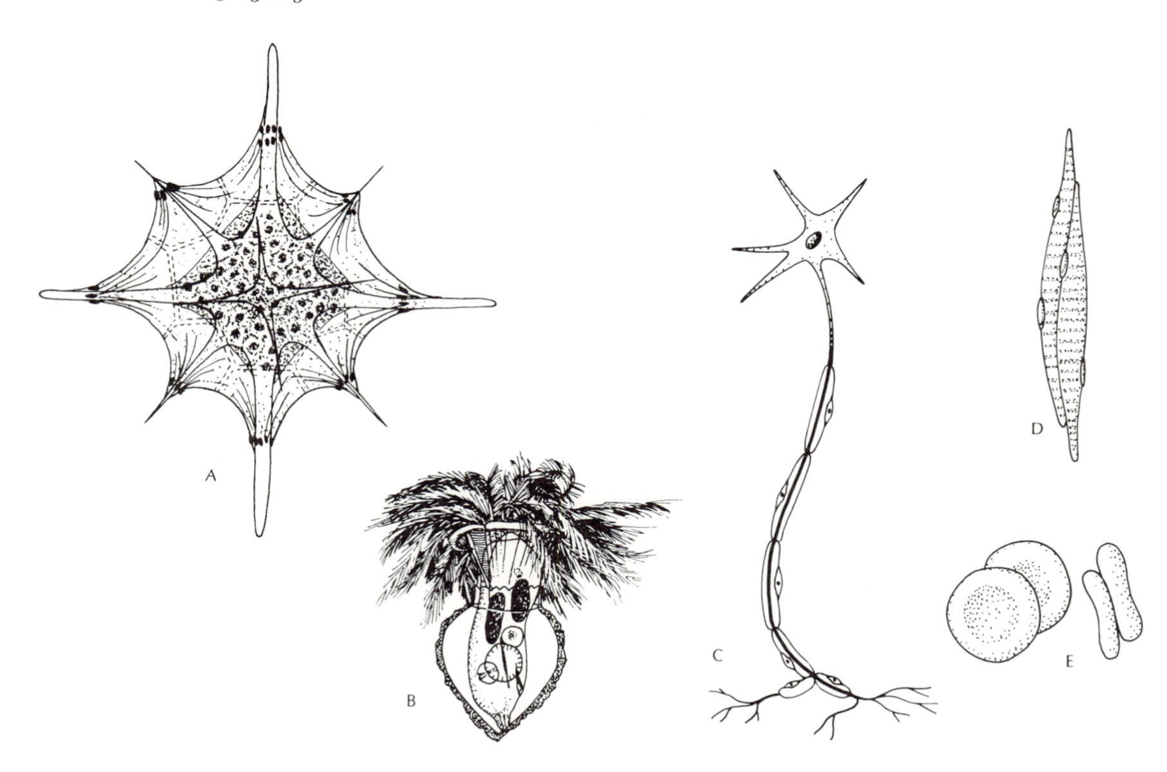

Fig. 9-6: Specialized eukaryotic cells. Free-living eukaryotic organisms include *Acanthostaurus purpurascens,* a floating ameboid protozoan (A), and *Tintinnopsis nucula,* a ciliated protozoan that lives in its test or shell (B). Examples of cells in tissues include a nerve cell (C), a striated muscle cell (D), and human red blood cells, viewed on their concave surface and from the side (E).

havior or sedimentation rates; and (3) biological traits, such as enzymatic abilities or immunological properties. All these have been discussed earlier in one context or another, except for the concept of homology, which we will discuss now.

Sequence studies are readily understood in the light of homologies. The sequence of amino acids along a polypeptide is a perfect example of point-to-point correspondence or its absence (Table 9–1). The same is true of nucleotide sequences. These latter sequences can be compared both in terms of actual bases or of hybridization. In this technique (Box 9–1), the similarity between DNA molecules or DNA and RNA molecules from different species is established by determining how complete a double helix is formed when single polynucleotide strands from both molecules being studied come into contact.

Determinations of physical characteristics is harder in terms of point-to-point comparisons. It is known, however, that the electrophoretic mobility of a molecule depends on its size and the charges on its surface. Similar mobility means similar size and charges (Box 4–3). Hence, there would be some correspondence beween molecules with similar mobility, but the

Table 9–1. Sequences of amino acids in cytochrome c.

										1										2										3					
	1	2	3	4	5	6	7	8	9	0	1	2	3	4	5	6	7	8	9	0	1	2	3	4	5	6	7	8	9	0	1	2	3	4	5
human	–	–	–	–	–	–	–	–	G	D	V	E	K	G	K	K	I	F	I	M	K	C	S	Q	C	H	T	V	E	K	G	G	K	H	K
rhesus monkey	–	–	–	–	–	–	–	–	(G ·	D ·	V ·	E ·	K ·	G ·	K ·	K ·	I ·	F)	I ·	M ·	K	(C ·	S ·	Z ·	C ·	H ·	T ·	V ·	Z ·	K =	G ·	G ·	K ·	H =	K
horse	–	–	–	–	–	–	–	–	G	D	V	E	K	G	K	K	I	F	V	Q	K	C	A	Q	C	H	T	V	E	K	G	G	K	H	K
pig, bovine, sheep	–	–	–	–	–	–	–	–	G	D	V	E	K	G	K	K	I	F	V	Q	K	C	A	Q	C	H	T	V	E	K	G	G	K	H	K
dog	–	–	–	–	–	–	–	–	G	D	V	E	K	G	K	K	I	F	V	Q	K	(C ·	A ·	Q ·	C ·	H ·	T ·	V ·	E)	K .	G	G	K	H	K
gray whale	–	–	–	–	–	–	–	–	G	D	V	E	K	G	K	K	I	F	V	Q	K	C	A	Q	C	H	T	V	E	K	G	G	K	H	K
rabbit	–	–	–	–	–	–	–	–	G	D	V	E	K	G	K	K	I	F	V	Q	K	C	A	Q	C	H	T	V	E	K	G	G	K	H	K
kangaroo	–	–	–	–	–	–	–	–	G	D	V	E	K	G	K	K	I	F	V	Q	K	C	A	Q	C	H	T	V	E	K	G	G	K	H	K
chicken, turkey	–	–	–	–	–	–	–	–	G	D	I	E	K	G	K	K	I	F	V	Q	K	C	S	Q	C	H	T	V	E	K	G	G	K	H	K
penguin	–	–	–	–	–	–	–	–	G	D	I	E	K	G	K	K	I	F	V	Q	K	C	S	Q	C	H	T	V	E	K	G	G	K	H	K
Pekin duck	–	–	–	–	–	–	–	–	G	D	V	E	K ·	G	K ·	K	(I ·	F ·	V ·	Q)	K	(C ·	S ·	Q ·	C ·	H ·	T ·	V ·	E)	K ·	G	G	K ·	H	K
pigeon	–	–	–	–	–	–	–	–	(G ·	D ·	I ·	E)	K ·	G	K ·	K	(I ·	F ·	V ·	Q)	K	(C ·	S ·	Q ·	C ·	H ·	T ·	V ·	E)	K ·	G	G	K ·	H	K
snapping turtle	–	–	–	–	–	–	–	–	G	D	V	E	K ·	G	K	K	I	F ·	V	Q	K	C	A	Q	C	H	T	V	E	K	G	G	K	H ·	K
rattlesnake	–	–	–	–	–	–	–	–	G	D	V	E	K	G	K	K	I	F	(T ·	M ·	K ·	C ·	S ·	Q ·	C ·	H ·	T ·	V ·	E ·	K ·	G ·	G ·	K ·	H)	K
bullfrog	–	–	–	–	–	–	–	–	G	D	V	E	K	G	K	K	I	F	(V ·	Q ·	K ·	C ·	A ·	Q ·	C ·	H ·	T ·	C ·	E ·	K ·	G ·	G ·	K ·	H)	K
tuna fish	–	–	–	–	–	–	–	–	G	D	V	A	K ·	G	K	K ·	T	F	V	Q	K ·	C	A	Q	(C ·	H)	T	V	E	N	G	G	K ·	H	K
dogfish	–	–	–	–	–	–	–	–	G	D	V	E	K	G	K	K	V	F	V	Q	K	C	A	Q	C	H	T	V	E	N	G	G	K	H	K
lamprey	–	–	–	–	–	–	–	–	G	D	V	E	K	G	K	K	V	F	V	Q	K	C	S	Q	C	H	T	V	E	K	A	G	K	H	K
fruit fly	–	–	–	–	G	V	P	A	G	D	V	E	K	G	K	K	L	F	V	Q	R	C	A	Q	C	H	T	V	E	A	G	G	K	H	K
screwworm fly	–	–	–	–	G	V	P	A	G	D	V	E	K	G	K	K	I	F	V	Q	R	C	A	Q	C	H	T	V	E	A	G	G	K	H	K
silkworm moth	–	–	–	–	G	V	P	A	G	N	A	E	N	G	K	K	I	F	V	Q	R	C	A	Q	C	H	T	V	E	A	G	G	K	H	K
tobacco horn worm moth	–	–	–	–	G	V	P	A	G	N	A	D	N	G	K	K	I	F	V	Q	R	C	A	Q	C	H	T	V	E	A	G	G	K	H	K
wheat	A	S	F	S	E	A	P	P	G	N	P	D	A	G	A	K	I	F	K	T	K	C	A	Q	C	H	T	V	D	A	G	A	G	H	K
Neurospora crassa	–	–	–	–	G	F	S	A	G	D	S	K	K	G	A	N	L	F	K	T	R	C	A	E	C	H	G	E	G	G	N	L	T	Q	K
baker's yeast	–	–	–	T	E	F	K	A	G	S	A	K	K	G	A	T	L	F	K	T	R	C	E	L	C	H	T	V	E	K	G	G	P	H	K
Candida krusei	–	–	P	A	P	F	E	Q	G	S	A	K	K	G	A	T	L	F	K	T	R	C	A	E	C	H	T	I	E	A	G	G	P	H	K

										7										8										9					
	1	2	3	4	5	6	7	8	9	0	1	2	3	4	5	6	7	8	9	0	1	2	3	4	5	6	7	8	9	0	1	2	3	4	5
human	K	N	K	G	I	I	W	G	E	D	T	L	M	E	Y	L	E	N	P	K	K	Y	I	P	G	T	K	M	I	F	V	G	I	K	K
rhesus monkey	K ·	N)	K	G	I	T	W	G	E	(D ·	T ·	L)	M	(E ·	Y)	L	E	N	(P ·	K ·	K ·	Y =	I ·	P ·	G ·	T ·	K ·	M =	I ·	F =	V ·	G ·	I ·	K)	K
horse	K	N	K	G	I	T	W	K	E	E	T	L	M	E	Y	L	E	N	P	K	K	Y	I	P	G	T	K	M	I	F	A	G	I	K	K
pig, bovine, sheep	K	N	K	G	I	T	W	G	E	E	T	L	M	E	Y	L	E	N	P	K	K	Y	I	P	G	T	K	M	I	F	A	G	I	K	K
dog	K	N	K	G	I	T	W	G	E	E	T	L	M	E	Y	L	E	N	P	K	K	Y	I	P	G	T	K	M	I	F	A	G	I	K	K
gray whale	K	N	K	G	I	T	W	G	E	E	T	L	M	E	Y	L	E	N	P	K	K	Y	I	P	G	T	K	M	I	F	A	G	I	K	K
rabbit	K	N	K	G	I	T	W	G	E	D	T	L	M	E	Y	L	E	N	P	K	K	Y	I	P	G	T	K	M	I	F	A	G	I	K	K
kangaroo	K	N	K	G	I	I	W	G	E	D	T	L	M	E	Y	L	E	N	P	K	K	Y	I	P	G	T	K	M	I	F	A	G	I	K	K
chicken, turkey	K	N	K	G	I	T	W	G	E	D	T	L	M	E	Y	L	E	N	P	K	K	Y	I	P	G	T	K	M	I	F	A	G	I	K	K
penguin	K	N	K	G	I	T	W	G	E	D	T	L	M	E	Y	L	E	N	P	K	K	Y	I	P	G	T	K	M	I	F	A	G	I	K	K
Pekin duck	K ·	N	K	(G ·	I ·	T ·	W ·	G ·	E ·	D ·	T ·	L ·	M ·	E ·	Y ·	L ·	E ·	N ·	P)	K ·	K	(Y ·	I ·	P ·	G ·	T)	K	(M ·	I ·	F ·	A ·	G ·	I)	K ·	K
pigeon	K ·	N	K	(G ·	I ·	T ·	W ·	G ·	E ·	D ·	T ·	L ·	M ·	E ·	Y ·	L ·	E ·	N ·	P)	K ·	K	(Y ·	I ·	P ·	G ·	T)	K	(M ·	I ·	F ·	A ·	G ·	I)	K ·	K
snapping turtle	K	N ·	K	G	I	T	W	G ·	E	E	T	L	M ·	E	Y ·	L	E	N	P	K	K	Y ·	I	P	G	T	K	M ·	I	F ·	A	G	I	K	K
rattlesnake	K	N ·	K	G	I	I	W ·	G	D	D	T	L	M	E	Y ·	L	E	N	P	K	K	Y ·	I	P	G	T	K	M ·	V	F ·	T	G	L ·	S	K
bullfrog	K	N	K	G	I	T	W	(G ·	E ·	D ·	T ·	L ·	M ·	E ·	Y)	L	E	N	P	K	K	Y	I	P	G	T	K	M	I	F	A	G	I	(K ·	K
tuna fish	K ·	S	K ·	G	I	V	W	(N ·	N ·	D)	T	L	M	E	Y ·	L	E	N	P	K	K ·	Y	(I ·	P ·	G)	T	K	(M ·	I)	F ·	A	G	I	K	K
dogfish	K	S	K	G	I	T	W	Q	Q	E	T	L	R	I	Y	L	E	N	P	K	K	Y	I	P	G	T	K	M	I	F	A	G	L	K	K
lamprey	K	S	K	G	I	V	W	N	Q	E	T	L	F	V	Y	L	E	N	P	K	K	Y	I	P	G	T	K	M	I	F	A	G	I	K	K
fruit fly	K	A	K	G	I	T	W	Q	D	D	T	L	F	E	Y	L	E	N	P	K	K	Y	I	P	G	T	K	M	I	F	A	G	L	K	K
screwworm fly	K	A	K	G	I	T	W	Q	D	D	T	L	F	E	Y	L	E	N	P	K	K	Y	I	P	G	T	K	M	I	F	A	G	L	K	K
silkworm moth	K	A	K	G	I	T	W	G	D	D	T	L	F	E	Y	L	E	N	P	K	K	Y	I	P	G	T	K	M	V	F	A	G	L	K	K
tobacco horn worm moth	K	A	K	G	I	T	W	Q	D	D	T	L	F	E	Y	L	E	N	P	K	K	Y	I	P	G	T	K	M	V	F	A	G	L	K	K
wheat	K	N	K	A	V	E	W	E	E	N	T	L	Y	D	Y	L	L	N	P	K	K	Y	I	P	G	T	K	M	V	F	P	G	L	K	K
Neurospora crassa	K	Q	K	G	I	T	W	D	E	N	T	L	F	E	Y	L	E	N	P	K	K	Y	I	P	G	T	K	M	A	F	G	G	L	K	K
baker's yeast	I	K	K	N	V	L	W	D	E	N	N	M	S	E	Y	L	T	N	P	K	K	Y	I	P	G	T	K	M	A	F	G	G	L	K	K
Candida krusei	K	R	A	G	V	E	W	A	E	P	T	M	S	D	Y	L	E	N	P	K	K	Y	I	P	G	T	K	M	A	F	G	G	L	K	K

Amino acid code

A, alanine
B, asparagine or aspartic acid
C, cysteine
D, aspartic acid
E, glutamic acid
F, phenylalanine
G, glycine
H, histidine
I, isoleucine
K, lysine
L, leucine
M, methionine
N, asparagine
P, proline
Q, glutamine
R, arginine
S, serine
T, threonine
V, valine
W, tryptophan

6	7	8	9	40	1	2	3	4	5	6	7	8	9	50	1	2	3	4	5	6	7	8	9	60	
T	G	P	N	L	H	G	L	F	G	R	K	T	G	Q	A	P	G	Y	S	Y	T	A	A	N	human
·T	·G	·P	·N	=L	·H	=G	·L	·F	=G	·R	·K	·T	=G	·Q	·A	·P	·G	·Y	=S	·Y	=T	·A	·A	·N ·	rhesus monkey
T	G	P	N	L	H	G	L	F	G	R	K	T	G	Q	A	P	G	F	T	Y	T	D	A	N	horse
T	G	P	N	L	H	G	L	F	G	R	K	T	G	Q	A	P	G	F	S	Y	T	D	A	N	pig, bovine, sheep
T	G	P	N	L	H	G	L	F	G	R	K	T	G	Q	A	P	G	F	S	Y	T	D	A	N	dog
T	G	P	N	L	H	G	L	F	G	R	K	T	G	Q	A	V	G	F	S	Y	T	D	A	N	gray whale
T	G	P	N	L	H	G	L	F	G	R	K	T	G	Q	A	V	G	F	S	Y	T	D	A	N	rabbit
T	G	P	N	L	N	G	I	F	G	R	K	T	G	Q	A	P	G	F	T	Y	T	D	A	N	kangaroo
T	G	P	N	L	H	G	L	F	G	R	K	T	G	Q	A	E	G	F	S	Y	T	D	A	N	chicken, turkey
T	G	P	N	L	H	G	I	F	G	R	K	T	G	Q	A	E	G	F	S	Y	T	D	A	N	penguin
(T	·G	·P	·N	·L	·H	·G	·L	·F	·G)	R	·K	(T	·G	·Q	·A	·E	·G	·F	·S	·Y	·T	·D	·A	·N)	Pekin duck
(T	·G	·P	·N	·L	·H	·G	·L	·F	·G)	R	·K	(T	·G	·Q	·A	·E	·G	·F	·S	·Y	·T	·D	·A	·N)	pigeon
T	G	P	N	L	N	G	L	·I	G	R	K	T	G	Q	A	E	G	F	·S	Y	T	E	A	N	snapping turtle
T	G	P	N	L	H	·G	L	F	·G	R	K	T	G	Q	A	V	G	Y	·S	Y	·T	A	A	N	rattlesnake
V	G	P	N	L	Y	G	L	I	G	R	K	T	G	Q	A	A	G	F	S	Y	T	D	A	N	bullfrog
(V	·G	·P	·N)	L	W	·G	L	F	·G	R	·K	T	(G	·Q)	A	E	G	Y	·S	Y	T	(D	·A	·N)	tuna fish
T	G	P	N	L	S	G	L	F	G	R	K	T	G	Q	A	Q	G	F	S	Y	T	D	A	N	dogfish
T	G	P	N	L	Q	G	L	F	G	R	K	T	G	Q	A	P	G	F	S	Y	T	D	A	N	lamprey
V	G	P	N	L	H	G	L	I	G	R	K	T	G	Q	A	A	G	F	A	Y	T	N	A	N	fruit fly
V	G	P	N	L	H	G	L	F	G	R	K	T	G	Q	A	A	G	F	A	Y	T	N	A	N	screwworm fly
V	G	P	N	L	H	G	F	Y	G	R	K	T	G	Q	A	P	G	F	S	Y	S	N	A	N	silkworm moth
V	G	P	N	L	H	G	F	F	G	R	K	T	G	Q	A	P	G	F	S	Y	S	N	A	N	tobacco horn worm moth
Q	G	P	N	L	H	G	L	F	G	R	Q	S	G	T	T	A	G	Y	S	Y	S	A	A	N	wheat
I	G	P	A	L	H	G	L	F	G	R	K	T	G	S	V	D	G	Y	A	Y	T	D	A	N	*Neurospora crassa*
V	G	P	N	L	H	G	I	F	G	R	H	S	G	Q	A	Q	G	Y	S	Y	T	D	A	N	baker's yeast
V	G	P	N	L	H	G	I	F	S	R	H	S	G	Q	A	E	G	Y	S	Y	T	D	A	N	*Candida krusei*

6	7	8	9	100	1	2	3	4	5	6	7	8	9	110	1	2	
K	E	E	R	A	D	L	I	A	Y	L	K	K	A	T	N	E	human
K	·E	·E	·R	·A	·D	·L	=I	·A	·Y)	L	·K	(K	·A	·T	·N	·E	rhesus monkey
K	T	E	R	E	D	L	I	A	Y	L	K	K	A	T	N	E	horse
K	G	E	R	E	D	L	I	A	Y	L	K	K	A	T	N	E	pig, bovine, sheep
T	G	E	R	A	D	L	I	A	Y	L	K	K	A	T	K	E	dog
K	G	E	R	A	D	L	I	A	Y	·L	·K	K	A	T	N	E	gray whale
K	D	E	R	A	D	L	I	A	Y	L	K	K	A	T	N	E	rabbit
K	G	E	R	A	D	L	I	A	Y	L	K	K	A	T	N	E	kangaroo
K	S	E	R	V	D	L	I	A	Y	L	K	D	A	T	S	K	chicken, turkey
K	S	E	R	A	D	L	I	A	Y	L	K	D	A	T	S	K	penguin
·K	(S	·E)	R	·A	D	L	I	A	Y	L	K	·D	A	T	A	K ·	Pekin duck
·K	(A	·E)	R	·A	D	L	I	A	Y	L	K	·Q	A	T	A	K	pigeon
K	A	E	R	A	D	L	·I	A	Y	·L	K	D	A	T	S	K	snapping turtle
K	K	E	R	T	N	L	·I	A	Y	·L	K	E	K	T	A	A	rattlesnake
·K	·G	·E	·R	·Q)	D	L	I	A	Y	(L	·K	·S	·A	·C	·S	·K	bullfrog
·K	G	E	R	·Q	D	L	(V	·A)	Y	·L	K	S	A	T	S	–	tuna fish
K	S	E	R	Q	D	L	I	A	Y	L	K	K	T	A	A	S	dogfish
E	G	E	R	K	D	L	I	A	Y	L	K	K	S	T	S	E	lamprey
P	N	E	R	G	D	L	I	A	Y	L	K	S	A	T	K	–	fruit fly
P	N	E	R	G	D	L	I	A	Y	L	K	S	A	T	K	–	screwworm fly
A	N	E	R	A	D	L	I	A	Y	L	K	E	S	T	K	–	silkworm moth
A	N	E	R	A	D	L	I	A	Y	L	K	Q	A	T	K	–	tobacco horn worm moth
P	Q	D	R	A	D	L	I	A	Y	L	K	K	A	T	S	S	wheat
D	K	D	R	N	D	I	I	T	F	M	K	E	A	T	A	–	*Neurospora crassa*
E	K	D	R	N	D	L	I	T	Y	L	K	K	A	C	E	–	baker's yeast
A	K	D	R	N	D	L	V	T	Y	M	L	E	A	S	K	–	*Candida krusei*

X, ?, Y, tyrosine Z, —, absent

Here the complete amino acid sequences for 26 different cytochrome c molecules are lined up for purposes of comparison (adapted from M. O. Dayhoff and C. M. Park, 1969. In *Atlas of Protein Sequence and Structure*. National Biomedical Research Foundation, Silver Spring, Md.).

Box 9–1. Hybridization of DNA from different species.

A. The polynucleotide strands of one DNA are separated and then embedded in agar or on nitro-cellulose filters. B. DNA from another source is degraded and the strands separated and added to the embedded DNA preparation. C. The small strands of DNA diffuse into the agar and combine, depending on their ability to hybridize, with the embedded DNA. The nonhybridized or nonhomologous material is washed away. Heating this preparation to 75 °C separates the homologous strands, which are then washed out and analyzed to determine the degree of homology.

One result of such a study on hybridized DNA is given below. Here a refinement of the technique just described was employed. Species A DNA was embedded in filters and a specified amount of radioactive species A DNA and varying amounts of DNA from one other species were added to each filter. After the

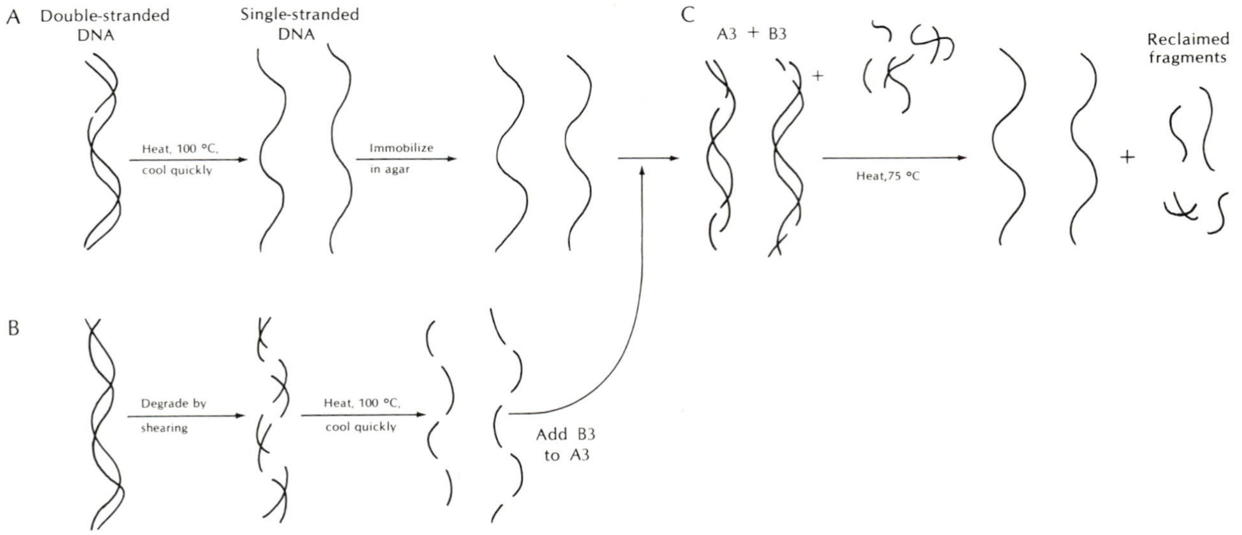

comparisons are clearly less precise than those from sequencing studies.

Such biological characteristics as enzymatic activity or antigenicity reflect quite specific properties of protein molecules. This is based on a precise knowledge of enzyme specificity and the precision of antigen-antibody reactions. These studies are nonhomologous DNA is washed out, the preparation is heated, and radiolabeled DNA from species A is recovered and the amount of radioactivity measured. The figure is the amount of homologous DNA from species A that hybridized to the embedded DNA. If the DNA from another species (B) is very similar to that of

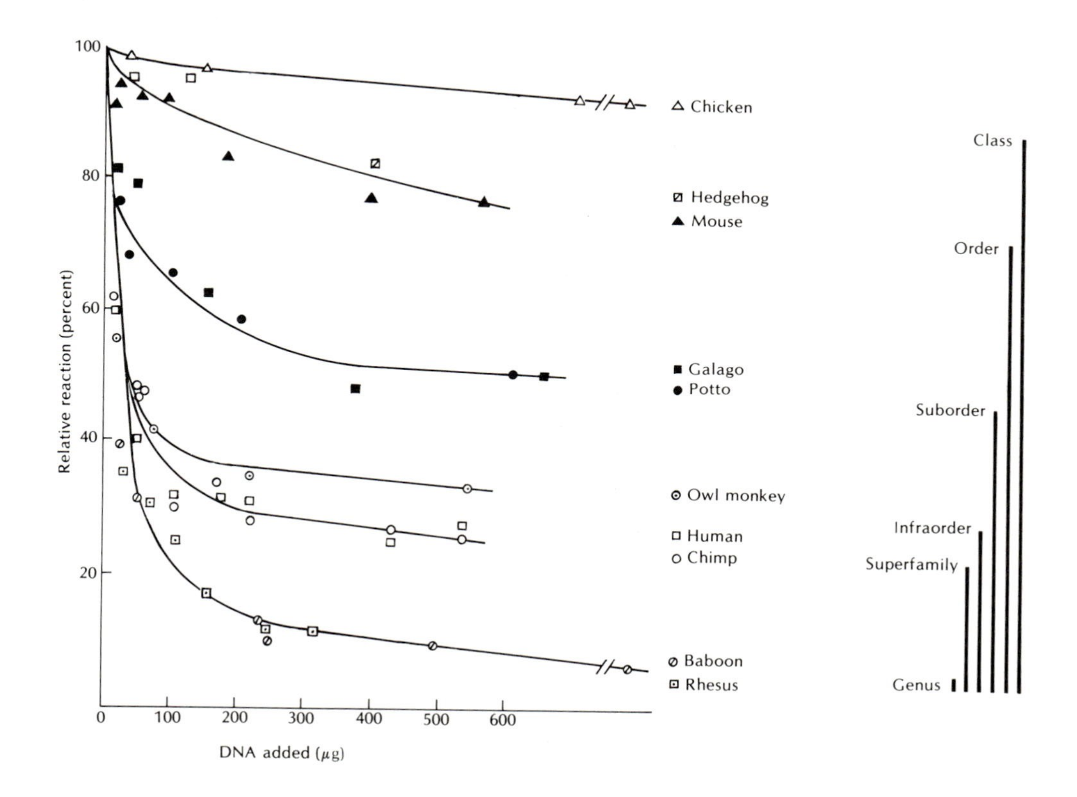

species A, then it will compete with the radioactive species A DNA for homologous sites on the embedded DNA. Therefore a low recovery of radioactive DNA means the DNA from species B has considerable homology with species A. Conversely, high recovery of radioactivity means little homology.

Another technique, based on hybridization, measures the proportion of nonhomologous DNA segments present, and thus, the stability of hybrid DNA molecules. Results here are comparable to those discussed above.

very useful in defining homologies between the molecules in question that function as enzymes or as antigens. They have specific point-to-point correspondences with substrates or with antibodies.

Homologous functions. Functions are perceived in the dimension of time. Structures are perceived in the three dimensions of space. Hence, structures offer more complicated systems for point-to-point comparisons than do functions. To compare functions effectively we must have a sequence of identifiable steps. Mitosis is a good example. Its various stages can be clearly identified, and there are sufficient features to be compared at each stage, such as the behavior of the nuclear membrane, the nucleolus, the spindle fibers, the centrioles, and the chromosomes themselves. It is possible to compare various kinds of mitosis in eukaryotic cells to see if they are or are not homologous processes.

Development is another good example. The egg is fertilized by a sperm, and cleavage follows, but the patterns of that process differ among different phyla. Gastrulation occurs by a variety of processes and germ layer formation can follow different steps in embryos from different phyla. We shall see that embryogenesis is an important phylogentic consideration, especially among invertebrate animals.

Among adult organisms we have already mentioned that such behavioral characteristics as hunting, feeding, and courtship patterns can be compared in the search for homologies. Such patterns have been used in studies among genera of the same family to help sort out phylogenetic relations. Here, as in mitosis and development, it is the sequence of events that is compared.

A last example of functional homologies comes from biochemistry and the patterns of biosynthesis or the role of certain molecules in certain functions. The role of adenosine triphosphate (ATP) or creatine triphosphate in energy storage has been one such study. The nature of the pigments used in sensory cells in the eyes—the visual pigments—is another. And the comparative biochemistry of the synthesis of various lipids and other compounds describes the steps of biosynthesis that allow the needed point-to-point comparisons to establish or deny homologies at the biochemical level.

SEMES: THE UNITS OF COMPARISON

The foregoing discussion has made it clear that homologies can be built on point-to-point comparisons. But we have not discussed, so far, exactly how similarities are established by point-to-point studies except in a general way, as in Fig. 9–5. We will now settle that issue by introducing the concept of a seme.

A *seme* is a character, or trait, that is sufficiently complex so as to allow meaningful comparison. It can be structural or functional. In terms of structure, a seme must have at least

Box 9–2. The basis for using four different points in three-dimensional space as the minimum number of points to establish homology.

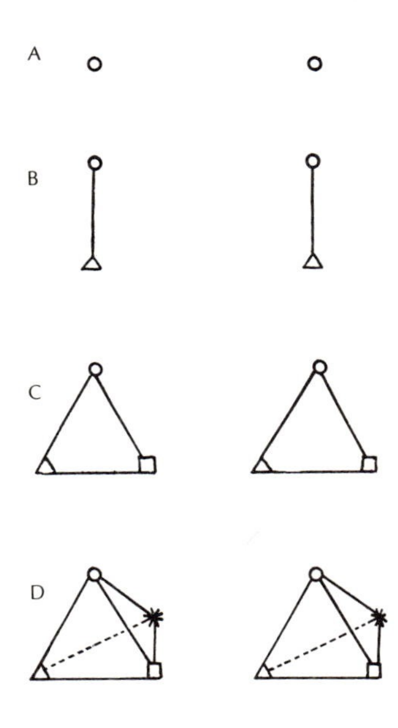

In each case we compare a set of points. A. With a single point in each set, it is obvious that one point can be superimposed on the other in an infinite number of ways. The apparent similarity is meaningless because there is no way to discriminate between the two points. B. Here there are two connected points in each set, and the two are different. But one set can be superimposed on the other endlessly by rotation around the long axis. Again the comparison is not informative except that circle lies on circle and triangle on triangle. (The circle and triangle only indicate that the two points are different in each set.) C. In the triangular sets, there exist only two possible point-to-point comparisons. One of these shows unambiguous superimposition on the other. However, it may be that one figure has been turned over, in which case one figure is the mirror image of the other. But here the ambiguity is reduced to one alternative from perfect similarity. D. Two tetrahedons provide sets with four different points. These are either similar, as shown here, or they are not. Ambiguity is absent, hence this is the minimum number of points needed for establishing homologies.

Remember, however, the examples given in Fig. 9–5, of aposemic traits. They illustrate the need to talk of *similarity,* rather than identity, in studying homologies. The transition from perfect similarity to general similarity (Figs. 9–3 and 9–5) to dissimilarity (the extremes in Fig. 9–4) is gradual. The logic behind these changes can be approximated mathematically in the following way.

There is one type of identity in mathematics called *transitivity,* whereby if $a = b$ and $b = c$, then $a = c$. The transitive relationship is of interest to a phylogenist in cases of limited or weak transitivity. Instead of assuming identity between the entities to be compared, let us assume degrees of similarity, given by $\sim_1$, $\sim_2$, and $\sim_3$. The first symbol $\sim_1$ expresses greater similarity than $\sim_2$, and both of them define greater similarity than $\sim_3$. Therefore, the following relationship can hold: $a \sim_1 b$ and $b \sim_1 c$, but $a \not\sim_1 c$. Namely, *a* can resemble *b* to a certain degree, and in that degree *b* resembles *c*, but *a* does not resemble *c* to that same degree. It may be that $a \sim_2 c$. This is another way of expressing the serial relationship. And greater degrees of divergence can be identified, i.e., $\sim_3$. The point is that limited transitivity may provide a logical language for expressing this most important homologous relationship of indirect or derived similarity.

four points in three-dimensional space to be uniquely recognizable and, therefore, comparable to another such structure (Box 9–2).

For functions it is more difficult to describe the minimal complexity needed to provide a meaningful comparison. One suggestion is that at least five different sequential steps are needed. The basic reasoning is that if there is an equal chance that any one point is or is not similar to its comparable point in another function, then the chance of all five points being similar is $(1/2)^5$ or 1/32. In statistics, if there is less than one chance in twenty that something will occur by chance alone, then something other than chance is credited as being the cause. Similarly here, if the correspondence of functions is not likely on a chance basis, then something else could cause it.

Fig. 9-7: A. In Mullerian mimicry, two distasteful species mimic each other, and so are thought to reduce predation on each one alone. Both butterflies are from the American tropics; the large one is *Heliconus xenoclea*, the smaller one *Actinote diceus callinira*. B. Mimicry between the reef fish *Calloplesiops altivelis* (top and middle) and the moray eel *Gymnothorax meleagoris* (bottom). In this Batesian mimicry, the reef fish is edible, but because it resembles its model, the eel, attacks by its predators are discouraged.

That could be either homology or convergence. To distinguish between them we must use Remane's criteria, as we said earlier. Before we turn to that point, one final comment is needed on functional semes. Note that in the probabilistic argument just given we have *assumed* that there is a 0.5 chance that two points will be alike. This is an arbitrary assumption. Unfortunately there is no good way to decide whether that value should be 0.5, 0.1, 0.9, or some other value. The value probably varies, depending on how closely related the species being compared are. But that relationship is precisely the phylogenetic problem our study of homology is trying to solve. In other words, the value cannot be determined at the time homologies are being identified. So, as was stated above, there is as yet no clear statement possible regarding the minimal number of points needing to be compared in functional homologies; five points is simply a reasonable guess.

CONVERGENCE

Unfortunately for phylogeny, similarities can occur for reasons other than descent from a common ancestor. In addition to convergence, there are also parallelism, analogy, mimicry, and chance resemblance. We can examine the last three rather briefly, since it will be apparent that they are less important than the first two.

Chance resemblance is just that, and usually a second look reassures the observer that the supposed similarity is entirely accidental. Examples are tree branches that look like deers' horns or driftwood that looks like a bone. Or an oval stone can look like an egg and a root might resemble a snake. These have nothing to do with phylogeny.

Mimicry can have a great deal to do with evolution, but little to do with phylogeny. A recently discovered example of mimicry is shown in Fig. 9–7. This is called Batesian mimicry; where the mimic is desirable to predators, but the model is not. The mimic is the reef fish *Calloplesiops altivelis* and the model is the moray eel *Gymnothorax meleagris*. There are reasons, from underwater observations, to believe that the similarity between these two have evolved so that predators will mistake the reef fish for the much larger and stronger eel and so leave it alone. Similarity was achieved by the action of natural selection, but has little to do with common ancestry. The same is true for many other examples of mimicry (Fig. 9–8).

Fig. 9-8: Imitation: This larva of the peppered moth *Biston betularia* (British Isles) (left) almost perfectly imitates, in shape, texture, and markings, a twig of the branch on which it rests.

Analogy refers to instances in which similar functions are performed by dissimilar structures. The wing of a butterfly and the wing of a bird are good examples. Superficially, the two wings are alike because they aid in flight. But in terms of structure the veined wing of an insect and the feathered wing of a bird are quite dissimilar, and phylogenetically distinct.

Parallelism comes about from similar selection pressures acting on different members of a group that are related by a common ancestor. It is very close to homology. In Fig. 9–9A are shown two species of porcupine. Present evidence shows that they evolved from a rodent ancestor. However, selection pressures were similar in South America and Africa, and in both areas, but quite independently of each other, a porcupine was evolved. There was parallel, but independent evolution of the porcupine characters in these animals who had a common ancestor.

An extensive set of parallels is found in the marsupials of Australia and placental mammals (Fig. 9–9B). Convergence differs from parallelism in that the shared ancestor may be very distant. So distant, in fact, that few of the shared characteristics are derived from that particular ancestry. Figure 9–10 gives several examples of convergence. The most striking examples are found in the lower part of the figure. In each pairing, one organism is unicellular and one is multicellular. Similar selection pressure has taken these very dissimilar ways of structuring an organism and produced quite similar phenotypes.

A

Fig. 9-9: Parallelism: A. Parallelism in porcupines (above, South American porcupine; below, African porcupine). B. Extensive parallelism is seen in a comparison (opposite page) between marsupial mammals (development completed, postnatally, in the mother's abdominal pouch) and placental mammals (development completed, prenatally, in the mother's uterus).

It is right here that we can look to Remane's compositional and serial criteria to differentiate convergences and homologies. Take, for example, the similarities between the ciliated protozoan *Stentor* and the invertebrate rotifer *Ptygura*. A close examination of the ciliated discs at the anterior end of each will show some real differences in the way they are formed. If we try to bridge these differences by looking for intermediates we find that with *Stentor* we get into ciliates more like *Paramecium*, which have no ciliated disc (Fig. 1–15); and with the rotifer we are not sure where to go—it may be some worm-like form. In other words the superficial similarities of body form disappear. What we find is that in terms of compositional relations (details of structure) and serial relationships (a series of intermediates that fulfill the positional and compositional relationships) the criteria of homology are not met. At best, convergences express only a positional relationship.

A look at the reptile, cartilaginous fish, and mammal in the upper part of Fig. 9–10 is very instructive in terms of homologies and convergences. All three organisms are vertebrates and so certain important homologies are present, i.e., the backbone, the dorsal nerve cord, and the gill slits at some stage of their lives. But what has caught the phylogenist's eye is the similarity in body form. It raises the question of common ancestry. Let us look for homologies and take three traits for study, i.e., the dorsal fin, the tail, and the paired anterior or pectoral fins or flippers, as they are called in the dolphin. The dorsal fin represents convergence in all three cases. In the ichthyosaur (reptile) it is supported by bony fin rays. In the shark (fish) it contains cartilaginous fin rays. And in the dolphin it is a fleshy extension supported by cartilage, but in a manner quite different from that found in the shark. There are

B
Placentals
Marsupials
Wolf
(Canis)
Tasmanian wolf
(Thylacinus)
Ocelot
(Felis)
Native cat
(Dasyurus)
Flying squirrel
(Glaucomys)
Flying phalanger
(Petaurus)
Ground hog
(marmota)
Wombat
(Phascolomys)
Anteater
(Myrmecophaga)
Anteater
(Myrmecobious)
Mole
(Taipa)
Mole
(Notoryctes)
Mouse
(Mus)
Mouse
(Dasycercus)

no intermediates to bridge these differences. In fact, as we go to forms closely related to dolphins, we emerge on land and there are no fins among these forms. The same is true of the ichthyosaur (Fig. 9–11). The shark dorsal fin goes back to primitive fossil fish only very indirectly related to the other two organisms.

The caudal fins are not homologous in the shark and ichthyosaur, but are parallel or convergent in their evolution. The intermediates include land-dwelling reptiles with tails, but these are also found in the amphibians from which the reptiles evolved. The amphibian tail is homologous with that of bony fishes from which they evolved, and the tail of bony fishes is homologous to that of the cartilaginous fishes. However, the caudal fin itself evolved independently in the icthyosaur after being lost in the amphibians. The dolphin tail is at best a poor convergence or parallelism. It is a horizontal rather than a vertical structure. (It is waved up and down to push the dolphin forward, rather than sidewise, as in the ichthyosaur and fishes.) The dolphin tail is supported by cartilage, which ex-

Fig. 9-10: Convergence: A. Convergence in vertebrates: an ichthyosaur (extinct marine reptile, reconstructed from the fossil record) (1); a sand shark (a cartilaginous fish) (2); a bottle-nosed dolphin (a mammal) (3); B. Convergence in unicellular and multicellular organisms: the flagellate *Leptodiscus* (1); the medusa *Homoeonema* (2); the ciliated protozoan *Stentor* (3); the rotifer *Ptygura* (4); the protozoan *Dendrosoma* (5); and the cnidarian polyp *Syncoryne* (6).

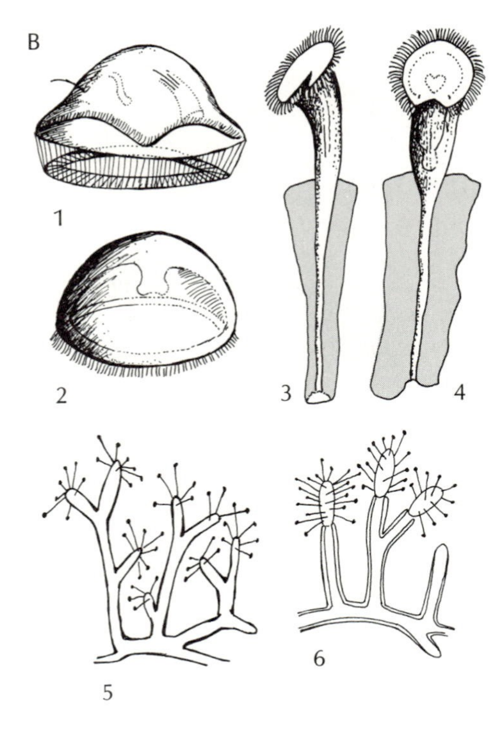

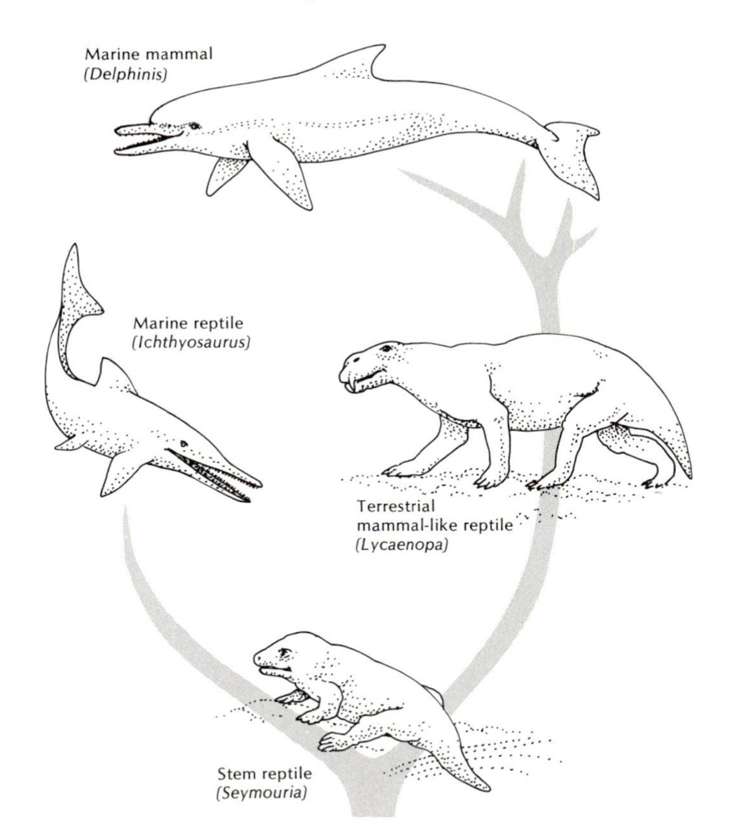

Fig. 9-11: Forms intermediate to the ichthyosaur and the dolphin.

tends laterally from the end of the spine. It is a brand new structure in these mammals.

The pectoral fins and flipper in all three organisms are homologous. There are intermediate forms that show positional and compositional criteria that bridge the differences among them (Fig. 9–11). A notable difference is the cartilaginous fin rays of the shark and the bones of the ichthyosaur and dolphin.

The overall body form of these three organisms reflects the fact that all three are surface-feeding organisms. There is a premium on the ability to catch swiftly swimming smaller fish as prey, hence the similar streamlined body. But that similarity is a convergence, since the compositional differences we find take us back to intermediate forms that lack the characters that define this body form except for the case of the pectoral fins. Some might argue that convergence in these three vertebrates is really parallelism. But that only raises the point of where to draw the line between the two. No explicit answer is possible except to say that parallelisms are usually restricted to members within the same order or to taxa below the ordinal level. At the level of classes, phyla, or kingdoms, we usually speak of convergences. Among some biologists there is a growing tendency to ignore parallelisms as a category and only speak of convergence. We tend to that point of view in this book.

Homology and phylogeny

Homologies have a clear and unambiguous definition and meaning and they can be distinguished from convergences even when both occur in the same organisms, as we just saw. There remains now the task of showing exactly how we can progress from homologies to phylogeny. What methodology allows us to read homologies so as to reconstruct and explain evolutionary history?

What follows now is a methodology published by Hanson in 1977. It is new and relatively untested, although it, in effect, integrates well-tested ideas. It brings together Remanian criteria, but reinterpreted more rigorously and extended to structures and functions at all biological levels. And it also uses Darwinian natural selection as interpreted through population genetics and the modern concept of the species. The newer aspects of this methodology lie in the concept of semes and in the quantitative measures of phyletic distance they generate. It is in those areas that future refinement of the ideas is certain to occur.

We perceive a species as a population of individuals. Thus, comparing species means comparing individuals from the species in question. That necessitates some words of caution. Organisms within a species can differ in terms of sex, stage of development, health, and the natural phenotypic variation of the species. Care must, therefore, be taken to compare forms that are indeed comparable. For example, healthy adult females must be compared with other healthy, adult females, or a larva with other larvae. Adults and embryos should not be compared. And there must always be due regard for possible variations.

When that is done, we can proceed to the phylogenetic analysis. The next step is to locate semic traits. We must have traits that allow for meaningful determinations of similarity.

Simple traits like color, weight, and length are not semes. It is often useful to consider organisms and list their organ systems and any other aspects, down to the molecular, which can provide semic information (Table 9–2). Such a listing has the important advantage that it guards against biased comparisons.

Quite understandably, a developmental biologist is especially familiar with larval stages, a biochemist with biosynthetic pathways, an anatomist with gross morphology, and so on. But the whole organism is liable to selection pressures and so the whole body, the whole life cycle, and all the bodily functions are products of evolution. To base evolutionary conclusion on just developmental, biochemical or anatomical facts is to read only part of the information that is available. The evolutionist, and in particular the phylogenist, must be a biologist.

When an array of semes that has been identified provides a good sample of the total biology of the species being compared, the search for homologies is initiated.

Table 9–2. Possible semes found in organisms.

1. Species semes
   a. Demic structure—genetic variability, etc.
   b. Breeding patterns
   c. Population size
   d. Migratory behavior
   e. Mutation rates
2. Semes found in organismic systems (multicellular and unicellular)
   a. Body organization
      (1) Size, shape, and symmetry
      (2) Special features—appendages, cavities, etc.
   b. Organ systems
      (1) Integumentary
      (2) Digestive (including nutritional needs)
      (3) Circulatory
      (4) Respiratory
      (5) Excretory
      (6) Muscular
      (7) Skeletal
      (8) Nervous
      (9) Endocrine
      (10) Reproductive
         (a) Mating systems
         (b) Development
3. Molecular semes
   a. DNA
   b. RNA
   c. Polypeptides and proteins
   d. Biochemical pathways and products

These semes, especially those referring to anatomy or major functions such as development, can be readily broken down into more than one seme.

PLESIOSEMES, APSOEMES, AND NEOSEMES

The discussion of homology has shown us that there are two kinds of homologies, those identified by direct evidence of similarity and those identified by indirect evidence. The former represent conservative semes, since they show little change; the latter are innovative. The former tells us that selection pressures have remained constant; the latter that selection pressures have changed.

Most importantly, then, when we put the two kinds of homologous semes together in viewing species, we immediately see the terms in which the species have evolved. Against a background of relatively stable or conservative traits, which are called *plesiosemes,* there stand out the innovative or changing ones, called *aposemes*. Sometimes one species will have a complex trait that is absent in another one. Such a trait is called a *neoseme*. (Consider the dolphins tail compared to the other tails in Fig. 9–11.) These three terms express the complexities inherent in the concept of homology. (Actually, only the plesiosemes and aposemes refer to homologies. A neoseme cannot be homologous, since it refers to a new trait present in one or more species and absent in others.)

The limitations inherent in comparing two species should now be mentioned. The chief one is that it may be impossible to determine the direction of change. That change has occurred in certain traits may be obvious. But did species A give rise to species B? Did the reverse happen? Or did both arise from a common ancestor, species C? Only if we have some added information, as could be supplied by fossil evidence or embryonic development or even distributional data regarding occurrence in different habitats, can we decide on the direction of change. However, when a group of species is being compared, as is usually the case, it is often possible to locate a form somewhat representative of the ancestral form. Given that information the flow of evolutionary change becomes much clearer.

IDENTIFYING THE PLESIOMORPH

The species closest to what might be considered an ancestral form to a number of species is called their *plesiomorph*. How do we identify it? Four kinds of information help us. They help us to varying degrees, depending on the species under consideration. We have mentioned three already, i.e., fossils, embryos, and distribution. The fourth is called phenoclines. Each will be discussed.

Fossil evidence. The plesiomorph can be a fossil species. If good fossil evidence is available then that provides an excellent reference point in time for studying subsequent changes. However as we shall see later (Chapter 10), fossil evidence is of varying degrees of usefulness. Some taxa are well represented as fossils and others are not because sometimes the fossil evidence is meager. Often fossils provide only a view of the external anatomy of an organism or its hard parts, such as shells or skeletal materials. In such cases fossil data are of limited usefulness. Usually the fossil record is used in conjunction with information from living forms.

Embryological evidence. Haeckel's famous saying that ontogeny recapitulates phylogeny is important here, but must be used in a very qualified way. What Haeckel claimed was that the evolutionary history of an organism is summarized in its development. But we now know his claim is not universally true. At best, it can be called a useful generalization. A fascinating review of this whole problem has been written by the American evolutionist S. J. Gould.

Earlier we pointed out (Fig. 1–5) the striking similarities in vertebrate embryos which all have gill-like structures, a notochord, and a dorsal nerve cord. In the fishes, the earliest vertebrates, the notochord is replaced in the adult by the vertebrae. The gills and nerve cord persist throughout the life of the fish. In the amphibia, again, the notochord is replaced by vertebrae, and the gills appear as functional structures in the tadpole or late embryonic stages and then disappear when their function is taken over by lungs. The nerve cord is always present. In reptiles, the notochord is replaced by vertebrae and the gills are only rudimentary in the embryo, while the nerve cord persists. And in birds and mammals, gill clefts are present for a short time early in the embryo. The notochord also has only a short existence before it is replaced, but the nerve cord is present throughout an individual's life.

What this brief survey of these three traits tells us is that once a basic character is present it can be changed. Even when the trait is no longer functional—for example, the gill clefts have no real role in respiration and another organ (the lung) has taken their place—they are still formed. In this sense development (ontogeny) continues to repeat the history of the group to which the organism belongs (phylogeny of the vertebrates).

The problems come when extra developmental steps are added to the historical sequence, when a step is finally

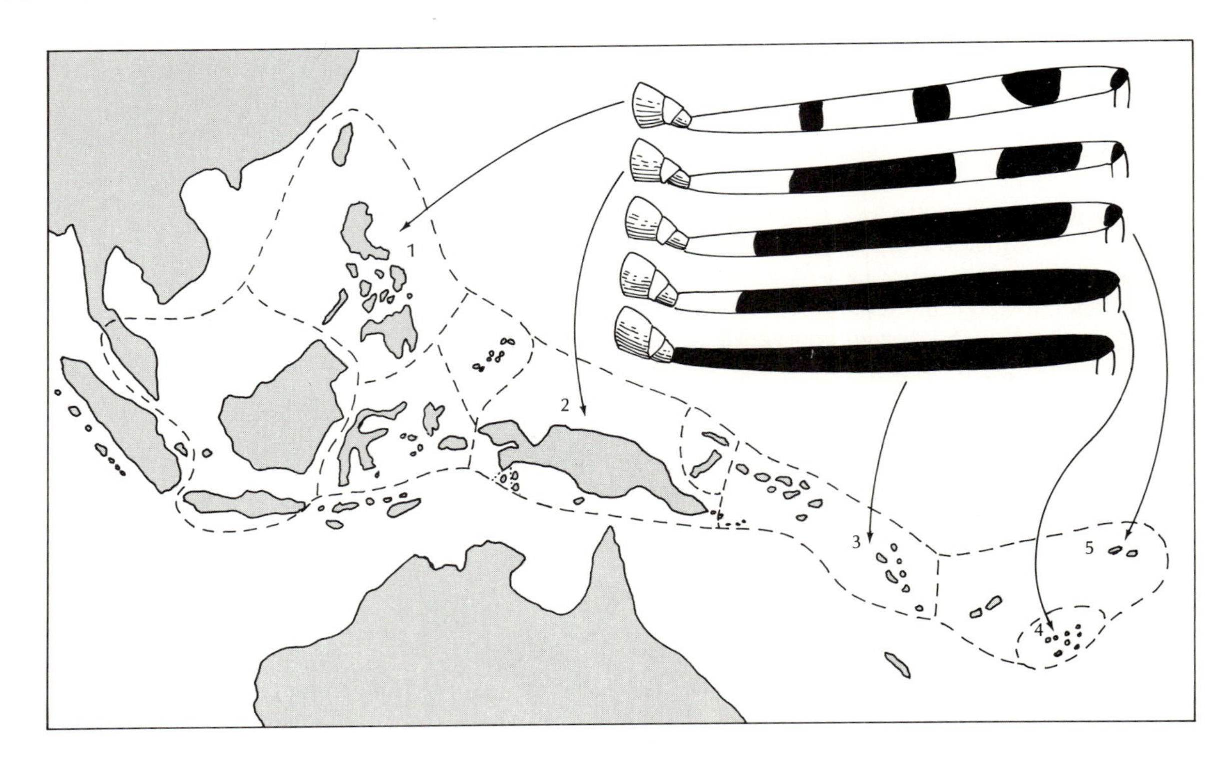

Fig. 9-12: An illustration of distributional data useful in identifying a plesiomorph. Note the difference in leg pigmentation (upper right) in five specimens of the insect *Mimigrella albimana* as the subspecies spread from Indonesia and the Phillipines (1) to the New Guinea area (2), to Samoa (3), to the Tonga Islands (4), and finally, to the New Hebrides (5). Though formally designated as subspecies, it may well be that the extreme forms in this distribution are reproductively isolated from each other.

dropped completely, or when development goes off in a new direction. Examples of all of these are known. They necessitate then a cautious reading of phylogeny from ontogeny.

Distributional evidence. This kind of evidence has been used successfully in some groups, but not all groups. In Fig. 9–12 we see one example of its use. When the geographic distribution of a group of species is known and when gradual changes in character are also present, an experienced researcher can often locate a plesiomorph. But where geographic variations are minimal or give contradictory stories—one character may vary in an east-west direction and another in a north-sourth one—interpretation is difficult. Also, distributional evidence, even when useful, is rarely extended to taxa above the familial level. It is most useful in species found among closely related genera.

As we shall see, the most intriguing and baffling phylogenetic questions lie among the very highest taxa, the phyla and kingdoms. We want to know in what ways the major groups of organisms are related to each other. Where did the protists come from and the land plants? Who were the ancestors of the multicellular animals and, within them, of the chordates? Fossils and embryos can tell us something on such questions, but distributional evidence is mute.

Phenocline evidence. This line of information is probably used the most in locating primitive forms and plesiomorphs, but unhappily, it is the most indirect. It is based on probabilistic arguments. Maslin has given us the clearest statement of these arguments and he coined the term morphocline. But since we are dealing with phenotypic characters, regardless of whether they are morphological or physiological, we prefer the term phenocline. Maslin's arguments are exemplified by the model systems in Fig. 9–13. Here the argument comes down to the most plausible explanation in terms of evolution by natural selection—by the accumulation of mutational differences between different species. The term phenocline, as shown in Fig. 9–13, refers to the phenotypic variation in a semic character or trait as it is found in various species. The distributional data, just discussed, represent a sort of phenocline study. In addition to clinal change, there is also spatial distribution to be considered. (In any case, we must not confuse clines, as gradual variations over distance, but *within* a species, with phenoclines. Phenoclines consider character or semic variations *between* species.)

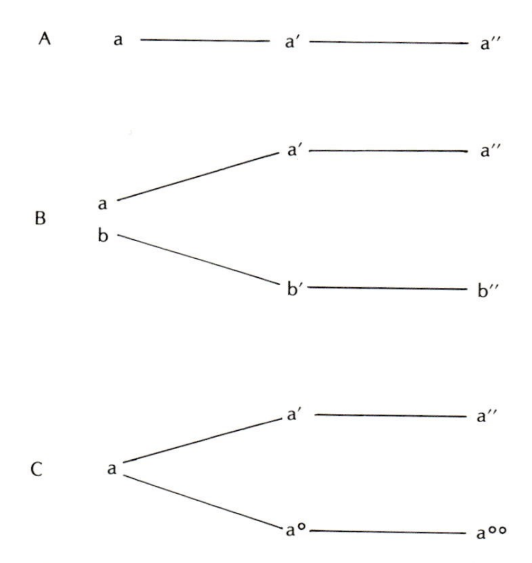

Fig. 9-13: Morphocline study based on Maslin's analysis. A. A semic character is found in three different species, but slightly modified in each: a, a′, and a″, with a and a″ being the most different states. If a lies in a species whose other characters are also less modified, then a is thought to be the conservative state of that character. B. Two different semes are present, each showing a morphocline: a, a′, and a″ and b, b′, and b″. The species containing the extreme of both morphoclines ab is thought to be the plesiomorph. C. Again, two morphoclines exist: a, a′, and a″ and a, a°, and a°°, but they share a common extreme state, a. The species with a is likely to be the plesiomorph.

In summary, regarding plesiomorph identification we can see that no simple definite procedure is available. One looks at all available information and from that obtains the best possible answer. In some instances, the chosen plesiomorph is convincing because several lines of evidence converge on it. In other instances, the evidence is inconclusive. This sort of ambiguity must, however, be expected in evolutionary studies. The very fact of evolutionary history drives us to try and understand it, to work out its details. On the other hand, evolution proceeds by replacing one species with another better adapted one. In that way the historical record is being continually erased, unless we are lucky enough to have a fossil record or embryonic characters that characterize the ancestral species.

Phyletic distance. The degree of difference or similarity, based on homologies, is termed *phyletic distance*. It is, also, a measure of evolutionary divergence. In recent years several methods have been proposed for expressing phyletic distance in quantitative terms. Just now we need to complete our analysis based on homologies. In particular, we will use plesiosemes, aposemes, and neosemes.

The plesiomorph allows us to read the evolutionary direction of changes that emerge with aposemes and neosomes. If a species shows the same semic trait that is found in the plesiomorph, that is then a conservative or plesiosemic trait. But if the trait differs significantly from what occurs in the plesiomorph, we have an aposeme. And if a brand new trait is present, i.e., it is absent in the plesiomorph, but present in another species, then it is a neoseme.

The quantitative use of these different semes, and the homologous relations they represent, is given in the following formula:

$$R = \frac{[-p + (2\,a)^2 + (3\,n)^2]}{t} + 1$$

The phyletic distance is given by R, in honor of Professor Remane. The different homologies are p for plesiosemes, a for aposemes, and n for neosemes. The total number of semic traits studied (the sum of p, a, and n) is given by t. What this formula says is that plesiosemes give no evidence of evolutionary development, but aposemes and neosemes tell us a great deal. Therefore, plesiosemes are given a negative value and the other two positive values. In fact, they are weighted to emphasize their importance; the aposemes are multiplied by two and squared and the neosemes, indicative of even more change, are multiplied by three and squared. What is the meaning of that final +1 at the end of formula? Suppose two organisms being compared happened to be members of the same species. In that case, unless they came from very different subpopulations, *all* the semic traits would be plesiosemes. Let us say there are ten semes under comparison, in which case the numerical values in the formula will be

$$R = \frac{(-10 + 0 + 0)}{10} + 1$$

(Remember there are no aposemes or neosemes in this case.) This means that $R = 0$ which is correct for members of the same species. The +1 is needed to change the −1, which comes from $-p/t$, to 0.

In Box 9–3, an analysis of six species has been set up. Though they are given as species A, B, C, D, E, and F, they do represent actual species. The analysis is given without reference to the actual organisms or to their specific semes to emphasize its logic. Note in particular, how the R values are generated for the phyletic distances between the different species. Given those values, we can construct a phylogenetic tree or dendrogram that summarizes the possible evolutionary relationships among these six species.

Dendrogram. The plesiomorph species D is placed at the bottom of the dendrogram and next to it is species E, with the lowest R value of the remaining species compared to D. And then next to E is the species showing the smallest phyletic distance from it, which is species B. From B there is an equal distance to species F, A, and C, but A and C are much more like each other than either is to F. This could mean parallel evolution of A and C from B or that both had a common ancestor in B or in some species, not considered here, but derived from B. On the basis of the evidence given here it is not possible to choose between these possibilities. More data are needed and in actual practice this would lead to the search for species possibly intermediate to B and to A and C.

It is now worth looking at the actual species used in this phylogenetic exercise. They are given in Box 9–4, along with the semes used. One can see here how these familiar animals give us the outlines of the well-known evolutionary tree of the vertebrates. Because this area is so well studied the analysis given in Box 9–3 is a bit superfluous, except it does make clear the details of the methodology. The technique is more useful when applied to less familiar areas, some of which will be considered in succeeding chapters.

METHODOLOGICAL OVERVIEW

The steps in deriving a dendrogram from homologies can now be summarized as follows:

1. Obtain a description of the species being studied that covers both structures and function.
2. Determine which traits are semes.
3. Determine which semes show homologies and identify plesiosemes, aposemes, and possible neosemes.
4. Identify the plesiomorph.
5. Calculate the phyletic distances.
6. Construct the appropriate dendrogram.

In addition to the above summary, there are three general rules worth keeping in mind, as the general context within which the above procedure takes on its phylogenetic meaning.

The first rule to remember is that species are the unit of evolution, and it is species that must be compared. Hence the examination of individual organisms is seen as the study of representatives of the species under consideration. The most

Box 9–3. A phylogenetic analysis of six species.

The first table is based on fairly extensive descriptions of 13 semes, labeled 1 through 13, for the species A, B, C, D, E, and F. The species and their semes are described in Fig. 9–16. Looking at seme 1 we see that it occurs in a state called 1A in species A and C; it is the same seme in both species. But among each of the remaining species, it exists in a recognizably different state. By contrast, seme 8 exists in only two states, 8A and 8D. Seme 11 shows information missing for species E. Actually, it is available, but is absent in the source consulted here. Its absence is included to show how occasional missing information can be adjusted for.

	Semes												
Species	**1**	**2**	**3**	**4**	**5**	**6**	**7**	**8**	**9**	**10**	**11**	**12**	**13**
A	1A	2A	3A	4A	5A	6A	7A	8A	9A	10A	11A	12A	13A
B	1B	2B	3B	4A	5B	6B	7B	8A	9B	10A	11B	12B	13B
C	1A	2A	3A	4A	5A	6A	7A	8A	9A	10A	11A	12A	13A
D	1D	2D	3D	4D	5D	6D	7D	8D	9D	10B	11D	12D	13B
E	1E	2B	3E	4E	5A	6E	7B	8A	9B	10B	?	12E	13B
F	1F	2F	3A	4A	5F	6F	7F	8A	9A	10A	11A	12F	13F

The next table shows a morphocline analysis for identifying the plesimorph. It is obvious that one species lies at one end of the morphoclines present in the majority of the semes. Where it is not so placed becomes clear (semes 1, 12, and 13) from looking at the actual semes (Box 9–4). Species D is the plesiomorph among these six species.

Semic morphoclines

1	2	3	4	5	6	7	8	9	10	11	12	13
A,F	F	A	A	F	A	F	A	A	A	A	F	A
				A								
	A	B		B?	E	A					B	
B			E		F?			B		B	E	F
D	B	E			B	B					D	
E	D	D	D	D	D	D	D	B	D	D	A	B

Species comparison	Semes 1	2	3	4	5	6	7	8	9	10	11	12	13	Plesiosemes	Aposemes	Neosemes	Total	Phyletic distance
D and A	*n*	a	a	*n*	a	a	a	a	a	a	a	a	a	0	11	2	13	41.00
B	*n*	a	a	*n*	a	a	a	a	a	a	a	a	*p*	1	10	2	13	34.46
C	*n*	a	a	*n*	a	a	a	a	a	a	a	a	a	0	11	2	13	41.00
E	a	a	a	*p*	a	a	a	a	a	*p*	?	a	*p*	3	8	1	12	22.83
F	*n*	a	a	*n*	a	a	a	a	a	a	a	a	a	0	11	2	13	41.00
A and B	*n*	a	a	*p*	a	a	a	*p*	a	*p*	a	a	a	3	9	1	13	26.38
C	*p*	*p*	*p*	*p*	*p*	*p*	*p*	*p*	*p*	*p*	*p*	*p*	*p*	13	0	0	13	0.0
E	*n*	a	a	a	*p*	a	a	*p*	a	a	?	a	a	2	9	1	12	28.58
F	*n*	a	*p*	*p*	*p*	a	a	*p*	*p*	*p*	*p*	a	a	6	6	1	13	12.31
B and C	*n*	a	a	*p*	a	a	a	*p*	a	*p*	a	a	a	3	9	1	13	26.38
E	*n*	*p*	a	a	a	a	*p*	*p*	*p*	a	?	a	*p*	5	6	1	12	13.33
F	*n*	a	a	*p*	a	a	a	*p*	a	*p*	a	a	a	3	9	1	13	26.38
C and E	*n*	a	a	a	*p*	a	a	*p*	a	a	?	a	a	2	9	1	12	28.58
F	*n*	a	*p*	*p*	a	a	a	*p*	*p*	*p*	*p*	a	a	6	6	1	13	12.31
E and F	*n*	a	a	a	a	a	a	*p*	a	a	?	a	a	1	10	1	12	35.00

We can now calculate R values of phyletic distances. First, the homologous relationships are determined between the semes of species D and each other species. Then the other species are compared with one another. The symbols p, a, n, and t refer to plesiosemes, aposemes, neosemes, and total semes, respectively. They are used in the formula

$$R = \frac{[-p + (2\,a)^2 + (3\,n)^2]}{t} + 1$$

to obtain values for R. Notice that species A and C appear to be identical. The reason for this is that the semes used refer to very broad categories, so broad that the do not discriminate between A and C (see Box 9–4).

From these values a diagram can be drawn showing all phyletic distances among the six species being studied here (below, left). That leads directly to the dendrogram showing possible evolutionary relationships among the species (below right).

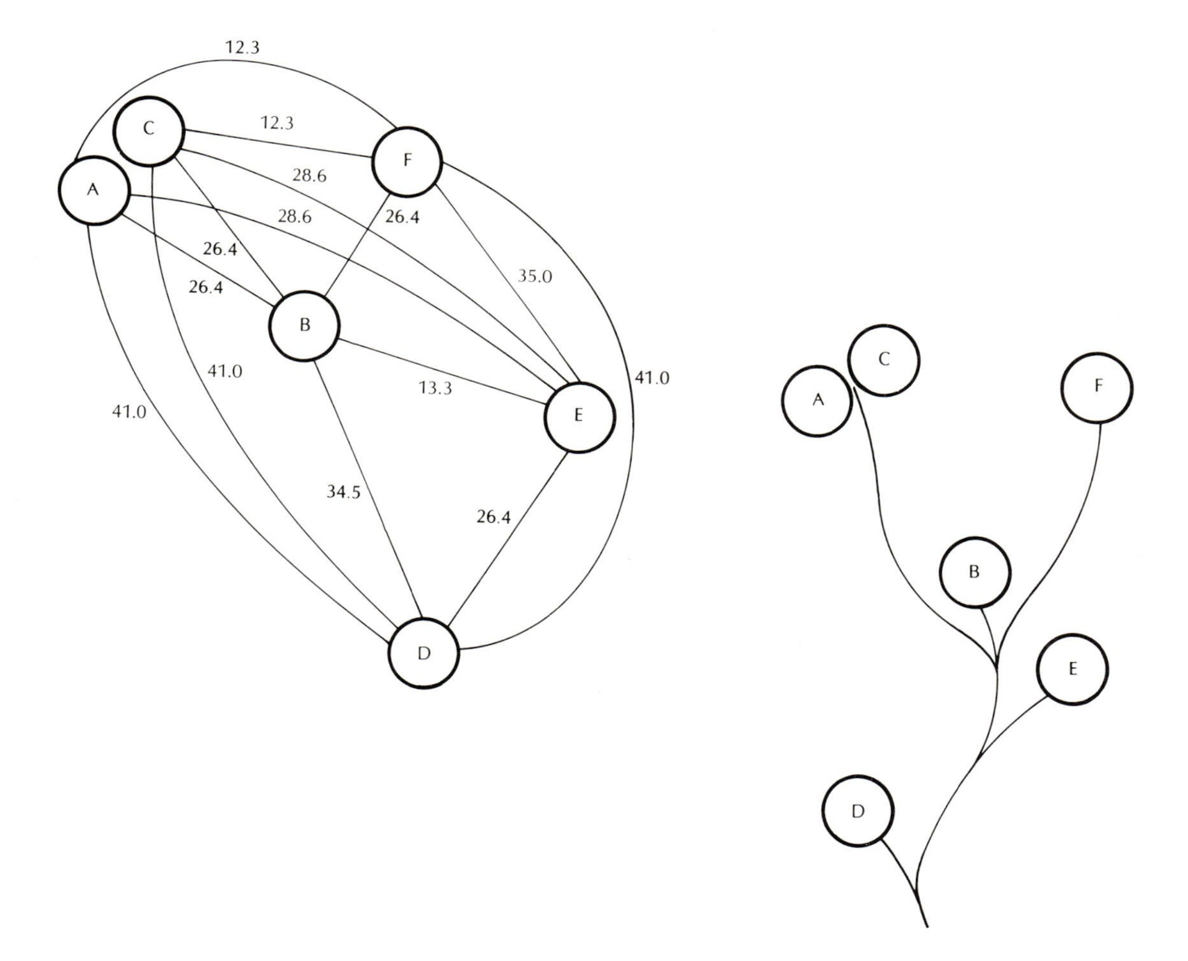

Note that no species is shown as directly derived from another. That is because we are dealing here with existing species, which could not be ancestral to one another. However, they do have common ancestors, which would lie at the point of divergence of evolutionary lines.

Box 9–4. The six species in the phylogenetic analysis summarized in Box 9–3.

The semic descriptions are taken from T. I. Storer *General Zoology* (1943) McGraw-Hill.

Species A, *Canis familiaris* domestic dog (mammal)
Species B, *Chrysemys picta,* painted turtle (reptile)
Species C, *Felis domesticus,* domestic cat (mammal)
Species D, *Perca flavescens,* yellow perch (fish)
Species E, *Rana pipiens,* grass frog (amphibian)
Species F, *Turdus migratorius,* robin (bird)

The semes used to describe these six species follow:
Seme 1. Integumentary system with emphasis on its specializations.
1A, hair; 1B, dry horny scales; 1D, mucous-covered, shingled scales; 1E, soft, moist, mucous-covered skin; 1F, feathers. (Note the difficulty in finding a morphocline.)
Seme 2. Digestive system with emphasis on the different organs.
2A, mouth, esophagus, stomach, small and large intestines, anus; 2B, cloaca inserted between large intestine and anus; 2D, short, highly coiled intestine; 2F, crop and gizzard added.
Seme 3. Circulatory system with emphasis on the heart.
3A, four-chambered heart; 3B, imperfect, four-chambered heart; 3D, two-chambered heart; 3E, three-chambered heart.
Seme 4. Respiratory system emphasizing presence or absence of functional gills and lungs.
4A, lungs; 4D, gills, 4E, gills in embryo, lungs in adult.
Seme 5. Skeletal system exclusive of skull (seme 6) and appendages (seme 7).
5A, vertebrae, ribs, pelvic and pectoral girdles; 5B, shell attached to vertebrae and ribs; 5D, many vertebrae with long spines; 5F, light bones, large breast bone.
Seme 6. Skull and its relation to vertebra.

important consequence of this rule, however, is to exclude any hypothetical form. There is much discussion in the phylogenetic literature of primitive forms, of ancestral and archetypal forms, of theoretical ancestors, and the like. The use of such figments of our imagination obscures careful analysis of semes and actually negates the concept of homology, which depends on relationships between *actual* structures or functions. The plesiomorph, as an actual species, may or may not be genuinely representative of an ancestral form. Probably no plesiomorph, unless it is a fossil, is really ancestral because any species shows certain peculiar adaptations not found in any other species. Some of these adaptations almost certainly are innovations among an array of otherwise conservative traits. But dealing with an actual species makes possible the needed critical analysis of actual homologies.

The second rule is to compare the total biology of each of the species under analysis. This almost always reveals gaps in

6A, teeth on jaw bones, two occipital condyles (sites of contact with vertebra); 6B, no teeth, one occipital condyle; 6D, teeth on various bones, skull rigid in relation to vertebra; 6E, fine teeth on various bones, two occipital condyles; 6F, beak (no teeth), one occipital condyle.

Seme 7. Appendages.

7A, long fore- and hindlimbs; 7B, short, stubby fore- and hind limbs; 7D, fins; 7F, wings and legs with four toes.

Seme 8. Musculature, considering only the organization of the body wall and disregarding specializations relating to skeletal attachments.

8A, voluntary muscles for various bodily movments. 8D, voluntary muscles in segments for side-to-side swimming movements.

Seme 9. Nervous system with emphasis on the brain.

9A, large cerebral hemispheres, folded cerebellum; 9B, medium-sized cerebral hemispheres, smooth cerebellum; 9D, small cerebral hemispheres, small cerebellum.

Seme 10. Nervous system with emphasis on the nerves coming from the brain (cranial nerves).

10A, twelve pairs of cranial nerves; 10B, ten pairs of cranial nerves.

Seme 11. Excretory system with emphasis on the size and shape of the kidneys.

11A, bean-shaped paired kidneys; 11B, paired, flat, lobed kidneys; 11D, long slender paired kidneys; 11E (no information in single source used here, but available elsewhere describing paired bean-shaped structures; omitted here to show how absence of data can be adjusted for);

Seme 12. Reproductive system with emphasis on the egg.

12A, minute eggs, little yolk; 12B, large yolk, leathery shell; 12D, small eggs, significant amount of yolk; 12E, medium-sized egg, large yolk; 12F, large yolk, brittle calcareous shell.

Seme 13. Reproduction with emphasis on where embryo develops.

13A, embryo in mother, parental care of young; 13B, embryo develops outside of mother, parental care usually absent; 13F, embryos develop outside mother, parental care of young.

available information. What is known in certain species may not be known in others. Embryological information very commonly is spotty from species to species. In instances where a group of closely related species is being studied it may be necessary to take some of the major semes and divide them into subsemes to uncover variations between the species. Again, embryology is a good example. Modes of fertilization, cleavage patterns, mechanisms of gastrulation could all be separate semes and thus allow for a more intensive analysis when appropriate.

The third and last rule is to remember that the dendrogram must be consistent with natural selection. We are, after all, studying evolution. At one time it was proposed that the chordates could be derived evolutionarily from the segmented worms or annelids. However, the different location of the nerve cord in the two phyla posed a problem. (It is ventral in the annelids and dorsal in the chordates.) This difficulty was thought to be resolved by turning the annelid over, somewhere along the evolutionary line to the chordates. Although such a solution may have looked neat in terms of drawings on paper, in terms of natural selection it was nonsense. One cannot flop a highly adapted organism on its back and expect it to go on competing effectively in its habitat. That would take an extraordinary macromutation or an extraordinary simultaneous occurrence of micromutations. Neither proposal is at all likely. The lesson is that the final dendrogram must contain evolutionary events entirely consistent with natural selection.

Now let us turn to another set of issues. There is a great temptation to bring systematics and phylogeny together in at least the description of organismic diversity. We shall look at certain current practices that do this, but first it is necessary to understand systematics and phylogeny on their own scientific terms.

Systematics. The goals of systematic work on which most workers can agree is that there be clear descriptions of species that can be stored and retrieved with efficiency. Where systematists disagree among themselves, and on occasions disagreement is vehement, is whether or not the stored information reflects evolutionary relationships.

Before describing briefly the positions taken by those with partisan viewpoints in this debate, let us review in more detail the work done by a systematist. George Gaylord Simpson, a leading paleontologist and evolutionist, has chosen to distinguish between systematics, taxonomy, classification, and nomenclature. He has given us the following definitions in his 1961 book, *Principles of Animal Taxonomy.*

"Systematics is the scientific study of the kinds and diversity of organisms and of any and all relationships among them (p. 7).

"Taxonomy is the theoretical study of classification, including its bases, principles, procedures, and rules (p. 11).

"Zoological classification is the ordering of animals into groups (or sets) on the basis of their relationships, that is, of associations by contiguity, similarity, or both (p. 9).

"Zoological nomenclature is the application of distinctive names to each of the groups recognized in any zoological classification" (p. 9).

Simpson refers only to animals in his definitions of classification and nomenclature because his book deals only with animals. It is fair to say that these definitions can be extended to all organisms, i.e., animals, plants, fungi, protists, and monerans. Furthermore we see that Simpson's definition of systematics is so broad that it really includes the other three. In that sense, then, systematics is composed of taxonomy, classification, and nomenclature. That is, a systematist develops the needed theory (taxonomy) and uses it to classify (classification) and to name what is classified (nomenclature).

There is good agreement today on the rules and procedures for naming organisms; it is a legacy of Linnaeus. The problems come in the areas of taxonomy and classification. On what basis do we classify? Which relationships are emphasized and which are ignored? Do we classify to assure, first and foremost, ease of storage and retrieval of named and classified organisms? Do evolutionary relationships impede or aid that process of storage and retrival? We will discuss possible answers after we quickly review the goals of phylogeny.

Phylogeny. Phylogenists depend on systematists for one primary set of information: good species descriptions. Given that information phylogenists can then describe the historical relationships between species and various groupings of species and explain those relationships. A systematist may want to describe groups of species as genera or families or classes, etc., but that is not a necessary part of phylogenetic work; at best it is a very useful by-product.

We should also remember that as one gets into those groupings called the higher taxa by systematists, that phylogenetic conclusions become less and less convincing. The greater the phyletic distance, the more difficult it is to read homologies. Therefore, phylogenetic conclusions range in certitude from very sound ones between species recently evolved from a common ancestor to very tentative ones regarding the possible relationships between different phyla and kingdoms.

Points of view. The "new systematics," as it was called when it was started in 1940, has been the most eloquent proponent of an evolutionary systematics. It argued that since differences arise and similarities remain between species because of natural selection, evolution must be integral to systematics. Theirs is a *phyletic approach to taxonomy.* Their case is well expressed by the German biologist Hennig in his book, *Phylogenetic Systematics.*

Others, most notably the numerical taxonomists in recent years, have emphasized what is called a *phenetic approach to taxonomy.* The basis for all categories is the phenotype of organism. The numerical taxonomists first list the characters of the organisms they are studying. They then determine, by careful mathematical analyses, the similarities among these characters. Organisms sharing groups of characters in terms of similarity are placed together in an *operational taxonomic unit* (OTU). And other organisms with other shared characters are placed in a different OTU. In this way the theoretical biases of the species concept are avoided and a purely objective taxonomy is generated. However, the founders of numerical taxonomy Sneath and Sokal devote a great deal of their book, *Numerical Taxonomy,* to showing that the hierarchies existing among their OTU categories can be interpreted phylogenetically. In fact, there is a good correlation between what they call an OTU and what an evolutionary systematist would call a species. The real advantage of numerical taxonomy seems apparent in descriptions of organisms for which no breeding analyses are available, which is the majority of the cases studied. Here their numerical analyses provide useful groupings of phenotypes and operationally sound evaluations of the descriptive data.

On the other hand, being aware of evolutionary principles, the evolutionary systematist recognizes the practical difficulties in specifying a reproductive community and tries to act accordingly. In the long run the recognition of evolution illuminates the many relationships apparent in the taxa being studied, i.e., homologies, convergences, and adaptations which are conservative and innovative. Numerical taxonomy pays a high price, it seems to many biologists, for its operationally defined units. Because it chooses to ignore the species concept it excludes one of modern biology's richest theoretical concepts. And in the process it ignores homologies and convergences and evolutionary innovation and conservatism. At best an OTU can identify phyletic relations, but it can never explain them, since it has no means of identifying plesiosemes, aposemes, and neosemes and since it ignores adaptations and natural selection.

Finally, we need to say something again about molecular phylogenies. The most successful ones are those based on amino acid sequences of proteins. One of the best known is shown in Fig. 9–14. This is based on the number of mutations needed to account for the amino acid differences between the cytochrome c proteins of different species (Table 9–1). By using the genetic code (Fig. 1–4), we can specify the minimum number of nucleotide changes needed to change one amino acid into another. A dendrogram, expressing phyletic distance as mutational distance, can then be drawn. This then

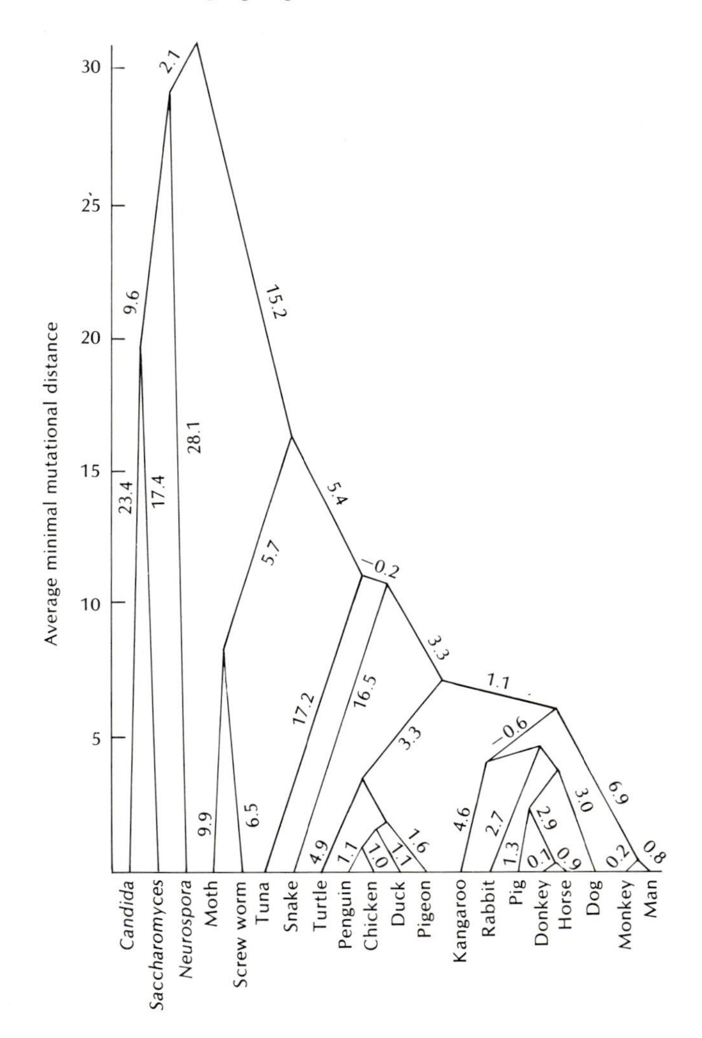

Fig. 9-14: A molecular phylogeny based on mutational distances between different cytochrome c molecules.

is a phylogeny based, remarkably, on one seme, i.e., the cytochrome c molecule! That it agrees as closely as it does with dendrograms built on a variety of traits is what is so remarkable. On the other hand, it tells us nothing that we did not already know about phyletic relations.

The potential of molecular phylogeny has not yet been fully exploited in phylogenetic studies. We will have occasions in later chapters to point to areas in which molecular and more conventional homologies can be used together to elucidate problems that neither category of homology can solve by itself.

The study of homologies, whether derived by careful use of Remanian criteria, by molecular sequence analysis, or very indirectly by numerical measures of similarity, is basic to clarifying evolutionary relationships. Whether those relationships are incorporated into systematics, or not, really depends on what we demand of systematics as a scientific discipline. The present tendency is to apply procedures that meet the job at hand, whether that be simple description or explanation of evolution. Since our goal in this book is to understand evolution, we must use the Remanian criteria of homology as a prerequisite to phylogeny. Only that approach fully exploits the information stored in reproductive communities and analyzable through the modern concept of direct and indirect homologies.

References

Bock, W. J., 1969. Discussion: The concept of homology. *Annals of the New York Academy of Sciences 167/1*: 71–73.

de Beer, G. R., 1971. *Homology, An Unsolved Problem*. Oxford Biology Readers. Oxford University Press, London.

Fisher, R. A., 1929. *The Genetical Theory of Natural Selection*. Clarendon Press, Oxford.

Gould, S. J., 1977. *Ontogeny and Phylogeny*. Belknap Press of Harvard University Press, Cambridge, Mass.

Hanson, E. D., 1977. *The Origin and Early Evolution of Animals*. Wesleyan University Press, Middletown, Conn.

Hanson, E. D., 1980. Phylogeny as description and explanation. *Scientia* (in press).

Hennig, W., 1966. *Phylogenetic Systematics*. University of Illinoi Press, Urbana, Ill.

Huxley, J. (ed.), 1940. *The New Systematics*. Clarendon Press, Oxford.

Maslin, T. P., 1952. Morphological criteria of phyletic relationships. *Systematic Zoology 1*: 49–70.

Mayr, E., 1963. *Animal Species and Evolution*. Belknap Press of Harvard University Press, Cambridge, Mass.

Mayr, E., 1969. *Principles of Systematic Zoology*. McGraw-Hill, New York.

Remane, A., 1971. *Die Grundlagen des natürlichen Systems in der vergleichenden Anatomie und Phylogenetik*. 2e. Auf. Koeltz, Königstein–Taunus. (This book is in German. The essentials of Remane's ideas on homology are given in Hanson, 1977, above.)

Simpson, G. G., 1961. *Principles of Animal Taxonomy*. Columbia University Press, New York.

Sneath, P.H.A., and R. R. Sokol, 1973. *Numerical Taxonomy*. W. H. Freeman, San Francisco.

Sonneborn, T. M., 1957. Breeding systems, reproductive methods, and species problems in Protozoa. In *The Species Problem*, E. Mayr (ed.), pp. 155–324. AAAS Publication, Washington, D.C.

Wright, S. S., 1978. *Evolution and the Genetics of Populations. Vol. 4. Variability Within and Among Natural Populations*. University of Chicago Press, Chicago.

4

Phylogeny

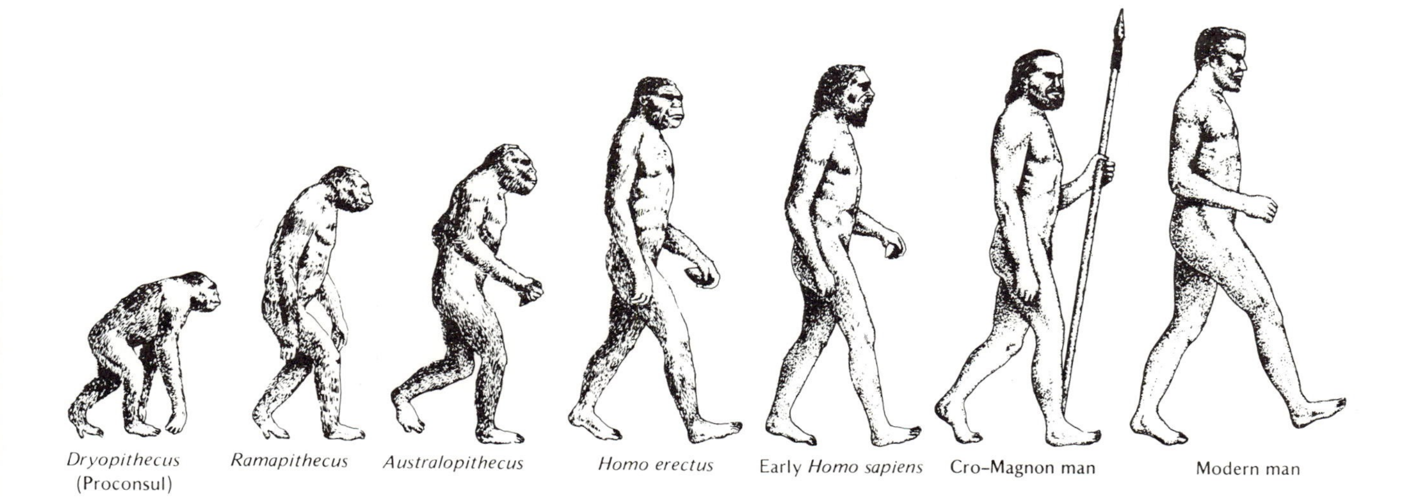

The ascent of man

An obvious place to start a critical review of phylogeny or evolutionary history is the origin of life. However, to study the origin of life we must start from what we know of life today and so work back from present-day living systems to their historical antecedents, and from that to the origin of life itself. This reverses the historical sequence.

In the context of this volume on evolution, the three preceding parts have made explicit our understanding of evolution through natural selection and the process of speciation. They have also informed us in considerable detail regarding present-day living systems. We are, therefore, ready to raise the question of how organisms arose in the first place. Chapter 10 will examine that problem. We will then proceed to consider various major groups of organisms to review what we know of their evolutionary history or phylogeny. To that extent we follow the chronology of the history of living things.

Evolvability is a major and distinguishing property of life. And as we have seen, it is a property that operates between the dual constraints of what constitutes a living system and the process of natural selection. Phylogeny is the expression of the potentials of evolvability. Hence the aim here must be twofold: to not only describe the history of life, but also, to explain why that history unfolded the way it has. Only by this twofold approach can we understand the potential of living things as evolvable systems.

TEN

The Origin of Life and the Fossil Record

AT THE VERY OUTSET there are four possible theories concerning the origin of life. The first of these considers life as a product of divine action. All cultures carry in their mythology such accounts. For example, the Judeo-Christian account is contained in the book of Genesis in the Old Testament. Such accounts are not scientific, nor are they amenable to scientific analysis and testing. The constraints on a scientific explanation for the origin of life are twofold. One is that nothing can be proposed that transcends our present understanding of physics, chemistry, and biology. The other is that whatever is proposed, within the limits of the natural sciences, must be testable. In particular, it must, in the most rigorous sense, be potentially falsifiable. Only then can we have any confidence that our tests are meaningful. Now let us turn to the remaining three possibilities for the origin of life.

The first of these possibilities suggests that life did not originate here on earth, but first appeared elsewhere in the universe. From that extraterrestrial origin something moved through space to infect our earth with life. This comes perilously close to being an untestable hypothesis because we have not yet found other life in the universe. As we shall see, probability favors the argument that life does exist elsewhere in the universe. The nearest possibility for such life to occur was thought to be Mars. It is now generally conceded, because of testing done by American and other scientists, that Mars is, in all probability, lifeless.

Radio-telescopes sweep outer space for signals revealing the presence of intelligent life elsewhere in the universe. They may or may not find evidence of such life, but if they do, that is just the first small step in analyzing the possibility of an infective origin of life on earth. There will still remain to be tested the mode of infection. It raises such questions as the following: Can life survive in outer space? How could it resist temperatures close to absolute zero? How long can it keep from drying out? What are the effects of irradiation? And so on. The difficulty in getting reliable answers to these questions suggests that the proposal of life originating elsewhere and then coming to earth is close to untestable.

What remains are the possibilities for a terrestrial origin of life. One of these says life is the result of highly improbable events. The other says that life is a probable outcome of terrestrial physics and chemistry. In the former viewpoint, a highly improbable configuration of atoms finally resulted in molecules that were life-like and from them all subsequent life evolved, including naturally, modern cellular forms of life. In the latter, molecular precursors of life were the expected outcome of natural events on earth and not to have found life would have been improbable.

Any further discussion of probabilities is not helpful at this point. Our discussion really focuses on three steps that must have occurred if life emerged from a non-living environment. These are

1. *The accumulation of precursors needed for life.* In particular this refers to molecular building blocks for proteins and nucleic acids. These two categories of macromolecules contain the information needed to specify living systems (see Chapter 1).

2. *The emergence of the first living thing.* There is sharp debate among specialists in this area. Many favor a DNA gene-like substance, as being the first form of life. Others argue that it was a cell-like structure composed largely of proteins—some with enzymatic capabilities.

3. *The evolution of the first living thing into cellular life.* Cells, whether prokaryotic or eukaryotic, are the basis of all life today. Somehow they came from the first living thing or things.

Let us now look at these three steps to see what can be said about a terrestrial origin of life.

The first life on earth

Recall that we have found it useful to define living systems as those that take matter and energy from their environment to maintain and reproduce themselves and at least some of their variations. All cells and viruses have these capabilities. Presumably so did the first living things. But what is the simplest system that could possess these capabilities?

THE SIMPLEST FORMS OF LIFE

Various candidates have been proposed as the simplest living thing.

1. *An inorganic autocatalytic reaction* is the basis of the first candidate. The following reaction between a hydrogen molecule and a copper ion has been put forward as at least an analog of a living system.

$$H_2 + 2\ Cu^{++} \xrightarrow{Cu^+} 2\ Cu^+ + 2\ H^+$$

Here the cuprous ion Cu^+ acts as a catalyst to form more of itself from the cupric ion Cu^{++}. How does this fulfill our definition of life?

First, we must realize that this autocatalytic system is not one entity, such as a single kind of atom, like the cuprous ion Cu^+. Rather the system is a collection of atoms that includes the hydrogen molecule H_2 and the hydrogen ion H^+. This set of inorganic substances and their interactions is thus the system.

We shall go on to evaluate this as a candidate for a living system, but first let us learn a lesson about viewing living systems in general. We expect living systems to have a certain structure—to be a cell, for example. But our definition of life does not necessitate such a viewpoint. Our definition is functional, not structural. It necessitates the *use* of matter and energy to *maintain* the system and to *reproduce* it and its variations. Anything that *does* these things can be considered to be alive. Therefore, the lesson is that the first living things may not have been cellular.

The example of copper ions is inorganic catalysis, in which the cuprous ion is the key substance that makes more of itself. In that sense it reproduces itself at the expense of certain materials in its environment, namely, H_2 and Cu^{++}. The energy used in the system is simply the ambient temperature. At absolute zero no reactions occur; at temperatures occurring on the surface of the earth there is a real probability, thermodynamically, that the reaction we are discussing can occur. But this example is not a living system because, essentially, no variations are possible. This reaction, in terms of modern chemistry and physics, will make more Cu^{+} and H^{+} ions and nothing else. Evolvability is utterly dependent on variation. If living things were devoid of mutations, they would not evolve. Hence, this reaction involving autocatalysis of inorganic substances is not a convincing candidate for the simplest living system.

2. In an *organic autocatalytic reaction,* there are instances when an organic compound formed by a reaction acts as a catalyst to form more of itself. In that proteins, as enzymes, are the most common example of catalysts, this proposal is suggesting that an autocatalytic enzyme is a possible form of early life. To evaluate this suggestion let us first look for concrete examples. The proteolytic enzyme trypsin is sometimes given as such an example.

$$\text{Trypsinogen} \xrightarrow{\text{trypsin}} \text{trypsin}$$

Trypsin, the product of the reaction is the enzyme catalyzing the reaction in which it is formed. This enzyme is found in many organisms. Trypsinogen is formed in the pancreas and initially is acted on by another enzyme, endopeptidase, which digests off a short section of the molecule. The remaining large portion is trypsin, which then also acts on trypsinogen to produce trypsin.

It is obvious, however, that this system has a long evolutionary history within living systems. Trypsinogen and endopeptidase are formed by the complex, gene-controlled machinery of multicellular organisms. No spontaneously formed autocatalytic enzymes are known today.

3. *A reflexively catalytic reaction* has been proposed as another possible primitive living system. There are no known examples of this, but nothing in this proposal contradicts our present knowledge of chemistry and physics. To the extent that no example is known, the proposal is untestable. However, since it is conceivable that an example of reflexive catalysis can be found and tested, let us look briefly at what is being proposed.

$$A \xrightarrow{X_n} X_1 \rightarrow X_2 \cdots \cdot \rightarrow X_n$$

Here a series of reactions occur such that a product, X_n, is eventually formed. This organic molecule catalyzes one of the reactions leading to its own formation. Presumably X_n is an enzyme and it and all its precursors, starting with substance A, are proteins. Protein A must be formed spontaneously from the environment. Proteins X_1 to X_n are able to fold into various configurations, some of them possibly enzymatic. But X_n is definitely enzymatic and catalyzes a reaction leading to formation of more of itself.

The environment as the source of matter and energy includes protein A and ambient energy. In catalysis, different amounts of energy, especially heat, will also affect the stability and conformation of these proteins. Reflexive catalysis is analogous to reproduction and the different foldings or conformations of proteins X_1 and X_n represent a kind of variation. In particular, if X_n takes on a changed conformation, X_n^1, which is a more efficient enzyme than X_n, then this is a variant of selective value if it thereby ensures more of X_n^1.

There is then in reflexive catalysis some clear parallels to living systems. The set of reactions that finally produces a protein catalyzing a reaction that makes more of itself comes the closest to meeting our definition of life of any of the proposals we have looked at so far. It emphasizes the need to find an actual example for experimental studies.

4. In *genic autosynthesis,* a gene is invoked as the simplest living system. Here the gene makes more of itself, not through catalysis, but through control of the synthesis of compounds needed to make another gene. Genes are self-replicating in a cellular environment, which they use to maintain and reproduce themselves and in which they mutate spontaneously. And the mutations can reproduce themselves. At first glance, then, the gene seems to fit very nicely our definition of a living thing and there are plenty of them around.

However, a gene maintains and reproduces itself *only* in a cellular environment or a close approximation of that environment. How could it have arisen in a noncellular environment? Here is where the origin of life as a highly improbable event comes into view. A gene is made up of thousands of nucleotides (Box 1–1) and each nucleotide has at least 33 atoms. What is the chance of a functional gene arising by chance alone? Put in those terms it is very remote. But those terms do not fairly describe the whole situation here, even though some specialists have used this as an argument against the spontaneous origin of life. More realistically we recognize that nucleotides are known to form spontaneously from mixtures of inorganic compounds containing water, carbon dioxide, ammonia, phosphate compounds, in the presence of an energy source such as heat or ultraviolet light. The real question is then, Do nucleotides spontaneously form gene-like aggregates of DNA? The answer, thus far, is no, they do not. But we are looking for conditions, which might have occurred in nature, under which DNA arises spontaneously from nucleotides and under which it might be able to reproduce itself.

The first result of our speculations regarding the first living thing is a negative one: we do not know what that living thing actually was. The second result is positive. Our speculations really center on two compounds—enzymatic proteins and a gene-like DNA. The problem comes down to seeing how those entities could have come into existence. Because once they were on the scene the emergence of cellular life as we know it is plausible.

THE PRECURSORS OF LIVING THINGS

For a long time there existed two somewhat contradictory beliefs in the Western world. One was the belief, reflecting the Judeo-Christian tradition, that all life was divinely created. The other was a belief in spontaneous generation. The latter never really included spontaneous generation of humans from nonliving material, but it did include the generation of many other animals (Box 10–1). Such a belief was finally put to rest by the scientific inquiry of Spallanzani and Redi and, especially, Louis Pasteur. Pasteur's experiments in the last century were the most convincing disproof of spontaneous generation.

There remained the idea of divine creation and probably because of it there was a special significance attached to all substances associated with life. Also, organic molecules were once thought to be so unique as to be formable only by living systems. The German chemist Wöhler dispelled that belief, in 1828, by his synthesis of urea, a nitrogenous waste product of many animals. Its structural formula can be shown as

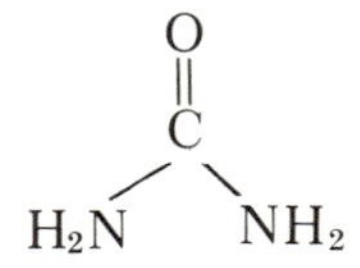

Wöhler and a contemporary of his, also in Germany, Justus

Box 10–1. Spontaneous generation.

The point of view of those who believed in spontaneous generation is seen in the following passages.

> Filippo Buonanni, the Jesuit scientist, surprised no one in his seventeenth-century world when he described a type of timberwood that, if soaked in the sea, produced worms that engendered butterflies which in turn became bright-colored birds. Not long before, the Belgian physician Jean Baptiste van Helmont had asserted that full-grown mice are produced in three weeks from wheat in a glass stopped with dirty linen and that scorpions are formed from the herb basil crushed in a cavity between two bricks. After all, it had been known since the time of the Egyptians that snakes rose from the mud of the Nile. Aristotle, greatest of the biologists, had stated that simple living things took shape from dust and mud—"from all dry things which become moist and all moist things which become dry." Even common folk were well aware that plant lice formed from dew drops, and snakes, of course, from a lady's hair dropped in a rain barrel. From Helena Curtis, "The Marvelous Animals," *Rockefeller University Review 3:* 13–21 (1965). Reprinted by permission.

And here is van Helmont's own recipe for generating mice.

> If a dirty undergarment is squeezed into the mouth of a vessel containing wheat, within a few days (say 21), a ferment drained from the garments and transformed by the smell of the grain, encrusts the wheat itself with its own skin and turns it into mice . . . . And, what is more remarkable, the mice from corn and undergarments are neither weanlings or sucklings nor premature but they jump out fully formed. (Translated from the original German.)

It was against this kind of superstition that the great French scientist Louis Pasteur performed his experiments. He was able to demonstrate convincingly that bacteria only grow in nutrient-rich containers when they get in from outside. They never arise spontaneously within such containers. He proclaimed, in 1864, to the French Academy that as a result of his work, "Never will the doctrine of spontaneous generation arise from this mortal blow."

Well, "never" is a difficult claim to establish in science. In effect, the origin of life on this earth revives the idea of spontaneous generation. Of course, it is being proposed in a form quite different from that envisioned by van Helmont and the others with whom Pasteur was arguing. The new guise of spontaneous generation was anticipated by none other than Charles Darwin. There is a passage in a letter to Hooker where Darwin says ". . . in some warm little pond, with all sorts of ammonia and phosphoric salts, light, heat, electricity, etc., present . . . a protein compound was chemically formed ready to undergo still more complex changes."

A tree stands by the water. Blossoms from the tree produce birds when they fall on land and fish when they fall in the water. This is from *Histoire admirable des Plantes,* by C. Duret, published in Paris in 1605.

Liebig, were the founders of organic chemistry. From that beginning it soon became clear that organic compounds also followed the laws of physics and chemistry. True, the highly reactive nature of carbon was a special property, but there was nothing mysterious about that. It reflected its place in the periodic table. True, also, that these carbon compounds could be of enormous size when compared to inorganic compounds.

But size meant complexity, not mystery. Hence, the chemical basis of life, no matter how complex, was clearly open to scientific inquiry.

Our emphasis on proteins and nucleic acids is justified from two points of view. One is the speculations we have summarized in an attempt to find an example of early life. The other and more important view is that proteins and nucleic acids are informed molecules, as the American biochemist Orgel has termed them. It is the informed molecules that determine the specificity of individual organisms. Such molecules are programmed by preexisting information in the cell, ultimately from genetic information within the nucleic acids themselves.

The concept of informed molecules is a neat way of encapsulating the chief problem in studying the origin of life. The question becomes How did informed molecules arise from uninformed ones? That question leads to another one: Do uninformed proteins and nucleic acids arise spontaneously? The answer is yes for proteins and no, so far, for nucleic acids.

Now we can pose our inquiry into the origin of life in these terms. First, regarding proteins: What are the precursors of proteins? How do they form uninformed proteins? How does an uninformed protein become an informed one? Similarly for nucleic acids: What are the precursors of nucleic acids? Can they form uninformed nucleic acids? How can informed nucleic acids arise? We shall now look at today's answers to these questions.

The origin of proteins. Protein precursors became an area of vigorous research when Stanley Miller, working with Harold Urey at the University of Chicago, published his experimental results in 1953. Miller placed CO_2, H_2O, NH_3, and CH_4 (methane) in a system of glass tubes such that by boiling the water the mixture of gases was forced past an electrical spark and returned to the water again (Fig. 10–1). When this system was allowed to operate for some days there appeared a dark precipitate in the water. Upon analysis this precipitate turned out to contain various amino acids.

Miller's basic experiment, including various modifications, has been repeated many times to reveal that not only can many essential amino acids be formed by this technique, but nonessential amino acids, significant amounts of proteins, and various other organic molecules. (Recall that the essential amino acids are those twenty that occur in informed proteins. See Box 1–1.)

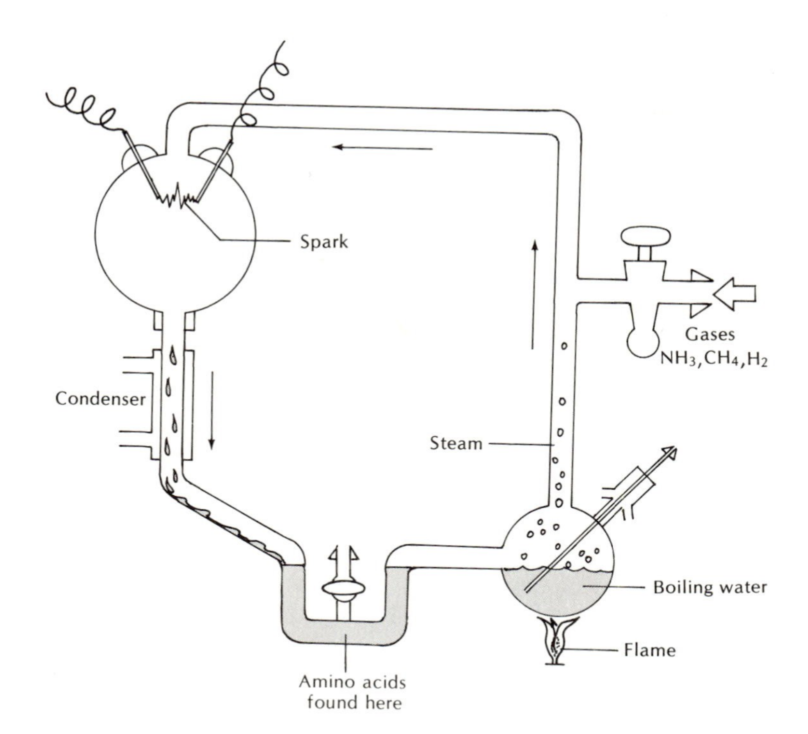

Fig. 10-1: A diagram of the apparatus used by Stanley Miller to illustrate spontaneous synthesis of amino acids. This sterile system contained water, ammonia, methane, and carbon dioxide. The water was heated to the boiling point, with steam and gases being forced past an electrical spark. These molecule and energy sources produced a dark precipitate of amino acids.

The answers to our first two questions regarding proteins are clearly in hand. Amino acids are the precursors of proteins, and they are readily formed spontaneously. And from them certain uninformed proteins are made. Now what can we conclude from this? Regarding the formation of amino acids there are two important points. First, it has become clear that an atmosphere in which amino acids will appear must be a reducing one (one in which H_2 is present). That was true of Miller's experimental set-up. In an oxidizing atmosphere (one in which O_2 is present) no amino acids were obtained.

Second, not all the essential amino acids are formed in this type of experiment and some amino acids were formed that are not found in organisms today. All of this is not so much a lesson as a puzzle. Why were only the 20 essential amino acids incorporated into living systems when some of them are not readily formed spontaneously, and why were others readily formed that are not incorporated? We cannot answer that question today.

And, then, the uninformed proteins require study. These proteins can be formed in large amounts from certain amino acids. That is, certain proteins are favored as are the amino

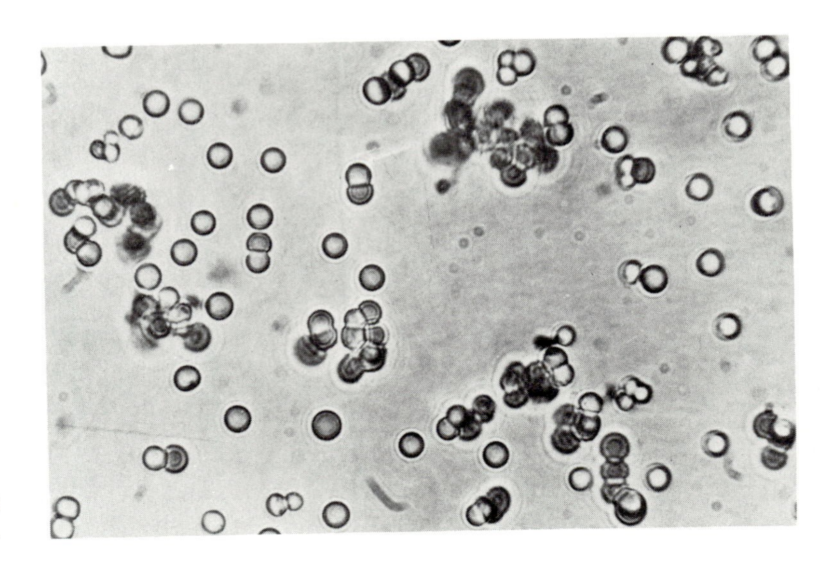

Fig. 10-2: Proteinaceous spherules obtained and studied by Sidney Fox and his co-workers.

acids that constitute them. None of them has been shown to have enzymatic activity. These proteins tend to aggregate and form small spheres and, most interesting, these spheres differentiate in a way that suggests cellular structure. In the electron microscope they show a dense outer layer; internally, there is a mass of material of differing densities. Biochemist Sidney Fox and his collaborators have been especially active in studying these proteinaceous spherules (Fig. 10–2). They suggest that these entities represent a structure that could anticipate cells at a later stage in the origin of life. Such microspheres are optimistically called protocells. A better term, coacervate, was first used by the Russian biologist Oparin, in 1924, in his pioneering examination of the origin of life. We will return to Oparin's coacervate after we have considered, in more detail, the transition from precellular living systems to cellular ones.

One further comment on the microspheres (coacervates) is needed. Amino acids and many other organic compounds tend to form microspheres in solution. The process is a kind of selective mutual adsorption in response to the surrounding charged molecules of water. In fact, the water molecules form a polarized film around the microsphere. This behavior is found in the hydrophilic and hydrophobic behavior of the protein and lipid components, respectively, of cell membranes. In other words, such orientation of molecules and formation of microspheres is common and may parallel the behavior of cells and their membranes. This observation is strengthened by one last one. The microspheres can accumulate certain substances from their watery environment by selective absorption. This, too, parallels the uptake of substances from their environments by cells through their semipermeable membranes.

Now what can be said regarding the emergence of informed proteins? Very little, and this is a major problem in the study of the spontaneous origin of life on earth. In addressing the problem this way, we have narrowed down the central issue. Instead of asking, How do we get living cells from a nonliving environment? we are asking, How do we get informed proteins from noninformed ones? This restatement of the problem helps us see the problem more clearly. Here the problem has two parts. First, we need to find how an uninformed protein can act like an informed one and, second, we need to find, then, a way it could be formed. The second part of the problem says that it must be formed from some kind of information-containing template. We will return to the problem of templates after first discussing the origin of nucleic acids and the emergence of cellular life. Returning to the first aspect of the problem, we can say that if a spontaneously formed, or uninformed, protein acts enzymatically, then it is acting like an informed protein. Very briefly, that brings us back to something like reflexive catalysis as a possible way both to generate such a protein and to advance a theory on the emergence of living systems as we know them today. But, as we have seen, there are serious problems with the experimental analysis of the proposal, so far.

The origin of nucleic acids. Now let us turn to the nucleic acids. Nucleotides are formed spontaneously in a Miller-type experiment. These nucleotides contain among them the all-important one adenine triphosphate or ATP. Its importance lies in the role it plays in living cells as a source of energy as well as a precursor for DNA. The other nucleotides needed for DNA, which contain cytosine, guanine, and thymine, as well as uridine for RNA, are also formed spontaneously. Also simple sugars have been found in these experiments.

The next question is whether or not uninformed DNA and RNA are formed. The answer really is no. Short lengths of nucleotide polymers, called oligomers, do appear. But so far no RNA with molecular weights similar to those of transfer or ribosomal or messenger RNA has been found. (These are, of course, the three main categories of informed RNA.) And nothing approaching a high molecular weight double helix of DNA has appeared spontaneously. The researchers in this area conclude, so far, there is no reason to suppose that nucleic acid molecules comparable in size to informed nucleic acids are formed spontaneously. It is always possible that tomorrow some researcher will find a way to effect the spontaneous formation of uninformed nucleic acids. And that may then provide the clue for the transition to informed molecules. But this is at present nowhere in sight. It forces us to think that perhaps informed nucleic acids did not arise spontaneously. That is, when they arose it was within living systems. Such living systems would have to be based on proteins. This, though somewhat implausible for the reasons given above, brings us back to a consideration of the origin of the first living system on the basis of proteins and, from that, the emergence of cellular life.

THE EMERGENCE OF CELLULAR LIFE

As problematical as the approach admittedly is, we must go back to the proteinaceous microspheres as possible protocells. We will approach them from the point of view of Oparin's coacervates, but adding that they contain uninformed proteins capable of some reflexive enzymatic activity. Enough has been said to show that such an approach is full of problems. But it is also clear why this seems to be the most promising approach. Frankly, at this point, we are speculating, but speculating within the rules of scientific inquiry; we are searching for testable explanations regarding the origin of life.

A coacervate containing reflexive enzymes can fulfill our definition of a living system. It could even be called an organism, since it could exploit its environment for the matter and energy to maintain itself. These would include spontaneously generated amino acids, nucleotides, sugar, and other organic compounds, including uninformed proteins, and as an energy source, heat or even possibly ATP. In the latter case one of the proteins might enzymatically split off the phosphate radicals and the released energy kept to help form more protein. This might or might not be part of the reflexive catalysis, but in any case, it contributes to formation of proteins making up the coacervate. Growth of the coacervate would come from the formation within it of various organic compounds, especially proteins. Large coacervates have been seen by Oparin and others to pinch in two when they reach a certain size. Apparently they are unstable above a certain size. This is growth and reproduction aided and abetted by the selective uptake of compounds and their enzymatically controlled reactions within the coacervate.

Extensive study of coacervates as proteinoid microspheres (Fig. 10–2) has come from the laboratory of Sidney Fox and his colleagues at the Institute for Molecular and Cellular Evolution at the University of Miami. They have confirmed and greatly extended Oparin's initial ideas. Variations in these microspheres might come about through variations in the folding of the proteins.

Another source of variations has been suggested by the distinguished Japanese protein chemist Akabori. He proposed that proteins absorbed on clays could undergo reactions to amino acid side chains. This would change the sequence of the amino acids and change the conformations of the proteins, as well. This is a very important suggestion in three respects. First, it points to how an uninformed protein could take on the functions of an informed one, if side-chain substitutions led to enzymatic activity. Second, it could help explain how amino acids not arising spontaneously could become part of a protein with informed activities. And, third, instead of absorbing on clays, perhaps these early proteins absorbed on structures within the microspheres and there underwent side-chain substitutions. And their new properties could be immediately advantageous within microspheres.

If, and we repeat, *if* such microspheres arose they could be regarded as living. And because life would be associated with a discrete, three-dimensional structure—not a set of reactions in an organic soup—such a system would also be recognized as an organism.

Many biologists today recoil at the notion of an organism without genetically functional nucleic acids. True, such an organism would be more efficient if it had genes. And genes would have to emerge before such a living microsphere could transform into cells as we know them today. However, it might just be that such a system would provide the possibility for the spontaneous formation of uninformed nucleic acids and their transition to informed ones. Let us speculate further on how that might occur.

Because these early organisms or protocells could be termed living, they would also be evolvable. That means natural selection would act on their variations and preserve the more advantageous ones. What would be the nature of such selection pressures? This can be made clearer by looking at a basic reality faced by all organisms. If a species is going to persist, the individual organism must survive long enough to form more of itself. Some rocks and other tough substances persist simply because they are durable. Organisms, by comparison, are not durable. In fact, they are comparatively delicate, but organisms that persist do so because they reproduce faster than they are destroyed. If we indicate the average period of duration of an individual as the survival interval, and the average period needed to form more of itself as the formative interval, then, for persistence, the survival interval must be greater than the formative interval. Therefore, *selection will favor anything that increases the survival interval or decreases the formative interval.* These selection pressures are unavoidably present; they arose with the first living thing and have been the basic reality of survival down to the present, and will continue to be.

What increases the survival period of the protocell? There are at least three conditions that increase survival: (1) an improved selective uptake of substances from the environment, (2) an increased concentration of reactants needed for the life of the protocell, and (3) stabilized functional (enzymatic) and structural (membranes, in particular) proteins. All three are, of course, interrelated.

What decreases the formative period? Two points deserve emphasis: (1) speeded up reactions needed for the life of the cell, which can be the result of increasing the concentration of reactants and improving the efficiency of enzymes, and (2) proper distribution of functional and structural components.

Increasing the survival interval and decreasing the formative interval make improving the proteins even more important. Their increased efficiency and stability as components of membranes and enzymes will have a very significant selective advantage as will their regular distribution to daughter protocells. At this point the advantage of informed proteins—proteins from information-containing templates—becomes obvious. Such templates can ensure selective formation of the needed proteins; this does not happen in reflexive catalysis. The latter process probably generates many useless proteins (in terms of cell survival) for each useful protein generated. Also, distribution of a few templates could be an efficient way of assuring that protocells can form needed proteins.

How could such a template system arise? We are still speculating here, but we can be guided by the known dependence of protein formation today on nucleic acid templates. Perhaps ribonucleic acid was the first template that absorbed the amino acids needed to form a selectively advantageous enzyme, for instance. If so, that nucleic acid would also be selected for. And if a DNA molecule assured formation of the needed RNA, it too would be selected for. Today these speculations are taking the form of research into the transfer RNA molecules as being the first informed nucleic acids.

We know that nucleic acids are maintained in cells. We also know that their precursors are concentrated and preserved in cells. The interior of protocells may, therefore, be the most likely place in which nucleic acids formed spontaneously, and in which if a nucleic acid that could act as a template were formed, it would be selected for. This point of view might get us around our present impasse of postulating the spontaneous formation of nucleic acids in a nonliving environment. Evolution of the template might have followed a path that is the reverse of today's protein synthesis. But that path would form the basis for understanding protein biosynthesis.

The resulting protocell, now endowed with informed molecules, would be very close to cellular life as we know it today. Arguably, it would have a real selective advantage over cells without genuinely informed molecules. The result would be replacement of the latter by the former. All living things would appear, then, to have arisen from an ancestor with nucleic acid (usually DNA) information storage that effects the formation of highly specific proteins. If these DNA molecules could also perpetuate themselves, we would have genes within cells as the model for an apparent unitary origin of life. Such a process, as formulated here, would have three stages. Stage one would involve living systems that are series of reactions; stage two, coacervates or microspheres with proteins that act as if they were informed; and stage three, microspheres that are

Table 10–1. Transition from reflexive catalysis to protocell during the early stages of the origin of life.

Living systems	Status of macromolecules	Level of organization
reflexive catalysis	spontaneously formed proteins, one or more with enzymatic activity	series of linked reactions, one product of which catalyzes a reaction leading to formation of more of itself
microsphere with reflexive catalysis	as above, but with a greater concentration of reactants and a more favorable environment	an organism; a three-dimensional structure resembling a protocell
protocell with nucleic acid-based genetic substances	nucleic acids and proteins are informed	cell-like structure with genes of DNA

transformed to protocells with nucleic acids emerging as the basis for all genuinely informed molecules (Table 10–1).

From stage three, we can discuss the evolution of what are essentially cellular systems. Such systems might be rather like today's prokaryotes. Hence, our discussion can, from this point on, rely heavily on known substances.

The early evolution of cells

We now look at the nutritional requirements of these early cells. What are they taking from their environments to maintain and reproduce themselves?

If an organism uses preformed organic compounds as building blocks for its macromolecules, and if it obtains the energy it needs from the breakdown of molecules, it is called a *chemoheterotroph*. These early cells were chemoheterotrophs. (So were all the living systems at that time.) Eventually, as these early chemoheterotrophs continued to multiply they inevitably used up the spontaneously formed precursors on which they depended. What then? The plausible answer is that they evolved biosynthetic pathways of ever increasing complexity. This answer was first developed by the American biochemist Horowitz, who has recently been concerned with the chemical testing for life on Mars.

THE EVOLUTION OF BIOSYNTHESES

Horowitz's proposal goes as follows. Suppose an early organism needs organic substance A as a precursor for a vital substance. Substance A eventually comes into short supply because many similar organisms are competing for it. Suppose, further, that A is being produced spontaneously from the following reaction, occurring outside the cell, $B + C \rightarrow A$. Any cell that can take up B and C from the environment and produce A now has a selective advantage over all the other cells that depend on an outside source of A.

The uptake of B and C may require, first, a changed membrane protein. This can occur through mutation. Proteins already in the cell might catalyze the reaction $B + C \rightarrow A$. Further, mutations could yield a more efficient catalyst. The cell with such new capabilities would now predominate in those parts of the early seas in which survival was possible. And then, in time, B or C or both would come into short supply. What then? A situation like that regarding the formation of A would occur. The cell or cells able to form B from its precur-

Box 10–2. Amino acids biosynthesis.

The biosynthetic pathway for two essential amino acids, valine and leucine, is shown here as it is known to occur in the bacteria. Each step in these reactions, indicated by an arrow, is catalyzed by a specific, gene-controlled enzyme. Note that the starting point is pyruvate, which contains three carbon atoms. This molecule, a common breakdown product of glucose metabolism, enters the citric acid cycle whereby carbon dioxide is released. In other words, amino acid metabolism is related to reactions that ultimately produce carbon dioxide, though normally these acids are not formed from that source directly but from the carbon dioxide that goes first to glucose, via the photosynthetic pathway to pyruvate. [From E. Umbarger and B. D. Davis. 1962 Pathways of amino biosynthesis. In *The Bacteria*, J. C. Gunsalaus and R. Y. Stainer (eds.). Vol. III. *Biosynthesis*. Academic Press, N.Y.].

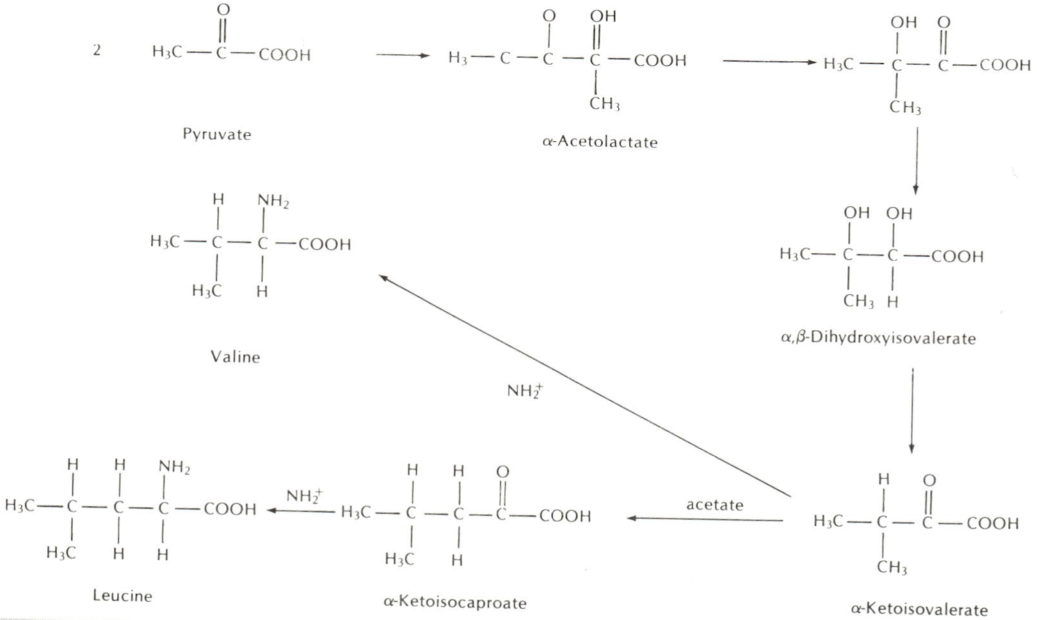

2 Pyruvate → α-Acetolactate → (α-keto-β-hydroxy acid) → α,β-Dihydroxyisovalerate → α-Ketoisovalerate

α-Ketoisovalerate $\xrightarrow{NH_2^+}$ Valine

α-Ketoisovalerate $\xrightarrow{acetate}$ α-Ketoisocaproate $\xrightarrow{NH_2^+}$ Leucine

sors and C from its precursors would be at a selective advantage. And this situation could repeat itself many times throughout the early history of the earth, which is now reliably estimated as being 4.6×10^9 years old.

The overall result of this early evolution would be the emergence of biosynthetic pathways. Such pathways would be like those known today, in which reactions are catalyzed by highly specific enzymes formed under the control of genes (Box 10–2). Also, of necessity, there would be a general direction in this evolution. The biosynthetic pathways would be lengthening in the direction of simpler and simpler precursors. Eventually, organisms would probably become so biosynthetically sophisticated as to require only such inorganic precursors as carbon dioxide, water, ammonia, sulfur, and phosphorus or their compounds or both. They might, also, by this time have evolved means for capturing energy from sunlight for biosynthesis. In short, nutritionally, they would have evolved from being chemoheterotrophs to being *photoautotrophs*.

THE INTERLOCKING OF NUTRITIONAL CYCLES

Microbiologists often say that if there is an energy-yielding reaction in nature, then there will be an organism that can make use of it. The relevance of that comment to the origin of life is that chemoheterotrophs evolved to various kinds of autotrophs. Some used sunlight, as various photosynthetic prokaryotes do today, but others used such energy-yielding reactions as the

Box 10–3. A nutritional classification of organisms.

The outline and table, given below, were put together by a group of distinguished microbiologists in 1946. It contains details not commonly used today, for example, lithotrophy and organotrophy as well as mesotrophy. It does illustrate, however, the logic inherent in such a classification, i.e., precise characterization with respect to the chief energy source as well as the ability to form essential molecular constituents [reprinted, by permission, from the Cold Spring Harbor Symposium of Quantitative Biology, Vol. XI, p. 302 (1946)].

I. Nomenclature based upon energy sources
- A. PHOTOTROPHY
  Energy chiefly provided by photochemical reaction
  1. *Photolithotrophy*
     Growth dependent upon exogenous inorganic H-donors
  2. *Photoorganotrophy*
     Growth dependent upon exogenous organic H-donors
- B. CHEMOTROPHY
  Energy provided entirely by dark chemical reaction
  1. *Chemolithotrophy*
     Growth dependent upon oxidation of exogenous inorganic substances
  2. *Chemoorganotrophy*
     Growth dependent upon oxidation or fermentation of exogenous organic substances
- C. PARATROPHY
  Energy apparently provided by the host cell
  1. *Schizomycetotrophy*
     Growth only in bacterial cells.
  2. *Phytotrophy*
     Growth only in plant cells.
  3. *Zootrophy*
     Growth only in animal cells

oxidation of sulfur or iron. However, these latter reactions would imply the presence of a significant amount of gaseous oxygen in the environment. We have mentioned, in connection with Miller's work, that the early atmosphere was a reducing one. The change to an atmosphere such as today's probably is the result of the release, by photosynthetic organisms, of large amounts of oxygen as a metabolic by-product. But nevertheless, in this early nutritional evolution we need not think of evolution as following a single path from chemoheterotrophy to photoautotrophy. Along the way there were opportunities for chemoautotrophs, as well as photoheterotrophs, to emerge. A summary of the various possibilities of nutrition is given in Box 10–3; it includes the use of other organisms as food sources.

Biogenic macromolecules. Up to this point we have only discussed spontaneously occurring molecules (organic and inorganic) as nutritional building blocks and sources of energy

---

II. Nomenclature based upon ability to synthesize essential metabolites
   A. AUTOTROPHY
      All essential metabolites are synthesized
      1. *Autotrophy sensu stricto*
         Ability to reduce oxidized inorganic nutrients
      2. *Mestrophy*
         Inability to reduce one or more oxidized inorganic nutrients, i.e., need for one or more reduced inorganic nutrients
   B. HETEROTROPHY
      Not all essential metabolites are synthesized i.e., need for exogenous supply of one or more essential metabolites (growth factors or vitamins)
   C. HYPOTROPHY
      The self-reproducing units (bacteriophages, viruses, genes, and so on) multiply by reorganization of complex structures of the host

Energy sources		Synthetic abilities		
		Autotrophy		Heterotrophy
		autotrophy S. S.	mesotrophy	
phototrophy	photolithotrophy	*Chlorella vulgaris*	no example given	*Rhodopseudomonas polustris*
	photoorganotrophy			*Rhodospirillum rub.*
chemotrophy	chemolithotrophy	*Thiobacillus dentrifricans*	*Escherichia coli*	*Saccharomyces cerevisia*
	chemoorganotrophy	*Pseudomonas fluorescens*		

along with sunlight or even heat. This means we have been ignoring another and by then extremely important source of molecules. These are the molecular products of organismic metabolism or biogenically produced compounds. If these compounds had not been utilized, the resources of the earth would have been eventually exhausted by living organisms. When, through natural selection, organisms began to utilize the products of other organisms, then recycling emerged on earth.

From this emergence of recycling, there appeared ecosystems, and new interrelationships of life on earth. In fact, organisms became interlocked in mutual dependence. Sometime at this point in evolutionary history, there were again many chemoheterotrophs. Organic compounds were again in an abundance, but now they were not formed spontaneously but as a result of biosynthesis, or they were formed biogenically. These secondary chemoheterotrophs were likely quite similar, nutritionally, to the prokaryotic or bacterial chemoheterotrophs of today. We shall call them chemoheterotrophs II to distinguish them from the earlier chemoheterotrophs I, which extend back (speculatively) to the origin of life. In between these two types of chemoheterotrophs, distinguished by the source of the organic compounds they use, there were the photoautotrophs and various combinations of chemotrophs, phototrophs, heterotrophs, and autotrophs (Box 10–3).

THE METHANOGENS

One of the most interesting evolutionary finds in recent years is the prokaryotic form, the methanogen. Their special evolutionary significance has been described by the American biophysicist and microbiologist Carl Woese and his colleagues. The methanogens are best classified as chemoautotrophs. They live in a hot, reducing environment. Their cell walls, their enzymes, and certain of their nucleic acids differ markedly from those of other prokaryotes, although their overall cellular organization is clearly prokaryotic. But especially because the nucleic acids of their ribosomes (essential in protein synthesis) are so different from any others known, Carl Woese and his colleagues have suggested that the methanogens might merit a kingdom of their own (Box 10–4). That suggestion has not been acted upon, but it is worth mentioning to highlight the special significance of the methanogens. For our purposes, they represent a group of extraordinary chemoautotrophs that depend on a reducing atmosphere. They could well

Fig. 10-3: The Gaia hypothesis is illustrated in these drawings showing the results of the absence and presence of living things on earth.

be living representatives of early chemoautotrophs—of those forms that first emerged when the ecosystems were evolving and that retain a need for a reducing atmosphere.

The Gaia hypothesis. This brings us to consider somewhat more carefully the changes that occurred in the atmosphere, hydrosphere, and the surface soils of the earth as life originated and evolved. The idea that life and its environment have co-evolved has been called the Gaia hypothesis by the British scientist James E. Lovelock. As the by-products of living systems accumulated on the surface waters and in the air and the soil of the earth, living things had to adapt to ever new surroundings (Fig. 10–3). We have already said that the appearance of oxygen as a by-product of photosynthesis changed the atmosphere from a reducing one to an oxidizing one. This made possible chemical respiration as we know it today. Most living things are aerobic and animals, in particular, depend on oxygen for life. Hence the change to an oxidizing atmosphere was an essential precursor to animal life and evolution.

Oxygen penetrates surface waters and enters the soil to some extent. These areas have also evolved oxygen-dependent life forms, but also important are the accumulated organic

Box 10–4. The methane-producing bacteria.

Methanogen is a general name applied to bacteria that grow anaerobically by oxidizing hydrogen and reducing carbon to methane (CH_4). As a result of analyzing the nucleotide sequence found in one of the ribosomal RNA (*r*RNA) molecules of these cells, it appears that the ten species of methanogens studies are separable into two groups and that both groups differ distinctly from all other prokaryotes. Earlier work has shown that the *r*RNA molecule in question, called 16S ribosomal RNA, is very conservative in that it shows sequences of nucleotides that are similar to those in all other prokaryotes in which it has been studied.

This *r*RNA is found in all prokaryotic cells, it is easily isolated, and can be cut up enzymatically (using RNase from a bacterial virus) into units with varying number of nucleotides, called oligonucleotides. The nucleotides from each species were compared to each other and to those from two species of bacteria and one species of blue-green alga. The result, expressed mathematically by an association coefficient, is a measure of homology. Similarities in oligonucleotides that contain six or more nucleotides represent the point-to-point comparisons necessary for identifying homologous structures. The number of identical nucleotides is divided by the number of nonidentical ones to give the association coefficient. A high value approaching 1.0, means a large proportion of identical oligonucleotides. Conversely, values for the coefficient that approach 0.0 mean that the proportion of identical nucletides is low. Higher values mean significant homology is present and phyletic distances are relatively small; lower values, the opposite.

The following table, published by George Fox, Linda Magrum, William Balch, Ralph Wolfe, and Carl Woese in this study on the methanogens, summarizes the association coefficients that were found. [Reprinted with permission from *Proceedings of the National Academy of Science 74:*4540 (1977.)]

Note that species 11 and 12 are typical bacteria and species 13 an unidentified member of the blue-green algae. Their 16S *r*RNA shows little homology to that of the methanogens. The two groups of methanogens referred to above, which show differences among themselves, are species 1 through 6, as one group, and species 7 through 10, as the other group.

The analysis was taken further. These researchers say that if methanogens and other prokaryotes are as different as the *r*RNA differences indicate other striking differences should also be evident. The following were found:

compounds that have dissolved in water and that continually enrich the soil. Such compounds transform a barren dust or sand to the rich topsoil that supports plant life and forms the basis for all food chains on land. Similarly, organic compounds are important to oceanic plankton and zooplankton. Plankton form the basis for life in the surface waters of the ocean. That life supplies not only organic compounds at the surface, but also contributes to the quiet rain of detritus that reaches the ocean floor and sustains life even at the greatest ocean depth. Such life lives in permanent darkness, with minimal oxygen and almost always low temperatures. But the fact that it exists has changed the ocean from an originally lifeless dilute soup of

---

1. The methanogens have at least two enzymes, related to methane metabolism, and not known to occur in other organisms.
2. The methanogens contain no cytochromes, the proteins widely used by other organisms for electron transport.
3. Most prokaryotes have peptidoglycan in their cell walls; Methanogens have none.
4. The transfer RNA (*t*RNA) of other organisms carries a distinctive squence TψCG in one part of each *t*RNA molecule. In its place, the methanogens have either ψψCG or UψCG. (ψ is a modified uridine.)
5. Changes occur in *r*RNA after it has been transcribed. (The change of uridine to "pseudo-uridine" or ψ, in 4 above, is an example.) Such changes in the methanogen 16S *r*RNA are very different from those found in other prokaryotes.

Altogether, then, significant differences are seen between the methanogens and typical prokaryotes. Whether this will justify a new kingdom of organisms, as hinted at by Carl Woese and his colleagues, remains to be seen. For the present it is convenient to refer to the methanogens as a very special group of prokaryotes.

Organism	Organism 1	2	3	4	5	6	7	8	9	10	11	12
1. *M. arbophilicum*	—											
2. *M. ruminantium PS*	.66	—										
3. *M. ruminantium M-1*	.60	.60	—									
4. *M. formicicum*	.50	.48	.49	—								
5. *M. sp. M.o.H.*	.53	.49	.51	.60	—							
6. *M. thermoautotrophicum*	.52	.49	.51	.54	.60	—						
7. Cariaco isolate JR-1	.25	.27	.25	.26	.23	.25	—					
8. Black Sea isolate JR-1	.26	.28	.26	.28	.27	.29	.59	—				
9. *Methanospirillum hungatii*	.20	.24	.21	.23	.23	.22	.51	.52	—			
10. *Methanosarcina barkeri*	.29	.26	.24	.24	.26	.25	.33	.41	.34	—		
11. Enteric-vibrio sp.	.08	.08	.11	.09	.09	.10	.05	.06	.07	.10	—	
12. *Bacillus* sp.	.10	.10	.14	.11	.11	.12	.08	.10	.10	.08	.27	—
13. Blue-green sp.	.10	.10	.10	.10	.10	.11	.08	.09	.08	.11	.24	.26

spontaneously formed compounds to a collection of complex ecosystems with interdependent inhabitants.

This awareness of co-evolution, the concept that life evolves along with the changes it produces in its environment, is emerging as a key concept in our search for life on other planets. This search is called exobiology—life beyond the limits of this earth, or outer life—and one of its leading practitioners in the United States, Cyril Ponnamperuma, of the University of Maryland, uses it as a criterion for predicting the probabilities of extraterrestrial life. In his laboratory Ponnamperuma mimics extraterrestrial conditions on the moon or on Mars, for example, so that he can study the spontaneous formation of molecules needed for life. When the first satellite descriptions were received on earth from Mars, Ponnamperuma set up his Mars atmosphere on earth and correctly showed that life on Mars was highly unlikely, a result confirmed by subsequent tests performed by the spacecraft that landed on the surface of Mars.

From the Gaia hypothesis we can expect to find a special kind of fossil on earth, the chemical fossil. These fossils are compounds the presence of which can only be accounted for as the product of living systems. It turns out that they have been found by Ponnamperuma and his colleagues in the oldest sedimentary rock known to us, that in Greenland, which is reliably dated as being 3.8×10^9 years old. They are the oldest evidence of early life on earth. With that, let us leave the many problems and speculations regarding the origin of life and turn our attention to the fossil record and the questions and answers contained there.

The fossil record

First we must be clear on answers to three questions. How are fossils formed? How are they discovered? And, most intriguing, how are fossils interpreted?

FOSSIL FORMATION

Any trace of life from the past can be called a fossil. This includes the molecular products of life as well as worm tracks in the mud of some bygone time. It also includes various kinds of hard parts, such as shells, bones, or teeth. It can include the impressions left by soft parts, which may be flattened and distorted by the pressures of the overlying mud that turned to rock or replaced by chemicals, which have no direct connection to living systems, but nonetheless are a record of life in the past. Perhaps the most extraordinary fossils are the living things that were frozen during an Ice Age. A mammoth, which is an extinct relative of the elephant, has been found which was trapped in the frozen Alaskan tundra some millions of years ago and preserved there. Discovered and recovered some years ago, it provides us with an extraordinary fossil. A sampling of fossils is given in Fig. 10–4.

Fossils are obviously formed in a variety of ways, and therefore, there is no one answer to our first question, How are fossils formed? Fossil formation is a chance event. It depends on living things, or any of their many products, being in a place where they are relatively undisturbed. Or, if they are disturbed, where the resulting changes still retain the record of the past. The examples of petrified wood (Fig. 10–4) show this dramatically. Here the original wood has been completely replaced by various chemicals from the environment. These latter were selectively deposited in various parts, even down to cellular details of the wood. Thus, despite replacement of the original living parts, a marvelously detailed record of that life is retained.

More typically, the fossilization of an animal, such as one of the ancestors of present-day horses, occurs along the following lines. The organism in question, say, for example, the grazing form called *Merychippus* that lived about 20 million years ago, died on the grassland where it lived. In a relatively short time the scavengers of that time—giant hyenas and vultures, and a whole array of other forms including beetles and flies and their larvae—would have consumed and dismembered the carcass of *Merychippus*. Such a fate would have precluded any fossil formation and was by all odds the most common situation. Another, less likely possibility was that a *Merychippus* was drowned while crossing a river. If the body sank and was covered by silt, the possibility of a fossil resulting is more likely. However, if the river dried up—perhaps our grazer was simply caught in a flash-flood—before the drowned animal was well buried, scavengers could have destroyed the carcass.

If, however, mud-burial was complete—the body was entombed and subsequent layers of mud and dirt accumulated on top of it and everything in time turned to rock—then a fossil would have been the result. The point here is that fossilization is the exception and not the rule. Most forms are destroyed before they can become fossils or fossils themselves

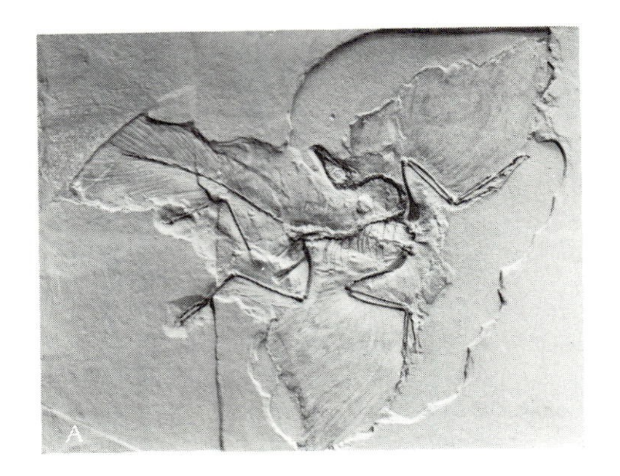

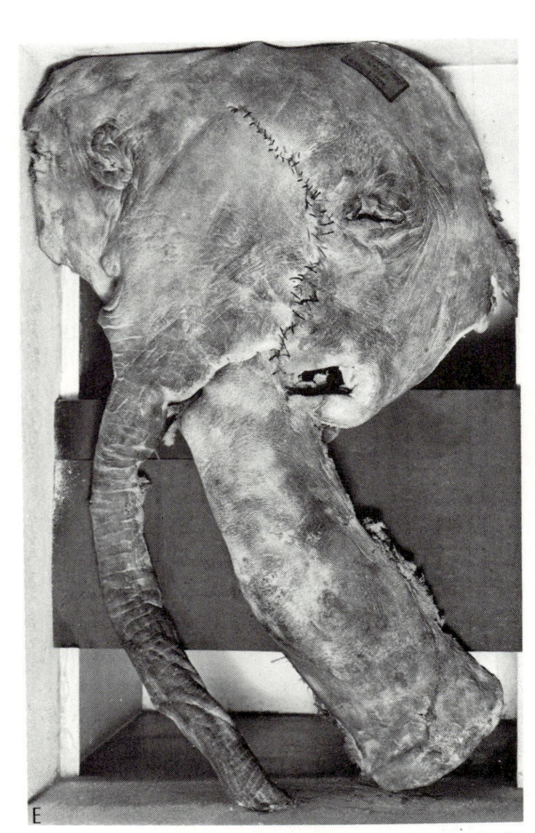

Fig. 10-4: Selected fossils. Prokaryote microfossils (A, arrows) from the Fig Tree Chert; a fossil protozoan (B), one of the foraminifera with a shell that was preserved in marine sediments; an imprint of a fossil leaf (C), so clear that the veins are visible; fossil fish (D), belonging to the herring family; an actual baby mammoth (E), retrieved frozen in Alaska.

may subsequently be destroyed. Those that are fossilized must have died in circumstances favorable to their being somehow preserved. (Organisms living on wave-swept, rocky coasts or in mountains where little sedimentation occurs, will rarely if ever appear as fossils.) Fossils represent a tiny and selected sample of life in the past. Their apparently great numbers tell us that life has been abundant, so much so that, despite the odds against fossilization, we have significant numbers with which to work.

FINDING FOSSILS

Fossils appear when erosion or digging uncovers them. Erosion goes on all the time, in total disregard of the presence or absence of fossils. Hence many fossils are uncovered and may be casually destroyed. The ideal fossil is one whose position in its surroundings is still identifiable; and it represents as complete a record as possible of the original organism. Isolated bones and teeth are important—they have told us much about human evolution—but complete skeletons are, of course, much more informative, though much less likely to be found.

Fig. 10-5: A fossil bed. Brachiopod fossil shells in limestone.

Sedimentary formations are naturally the richest source of fossils for the paleontologists who seek them. Here fossils were formed and preserved through deposition of muds and silt. When such a bed of fossils is found (Fig. 10–5) great care is taken in the removal of specimens. Often, chunks of the sedimentary material are removed and worked on with special tools in a laboratory so as to assure intact specimens. Depending on the size and nature of the fossil, the paleontologist's tools are appropriately varied. In looking at microscopic fossils, the techniques of a gem cutter and polisher are utilized. This assures thin, transparent preparations for the microscope. Larger fossils may require accurate splitting of sedimentary layers followed by the addition of preservatives, e.g., plastics, to protect the fossil remains. And large dinosaur bones are often covered with plaster at the site of their excavation to protect them from rain and other wear and tear.

Most fossils are destroyed before they catch the eyes of fossil-hunters. Sedimentary rocks are changed through movements of the earth's crust or through heating, and this damages or destroys fossils. When we put this fact together with the realization that relatively few organisms are successfully fossilized, we can conclude that the fossils we have recovered are a very limited sample of the life of the past. But that conclusion protects against the extremes of wild speculation or pessimistic conservatism in interpreting the fossil record. Neither the extreme of assuming we have a complete record of the past nor the extreme of assuming we can say nothing of the past is justified.

INTERPRETING FOSSILS

There are two aspects to learning what fossils can teach us. The first comes from the direct observation of fossil evidence. The second is less direct because it is concerned with the relative position of fossils in the rocks where they are found.

Direct observation of fossils. Except for rare cases in which organisms are immediately preserved, e.g., the frozen mammoth or, another case, insects trapped in amber, fossil remains present us only with a part of some organism. So, obviously, one problem in interpretation is the reconstruction of the whole organism. Perhaps the most famous practitioner in this area was Cuvier (see Box 2–1), the father of paleontology. A fossil tooth was a fascinating clue to Cuvier. From it he could make educated guesses as to the size of the original organism and its age and its food. The latter then allowed cautious speculation as to the nature of such internal organs as the digestive system. This led to more information on body size and shape. And so on. Comparable work is being done today and the reconstruction of whole humans from parts of their skull is a case in point (Fig. 10–6). Such reconstructions can be extremely useful in casting light on the creatures of bygone times. In some cases specially preserved fossils, such as the impression of dinosaur skins, allow us to reconstruct details that we would otherwise have had no knowledge of.

More difficult to interpret are the traces left by organisms, such as worm tracks left in the mud of ancient seas or the molecular fossils, mentioned earlier. In these instances we may never know the exact nature of the organism that left such evidence of its existence. Nonetheless the information we do have is useful. It helps complete the picture of what type of organism was alive at a certain time in the past history of this earth. Worm tracks, for example, document, at the least, that worms were present. Even though we may not be able to say which species or class or maybe even which phylum, we do know that crawling things of a certain approximate size were part of the ecosystem under study. That enhances, though admittedly in a limited way, our understanding of past life.

Molecular fossils are even more ambiguous. They tell us practically nothing about the size and shape of what produced

Fig. 10-6: These skulls (A) and skeletal remains (B) were photographed in a cave at Shanidar, Iraq. These human-like fossil fragments, in particular, give us an insight into the cultural life of Neanderthal-like persons, since remains of flowering plants were discovered with the fragments. From this, an artist has recreated a scene from Neanderthal times (C).

them. They can, however, clearly indicate the biochemical capabilities and functions of what produced them. There are, as we know, a variety of organic molecules synthesized by living systems, but they are stable to very different degrees. Proteins and their constituent amino acids would be fascinating fossils, but proteins are essentially non-existent as fossil molecules and extensive searches for amino acids have led to the following conclusions. Amino acids are present in ancient rocks and the majority are in the L-form, which is characteristic of life

Box 10–5. The occurrence of D- and L-molecules in living systems.

It is an unexplained fact that, except for a few D-amino acids in certain bacterial cell walls, all amino acids in organisms now living are in the L-form. On the other hand, sugars exist in the D-form in organisms. The possibility of D- (dextrorotatory) and L- (levorotatory) forms arises when a carbon atom is covalently bonded to four different atoms or sets of atoms. The differences in three-dimensional structure are comparable to our right and left hands, which are almost identical except that they are mirror images of each other. Consider the middle carbon atom in an amino acid with the structure

$$\begin{array}{c} NH_2 \\ | \\ R—C—COOH \\ | \\ H \end{array}$$

(R refers to the rest of the molecule, which differs among the different amino acids. See Box 1–1.)

Now imagine that the middle carbon sits at the centers of a tetrahedon and at the four apices of the tetrahedon there are the four different atoms or groups of atoms to which it is bonded. Such a molecule can be drawn two ways, as follows:

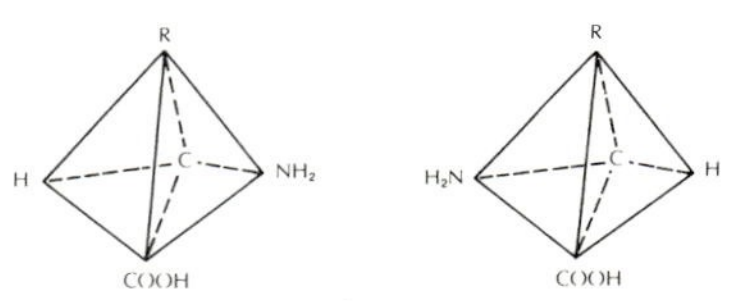

Note that these are two different molecules structurally, although their atomic constituents are identical. The one on the left is the L-form.

(Box 10–5). However, in rocks over 60 million years old, the L- and D-forms occur in about equal amounts. Hence, it appears that L-amino acids, when associated with ancient rocks (Precambrian, over 600 million years old), are probably of relatively recent origin.

The most useful, it now seems, of the fossil molecules are the straight-chain hydrocarbons (Fig. 10–7). These are very stable compounds and are easily recovered from ancient rocks by washing in benzene and methane. The occurrence of pristane and phytane, products of the chlorophyl molecule, can be used to infer the presence of organisms capable of photosynthesis. (Cuvier would have appreciated this detective work.)

However, there is still lively debate regarding the interpretation of molecular fossils. The work on the generation of organic compounds in the absence of life has shown, as we discussed earlier, a quite extraordinary array of spontaneously formed substances. Amino acids were among them. Some organic geochemists suggest that the long-chain hydrocarbons should be added to this list. And so the search is still underway for unambiguous fossil molecules. The field is, nonetheless, considered to be a promising one by many researchers.

Finally, under problems of direct observation of fossils, we must consider the species problem. Unavoidably, fossil specimens are named and classified and this places them beside extant or neontological forms. The problem now is this: Are the criteria for species identification the same for paleontological and for neontological material? We raise this question because both are used in the study of evolution. And when evolutionary sequences are constructed from extinct forms leading to modern extant forms, we might wonder if we are using comparable terms for both forms. In brief, the answer is that we are not. The problem is not a serious one if we keep in mind certain precautions.

To describe apparently related fossil forms as members of different species we apply the kinds of differences apparent between comparable extant species. Some workers prefer to work only with differences that justify generic names. This degree of caution is not always necessary, but it is more desirable than the reverse case in which every little difference between two fossils is used as an excuse for describing new species. The necessary precautions come down to an awareness of the natural variation that occurs in any species, a sense of the genetic basis for such variation, and then a cautious usage of species designations. In this way we can quite confidently combine our data on extinct and extant forms and expand the data base used for evolutionary studies.

The relative positions of fossils: Chronology. The most informative aspect of the fossil record is that it allows us to perceive evolutionary history directly. This can happen, however, only when we have determined the relative ages of fossils correctly. The basic rule is that the oldest fossils are toward the bottom of the pile. The pile we refer to is the bedded sequence of fossils. The oldest fossil beds are laid down first, the next oldest on top of them, and so on up to the surface where today's forms are living. Ideally, then, the evolutionary sequence is simply read from bottom to top. But things are rarely ideal. Two phenomena, at least, can confuse the ideal state of affairs. One of these is that fossil beds can be moved. Geological folding and uplift, which can obscure an otherwise quite readable story, occur. Folding often reorients layers or strata of fossil beds relative to each other and uplift in one area

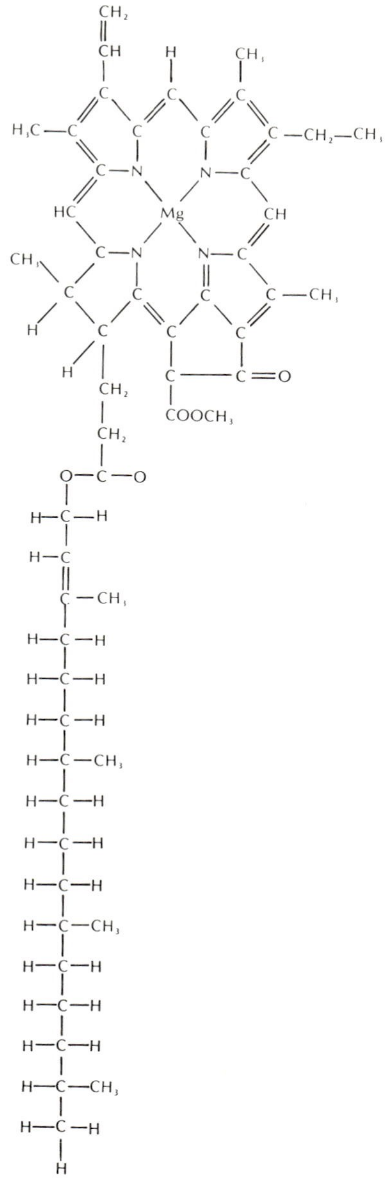

Fig. 10-7: Straight chain hydrocarbons as molecular fossils. The chlorophyll molecule has attached to it a sixteen-carbon chain after the oxygen. This carbon chain can be hydrolized at the oxygen bond to give rise to two very stable compounds, pristane and phytane.

can raise a bed that was at the bottom, and quite old, in another area. Another troublesome phenomenon is the loss of strata or beds. Erosion can remove millions of years of sedimented history and thus produce gaps in the fossil record (Fig. 10–8). Such gaps, depending on the amount of discontinuity they cause, can pose genuine problems in our understanding of how one form evolves into another.

The answer to the foregoing largely depends on common sense as well as the competence of the geologist and the paleontologist. Common sense tells us to go slow in interpreting sequences from fossil beds until the geologist can determine the degree of folding, uplift, and erosion. The paleontologist can then compare the results from one stratum with those from a comparable stratum elsewhere (Fig. 10–8) and thus carefully reconstruct historical sequences.

Ultimately, when we deal with history, we look for absolute measures of time, not just relative ones. We want to know precisely how long ago a certain event occurred and not just that it occurred earlier or later than other events. The most informative method in this regard is an analysis that depends on the amount of specific radioisotopes present in a fossil bed or a fossil.

The key to this method is what is called the *decay rate*. This refers to the rate at which a radioactive isotope changes spontaneously to another, usually nonradioactive form. Such a rate is expressed most commonly as a half-life, namely, the time during which one-half the radioactive element decays, or is changed spontaneously to its decay product. If we know the half-life of a given element and can measure the amount of ra-

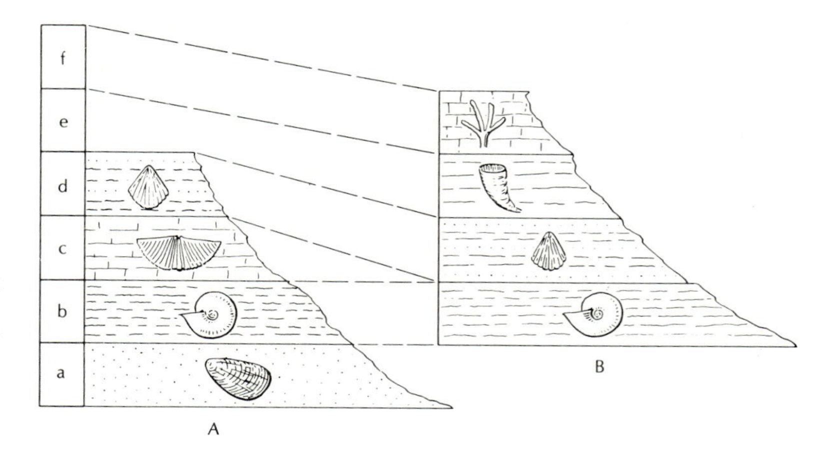

Fig. 10-8: Problems in correlating fossils in different localities. Fossils from strata (a through f) in fossil beds from different localities (A and B) are compared. But stratum c is missing from B, presumably because it was eroded before stratum d was laid down. And strata e and f are missing from A. Note that strata are characterized by all the fossils present, not just one.

Box 10–6. Radioactive dating.

Radioactive decay is used as an absolute measure of the age of fossils or of the rocks that contain fossils. The half-lives of various isotopes are as follows:

Radioactive isotope	Decay product	Isotope half-life (years)
carbon-14	carbon-12	573×10^1
uranium-235	lead-207	713×10^6
potassium-40	argon-40	130×10^7
uranium-238	lead-206	451×10^7
rubidium-87	strontium-87	470×10^8

Obviously, the last four decay more slowly than carbon-14 (^{14}C). For relatively recent events (up to about 40,000 years ago), ^{14}C is a useful isotope. It is found in fossils because ^{14}C, like its stable product ^{12}C, enters living things as CO_2 and is assimilated into the very structure of living matter. In 40,000 years, ^{14}C has gone through about seven half-lives and, therefore, the amount of the original isotope has been reduced to less than 1 $[(\frac{1}{2})^7 = \frac{1}{128}]$ and can no longer be reliably detected.

Of the other isotopes, uranium (^{235}U or ^{238}U) is also incorporated into living systems, but not always in amounts sufficient for dating analyses. The remaining isotopes in the table are used to date rocks directly. That date is then applied to the fossils they contain, if possible. We say "if possible" because dating of rocks usually applies to igneous rocks, which were once molten and have cooled; they are not sedimentary, fossil-containing rocks. To date sedimentary rocks, the association between them and datable igneous rocks must be carefully studied. This can be done. One result is that we have documented, a few, very old sedimentary beds with ages that exceed 3×10^9 years. These are the ones most avidly screened for fossil molecules, since they are the oldest known possible storehouses for this kind of fossil.

dioactive isotope and its decay product in a rock, we can then accurately estimate the age of that rock (Box 10–6). It is important to realize that radioactive decay is an atomic phenomenon that is unaffected by such external factors as weather, temperature, and sunlight. It depends solely on subatomic events determining atomic structure. It is this independence from environmental vagaries that makes radioactive decay such an indispensable clock for measuring absolute time in the fossil record.

MAJOR FEATURES OF THE FOSSIL RECORD

To determine the more important features of fossil history we proceed in two steps. First let us look at the major periods of that history and glimpse the geological events that occurred throughout that history, and then let us turn to the fossils themselves. Figure 10–9 summarizes currently recognized eras, periods, and epochs of the fossil record and gives the time spans associated with each one. Also given are some of

TIME SPANS (Rough estimates of years × 10⁶) Years ago	DURATION Eras	DURATION Periods	DURATION Epochs	GEOLOGICAL PERIODS Eras	GEOLOGICAL PERIODS Periods	GEOLOGICAL PERIODS Epochs	GEOLOGICAL EVENT
0.025	70	1.5	0.025	Cenozoic	Quaternary	Recent	Glacial conditions, followed by recent times
1.5			1.475			Pleistocene	
12		68.5	10.5		Tertiary	Pliocene	Warm climates, gradually cooling; continental areas mainly free of seas; continued growth of mountains, including Alps and Himalayas
25			13			Miocene	
34			9			Oligocene	
60			26			Eocene	
70			10			Paleocene	
132	155	62		Mesozoic	Cretaceous	(Epoch divisions not necessary for present purposes)	At first great swamp deposits, followed by birth of Rocky Mountains and Andes, and cooling of climates
180		48			Jurassic		Much of continental lowlands are near sea level
225		45			Triassic		Widespread desert conditions
275	375	50		Paleozoic	Permian		Continued mountain-building, and variable climates including aridity and perhaps glaciation
310		35			Pennsylvanian		Lands low and warm with seas over much of continents, at the beginning; coal swamps, from which come the greatest of our coal deposits; mountain-building toward the end
350		40			Mississippian		
405		55			Devonian		Still considerable portions of land below water; evidences of aridity and continental areas
430		25			Silurian		Much of land below the sea at first, followed by mountain-building at the end
485		55			Ordovician		Great submergence of lands
600		115			Cambrian		Lowlands and mild climates; first abundantly fossiliferous rocks
4500?		3900?		Precambrian	(Period divisions not well established)		

Fig. 10-9: Periods in the fossil record and related geological events.

the geological changes that have occurred on the earth's surface.

The major lesson regarding geological changes is the one Lyell taught us (Box 2–3), namely, the earth's surface is always changing. When we look past time spans of a human lifetime, past the birth and death of nations, and past recorded history, we proceed from periods measured in decades to those whose duration is measured in centuries and in millennia. But the fossil record covers millions and hundreds of millions of years. We must expand our imaginative powers to sense these dimensions of time. This is the duration of natural selection.

And it is quite clear then why natural selection must be a process that preserves organisms able to respond to change. The environment provided by the earth is changeable and always in flux. Not only are there daily and seasonal changes, and cycles of wet or dry years and winters and summers of differing severity, but—again in terms of geological time spans—mountains have been rising and falling, oceans advancing and receding. In brief, the whole surface, what geologists call the crust, is in continuous dynamic motion. And we living things survive as best we can under such conditions and add our own changes (remember, the Gaia Hypothesis) to those of the earth itself.

As to the geological time spans, one place to start is where the fossil record becomes most clear. This is at the beginning of the Paleozoic era, about 570 million years ago. The oldest period in the Paleozoic is the Cambrian and so the time prior to that is, logically enough, the Precambrian. The Precambrian covers something like 4000 million years (4×10^9). It is in this time span that we think life started on earth and gave rise, successively, to prokaryotes and then to eukaryotes. The latter evolved into various types of unicellular producers, consumers, and decomposers. And from each of those categories multicellular forms emerged. In the Cambrian, we find fossil evidence of all major groups of life, except for the land plants and the Chordata. But within the Paleozoic, these groups also appear so that by around the Devonian period, about 480 million years ago, we can say all present-day major taxa had appeared. The subsequent history is one of continued change with these taxa; new forms appear and many other forms disappear. In fact, extinction appears to be the fate of many forms. The distinguished vertebrate paleontologist and evolutionist Alfred Romer has estimated that over 95% of the tetrapods (four-legged vertebrates) have become extinct. That means our present enormous diversity of amphibia, reptiles, birds, and mammals is but a small sample of the diversity that has been produced over time in these groups. The same conclusion may be drawn regarding other groups of animals, plants, protistans, and monerans. Now let us survey in detail what we find in the fossil record of the various eras, periods, and epochs. We shall work forward from the Precambrian.

Precambrian life. The oldest fossils we have of cellular life are from Western Australia. In that area, the North Pole Dome, there is a black chert (a kind of rock) reliably estimated to be 3.5×10^9 years old. Four researchers, Awramik, Schopf,

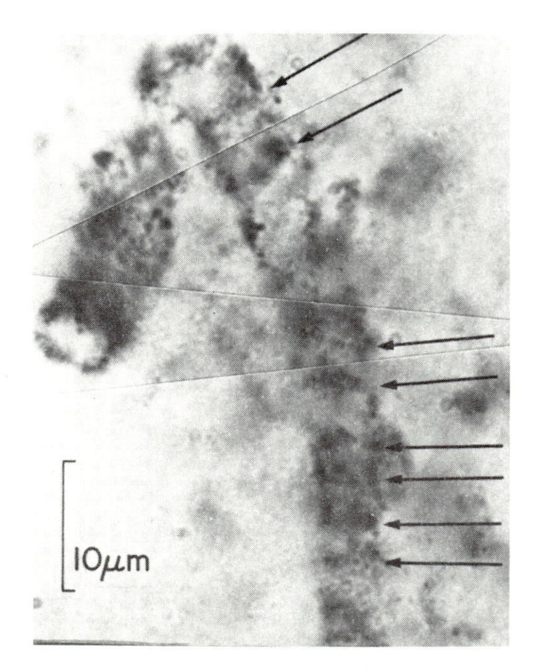

Fig. 10-10: Prokaryote microfossils. This material, which appears to be filamentous, was prepared from Precambrian rocks estimated to be 3.5×10^9 years old. These rocks are in Western Australia. The two topmost arrows point to what can be cross-walls in the filament. The other arrows indicate other structures, which appear to be cell walls.

Walter, and Buick, prepared very thin slices of this rock for microscopic examination and found structures that look convincingly like filaments of prokaryote cells (Fig. 10–10). In older rocks from Greenland, i.e., 3.8×10^9 years old, Ponnamperuma found molecular fossils, as we noted earlier.

The special importance of the North Pole Dome finds, other than their great age, lies in the filamentous structure. Previously, what were thought to be old cellular fossils were round. Were they prokaryotes or microspheres? Were they alive? Were they what we have called protocells—something between microspheres and prokaryotes? Recent work has shown that microspheres impregnated with silicon are indistinguishable from the spherical structures suspected to be fossils (Fig. 10–11). But the presence of filamentous structures is strong evidence in favor of prokaryote fossils.

Other reliable identifications of microfossils come from rocks about $1.6–2 \times 10^9$ years old. In the Gunflint chert of Canada, there are structured forms that are very similar to the blue-green algae known today. And such similarity to modern forms is especially apparent in the fossils from the Bitter Springs rocks of Australia. These rocks, which are about 9×10^8 years old, have revealed some 30 different species of fossils. (Remember, species are here described in morphologi-

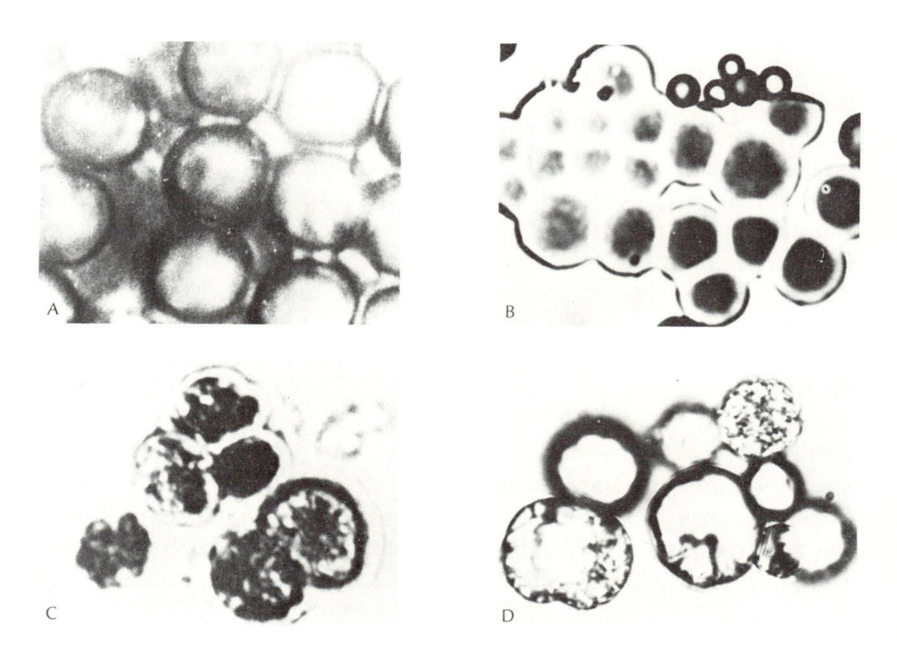

Fig. 10-11: Ancient microfossils and artificial "fossils." The tiny, cell-like structures found in ancient sedimentary rocks (A and C); the fossilized microspheres manufactured in the laboratory (B and D).

cal terms.) Of these, nineteen appear to be blue-green algae, with six or seven similar in structure to present-day blue-green algae. Thus, we can deduce that prokaryotes were well established and quite diverse by the middle of the Precambrian period. Probably photosynthesis (photo-autotrophy) had evolved by that time and the earth's atmosphere was changing from a reducing to an oxidizing one.

Sometime in the more recent Precambrian, perhaps $1\text{–}1.5 \times 10^9$ years ago, the first eukaryotes appeared. We have found fossil fungi from that period and from what we can infer today of their origin (see Chapter 12) they were probably descended from eukaryotic algae. Only the discovery of unequivocal eukaryote fossils will help us determine the actual time of their origin. But because we find abundant fossils of most groups of multicellular plants and animals by the Cambrian (570×10^6 years ago) we must conclude that the recent 1000×10^6 years of the Precambrian was a period of extraordinary evolutionary change, which saw unicellular eukaryotes diversfiying into a variety of multicellular forms. These forms were the origins of the dominant phyla of recent times, namely, the Metaphyta, Fungi, and Metazoa (see Fig. 1–21).

Paleozoic life. As can be seen in Fig. 10–12, all the major animal phyla are represented by fossils in the Paleozoic. Estimates regarding the abundance of these phyla come from the frequency of appearance of their fossils. This may have little relation to the actual abundance of these forms. (Recall our earlier remarks on the formation and recovery of fossils as being a limited sample of past life.)

The first chordate fossils appear in the Ordovician period of this era. These are fossils of the earliest vertebrates—the jawless fishes or Agnatha (see Appendix I). From then on, other groups of vertebrates appear in succession (Fig. 10–13). This history of the vertebrates illustrates two important points we will have to remember as we continue to discuss the phylogeny of living things. The first point concerns the origin of the phylum Chordata. According to the data in Figs. 10–12 and 10–13, this great phylum makes its first appearance fully developed in the fossil record. In the Cambrian there are no animals with a notochord in the fossil record and, then, there they are in the Ordovician record. And when we do find them they are recognizable as primitive jawless fish. It is these kinds of data that make evolution, at least of the higher taxa, appear to proceed by sudden jumps or saltations. Such a view is quite different from Darwin's idea of evolutionary change

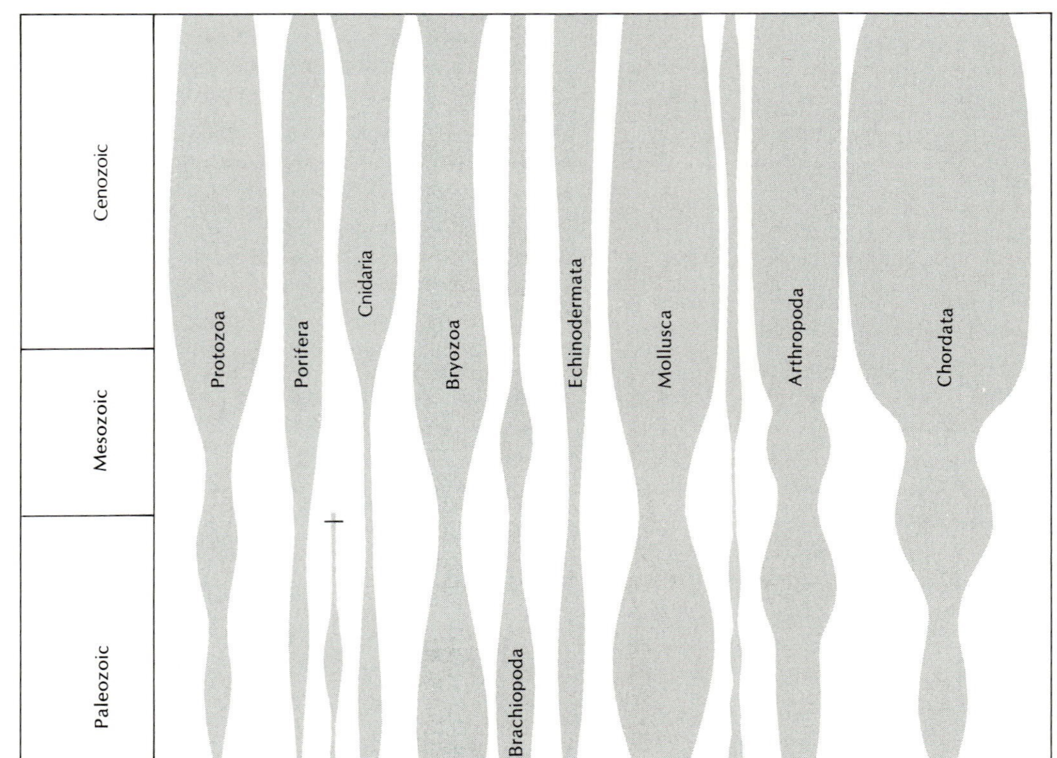

Fig. 10-12: The fossil history of major animal phyla.

through the slow accumulation of small mutational events. The problem is usually thought to lie with the quality of the fossil record; since it is a partial record, only the most abundant forms living where fossilization is possible are apt to appear. However, another view of gaps in the fossil record is that of the American paleontologists Eldredge and Gould. They propose that, instead of the "phyletic gradualism" we expect as a legacy from Darwin's view of evolution as a slow accumulation of small changes, "punctuated equilibria" better describes what we can expect. They argue that speciation through spatial isolation produces new gene pools that differ sharply from parental pools—the new pools punctuate the history of old pools by establishing their own tendencies toward a new adaptive phenotype or a new epigenotypic equilibrium. Hence, of necessity, there is a gap between old and new phenotypes and this will appear in the fossil record. Whether punctuated equilibria or simply incompleteness of the fossil record will account for the apparently sudden appearance of the vertebrates in the fossil record awaits further work.

The second point is related to the first one, namely, the ancestors of each group within the vertebrates are obscure. We have no good documentation, step-by-step (which means, ide-

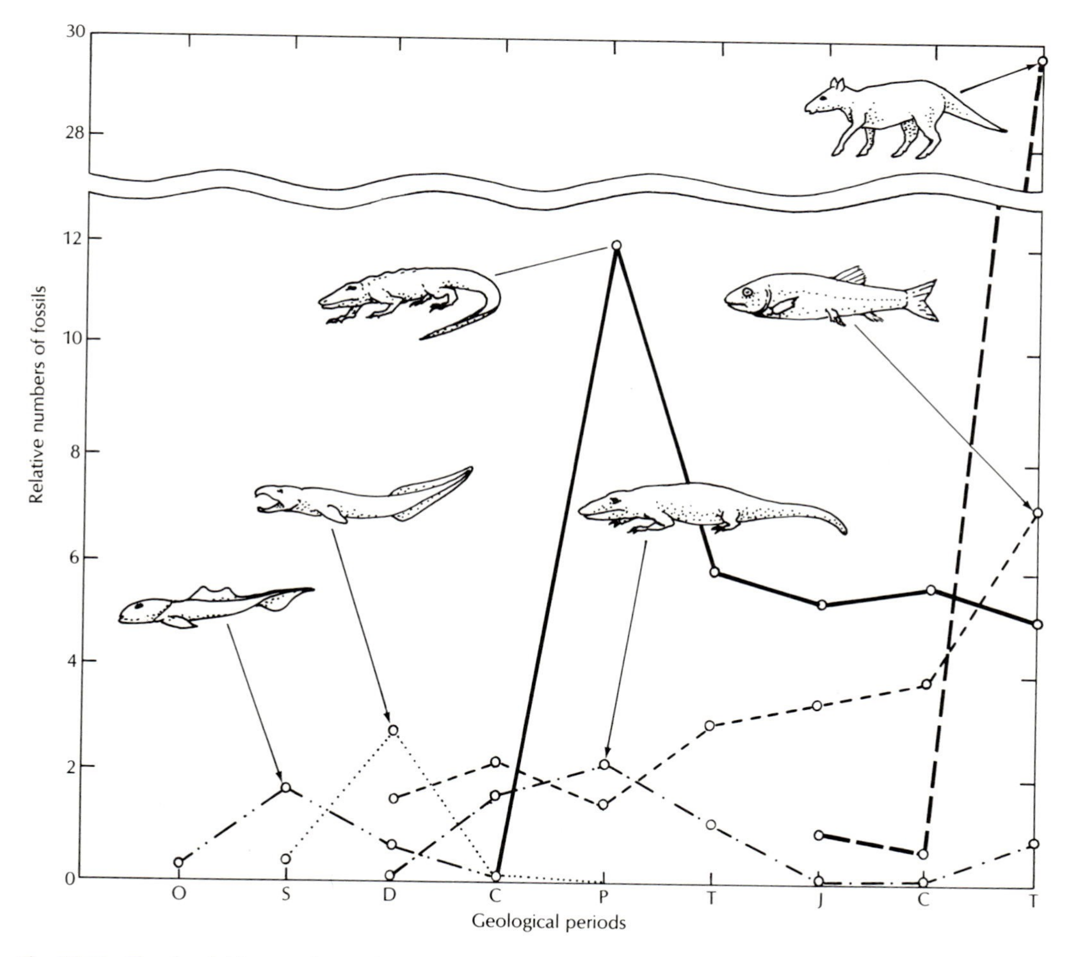

Fig. 10-13: The fossil history of vertebrates, extending from the Ordovician to the Tertiary (see Fig. 10-9).

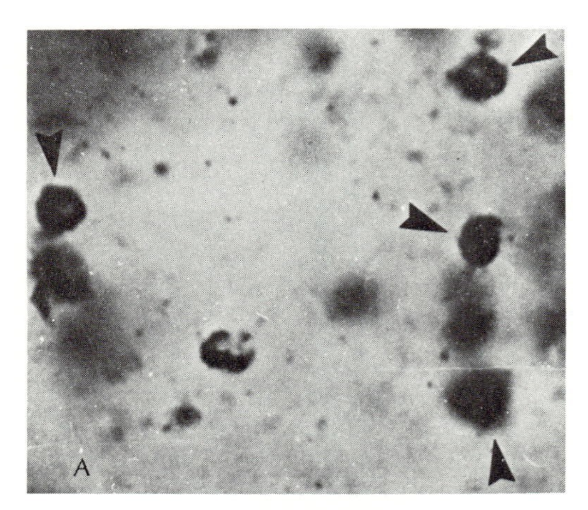

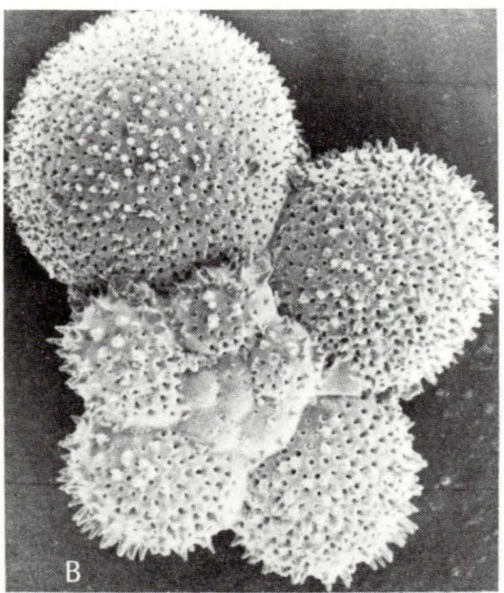

Fig. 10-14: Primitive birds: The fossil (A) was the model for the reconstruction (B) of *Archaeopteryx*. John Ostrom, a Yale University paleontologist, proposes that feathers were initially an adaptation for bipedal reptiles in the capture of prey (C). Later on, this led to gliding (D), and subsequently, flying evolved.

ally, species-by-species) of the changes giving rise to the various vertebrate orders. And the reason is, again, either the limitations of the fossil record or evolution as punctuated equilibria, i.e., significant departures from antecedent forms.

Does this mean that the origins of major groups are forever lost to us and that all we can do is speculate on these bygone events? Yes and no. Yes, in the sense that actual forms documenting the transition from one major taxon to another are in all probability not available. [New fossil finds are always possible, but key missing links, as they are called, are rare. One good example is *Archaeopteryx*, which documents the transition from reptiles to birds (Fig. 10–14).] No, because the phylogenetic methodology available to us (Chapter 9) allows us to reconstruct, more or less reliably, the transitional events that bring about evolutionary change.

Last, regarding major evolutionary events in the Paleozoic, we should report on the invasion of land by living forms. In the Devonian period there is clear evidence that amphibians and vascular plants became established as terrestrial organisms. Insects were also on land by that time and they gave rise to airborne forms by the end of the Paleozoic. Flying vertebrates appeared only in the next era, the Mesozoic. So by the time the Paleozoic closes with the Pennsylvanian and Permian periods, the present-day diversity of life was very much in evidence. The Pennsylvanian period is remarkable for its plant life, which provide us with major coal beds that are, in reality, fossil plants.

Mesozoic life. This era is often called the Age of Reptiles. It is better thought of as the Decline of Reptiles. These vertebrates were numerous and highly evolved in the Permian period and then their numbers decreased. With the appearance of birds and mammals in this era, their numbers went into a further decline. This inverse correlation between changes in numbers of various taxa—one decreases in number as others increase in number—suggests a causal connection between the two sets of events. The question might be posed this way: Did the reptiles decrease in number as a result of unsuccessful competition with the warm-blooded vertebrates? Unfortunately the fossil record does not really answer such questions.

The extinction of those great reptiles, the dinosaurs, is, in fact, one of the great puzzles of the fossil record, and scientists are still working on possible solutions. Dinosaurs (Fig. 10–15) dominated the land in the early and mid-Mesozoic and then

Fig. 10-15: Selected dinosaurs from a Mesozoic landscape. The large form in the center is *Tyrannosaurus rex,* a carnivore. The other dinosaurs are herbivores. A flying reptile is also shown.

declined and largely disappeared by the end of that era. Much effort has been expended on determining the cause of their decline. During this era, there was significant geological activity (Fig. 10–9), and perhaps dinosaurs could not evolve as fast as their environment was changing. Perhaps the first mammals, being small, scurrying forms, found reptilian eggs an abundant and accessible food source. Perhaps the reptiles actually did evolve fast enough to survive, but their evolutionary changes were so great that what remains today is not what we recognize as reptilian, but is now avian—the birds.

All the foregoing have been cited as factors in understanding the fate of the dinosaurs. The last idea is a very recent one and is summed up in the notion that birds are, in effect, warm-blooded reptiles. A careful reconstruction of the biology of reptilian life from their fossils tells us that the reptiles were probably evolving as many warm-blooded species. But why that should lead to a wholesale loss of many cold-blooded species is not entirely clear. The story of dinosaur extinction is probably a complex one. It is not unlikely that all the causes proposed thus far contributed to their decline.

Cenozoic life. This era has been called the Age of Mammals. That name tells us that mammals occupied a dominant position as land animals during the 65 million years of this era. But it ignores the fact that birds too were well established and that the seas continued to flourish as a home for the other great animal phyla. Also obscured is the important evolution of plant life during this time. Seed plants are clearly present in late Paleozoic fossils. The angiosperms, which include flowering plants, appear in the fossil record toward the end of the Mesozoic era, namely, in the Cretaceous period. Today flowering plants are the dominant land plants and so the Cenozoic could also be called the Age of Flowers.

It is from the Cenozoic that we have obtained one of the finest chapters in the fossil record. This is the carefully documented evolution of the modern horse. The major finds in that evolutionary history appear in Fig. 7–1. The story starts with *Hyracotherium,* formerly called *Eohippus,* an animal the size of a small dog, which fed by browsing on leaves of bushes. Its front feet had four toes resting on the ground, its back feet three functional toes. Its chewing teeth were differentiated into premolars and molars. All of this changed as *Hyracotherium* evolved into modern *Equus* over a period of about 60 million years. These changes affected especially the size of the body, the structure of the legs, and the structure of the skull and its teeth. The small bush-dwelling, browsing *Hyracotherium* became an inhabitant of grasslands. It became a large-bodied, swiftly running grazer. The changes in the lower bones of the legs are shown in Fig. 7–1, as are the changes in the skull and the teeth. As dependence on grass as food gradually emerged, the teeth became larger and longer, with heavier crowns, so as to withstand the grinding necessary to eat grass. The difference between premolars and molars was minimized—they all served as grinding molars—and the jaw increased in size. Note in Fig. 7–1 how the muzzle extended in length. The eye socket, for example, instead of lying midway between the nose and the base of skull, as in *Hyracotherium,* is, in *Equus,* only about one-third of the way from the base of the skull to the nose.

This fossil record has been interpreted by experts as the phylogeny shown in the next figure (Fig. 10–16). Note here that, though there is a convincing series of intermediate forms joining the browser *Hyracotherium* to the grazer *Equus,* much else was happening in the history of this group called the horse family or the Equidae. With the appearance of *Miohippus* we find a form that has a significant evolutionary potential. From it, in the early Miocene, about 26 million years ago, we see the beginning of an *adaptive radiation.* The genus *Miohippus* gave rise to a variety of forms capable of adapting to the grassland habitat. The majority of them have become extinct, but in the Pliocene they were abundant and diverse enough to leave a fossil record of many genera.

TRENDS

Now, let us stop and determine what it is we can learn from the fossil record.

This analysis of the fossil Equidae is a convincing example of what the fossil record can tell us, despite the obvious pitfalls and limitations inherent in it. It is by no means the only example that could be cited. The records of invertebrate animals and plants provide comparable stories. Paleontology is a unique part of evolutionary studies in that we see changes in time or trends more clearly here than in any other evolutionary study. Trends can be inferred, as we will see, from phylogenetic analyses, but *direct* observation of trends is possible with fossils.

Furthermore, when the record is as complete as it is with the Equidae, we find that it confirms our ideas that evolution proceeds by Darwinian natural selection. The actual rates of

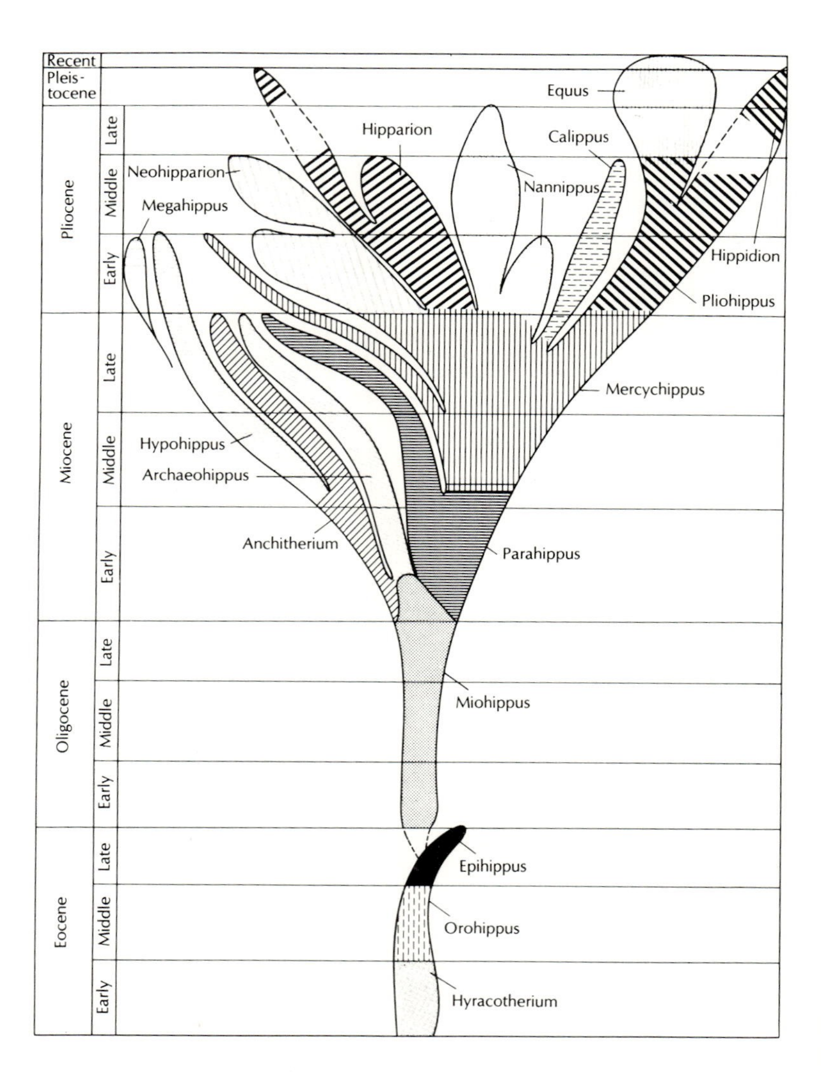

Fig. 10-16: Horse phylogeny showing the adaptive radiation of this genus, starting from browsers and branching into various grazers. Various fossil forms have been given different generic names.

change are very hard to determine in the sense of the numbers of mutations that separate *Hyracotherium* from *Orohippus* and *Orohippus* from *Mesohippus* (Fig. 10–16). Though we have good estimates of the ages of these fossils, we have no good estimates of the genetic changes underlying the differences between these apparent genera. This is another example of the problem discussed earlier (Chapter 8) of the actual genetic differences between a parent gene pool and its progeny species.

Trends, however, can be viewed another way. Rather than looking at them from the end of the spectrum involving species differences, let us look from the other end of the spectrum. What about the time taken to achieve some of the major distinctions among living things? These have been brought together in a useful way by Ponnamperuma (Fig. 10–17). Note, for example, that the bacteria probably originated at least 3×10^9 years ago, the eukaryotes about 1.5×10^9 million years ago. Chordates, we saw, first appeared about 5×10^8 years ago. This suggests clearly that the prokaryotes have been on earth two or three times longer than eukaryotes and about six times longer than the chordates. From data such as these it

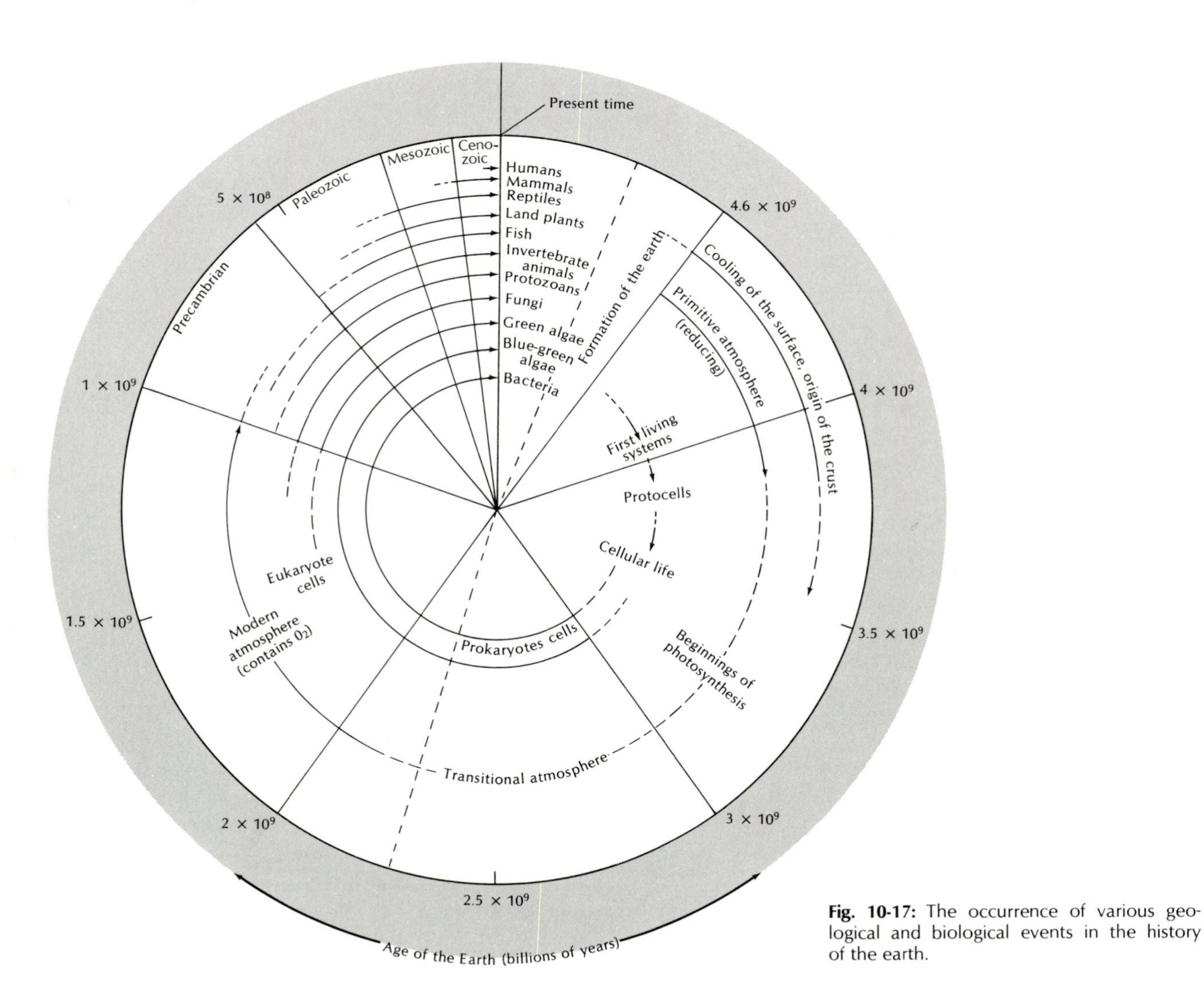

Fig. 10-17: The occurrence of various geological and biological events in the history of the earth.

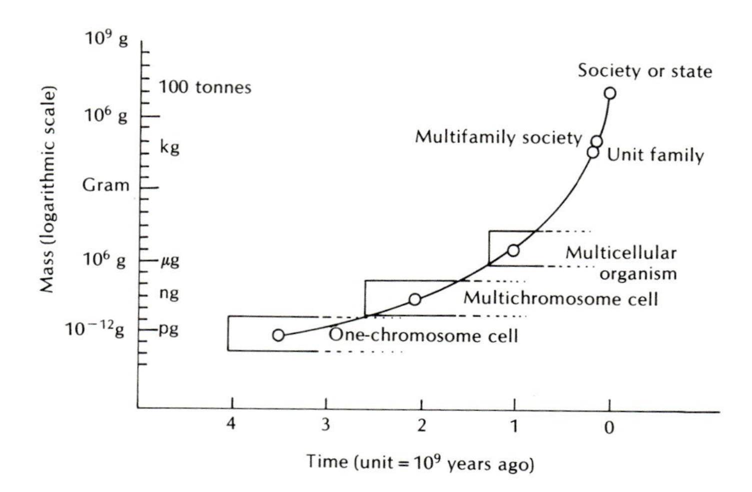

Fig. 10-18: The continuously increasing rate of evolutionary change. Here change is measured as increase in mass of selected biological units. Note that mass (vertical axis) is on a logarithmic scale, whereas time (horizontal axis) is linear.

has been argued that evolution is a continuously accelerating process (Fig. 10–18).

Reconsidering the details of the trends, we realize that evolutionary history is really a combination of different trends. The broad trends we have just been discussing are an average of a rich blend of detailed histories of many different organisms. Some of these show remarkable innovation, others are conservative, and then there are many others between these two extremes. George Gaylord Simpson, a great paleon-

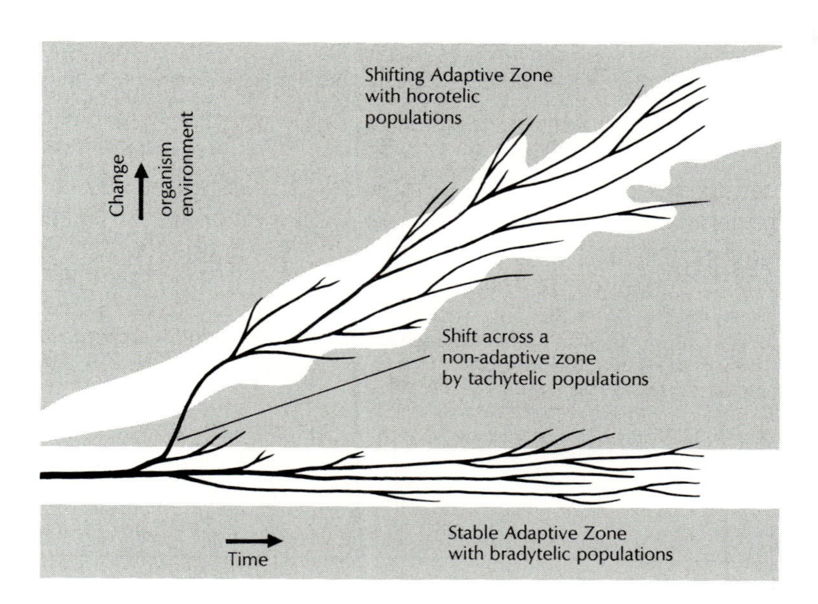

Fig. 10-19: Different rates of evolutionary change and their relation to adaptive zones.

tologist and profound student of evolution, has summarized some of his views on evolution in terms of groups that are *bradytelic* (changing little), *horotelic* (showing steady change), and *tachytelic* (changing rapidly) (Fig. 10–19). This view of evolutionary change is also related to the adaptive zone in which the organisms live. Simpson does not precisely define an adaptive zone—it is broader than an ecological niche and less specifically bound than a biome. (A *biome* is such a geographical-botanical area as the hardwood forests, the coniferous forests, or the prairies of North America.) The grasslands invaded by *Mesohippus* represent an adaptive zone. The most rapid evolution occurs when members of a species leave one adaptive zone and invade a new one. Rapid (tachytelic) adjustment to the new zone in terms of losing unfit genes and selecting for new ones occurs at this time. This is essentially identical to Mayr's genetic revolution, when a small sample of a parental gene pool breaks away (founder principle) to establish its genetic resources as a new species or to the punctuated equilibria of Eldredge and Gould. Simpson adds to this the longer view of the paleontologist and anticipates survival in a new adaptive zone with subsequent adaptive radiations and their attendant innovations and also their extinctions. Horotelic populations are those showing continued change after the period of tachytelic evolution. Some populations, however, end up in stable environments. These are the bradytelic ones that show little change. Evolution is a mosaic of these tendencies and the fossil record is the historical document of that flow of biological events.

References

Awramik, S. M., J. W. Schopf, M. R. Walter, and R. Buick, 1980. Filamentous fossil bacteria 3.5 Ga-old from the Archean of Western Australia. *Science* (in press).

Eldredge, N., and S. J. Gould, 1972. Punctuated equilibria: An alternative to phyletic gradualism. In *Models in Paleobiology,* T.J.M. Schopf (ed.), pp. 82–115. Freeman, Cooper and Company, San Francisco.

Fox, G. E., L. J. Magrum, W. E. Balch, R. S. Wolfe, and C. R. Woese, 1977. Classification of methanogenic bacteria by 16 S ribosomal RNA characterization. *Proceedings of the National Academy of Science, U.S.A., 74:* 4537–4541.

Fox, S., 1980. New missing links. How did life begin? *The Sciences 20:* 18–21.

Fox, S. W., and K. Dose, 1977. *Molecular Evolution and the Origin of Life*. Revised ed. Dekker, New York.

Gould, S. J., and N. Eldredge, 1977. Punctuated equilibria: The tempo and mode of evolution reconsidered. *Paleobiology 3:* 115–151.

Hanson, E. D., 1966. Evolution of the cell from primordial living systems. *Quarterly Review of Biology 14:* 1–12.

Horowitz, N. H., 1965. The evolution of biochemical synthesis—retrospect and prospect. In *Evolving Genes and Proteins,* V. Bryson and H. J. Vogel (eds.), pp. 15–23. Academic Press, New York.

Margulis, L., 1979. *The Evolution of Cells*. Harvard University Press, Cambridge, Mass.

Margulis, L., and J. E. Lovelock, 1975. The atmosphere as circulatory system of the biosphere—the Gaia hypothesis. *Co-Evolution Quarterly* (Summer), pp. 31–40.

Lovelock, J. E., 1979. *Gaia: A New Look at Life on Earth.* Oxford University Press, London.

Madison, K. M., 1953. The organism and its origin. *Evolution 7:* 211–227.

Miller, S. L., and L. E. Orgel, 1974. *The Origins of Life on Earth.* Prentice-Hall, Englewood Cliffs, N.J.

Oparin, A. I., 1968. *Genesis and Evolutionary Development of Life*. Academic Press, New York. (Translated by Eleanor Maass.)

Orgel, L. E., 1973. *The Origins of Life: Molecules and Natural Selection.* Wiley, New York.

Ostrom, J., 1979. Bird flight: How did it begin? *American Scientist 67:* 46–56.

Ponnamperuma, C., 1972. *The Origins of Life*. Thames and Hudson, London.

Ponnamperuma, C., A. Shimoyama, M. Yamada, T. Hobo, and R. Pal, 1977. Possible surface reactions on Mars: Implications for Viking biology results. *Science 197:* 455–457.

Ponnamperuma, C. (ed.), 1977. *Chemical Evolution of the Early Precambrian*. Academic Press, New York.

Peterson, M., 1978. Acceleration in evolution, before human times. *Journal of Social and Biological Structure 1:* 201–206.

Schopf, J. W., and D. Z. Oehler, 1976. How old are the eukaryotes? *Science 193:* 47–49.

Simpson, G. G., 1953. *The Major Features of Evolution*. Columbia University Press, New York.

Simpson, G. G., 1967. *The Meaning of Evolution*. Revised ed. Yale University Press, New Haven, Conn.

Woese, C. R., and G. E. Fox, 1977. The concept of cellular evolution. *Journal of Molecular Evolution 10:* 1–6.

ELEVEN

The Monera and Protista

WHAT HAS DRIVEN LIFE to increasingly complex organisms? And with the production of complex forms, why is it that simpler forms persist? Why do we have amebas and humans, bacteria and flowering plants? These questions highlight one of the most intriguing aspects of the diversity of life, but they are all misleading. The real question is not degree of complexity, but simply, survival. And the answer to that is adaptation.

As we progress through increasingly complex organisms, we must realize that progression does not necessarily mean progress in the sense of achieving some particular good. Many of the most complex forms are now extinct—recall the dinosaurs—and yet many bacteria and their relatives and many unicellular relatives of the ameba survive. Our continuing question, as we survey the Monera and Protista and then the Metaphyta, Fungi, and Metazoa, is In what ways are these groups of organisms adapted to survive? And since that question is posed in the context of the history of life, in answering it we are also answering the most significant questions of phylogeny, namely, the reasons why the history of life has unfolded the way it has.

The Monera

We start with the monerans because the fossil record and our speculations regarding the origin of life both indicate that they are the first organisms.

The most interesting thing about the monerans is their unusual combination of biosynthetic sophistication and morphological simplicity. Structurally they are prokaryotes, which means a rather small cell with a cell membrane, only a few cytoplasmic organelles, and a nucleus that is a highly coiled circular length of DNA (Fig. 1–1). Functionally, some of the bacteria are extraordinary because, as autotrophs, they make more of themselves from simple organic compounds. Their energy can come from the oxidation of iron or sulfur or from sunlight (they are chemoautotrophs and photoautotrophs, respectively). It is this biochemical sophistication that is one key to the success of the Monera.

Here a word of warning must be interjected. In our surveys of the different kingdoms, we will have to generalize broadly about groupings. That necessarily reduces the amount of detail presented, and therefore, groups will appear more homogeneous than they actually are. The monerans contain forms as diverse in function as methanogens, blue-green algae, free-living photoautotrophic bacteria, and pathogenic chemoheterotrophic forms. So when we speak of the biochemical sophistication of these prokaryotes, we refer not to just one type, but to a great many types of biochemical capabilities. In the aggregate, and in terms of the chemoautotrophs, especially, the range of biochemical capabilities of these tiny cells is truly staggering.

Another aspect of their biochemical capabilities is a high metabolic rate. In particular, consider a cell that can reproduce itself in about 20 minutes. Most bacteria have longer cell cycles, say about 30 minutes, but a 20-minute cell cycle does occur. In such a rapidly metabolizing cell, it has been calculated that 40,000 amino acids of one kind can be synthesized each second or that 150,000 peptide bonds are formed each second. The lactose-fermenting bacteria can break down 1,000 to 10,000 times their own weight in lactose in 1 hour. (Humans, by contrast, need half a lifetime to metabolize material 1,000 times their weight.)

Perhaps most dramatic is the reproductive capabilities of a prokaryote that divides once every 20 minutes, when we consider that rate as sustained over 48 hours. There are 144 periods of 20 minutes each in that time, which means one original cell could produce by fission 2^{144} progeny, or 2.2×10^{43} cells. Each cell weighs close to 10^{-12}g. The total weight of all the cells is, therefore, 2.2×10^{31}g. That is 24×10^{24} tons or 4,000 times the weight of the earth.

What does all this mean in terms of adaptation?

THE IMPORTANCE OF SMALL SIZE

Paradoxically, because the prokaryotes are relatively small, as cells go, they can take maximum advantage of their small size. This is made clearer by first looking at the surface to volume ratio of a prokaryote cell. Calculations show that the surface/volume value for a 200-pound human is around 0.3; for a hen's egg, it is 1.5. For an ameba, whose diameter is 300 μm, the same value is 400. And for a bacterium (diameter 0.5 μm), it is 120,000. These different values reflect the simple fact that in a sphere the volume increases with the cube of some linear dimension, such as the radius, and the surface with the square. Therefore, the volume is proportionately much larger relative to the surface in large organisms and much smaller in small ones.

This has two important consequences for bacteria. One is that metabolites can be moved rapidly into and out of the cell. The second is that heat is rapidly dissipated. If humans metabolized—either to build proteins or to break down sugars—at the rate bacteria do, we would release so much heat energy internally that high fever and death would quickly ensue. But bacteria, which live in a watery environment, whether inside a host or free-living, can lose heat rapidly to their environment. They cannot overheat, metabolically, when they have a high surface/volume value that immediately rids them of heat. To sustain a high metabolic rate, the cell must be able to transport precursors and waste products (both are metabolites) at a high rate. This also requires a large surface relative to volume, which these small monerans possess.

STRUCTURE AND FUNCTION

Now let us look at high metabolic rates relative to prokaryote cellular organization. As we have noted many times, the prokaryotes are much simpler than the eukaryotes (Fig. 1–2). What does this mean in terms of adaptation? If it is adaptively advantageous for a cell to divide rapidly, it will also be advantageous—and even necessary—to keep cellular structures sim-

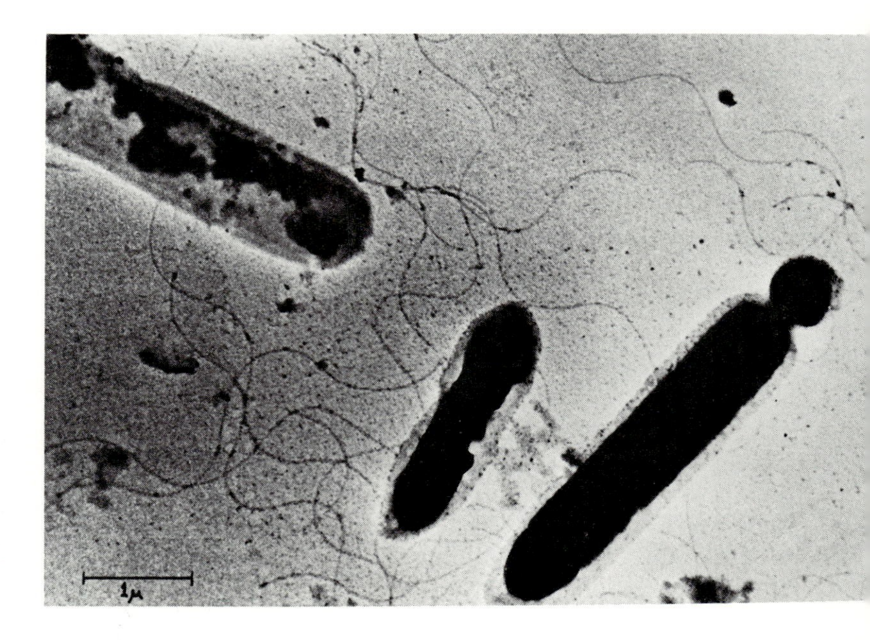

Fig. 11-1: Spore formation in a bacterial cell (right). These bacteria, *Clostridium tetani,* can cause lockjaw (tetanus).

ple and to a minimum. Complexity and multiplicity of cell organelles complicate cell division. To exaggerate for a moment, we can argue that a bag of marbles is more readily divisible into two comparable bags than a tool kit is readily divisible into two kits, each with its complete set of wrenches and other tools. The prokaryote parallels the bag of marbles if we replace the marbles with populations of different enzymes and add a genetic apparatus that ensures regulation of protein formation. And a eukaryote cell, with its many specialized cytoplasmic organelles and nucleus, parallels the tool kit.

We need to look at the genetic apparatus of the prokaryote cell more closely here. It consists only of a naked, Watson-Crick double helix of DNA, which is continuous or circular. Essentially, bacterial cells are haploid, for even when they are multinucleate, each nucleus is haploid and, except of rare mutations, identical to the other nuclei in the cell. In haploid cells, mutants are expressed as soon as the cell contains only mutant nuclei. This, combined with large numbers of cells, can be considered adaptively advantageous. Rapid proliferation of bacteria will occur, in a suitable environment, whenever foodstuffs are available. Geometric increase through fission assures such proliferation. When the food is exhausted toxic metabolites may be present, and new resources will be required. Mutation now becomes important as do two other aspects of prokaryote biology, sexuality and spore formation. We need to comment on each of these before continuing with our discussion of moneran adaptations involving structure and function.

Spore-formation (Fig. 11–1) will occur in many bacterial species as a response to adverse or life-threatening conditions. Spores are quite resistant to drying and to the more common toxic substances that occur in nature. When favorable conditions reappear or when the spore is transported to a place where conditions are favorable to growth, it germinates and starts a new period of vigorous growth. Spores, then, allow a bacterium to survive conditions under which regular cells die and are also a means of dispersal, or preservation. Spore-formation is triggered at the appropriate time to ensure further survival. Many species of bacteria and blue-green algae are known to be spore-formers.

Sexuality occurs as three different processes in the bacteria (Box 11–1). It has not yet been described as occurring in the blue-green algae. From a genetic-evolutionary viewpoint, any process that has the potential for new genotypes as the result of bringing together separate genomes, or part of them, is a sexual process. Hence, the process whereby DNA from one cell enters another cell and changes its phenotype (transformation) can be thought of as sexual. Similarly, transfer of genetic material by a virus (transduction) is also a kind of sexual process. Conjugation between two different cells and the transfer of genetic material from one to the other is the process most like the sexuality of eukaryote systems. Transformation is not common among bacteria. Transduction may be fairly widespread as may conjugation. From what we know now, it is important to note that conjugation occurs most readily in bacterial cultures that have exhausted their food supply. In other words, when there may be a survival advantage for new genotypes, sexuality tends to appear.

We can now present an overview of the adaptive nature of the prokaryotes: They have a kind of feast or famine way of surviving. Readily available nutrients and energy sources are rapidly exploited to produce large numbers of cells in a relatively short time. These large numbers and their essentially haploid genomes ensure that a significant number of new mutations will arise and will be expressed. That plus new genotypes from sexual processes optimize the chances of taking advantage of new resources. But if famine is unavoidable, many prokaryotes can form spores as a survival mechanism. This necessitates our viewing the Monera not as relics of bygone life forms that somehow have managed to survive as evolu-

Box 11–1. Sexual processes in bacteria.

Sexuality in bacteria is seen in the three processes described below. To understand why they are described as sexual it is necessary to ignore the concepts of male and female, and even those of sperm and egg, found associated with sex in higher plants and in animals. The point of view here is a genetic-evolutionary one that identifies a process as sexual if genetic materials of separate origin—from two different genomes—contribute to the formation of another genome. The essential point is the potential for new genotypes that differ from those in either parental genome.

Bacterial transformation. This type of sexuality was first discovered by Griffiths, in 1928, in *Diplococcus pneumoniae* (originally called *Pneumococcus*). He found that dead bacteria of one phenotype could release a substance that permanently changed the phenotype of other strains of the same bacterium. In 1944, Avery, McLeod, and MacCarty, of the Rockefeller Institute in New York, further showed that DNA caused the hereditary change. Although the details of the process are still not clear, we now know that in various species of bacteria the DNA of one strain can enter other strains of that species and somehow become incorporated in the recipient's DNA. The whole genome of the donor strain is not transferred to the recipient, only parts of it are. But these parts can come from anywhere in the donor genome; they are not gene-specific. This produces a variety of new genotypes, which are expressed in the recipient. Hence, it is classified as a sexual process.

Transduction. Here a virus is the physical agent of transmission of genetic material from one bacterium to another. This was described by the American microbiologists Lederberg and Zinder for the bacterium *Salmonella typhimurium*. Since then, it has been found to occur in other species of bacteria, too. By using genetically different strains separtated by a filter that viruses, but not bacteria, could cross, they were able to identify, first, the origin of new genotypes and, then, the virus responsible for transduction. The virus encloses part of the host genome in the viral protein coat. The enclosed viral and bacterial DNA can enter a new host and there the introduced bacterial DNA can somehow become incorporated into the host genome, giving rise to a new genotype.

Bacterial conjugation. This is thought to be the most common kind of bacterial sexuality, although it has not been reported in many species. Conjugation requires physical contact between complementary strains of a species. In *Escherichia coli*, the complementary strains are F^+ and F^-. The movement of genetic material is from an F^+ cell to an F^- one. The circular DNA in the F^+ cell is broken; it then migrates through a cytoplasmic tube connecting the two cells. The amount of DNA transferred depends on how long the tube between the conjugants remains intact. Connections of shorter duration are more probable than longer ones. Hence shorter segments of DNA more often enter the F^- cell than long segments. Nonetheless, the transferred DNA becomes incorporated in the F^- genome, to produce a new genotype.

tionary change has swept past them and on to the eukaryotes, but rather as highly successful forms in terms of their own strategy and tactics for survival. The strategy is to exploit faster than anyone else the available environmental resources. The tactics used to implement that strategy depend on small size in association with biochemical sophistication, rapid metabolism, simple cellular organization, and genic variability.

Such a viewpoint not only gives us a better appreciation of the prokaryotes as beautifully adapted organisms, but it raises two intriguing questions! How much diversity can we expect in this group that, by traditional taxonomic standards, is said to contain a few thousand species? What is the lower limit of cell size? Related to that last question is the broader issue of the ideal size for a prokaryote and its upper and lower limits in size.

MONERAN DIVERSITY

Special forms of sexuality and a lack of knowledge of their occurrence outside the laboratory combine to confuse the species concept in the Monera. A long-standing reference on bacterial classification, *Bergey's Manual of Determinative Bacteriology*, has been criticized from several points of view. One criticism is the extensive revision that occurs in each edition. Much of that is unavoidable, even though it is a genuine source of confusion when taxonomic names are not stable. More important is the criticism leveled at criteria for species differences. In some cases, such differences come down to the presence or absence of a single enzyme, which can reflect a single gene mutation. Such a difference is wholly inconsistent with the usual concept of species. And, the bacteria are given a hierarchical classification into species, genera, families, orders, and so on. The desire to parallel the systematics of other groups such as plants and animals is understandable. However, in plants and animals, as we shall see, the systematic schemes justifiably contain more or less explicit reference to phylogeny. The eminent bacteriologist van Niel has vigorously criticized such an attempt on the part of bacteriologists. He has urged a perspective that is a common issue in systematic work these days, namely, rather than use systematics to classify and show evolutionary relations, let it stick to classification. In his view, bacterial systematics is strictly taxonomic. It is a device for efficiently classifying bacteria and for storing and recovering that information as needed.

This point of view has been called phenetic. It argues that rigorous attention to the organismic phenotype (structural and functional) is the essential concern of taxonomists and they could even ignore the species concept. By proper use of statistical tools, developed especially by Sokal and Sneath, organisms will be grouped by the phenotypic characters they share. Such a grouping is called an *operational taxonomic unit* (OTU). (See Chapter 9.) This approach to grouping has the great advantage of a relatively objective specification of the characters studied and their explicit usage in defining the OTU. Nevertheless these workers, collectively referred to as numerical taxonomists, then go on to develop phylogenetic schemes based on their taxonomy, the latter being developed with the recognition that their OTUs coincide well with species defined as gene pools.

The numerical taxonomists have applied their techniques to the bacteria and have achieved useful classifications thereby. But we need not pursue further the debate regarding the proper role of systematics, i.e., it should be purely taxonomic or it should combine taxonomy and phylogeny. We will now turn to phylogeny and to the use of widely recognized groups of bacteria without entering into a discussion as to their taxonomic designation as a species, a genus, or whatever. A possible phylogenetic scheme is given in Fig. 11–2.

This scheme was not developed using the explicit phylogenetic analysis given in Chapter 9. It does, however, use the concepts of homology, phenoclines (including what we

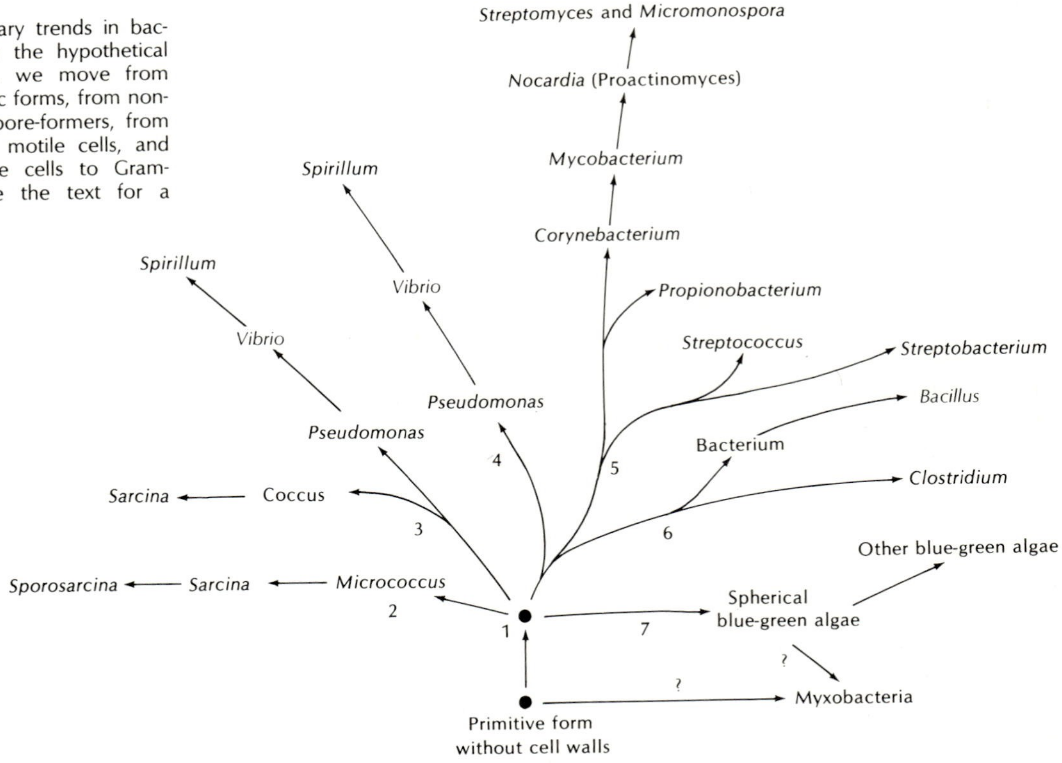

Fig. 11-2: Evolutionary trends in bacteria. Starting from the hypothetical ancestral form (1), we move from anaerobic to aerobic forms, from non-spore-formers to spore-formers, from nonmotile cells to motile cells, and from Gram-negative cells to Gram-positive cells. (See the text for a fuller discussion.)

can call physiological changes or physioclines), and a plesiomorph, but intuitively rather than explicitly. It was felt that the ancestral bacterium would be a round or coccoidal form, without a cell wall. It would also be anaerobic, nonspore-forming, and nonmotile. In fact, such a form, shown at (1) in Fig. 11–2, is not a plesiomorph, but a hypothetical ancestral form. (Remember that a plesiomorph must be an actual organism, the one that most closely resembles the ancestral form of the group under study.) However, coccoidal forms are known that could serve as the plesiomorph, notably *Micrococcus,* or even one of the methanogens.

From such a plesiomorph, six evolutionary trends can be identified. These are based on structural and functional phenoclines. In the first one, trend (2) in the figure, we find coccoidal forms becoming motile spore-formers in *Sporosarcina.* In trend (4) the original coccoidal form becomes a curved rod, at first C-shaped (*Vibrio*) and then S-shaped (*Spirillum*). In this evolutionary line the motile cells are propelled by polar flagella (tiny whip-like organelles at the ends of the cells). *Vibrio* has one flagellum at one end; *Spirillum* has one at both ends. Trend (3) parallels the trends in lines (2) and (4), and in addition, these cells are capable of photosynthesis. Collectively, they are called the red bacteria or Rhodobacteriales. In trend (5) the cells are straight rods and largely nonmotile. The extreme forms, the Actinomycetales, form long filaments. In *Nocardia,* the filaments have cross walls and are, therefore, chains of cells. In *Streptomyces* and *Micromonospora,* cross walls are absent and the filaments are often elevated (aerial) and suggest small fungi. They form special spores called conidia. The sixth line also consists of rods. But whereas the rods in trend (5) were nonmotile, the rods in trend (6) are multiflagellated. The evolutionary line (7) that gives rise to coccoidal blue-green algae could be the ancestor for the rest of those forms. They have filaments and spores, but no flagella.

Other trends present in these evolutionary lines are a tendency to be aerobic, and in some, the more highly evolved forms can be stained with the Gram stain. (They are Gram-positive, the others are Gram-negative.) This stain is extremely useful in that it is, in most cases, a clear-cut character. Gram-positive organisms stain because of the presence of carbohydrates, protein, and RNA. The latter is especially important in the staining reaction. Just why the apparently more highly evolved bacteria are Gram-positive is not entirely clear. It does, nonetheless, represent another useful evolutionary trend.

Bacteria in Fig. 11–2 not contained in the Rhodobacteriales, the Actinomycetales or the blue-green algae, are Eubacteriales. This covers the majority of the Monera, leaving out only two other groups, the myxobacteria and the spirochaetes. These last two groups do not fit readily into this scheme, although Stanier and van Niel suggest that the myxobacteria might have arisen from a primitive coccoid lacking a cell wall or a coccoidal blue-green alga that lost its photosynthetic pigments.

The diversity, then, in the Monera is considerable. Certain evolutionary trends can be displayed in a phylogenetic tree such as that just discussed. Even though the prokaryote cell is simply organized, it does show specializations of cell form, flagellar motility, and spore-formation. The structural diversity is exceeded by the biochemical diversity. Figure 11–2 does not adequately show that aspect of moneran evolution. The special biochemical capabilities of this group are, as we have emphasized, the adaptations that make these small, simple cells the giants of evolutionary history in terms of period of survival on earth.

THE SMALLEST CELL

Cellular organization is the foundation of all life. Even noncellular viruses survive only by reproducing within cells. We have repeatedly referred to prokaryotic cells as relatively small and relatively simple. But let us look into the lower limits of size and organization found in the prokaryotes. Such an inquiry will enhance our appreciation of cells and the limits within which they function and successfully evolve.

The nature of a Mycoplasma *cell.* The smallest known

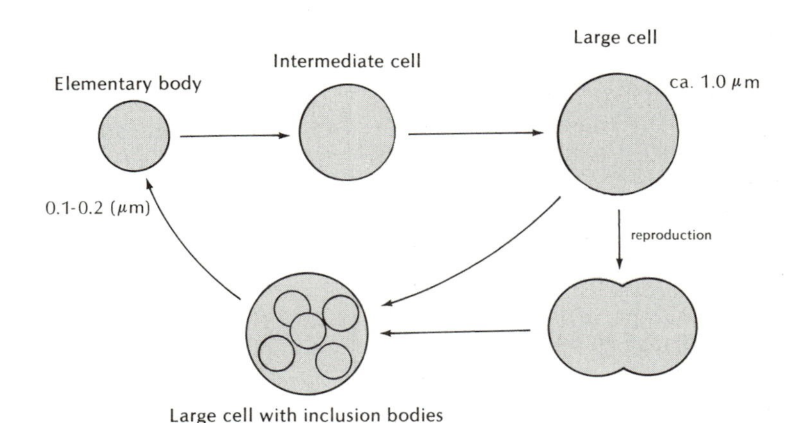

Fig. 11-3: A simplified diagram of the life cycle of *Mycoplasma laidlawi,* one of the PPLO. (See Fig. 1-2A for details of the intermediate cell.)

Table 11–1. The molecular composition of a PPLO cell.

Cell size	Atoms in dry matter	Molecular weight of DNA (4% of cell contents)	Number of monomers (amino acids and nucleotides	Number of large molecules
0.1 μm	12,000,000	2,880,000	600,000	1,200
0.25 μm	187,500,000	45,000,000	9,375,000	18,750

From H. Morowitz and M. E. Tourtelotte (1962) *Scientific American* (March).

cell has a diameter of 0.1 μm. It is one stage in the life cycle of *Mycoplasma laidlawi,* one of the pleuropneumonia-like organisms (PPLO). The life cycle of these cells is shown diagrammatically in Fig. 11–3. The small elementary body can grow into a cell about 1.0 μm in diameter and that cell either divides by fission or forms a larger cell, sometimes 10 μm in diameter, with inclusion bodies. From that larger cell, elementary bodies can again be formed. The PPLO cells usually occur as animal parasites, but they can also be recovered from sewage. Nutritionally, they are chemoheterotrophs; and they are Gram-negative. The composition of PPLO cells is given in Table 11–1. To better understand the DNA content we need to remember the following points. A protein that contains 100 amino acids will need at least 300 nucleotides to specify it. Since DNA is a double helix, each length of 300 nucleotides requires that 600 complementary pairs of nucleotides be present. Further, the average molecular weight of a nucleotide is around 312, and so each length of DNA that specifies a protein of 100 amino acids must have a molecular weight of 1.87×10^5. Therefore, the smallest PPLO cell has enough DNA to specify only about 15 proteins if its genes do not overlap. If the genes do overlap (Fig. 3–8), then as many as 40 to 50 proteins may be specified. However, considering the organism in which overlap is best studied, e.g., the bacterial virus ϕX174, we see that 20 or 25 proteins seems to be a maximum number for PPLO. This gives us a first glimpse into problems of small cell size when approached genetically. Are 20 to 25 different proteins enough for cell survival?

The biophysicist Harold Morowitz has studied small cells in detail. He suggests that most cells need about 100 different proteins. These are largely enzymes, to break down sugars and for cellular respiration, to synthesize more DNA and the different kinds of RNA. Other proteins are needed for cellular structure to form membranes and other cellular organelles. Clearly, then, the small PPLO cell we are looking at is somehow getting along with a very minimal number of proteins. This is perhaps understandable in the light of its nutrition, which is chemoheterotrophic. As an animal parasite it obtains much of its food already broken down by the host. This means it needs fewer enzymes than it would if it were free living.

To digress briefly, we can state that the smallest genome now known, that of the RNA virus of *Escherichia coli* MS2, has only three genes that produce three proteins. These are a coat protein surrounding the mature virus, a protein for assembling the virus from RNA and coat protein, and an enzyme for directing synthesis of more RNA. Quite obviously this organism depends almost entirely on its host for a variety of precursors and their formation into the needed minimum of RNA and protein that will constitute a new organism. Although PPLO is not as minimally organized as MS2, living at the expense of a host organism can minimize its survival needs. We do not know the exact requirements for PPLO, as we do for MS2, but they are considerably fewer than those for most cells.

A physicist's view of a small cell. Returning now to the problem of the needs of the smallest cell, we examine that problem from another point of view. This comes from considerations raised by the great physicist Erwin Schrödinger, who was also a profound student of molecular biology. He pointed out that small size has a very important implication in terms of the errors that are permissible. Statistically, errors are expressed as *standard errors* and a general estimate of a standard error is plus or minus the square root of some average value. If n is the average value, the standard error is $\pm\sqrt{n}$. Now, as Schrödinger points out, when n is large, for example, 6×10^{23}, the standard error is *proportionately* small, i.e.,

$$\text{Standard error} = \sqrt{6 \times 10^{23}} = 8 \times 10^{11}$$

This consideration can be applied to the occurrence of atoms of a certain type in a certain space. Compared to 6×10^{23}, the error of 8×10^{11} is a fraction of a percent. But when n is smaller the standard error is proportionately larger.

n	*Standard error*	*Error (%)*
10^{10}	$\pm 10^5$	0.001
10^6	$\pm 10^3$	0.1
10^4	$\pm 10^2$	1.0

Now let us look at the situation involving hydrogen ions.

The concentration of these atoms determines the acidity of a solution. Most bacteria live under conditions where the hydrogen ion concentration is 10^{-7} grams per liter, or pH 7. If the pH is the same inside and outside the cell (hydrogen ions are small and can move freely across a membrane; work must be done to maintain different concentrations on either side of a membrane) then in a coccoidal bacterium 0.5 μm in diameter there would be only about 3.6 hydrogen ions. The calculations for this figure are given in Box 11–2. With an n this low, statistically, a significant number of the cells will have no hydrogen ions. (According to the Poisson distribution, discussed in Chapter 6, we should expect, on the average, that this will be 0.027 or about 3% of the cells.) The absence of these ions will have a profound effect on the chemistry of these cells. If the size of the cell decreases, the probability of the absence of hydrogen ions increases dramatically. In a cell with a diameter of 0.1 μm, the possibility of a hydrogen ion being present is 0.03; thus, most cells will have no hydrogen ion (Box 11–2).

This suggests that the PPLO elementary bodies do work to retain their hydrogen ions, that they somehow do not depend on hydrogen ions, or that most small PPLO cells do not survive. Which of these is correct is at present not clear.

Box 11–2. Hydrogen ion concentration in prokaryotes.

A coccoidal cell whose diameter is 0.5 μm has a volume of 6×10^{-14} cubic centimeters (cm^3). The calculation is as follows:

The volume of a sphere is $V = 4/3\ \pi\ r^3$, where π = a constant = 3.17. For the coccoidal cell, $r = 0.25\ \mu$m or 2.5×10^{-5} cm. Hence, the volume of the cell is close to 6×10^{-14} cm^3.

In one liter (1,000 cm^3), an aqueous solution at pH 7 contains 10^{-7} gram-ions of hydrogen. In the volume of our coccoidal cell, there will therefore be 6×10^{-24} gram-ions ($6 \times 10^{-14} \times 10^{-3} \times 10^{-7}$). Since in one gram-ion there are 6×10^{23} ions (Avogadro's number), then the coccoidal cell contains 3.6 ions, i.e. $(6 \times 10^{-24})\ (6 \times 10^{23})$.

To calculate the proportion of cells with no hydrogen ions, we can use the first term of the Poisson series

$$P_0 = e^{-m}$$

where e is the base of natural logarithms, P_0 the proportion with no hydrogen ions and m the average number of ions per cell. We solve for P_o as follows:

$$\log_e P_0 = -m$$

where P_0 is the antilog of -3.6. That value is 0.0273.

For a cell whose size is that of the PPLO elementary body (0.1 μm), the volume is close to 5×10^{-16} cm^3. This means that the gram-ion concentration is 5×10^{-26} and, therefore, the number of hydrogen ions per cell is 0.03.

The essential point in Schrödinger's discussion is that cells must be large enough so that statistical errors are not life-threatening. Cells cannot afford to have just a few copies of essential atoms or molecules unless there are also mechanisms to correct errors. One copy of each gene is protected by a stable means of reproduction, of error correction, and of precise distribution during fission. Also, genes can be transcribed and translated into multiple copies. The provision for copies is an important safeguard against error, since it ensures that at least one good copy of the needed enzyme is present in each cell, a copy that can be used many times to catalyze reactions.

From these considerations we see that PPLO cells, in particular, and prokaryotes, in general, are close to the lower limit of viable cells. The actual lower limit will depend on nutritional requirements, stability of genetic information storage, and accuracy in expressing that information so as to ensure that needed copies of functional proteins are produced. Minimal information needs and the mechanisms needed to assure essential redundancy are thus balanced.

From the foregoing discussion, we realize that the first organisms could not have been very small, since, with a limited ability to form informed macromolecules, many errors would be probable and a high redundancy would be needed to compensate for the effects of those errors. And then, thinking of eukaryotes, to which we turn next, their relatively large size ensures that these cells can do more than just compound a limited amount of information (redundancy); they are large enough to carry a variety of information. We expect, in fact, we predict the eukaryotes to be truly complex.

The Protista

The Protista are predominantly unicellular forms, although colonial aggregations of various sorts are not uncommon. All are eukaryotes, and they may be either photoautotrophs or chemoheterotrophs. Perhaps the most fascinating example of the range of functional capabilities found in these cells are those displayed by *Ochromonas malhamensis* (Fig. 11–4), a member of the golden algae or Chrysophycophyta. These algae are so named because their photosynthetic pigments are golden-yellow, as compared to the pigments of the green algae or Chlorophycophyta, for example.

When grown in sunlight, *O. malhamensis* carries out pho-

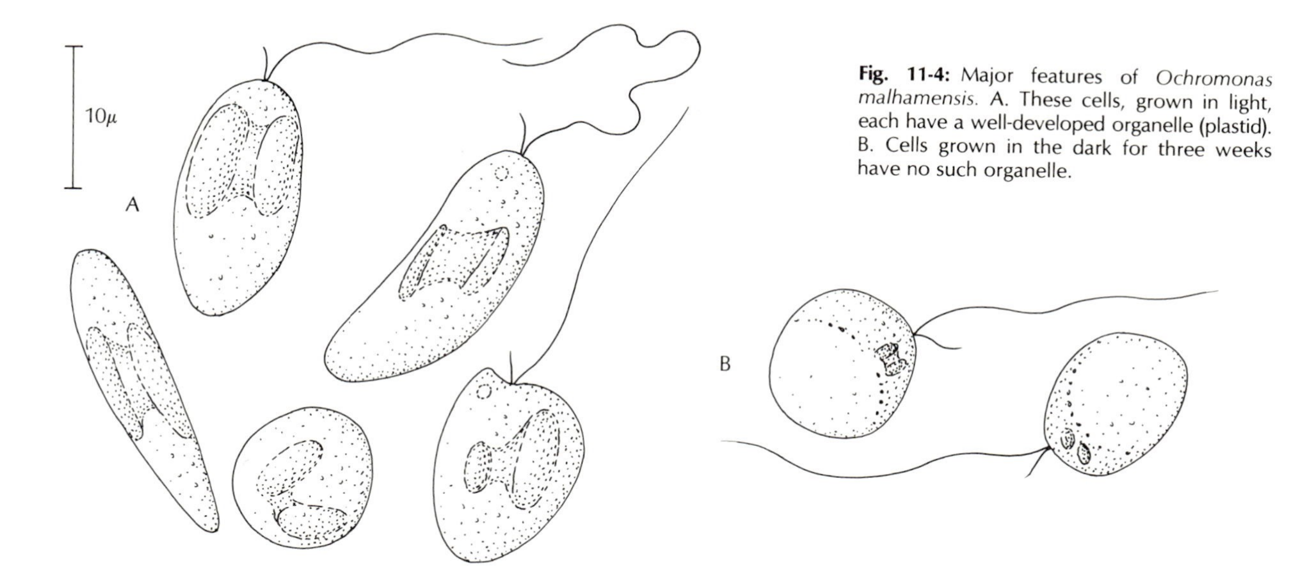

Fig. 11-4: Major features of *Ochromonas malhamensis*. A. These cells, grown in light, each have a well-developed organelle (plastid). B. Cells grown in the dark for three weeks have no such organelle.

tosynthesis and uses very few preformed molecules in its nutrition. It is very close to being a photoautotroph. When deprived of sunlight, but provided with such organic molecules as the proteins, nucleic acids, and certain lipids and carbohydrates found in a nutrient broth, it is a chemoheterotroph. It releases enzymes into its environment to break down macromolecules to amino acids, nucleotides, sugars, etc., which then are moved into the cell to be utilized. Finally, when *O. malhamensis* is denied light and a nutrient broth, but is provided with bacteria in its environment, it uses them as food. Here, too, it is a chemoheterotroph, since it takes the bacteria into food vacuoles where digestion occurs.

These three nutritional activities must be viewed ecologically to be properly appreciated. In effect, *O. malhamensis* can function as a producer (a photoautotroph), a decomposer (a chemoheterotroph living off organic macromolecules), and a consumer (a chemoheterotroph living off ingested food organisms). We see in these cells the capabilities that finally emerge in the three great kingdoms of multicellular organisms, i.e., the Metaphyta as producers, the Fungi as decomposers, and the Metazoa as consumers. The protistans represent the evolutionary playground where these three major trends were sorted out. That, combined with the emergence of multicellularity, is an important way to view the evolutionary developments within the Protista. Here we can introduce two useful terms, the protophyta and the protozoa.

Protophyta is not a taxon used in any systematic schema. The term covers the unicellular eukaryotic algae and refers to those protists that are producers. Comparably, we have the protozoa, which are the protistan consumers. They have been treated as a taxon, but the diversity within the protozoans is better viewed in terms of several different protistan phyla. The term is nonetheless useful to indicate all those eukaryote cells that show animal tendencies, i.e., that are consumers or ingestors.

There is no group easily identified as the protofungi or protomycota. This is not because unicellular decomposers do not exist. They do exist, but they are usually included in with the rest of the fungi. It seems that among the eukaryotes, decomposers are more successful as multicellular forms than as unicellular ones. Possibly because many prokaryotes are very successful as decomposers, protistan decomposers are not common. If we next compare eukaryotic unicellular producers and prokaryotic producers, the success of the protophyta as compared to the photosynthetic bacteria may depend on differences in the organization and function of the photosynthetic organelles of the two groups; both are successful, but in somewhat different ways. Last, there are no consumers among the prokaryotic cells, and hence, the protozoa are a unique evolutionary development. Let us look more closely at the Protista by reviewing the protophyta and the protozoa.

THE PROTOPHYTA

These algae are easily introduced by looking at the organization of *Chlamydomonas reinhardi*. The genetics, physiology, and fine structure of this member of the Chlorophycophyta have been intensively studied. What we find is a complex cell with well-defined cytoplasmic organelles. It reproduces asexually, undergoes sexual processes, and has various other complicated processes in its life cycle. Asexual reproduction ensures a relatively rapid increase in numbers under suitable conditions, namely, the availability of simple nutrients and energy (adequate sunlight). The sexual process, in which whole cells fuse into a nonmotile zygote, leads to meiotic cell division and the segregation of new genotypes in the progeny cells. So we find that this species takes advantage of asexual reproduction for rapid increase in number of haploid cells, parallel to the situation in bacteria, and also uses sexuality to gen-

Table 11–2. Characterization of the major phyla (divisions) of protophyta.

Phylum	Photosynthetic pigments	Photosynthetic carbohydrate product	Cell wall
Chlorophycophyta	chlorophyll a, b	starch	cellulose
Chrysophycophyla	chlorophyll a, some have c, some e	oil, leucosin	pectin and silicon dioxide
Pyrrophycophyta	chlorophyll a, b	starch	cellulose; or wall absent
Euglenophycophyta	chlorophyll a, b	paramylon	absent

erate new genotypes. Interestingly, fusion of complementary mating types is brought about by mild starvation, again paralleling the case in bacteria in which sexual reproduction most commonly occurs when food resources are depleted. Last, we observe that a parallel to bacterial spores is also found here in that chlamydomonad cells can encapsulate in a jelly-like secretion in which nonmotile, thick-walled spores are found. This occurs under adverse growth conditions.

Major protophytan phyla and their evolutionary trends. Thus far we have mentioned two phyla among the protophyta, the Chlorophycophyta and the Chrysophycophyta. Two more deserve mention. (We speak of phyla here, but botanists also refer to them as divisions. The terms are synonymous.) These are the Pyrrophycophyta and the Euglenophycophyta. (Those interested in more detail are referred to Appendix I.) These phyla are briefly characterized in Table 11–2. They differ in their cell structure, photosynthetic pigments, and storage products resulting from photosynthesis. Figure 11–5 provides examples of the Pyrrophycophyta and the Euglenophycophyta.

Two evolutionary trends should be considered: (1) colony formation and (2) loss of photosynthesis or the appearance of colorless forms. Colony formation is common in the Chlorophycophyta and the Chrysophycophyta. It can take the form of spherical aggregates of cells and of chains of cells forming filaments, plates of cells, branching colonies, net-like aggregates, and even other variations. For our purposes the most interesting examples are those found in the Chlorophycophyta. Details of the three major evolutionary trends are given in Box 11–3. Here we emphasize the outcome of these trends.

The volvocine line of evolutionary development culminates

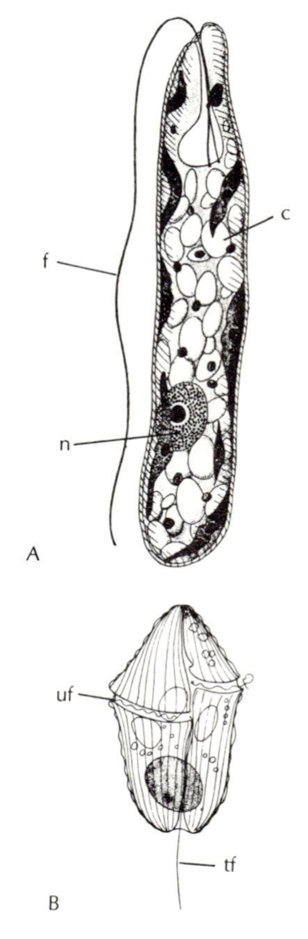

Fig. 11-5: Euglenoid (Euglenophycophyta) and dinoflagellate (Pyrrophycophyta) cells. A. *Euglena gracilis* showing flagellum (f), nucleus (n), chloroplasts (c), and other structures. B. *Gymnodinium pavillardi* showing its trailing flagellum (tf) and its undulating flagellum (uf) lying in a groove in the cell wall.

Box 11–3. Evolutionary trends in the green algae (Chlorophycophyta).

The volvocine trend is to the left, the siphonalian trend to the right. Tetrasporalian evolution is in the center of the figure.

To make clearer the events in these three evolutionary trends, the following comments are worth noting. The various events in the life cycle of the plesiomorphic *Chlamydomonas* are at the bottom center of the figure. Its cycle of *asexual reproduction* is shown to the right of the chlamydomonad cell, and is represented by the following stages: A spore or cyst within which two mitotic cell divisions occur and the products develop into a mature cell. *Sexual processes* are shown to the left of the chlamydomonad cell: Again there is a spore or cyst and within this as many as three or four mitotic divisions occur. The products, biflagellated gametes, can fuse with gametes from other cysts of the complementary mating type. The diploid zygote that is formed then becomes a zygospore and undergoes meiosis. The four haploid products of meiosis emerge as mature cells, two of each mating type. Then there is a special *spore stage* shown above the chlamydomonad cell: Under adverse growth conditions, spores are formed and a gelatinous matrix is secreted around the spores. Upon the reappearance of favorable growth conditions, mature chlamydomonad cells emerge.

The *volvocine* line of evolution starts with the species *Gonium sociale,* which is represented by four chlamydomonad cells attached in a cluster, their flagella all extending in much the same direction. *Gonium pectorale* is a flat plate of sixteen cells; with *Pandorina,* a ball of cells emerges. Note here the slight difference in the size of the gametes. The sexual cycles are given to the left of each volvocine species or genus. Anisogamy (morphological differences between the gametes) is found in all forms from *Pandorina* to *Volvox.* In the latter, it reaches the stage of gamete differentiation called oogamy—there is an active sperm-like gamete and a nonmobile, large, egg-like gamete. Note also the differences between members of the same colony in the two *Pleodorina* species. Here the members of one colony (coenobium) are differentiated into two different cell types. In *Volvox,* the smaller spheres within the larger one are daughter coenobia, the result of asexual reproduction.

The *siphonalian* line, at the right, shows various evolutionary experiments in addition to the one starting at *Protosiphon* and going on to the coenocytic filaments of *Vaucheria.* Anisogamy is found here, and even oogamy, in *Vaucheria.* Differentiation of these algae appears in the colorless holdfast and the pigmented upper parts.

The *tetrasporalian* evolutionary line also shows a variety of forms before such complex forms as *Draparnaldia, Coleochaete,* and *Ulva* emerge. In these forms, the gametes show little tendency to anisogamy. However, there is extensive differentiation into holdfasts and a thallus of different kinds. Not shown here is the important fact of alternation of generations in *Ulva* (see Fig. 11–6).

These three lines of evolutionary development are analyzed by the use of homologies, largely, structural homologies. The shape of the individual motile cell is fairly conservative or plesiosemic—compare the free-swimming cells of *Coleochaete* with *Chlamydomonas*—but there are aposemic changes in coenobial evolution and in the appearance of coenocytic and cellular filaments. The latter are seen in *Ulothrix* and the complex tetrasporalian forms. A neosemic trait is the alternation of generations in *Ulva.* What is still missing in this analysis is an explanation of these changes. That necessitates identification of the selection pressures that provide a selective advantage to coenobial or filamentous organization. In other words, we need to learn how these different structures provide functions that are selectively advantageous.

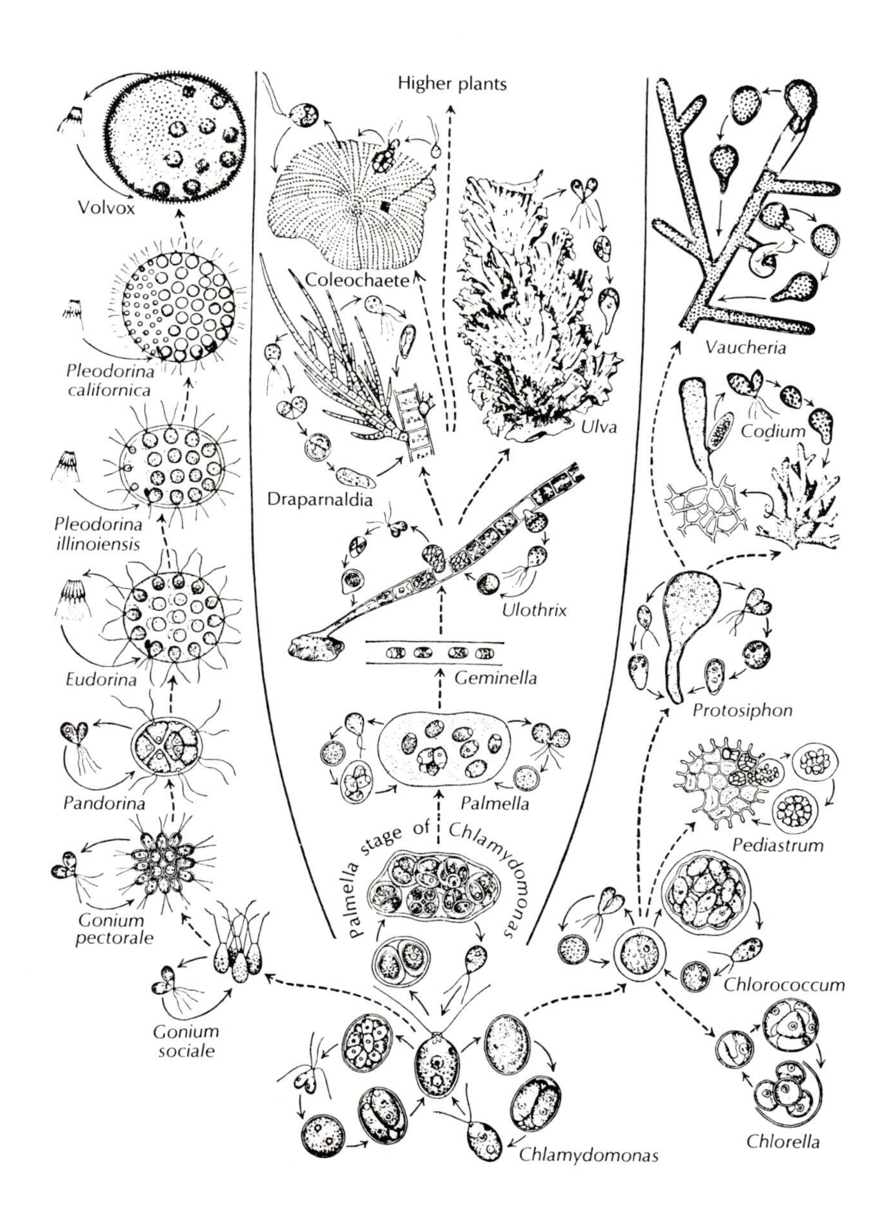
Higher plants
Volvox
Coleochaete
Ulva
Vaucheria
Pleodorina
californica
Draparnaldia
Codium
Pleodorina
illinoiensis
Ulothrix
Eudorina
Geminella
Protosiphon
Pandorina
Palmella
Palmella stage of Chlamydomonas
Pediastrum
Gonium
pectorale
Chlorococcum
Gonium
sociale
Chlamydomonas
Chlorella

in species of the genus *Volvox*. These colonies are composed of a thousand or more cells, each one like *Chlamydomonas;* additionally, each one is connected to adjacent cells by cytoplasmic strands. This hollow sphere of cells, which contains a rather precise number of cells, is called a coenobium. The coenobium reproduces sexually when certain of its cells differentiate as slender, flagellated gametes and others develop into large, nonmotile bile eggs. Fertilization here is called oogamy because of the differentiation of the gametes. Specialists in the algae consider the volvocine line of evolution an evolutionary dead end. There are no rigorously demonstrated homologies between the coenobia, with their chlamydomonad cells, and other multicellular forms. In the last century, the father of phylogeny, Ernst Haeckel, seriously proposed that coenobia were the evolutionary antecedents of the hollow blastula of certain lower invertebrates. In other words, Haeckel derived the animals from these green algae. What Haeckel overlooked is that no colorless volvocines are known (they never show animal-like nutrition), chlamydomonad cells are plant cells and not animal cells, and the lower invertebrates, except for certain sponges, do not develop from hollow blastulae. This problem will be examined again in Chapter 13, but the conclusion of the experts in this field bears repeating, namely, volvocine evolution is an evolutionary dead end. The remaining two trends in the green algae are not dead ends. The siphonalian line of evolution has given rise to several evolutionary experiments. Those giving rise to plates of cells, such as seen in the genus *Pediatrum,* are probably going no further, evolutionarily speaking, but a form such as *Vaucheria,* when it becomes colorless as in *Saprolegnia,* is thought to be ancestral to certain water molds. That is, from it there evolved certain of the fungi. *Vaucheria* is coenocytic (Box 11–3); it is one large multinucleate organism, since the filaments have no cross walls. Its highly specialized gametes no longer show a chlamydomonad-like structure. (The homologies between *Vaucheria* and *Chlamydomonas,* the plesiomorph of the green algae, can only be established by the serial criterion. This criterion establishes indirect similarities through the use of intermediate forms that here constitute the siphonolian line of evolution.)

The tetrasporalian line of evolution is thought to be ancestral to the green plants that invaded land and make up today's mosses, ferns, trees, shrubs, and grasses. At the peak of tetrasporalian evolution, both tissue formation and alternation of generations in the life cycle occur. Tissue formation is seen in the specialized mass of cells that make up the holdfast that attaches the plant to the substratum and in the rest of the plant, the thallus, that extends into the watery environment.

In the sea lettuce *Ulva,* the thallus has not only multicellular height and breadth, but it is also at least two cell layers thick. *Ulva* shows both a sporophyte and a gametophyte generation (Fig. 11–6). These two alternating generations played an important part in the evolution of the land plants, as will be seen in Chapter 14.

In brief, colony formation is the precursor of the multicellularity found in the Metaphyta and in the Fungi.

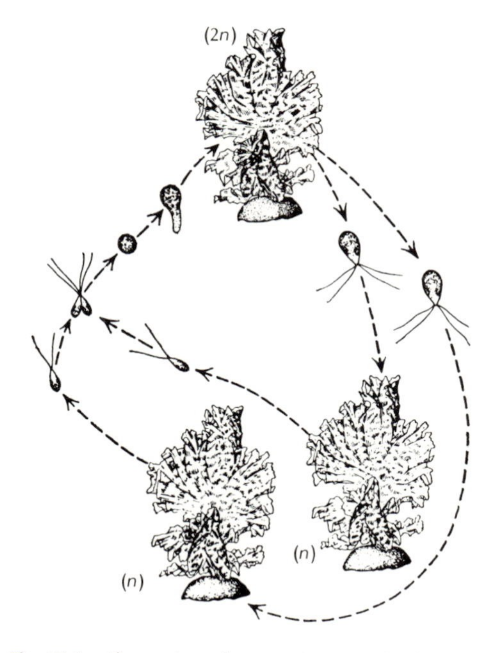

Fig. 11-6: Alternation of generations in *Ulva lactuca.*

Turning to the other trend—the appearance of colorless forms—we find the precursors of certain other fungi and the protozoa. From *Vaucheria,* as we saw, loss of plastids can result in such coenocytic colorless forms as *Saprolegnia.* But most fungi are not coenocytic. Though their origins are still obscure, it is likely that they are polyphyletic. That is, they arose several different times from colorless green and golden algae and perhaps from other protophyta, too. It is important to note that all the major groups of the protophyta contain colorless forms (Table 11–3). If loss of pigment occurs in forms that can also function as decomposers, then, if further evolution is successful, a fungal form can evolve. If, on the other hand the colorless form can function successfully as a consumer, then protozoan evolution is indicated.

There are many ways in which photosynthetic pigments can be lost. Mutations in the nucleus or in plastid genes can result in the loss of plastids or in the loss just of the photosynthetic compounds. Abnormal cell division can produce cells with no plastid. Many of the protophyta have one or a few plastids, which must divide when the cell divides and whose products must segregate to each daughter cell to ensure at least one plastid in each cell. Experimentally, various antibiotics or irradiation or extended growth in the dark also produce colorless cells. How important these factors are in natural populations is not known. Nevertheless, the descriptive facts are clear (Table 11–3): unpigmented cells that otherwise resemble the protophyta are numerous. Studies such as the study described earlier on *Ochromonas malhamensis* show that the colorless state can lead to cells that function as do fungi or protozoa.

Table 11–3. A listing of photosynthetic algae also shows closely related, colorless forms.

Form	Pigmented	Colorless
single cells	*Ochromonas* *Chromulina* *Mallomonas*	*Monas* *Oikomonas* *Mallomonas apochromatica*
spherical colonies	*Synchromonas pallida*	*Anthophysa vegetans*
branching colonies with stalks	*Chrysodendron ramosum*	*Dendromonas virgaria* *Monadodendron distans*
cells in tests	*Epipyxis* *Stylopyxis* *Poteriochromonas*	*Stokesiella lepteca*
rhizopodial	*Chrysarachnion insidians* *Rhizochrysis crassipes*	*Leukochrysis* *Amoeba stigmatica*
rhizopidial, with tests	*Eleuthropyxis* *Kybotion eremita* *Lagynion*	*Leukopyxis asymmetrica* *Platytheca micropora* *Heterolagynion*
Bicoecaceae, cells with tests and one flagellum	*Kephyrion* *Conradocystis dinobryonis*	*Donatomonas cylindrica* *Codonodendron ocellatum*
Craspedomonadaceae, with collars and one flagellum	*Stylochromonas minuta*	*Monosiga* and others

Note that the colorless forms occur in all major divisions of the protophyta. From Hanson (1977) *The Origin and Early Evolution of Animals*, Wesleyan University Press and Pitman Publishing. Based on data from Pringsheim (1963) *Die Farblosen Algen*, Springer Verlag, Stuttgart.

THE PROTOZOA

The protozoa are more diverse than the protophyta, including ameboid, flagellated, ciliated, and symbiotic forms. The latter include some of the most devastating parasites of humans, such as the causative agents of malaria and sleeping sickness. A more critical review of the diversity of the protozoa starts with the realization that they, in all probability, arose from colorless protophyta. This also means that there has been a multiple or polyphyletic origin of the protozoa; they are derived, at least, from the Chrysophycophyta, the Euglenophycophyta, and the Pyrrophycophyta. This polyphyletic origin is the main reason for no longer using the term protozoa as a taxonomic designation. Within the protozoa, the symbiotic forms are polyphyletic, since there are ameboid, flagellated, and ciliated cells that live in or on host organisms. The association may (1) mutually benefit host and symbiont (mutualism), (2) damage

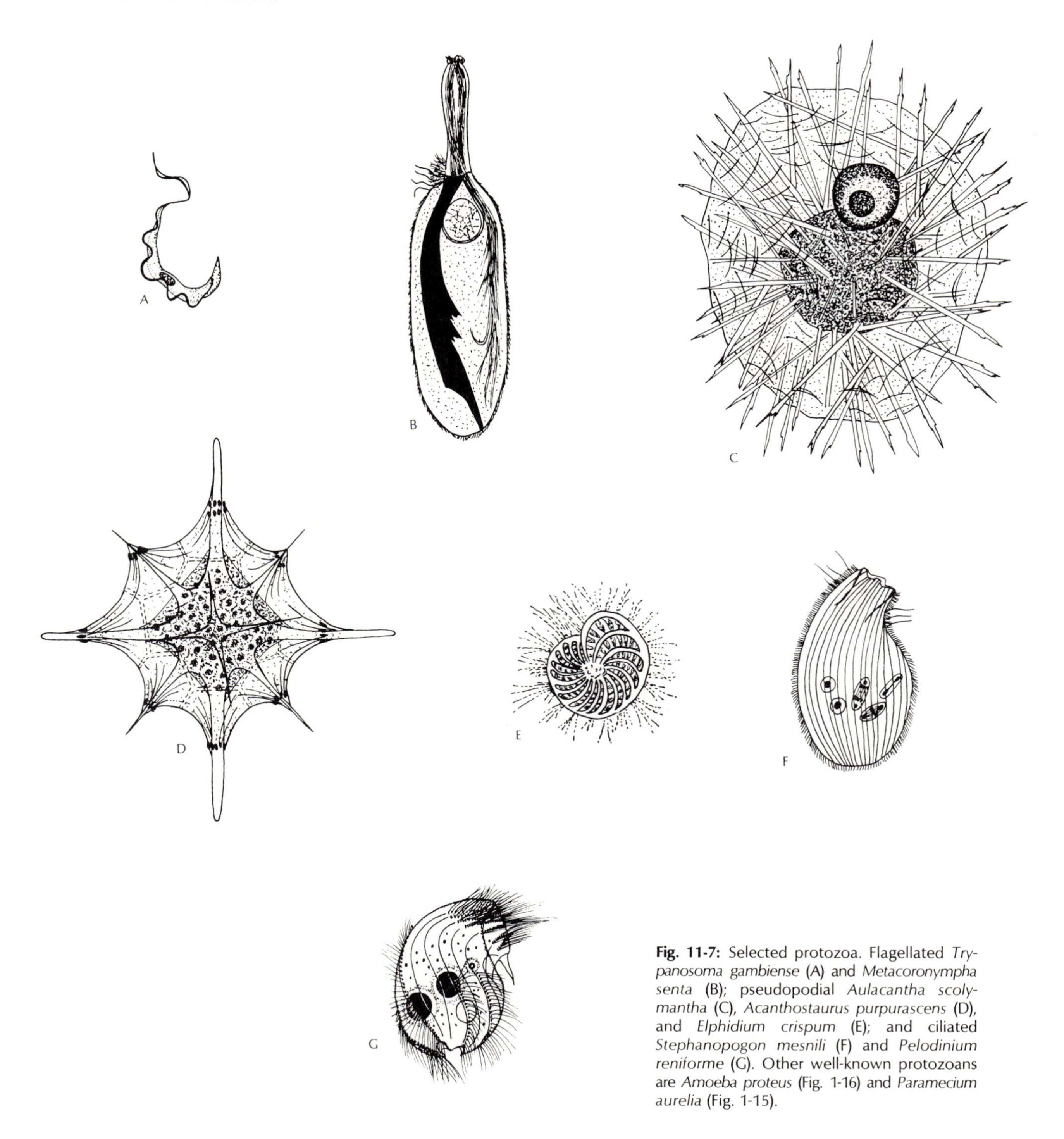

Fig. 11-7: Selected protozoa. Flagellated *Trypanosoma gambiense* (A) and *Metacoronympha senta* (B); pseudopodial *Aulacantha scolymantha* (C), *Acanthostaurus purpurascens* (D), and *Elphidium crispum* (E); and ciliated *Stephanopogon mesnili* (F) and *Pelodinium reniforme* (G). Other well-known protozoans are *Amoeba proteus* (Fig. 1-16) and *Paramecium aurelia* (Fig. 1-15).

the host but benefit the symbiont (parasitism), and (3) neither benefit nor damage either host or symbiont (commensalism).

Regardless of whether the protozoa are free-living or symbiotic, they occur in three broad categories or types. They are flagellated, ameboid, or ciliated (Fig. 11–17). However, the evolutionary trends are twofold, i.e., kinetidal and pseudopodial.

Pseudopodial evolution. The pseudopod, or false foot, a cytoplasmic extension of the protozoan cell, is used for locomotion or food capture or both. All ameboid protozoa have pseudopodia, but these can be quite different in different species. One species can have one blunt pseudopodium and another many. The pseudopodium can be thin, and hundreds can extend in all directions from a spherically symmetrical cell. The pseudopodia in some forms can interconnect to form net-like structures.

The pseudopodial or ameboid cells are now viewed as all evolving from originally flagellated cells. In some cases, the flagellated form was a pigmented protophyte that showed ameboid tendencies. With loss of plastids it became an ameboid protozoan. Others lost their pigment, but remained flagellated, and then became ameboid.

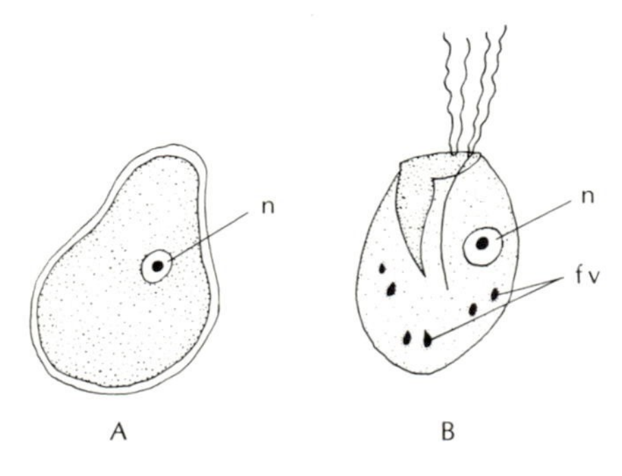

Fig. 11-8: Interchangeability of pseudopodial and kinetidal forms in the ameboflagellate *Tetramitus rostrata*. These environmentally determined transformations are not the same as DNA-determined transformations in bacteria. In salt solution, the ameboid stage (A) occurs at a one-eighth dilution from normal, the flagellated stage (B) at a one-twentieth dilution from normal (fv, food vacuoles; n, nucleus).

The close relation between flagellated and ameboid protozoans is seen in the life cycle of the ameboflagellates (Fig. 11–8). These protozoa can change from ameboid to flagellated cells and back again, depending on culture conditions. Even in those ameboid forms in which the pseudopodial stages are dominant, one can still find flagellated reproductive stages (zoospores or gametes). These are apparently retained as an adaptation to ensure dispersal and union of gametes from different sources. However, many species of amebae are entirely without flagella. And, curiously enough, many of these permanent amebae have also dispensed with sex. Since no known sexual process occurs in these amebae, it therefore renders problematical their designation as a species, in the sense of sharing a common gene pool. They have been described as an ecospecies, which means a group of organisms driven by selection to share a common adaptive peak. Presumably they are so well adapted to that peak that sexual processes and the genotypic variety they generate are selected *against*. It is a situation deserving much more study, as we stated in Chapter 8 when we first considered the phenomenon of ecospecies.

Evolutionary trends among the pseudopodial forms are summarized in Table 11–4. The extremes in such an evolution

Table 11–4. Trends in protozoan evolution. Examples of these forms are shown in parts C, D, and E (pseudopodial), A and B (kinetidal flagellated protozoa), and F and G (kinetidal ciliated protozoa) of Fig. 11–7.

Features	Pseudopodial forms	Kinetidal forms
size	increase from small amebae (ca. 10 μm) to testate forms (ca. 1,000 μm)	increase from slender zooflagellates (ca. 10 μm) to large ciliates (ca. 1,000 μm)
symmetry	many asymmetrical, others with spherical or radial symmetry	radial to bilateral symmetry with anterior-posterior polarity
nuclear organization	single to many nuclei, usually haploid, occasional polyploid nuclei	in the zooflagellates, single to many haploid nuclei, occasionally diploid; in the ciliates, single to many diploid nuclei and single to several polyploid nuclei
cytoplasmic organization	specialized pseudopodia; flotation mechanisms, tests	compounding and complexing of kinetides and associated structures; cortical differentiations, especially the cytostome
life cycles	extracellular flagellated gametes develop into pseudopodial adults	intracellular nuclear gametes, adult differentiation persists; sexuality and permanent diploidy common in ciliates

From Hanson 1976 *J. Protozool. 23:*1–12.

are the large, often spherically symmetrical marine forms with tests or shells (the Radiolaria and Acantharia, especially, but also the Foraminifera) (see Appendix I). There is no evidence that they have given rise to other protozoa or to multicellular metazoans. They are an evolutionary dead end.

Kinetidal evolution. The protozoa with kinetides are arguably the ancestors of all multicellular animals. A kinetide is composed of a flagellum or cilium, its basal granule or kinetosome, and associated structures. The latter include microtubules and microfilaments organized in various ways (Fig. 11–9). The kinetide has many functions. It is used for locomotion. Associated with that function is the capture of food or the location of sexual partners. The kinetide also plays a central role in regulating cellular form. To better understand these multiple functions—and sometimes all functions are performed by the same kinetides—we now describe the organizational diversity of kinetides.

There can be one kinetide per cell or there can be thousands of kinetides (Fig. 11–9). The single kinetide can be compounded into rows or joined into membranelles or special

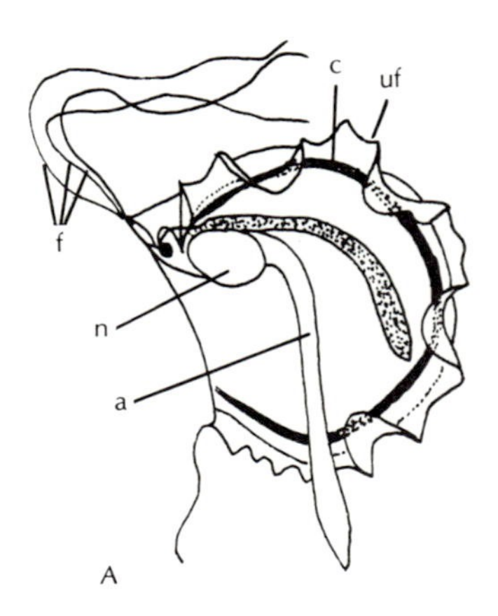

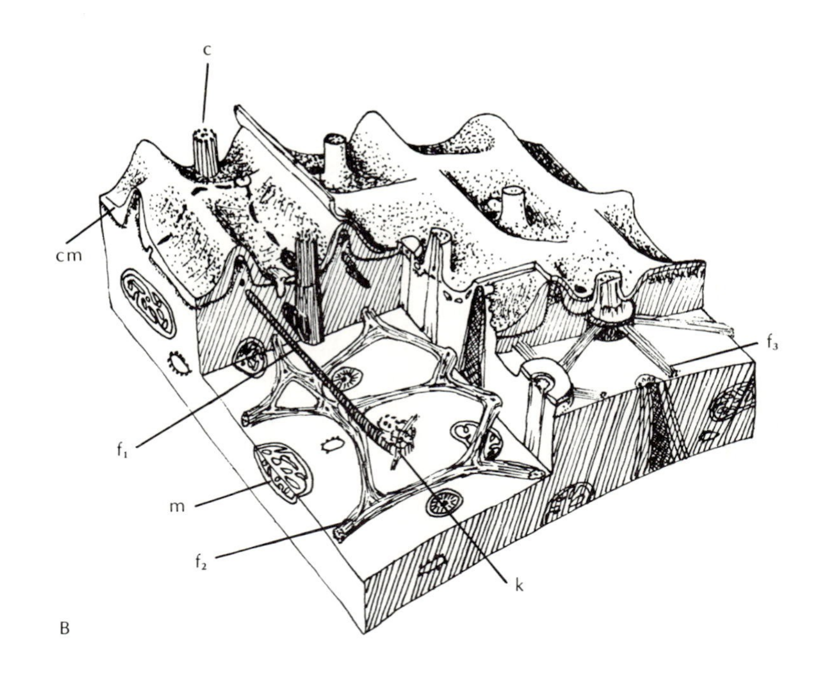

Fig. 11-9: Kinetides and associated structures. A. The flagellate *Trichomitopsis,* an unidentified species (a, axostyle; c, costa; f, flagellum; n, nucleus; uf, undulating flagellum). B. A three-dimensional reconstruction from electron-microscope studies of the surface structures of *Paramecium caudatum* (c, cilium; f_1, f_2, and f_3, different fibrillar systems; k, kinetosome or ciliary base; m, mitochondrion).

tufts. In the latter two conditions, new organelles with capabilities beyond those of single kinetides are formed. In brief, single kinetides become compound and then complex as they become new organizational units. Furthermore, the kinetosome of the kinetide can be associated (especially in the flagellates) with the nucleus. It can organize the mitotic spindle of the cell and play a key role in cell division. In the ciliates, kinetosomes are not associated with the nuclei, but they are organizing centers for the complex architecture of the outer layers of the cell (Fig. 11–9B) and detailed studies have shown us how the ciliate cortex plays a role in regulating development and in maintaining cell structures.

The kinetide is an organelle concerned with not only the immediate survival needs of the protozoan—locomotion and food—but also with its organization and development. Research has shown that its role in development is essential to proper expression of the genes. Thus, the kinetide increases the evolutionary potential of the flagellated and ciliated protozoa.

What should be emphasized in the evolution of kinetidal forms are those trends that lead to multicellular animals. As we have said, all animals are consumers—they ingest their food. That food can be other organisms or parts thereof. Whatever the case, the food is presented as chunks, which are ingested and then digested internally, either in food vacuoles or in digestive cavities. The protozoa use food vacuoles, and ingestion is by phagocytosis. Phagocytosis is usually thought of as occurring in amebae (Fig. 11–10A), in which the food particles are engulfed and then located internally in a food vacuole. The same process, essentially, occurs in the ciliates (Fig. 11–10C). But here food is ingested at a special cortical organelle—the cell-mouth or cytostome. The ciliates are the only protozoa with a permanent mouth. All the others ingest through temporary mouths, which are induced when phagocytosis occurs.

Note also that the role of consumer or ingestor is also that of predator, which usually means that the predator must be larger than the prey; otherwise, ingestion is impossible. This puts a selective advantage on increase in size, at least for certain predators. That has many functional as well as evolutionary complications. Increase in size means new solutions to locomotion. The kinetide solves that by becoming compound and then complex in response to selection pressures. Increase in size means an adjustment of nuclear content to cells needs. Various solutions are found. Large flagellates (and ameboid forms) become multinucleate. Large ciliates develop a macronucleus, a highly polyploid structure. Large size also demands coordination of various parts of the cell body. The microtubules and microfilaments of the kinetide as well as associated membranes seem to act as coordinators. In addition to coordination of body parts, a predator must coordinate sensory input with body function. Predators must be able to locate prey, capture it, and ingest it. In the ciliates, especially, the kinetides aid in all this. Ciliates do respond chemotactically and thigmotoctically (by touch) to prey in their vicinity. They swim toward their prey. Prey is swept into the cytostome or captured by other special cortical organelles, and then ingested.

This predatory behavior puts a premium on body specialization, which includes sensory, locomotory, and ingestatory apparatuses, all within a single cell. It demands a complexity of the cell body not found in any other living thing. The protozoa are the most complex of all cells, and the ciliates are supreme among the protozoa. We see in them the consequences of the selective pressures arising from a predatory mode of life. In the multicellular animals this means bilaterally symmetrical forms with specialized anterior ends. The

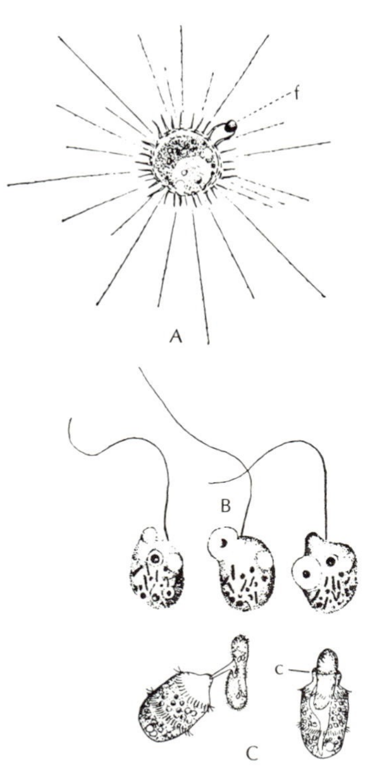

Fig. 11-10: Ingestion in the protozoa (see also Fig. 1-16). A. The spherical ameba *Raphidiophrys elegans* capturing a food particle (f), which will be phagocytized. B. Phagocytosis in the flagellate *Oikomonas termo*. C. The ciliate *Didinium nasutum* capturing and ingesting a paramecium through its well-developed, permanent cytostome (cell mouth) (c).

anterior ends typically carry special sensing devices—eyes, ears, noses, taste buds—and ingestatory structures—jaws and mouths. Bilateral symmetry allows these predators to develop the specific orientation—right and left, up and down—necessary for predation. All these evolutionary innovations are anticipated by the ciliates. They will be discussed in further detail when we come to metazoan origins (Chapter 13). Now let us make some concluding comments on kinetidal protozoa.

As was summarized in Table 11–4 and Fig. 11–9, the flagellates and ciliates get larger, develop more complex kinetidal and nuclear apparatuses, and become more complex; this complexity includes permanent diploidy and polyploidy in the ciliates. The key step, in going from the zooflagellates to the ciliates, is the emergence of a permanent mouth, a hallmark of predation. This neosemic character may well explain why most of today's zooflagellates are symbiotic and why the ciliates are mostly free swimming. The now more efficient predators, the ciliates, eliminated their ancestors, the zooflagellates, except in cases in which the latter had invaded specialized niches. These niches appear today as host organisms. But note, when the host organisms are multicellular animals, those hosts evolved *after* the appearance of the protozoa. Presumably the zooflagellates survived in special, free-living niches and then when opportunities for symbiosis arose, they took them. Extant symbiotic zooflagellates can therefore be viewed as relict populations of once widespread free-living species, now replaced by the ciliates.

Ciliates also differ from zooflagellates in that the former typically have a macronucleus, whereas ciliate plesiomorphs, e.g., *Stephanopogon mesnili* (Fig. 11-7F), do not. Instead, they have many diploid nuclei. A few zooflagellates are also known to have many diploid nuclei. It, therefore, seems likely that ciliates with cytostomes arose from zooflagellates and then evolved. That further evolution included the appearance of the polyploid macronucleus and refinements of the kinetide and ciliate cortex.

Protistan origins

Let us now leave our consideration of trends within the Protista and turn to a major phylogenetic problem, the origin of the Protista. Before discussing that problem, we can prepare for it by returning briefly to the Monera. The origin of the Monera is most precisely formulated by looking into the origin of the moneran plesiomorph. That plesiomorph we identified as *Methanococcus,* or one of the methanogens. Its origin was discussed in Chapter 10 in the context of the emergence of cellular life from protocells. What we have to realize is that the Protista probably arose from the Monera. We are forced to this conclusion because, based on present evidence, the Monera are the first cellular forms of life, and the genetic code is universal. The latter says that the eukaryotes did not have a separate origin, the former that the Monera came first. So our problem regarding the origin of the Protista comes down to these questions: What is the protistan plesiomorph? Does it share any homologies with monerans? If it does, what can we say regarding its evolutionary development?

THE PROTISTAN PLESIOMORPH

There are as many candidates for the protistan plesiomorph as there are protophyte phyla, since we cannot decide which phylum is plausibly ancestral to the others. Although a species of *Chlamydomonas* would be an excellent plesiomorph for the Chlorophycophyta, and a species of *Euglena* the plesiomorph for the Euglenophycophyta, we have no good evidence for deriving an euglenoid cell from a chlamydomonad cell or vice versa; and similarly for a chrysomonad or pyrrophycophytan (dinoflagellate) plesiomorph. All are equally good or equally bad as a protistan plesiomorph. This situation can be explained in two ways: (1) we have as many independent origins of the protophyta from the Monera as we have separate plesiomorphs, and (2) there was a unitary origin of the protophyta, but the rapid, tachytelic evolution (Fig. 10–19), in going from the adaptive zone of the metabolically specialized Monera (prokaryotes) to that of the structurally complex Protista (eukaryotes), resulted in rapid adaptive radiation into various protophyte phyla. In both cases we can argue that many intermediate forms, i.e., those that are neither good prokaryotes nor good eukaryotes, were lost and, hence, there is a considerable evolutionary gap. We are trying to peer across this gap; we are trying to reconstruct, conceptually, a phylogenetic bridge. Our phylogenetic methods tell us to look for serial relations to bridge such a gap. But that may well be futile in terms of cellular structure. Researchers have been aware of the difference between prokaryotes and eukaryotes for decades and have looked in vain for intermediates or missing links. They may yet turn up as fossils, but it is questionable that the needed fine-structural detail will be preserved.

The best remaining possibility is molecular data. We need to compare conservative or plesiosemic molecules in protophyte plesiomorphs with each other and with comparable molecules in the monerans. Some such comparisons have been made, as we will see. Before turning to them, it is worth mentioning the work of the German botanist Pascher, who was a profound student of the algae in the early part of this century.

Pascher, influenced by Haeckel's phylogenetic speculations, constructed a hypothetical ancestor of the Protista. It was a unicellular, photosynthetic cell with two flagella—a phytoflagellate. It had one plastid and a cell wall. From such an ancestor, Pascher believed there could be derived all the algae and from them the higher plants, fungi, and protozoa. In other words, this was the start of eukaryotic evolution. This hypothetical ancestral phytoflagellate of Pascher illustrates the weakness of this kind of phylogenetic speculation. No experiments can be done with it, since it does not exist. Rigorous comparisons for homology are impossible, since it does not exist. The only thing in its favor is that it alerts us to the phylogenetic problems in this area by saying that a biflagellate, unicellular eukaryotic producer is the kind of cell we are looking for. But the worst effect of this kind of thinking is that it is prejudicial to any other theory. The implication is that there was a unitary origin of the Protista. But was there? What is the evidence? It suggests that a certain moneran species evolved into this type of protistan. Did it? We will see shortly that some people favor a startlingly different point of view. In summary, Pascher's suggestion and others like it must be held at arm's length to avoid the distortions they can introduce. Perhaps better, we should ignore them altogether and simply work from plesiomorphs.

Working from plesiomorphs, we immediately face the situation already described, namely, there are many plesiomorphs and there is a huge gap between the plesiomorphs and the prokaryotes. The next step is to look for homologies, which, as we have also noted, will probably have to be molecular. Are there any? The growing answer is yes, there seem to be some, but the story is not yet entirely convincing.

MOLECULAR HOMOLOGIES

A recent report from the laboratory of Margaret Dayhoff and her colleagues at the National Institutes of Health (U.S.) shows that the respiratory protein cytochrome c and the small, iron-containing proteins, the ferredoxins, as well as the 5S ribosomal RNA molecules, all show homologous similarities among monerans and protistans. Of particular interest is the fact that in the eukaryotes the cytochromes are located in the mitochondria and the ferredoxins are obtained from plastids (they are involved in photosynthesis). Similarly other workers have compared the RNA from the plastids of *Euglena* with moneran RNA and have come up with good evidence of homology. All these molecular studies are based on point-to-point similarities of amino acid sequences in proteins and of nucleotide sequences in RNA.

This kind of evidence strongly supports what is now called the serial endosymbiont theory of protistan origins. This theory was first advanced in the nineteenth century and given its modern form by Lynn Margulis, of Boston University, in 1970. This theory argues that protists are symbiotic associations of monerans. Or, stated more exactly, prokaryotes enter into cooperative associations to form eukaryotes. Margulis and others differ somewhat on the exact details of how this occurred, and those differences are being investigated. Margulis proposes that plastids, mitochondria, and the microtubular structures of eukaryotes are derived from different prokaryotes. For example, a host cell engulfed a blue-green alga that evolved into a plastid. An engulfed aerobic prokaryote could have provided a mitochondrion. And association with a motile prokaryote like a spirochaete could have given rise to flagella, kinetosomes, and other microtubular elements, in short, the primoridal kinetide. In fact, Margulis has proposed that a colorless ameboid prokaryote was the host cell and phagocytosed other prokaryotes of the sort just mentioned. This point of view suggests that protophyta and protozoa, both, are products of serial endosymbiosis.

For reasons given above regarding the homologies between protozoa and colorless protophyta, it seems unnecessary to suggest that the protozoa originated by endosymbiosis. But the suggestion that the protophyta originated by endosymbiosis is attractive and is supported by the molecular evidence given above.

There are, however, other points of view. The American microbiologists Mahler and Raff concentrated on the origins of the plastids and mitochondria, since these are similar to prokaryotes. In addition to the above descriptions, it is important that both sets of organelles contain DNA that is a circular double helix of naked DNA. Also, antibiotics that inhibit RNA and protein metabolism in bacteria inhibit the same activities in plastids and mitochondria. Mahler and Raff argue that, rather than endosymbiosis, the extranuclear DNA of prokaryotes, called episomes, could be associated with membranes and eventually segregated off into organelles. This is a type of species transformation wherein, given enough time and appropriate selection pressures, one species slowly evolves into another. It is a well-established process in evolution. The lack of

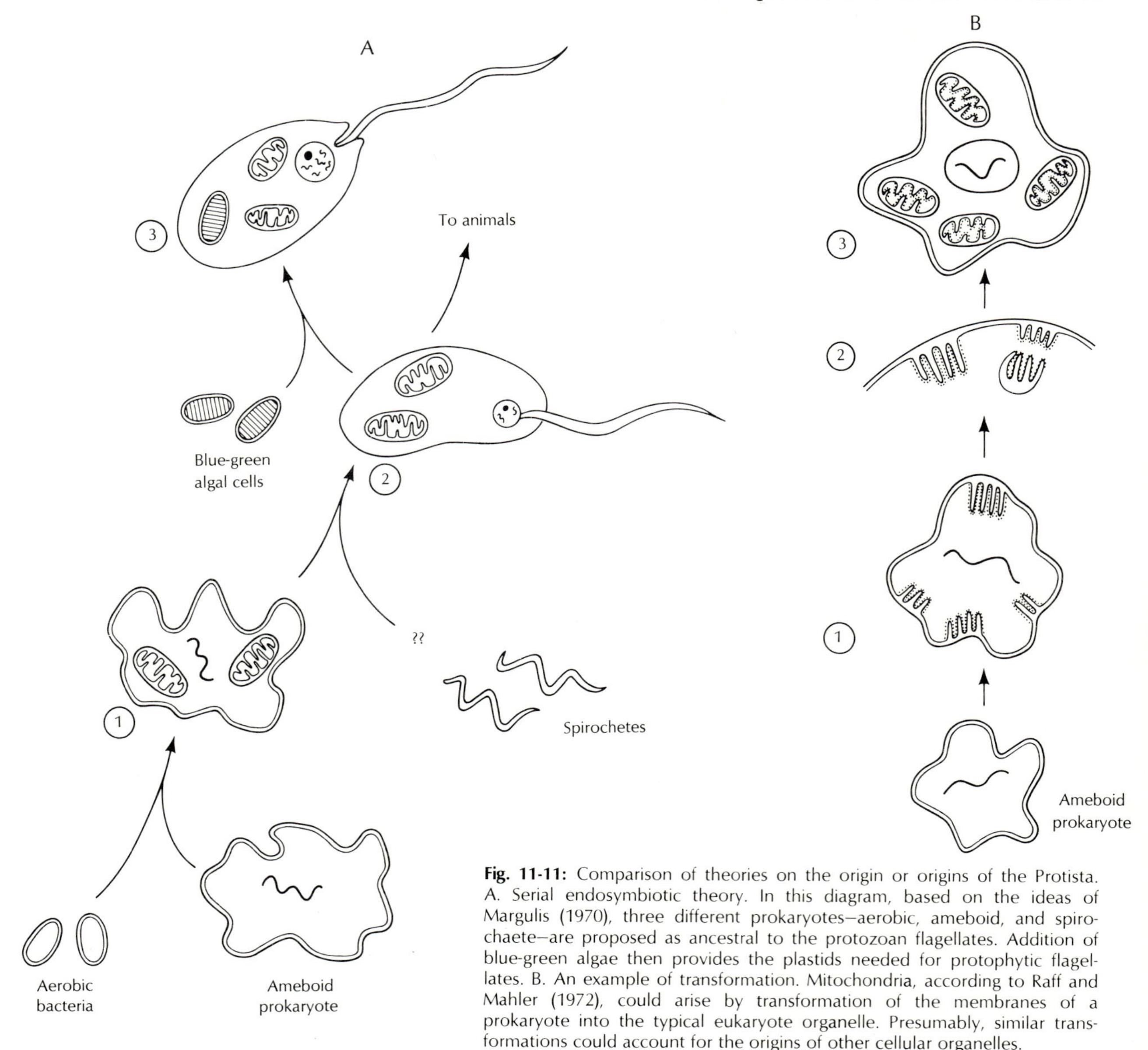

Fig. 11-11: Comparison of theories on the origin or origins of the Protista. A. Serial endosymbiotic theory. In this diagram, based on the ideas of Margulis (1970), three different prokaryotes—aerobic, ameboid, and spirochaete—are proposed as ancestral to the protozoan flagellates. Addition of blue-green algae then provides the plastids needed for protophytic flagellates. B. An example of transformation. Mitochondria, according to Raff and Mahler (1972), could arise by transformation of the membranes of a prokaryote into the typical eukaryote organelle. Presumably, similar transformations could account for the origins of other cellular organelles.

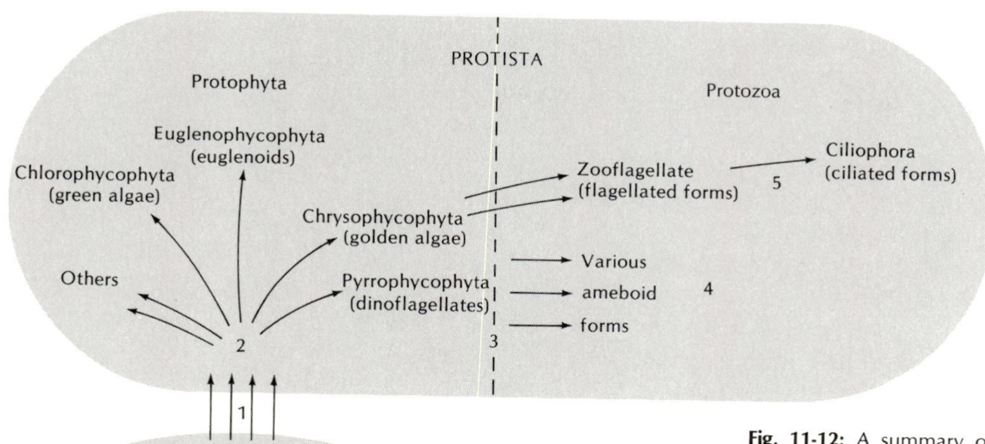

Fig. 11-12: A summary of protistan phylogeny. Note that other views on protistan evolution differ from this one. Five problem areas can be mentioned: (1) Origin of the protists. This could be by endosymbiosis or by transformation. (2) Diversification of the protophyta. This could be the result of a polyphyletic origin of the algae from the Monera. (3) Origins of the protozoa. There is convincing evidence for a polyphyletic appearance of animal protists (protozoa) from certain colorless algae. (4) Pseudopodial evolution. Many different evolutionary experiments among ameboid forms. (5) Kinetidal evolution. Polyphyletic zooflagellates were presumably replaced as free-swimming forms by ciliated cells.

intermediates would be due, as we said earlier, to tachytelic evolution, in which prokaryotes rapidly became eukaryotes. This theory and the endosymbiont theory are summarized in Fig. 11–11.

PRESENT STATUS OF PROTISTAN ORIGINS

We have then two competing views of the origin of the protistan eukaryotes. Both views agree that the protists arose from the prokaryotes; both can account for the observed molecular homologies; and both can explain the morphological gap that separates the prokaryotes and eukaryotes. Also, both suggest that there was a multiple or polyphyletic origin of the Protista. (They both disagree with Pascher.) In other words, both can explain the same set of phylogenetic data. To determine which view is the correct one, other predictions from the two hypotheses will have to be made and tested. This has not yet been done and therefore we are left with both views as options. One general summary of protistan phylogeny is shown in Fig. 11–12.

In closing, one more complication should be noted and that pertains to the validity of macromolecular homologies. Usually the argument against homology is stated in terms of random events producing comparable arrays of amino acids or nucleotides, and the argument for homology is stated as due to similarity from a common ancestry. That is not correct. It is really convergence versus homology, not random chance versus homology. And we really do not yet know how precise macromolecular convergence can be. The fact that hemoglobin appears in such diverse animals as flatworms, annelids, certain molluscs and insects, some echinoderms, and vertebrates has been widely attributed to convergence. Unfortunately, only the sequence of the vertebrate hemoglobins has been studied; therefore we cannot determine the nature of molecular convergence between different phyla precisely. But it is conceivable that there are only a limited number of ways to build an oxygen-transport molecule. Perhaps only a rather specific sequence of amino acids will be functional, and hence, similar sequences will appear due to similar selection pressures, a process quite different from chance. The same might also apply to similar RNA molecules—their role in ribosomal function might well be so precisely defined as to select for highly similar nucleotide sequences. Evolution produces remarkably similar, complex organs of sight in the cephalopod molluscs (octopus and squids) and vertebrates, which are convergent, not homologous. Perhaps less complicated structures, such as certain functionally identical molecules, will also be precisely convergent. It will be of great theoretical interest to see if the Remanian criteria for homology can be used to detect convergence of molecules.

When, finally, the evolutionary path from prokaryotes to eukaryotes is drawn with clearer lines, then, too, we will see the selection pressures that were in force. That should provide a better understanding of how the metabolic specializations of the prokaryotes evolved into the structural complexities of the eukaryote cells.

References

Breed, R. S., E.G.D. Murray, and A. P. Hitchens, 1948. *Bergey's Manual of Determinative Bacteriology.* 6th ed. Williams & Wilkins, Baltimore.

Frederick, J. F., and R. M. Klein (eds.) 1980. *Origins and Evolution of Eukaryotic Intracellular Organelles. Annals of the New York Academy of Sciences.* (in press).

Hanson, E. D., 1977. *The Origin and Early Evolution of Animals.* Wesleyan University Press, Middletown, Conn.

Margulis, L., 1970. *Origin of Eukaryotic Cells.* Yale University Press, New Haven, Conn.

Mahler, H. R., and R. A. Raff, 1975. The evolutionary origin of the mitochondrian: A non-symbiotic model. *International Review of Cytology 43:* 1–124.

Morowitz, H., 1966. In *Principles of Biomolecular Organization,* G.E.W. Wolstenholm and M. O'Connor (eds.), pp. 336–462 Churchill, London.

Ragan, M. A., and D. J. Chapman, 1978. *A Biochemical Phylogeny of the Protists.* Academic Press, New York.

Schwartz, R. M., and M. Dayhoff, 1978. Origins of prokaryotes, eukaryotes, mitochondria and chlorophasts. *Science 199:* 395–403.

Sneath, P.H.A., and R. R. Sokal, 1973. *Numerical Taxonomy.* W. H. Freeman, San Francisco.

Symposium "Early Evolution of Protists." (Sponsored by the Society of Protozoologists, the Phycological Society of America, and the Society for Invertebrate Pathology.) The following papers were published in the *Journal of Protozoology 23:* 4–56 (1976).

- Hanson, E. D. Major evolutionary trends in animal protists. (pp. 4–12)
- Loeblich, A. R. III. Dinoflagellate evolution: Speculation and evidence. (pp. 13–28)
- Taylor, F.J.R. Flagellate phylogeny: A study in conflicts. (pp. 28–40)
- Buetow, D. E. Phylogenetic origin of the chloroplast. (pp. 41–47)
- Hutner, S. H., and J. O. Corliss. Search for clues to the evolutionary meaning of ciliate phylogeny. (pp. 48–56)

Taylor, F.J.R., 1974. Implication and extensions of the serial endosymbiosis theory of the origin of eukaryotes. *Taxon 23:* 229–258.

Thimann, K. V., 1955. *The Life of the Bacteria.* Macmillan, New York.

TWELVE

The Metaphyta and Fungi

In both the kingdom Metaphyta, which are multicellular producers, and the kingdom Fungi, which are multicellular decomposers, the cells are eukaryotic. We shall, in both kingdoms thus defined, first characterize the diversity that occurs there and then discuss evolutionary trends and what we know of the origins of each kingdom.

The Metaphyta

The number of described metaphytan species is something in excess of 300,000 (Table 12–1). The species range from tiny, microscopic members of the aquatic red algae to redwood trees—terrestrial giants over 100 m in height. More impressive than the size range is the diversity of habitats and niches successfully invaded by these plants. They occur in salt water and freshwater, and on land they are distributed everywhere that is not excluded by freezing temperatures, lack of water, or other extreme conditions. Except for very rare exceptions, all are photosynthetic, i.e., they are phototrophs. They have varying needs for salts and a few organic nutrients, but because they can produce more of themselves from largely inorganic raw materials they are characterized ecologically as producers.

Table 12–1. Estimated numbers of described living metaphytan species.

Species	Compiled by T. Dobzhansky (1970)	Compiled by H. C. Bold (1973)
flowering plants	286,000	ca. 300,000
evergreens and related forms	640	720
ferns and related forms	10,000	ca. 11,000
mosses and related forms	23,000	20,000–32,000
red algae	2,500	ca. 3,900
brown algae	900	ca. 1,500
total	ca. 323,000	ca. 337,000–349,000

Together with the photosynthetic monerans and the protophyta, the Metaphyta represent the starting point of all food chains. Without them other living things—consumers (animals) and decomposers (fungi)—could not survive.

MULTICELLULAR ALGAE

The green algae (Chlorophycophyta, see Chapter 11) have sometimes been classified among the Metaphyta for the obvious reason that many of their member species are organized as highly integrated colonies or even as multicellular plants. We have treated them as protistans because of their obvious derivation from unicellular algae. By contrast, the brown algae (Phaeophycophyta) show no unicellular forms except for gametes, and in the red algae (Rhodophycophyta) there are only a few unicellular species.

Multicellular plants typify the red and brown algae (Fig. 12–1). They are differentiated into the same general structures we encountered in the green algae, i.e., tissues specialized as holdfasts to anchor the plant to the substratum, and above that a buoyant thallus or vegetative structure. The thallus varies enormously in size and shape. The largest ones, found in the California kelps, are long blades of brown algae 50 m or more in length. Furthermore, the thallus can be organized variously into unbranched or branched structures and can be flattened blades or cylindrical stems. Often in these algae reproductive structures form another kind of tissue specialization with male and female gametes being produced in different parts of the plant. Finally, alternation of generation occurs. In the red algae there can be three different generations of plants before a life cycle is completed. This involves various combinations of haploid and diploid stages (Fig. 12–2).

No motile cells have ever been found in the red algae, not even among their gametes. The sexual structures (Fig. 12–2)

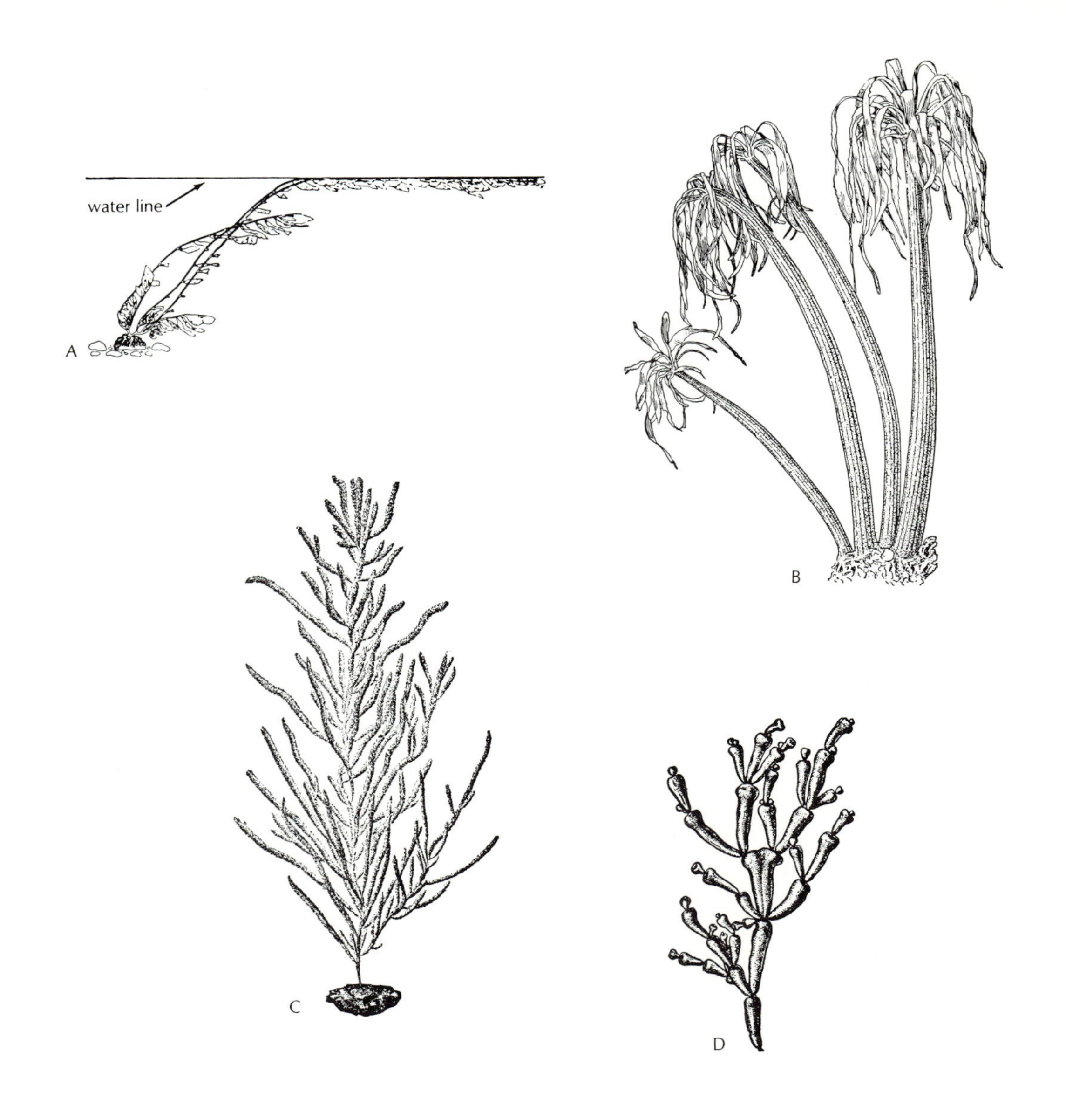

Fig. 12-1: Selected brown and red algae. A. *Macrocystis pyrifera,* the largest brown alga, attains a length of 50 m. B. *Postelsia palmaeformis,* a brown algal form, grows in the tidal zone and can withstand severe wave action. C. *Dasya elegans,* a beautiful feathery red alga, is highly branched. D. *Corallopsis salicornia,* a red alga a few inches long, is often encrusted with calcium carbonate.

typically produce well-differentiated eggs and sperm—so well differentiated, in fact, that fertilization is referred to as oogamy. The male gamete is released into the watery environment of these plants and apparently, through random motion, comes into contact with and fertilizes the egg. The brown algae, however, have flagellated gametes and zoospores. Fertilization of gametes produces a zygote that germinates and develops into a sporophyte—the diploid phase of these plants. Meiosis occurs in the special cells that form zoospores, and these flagellated cells swim free and attach to the bottom where they then develop into haploid gametophytes.

There are various parallels between the organization of the

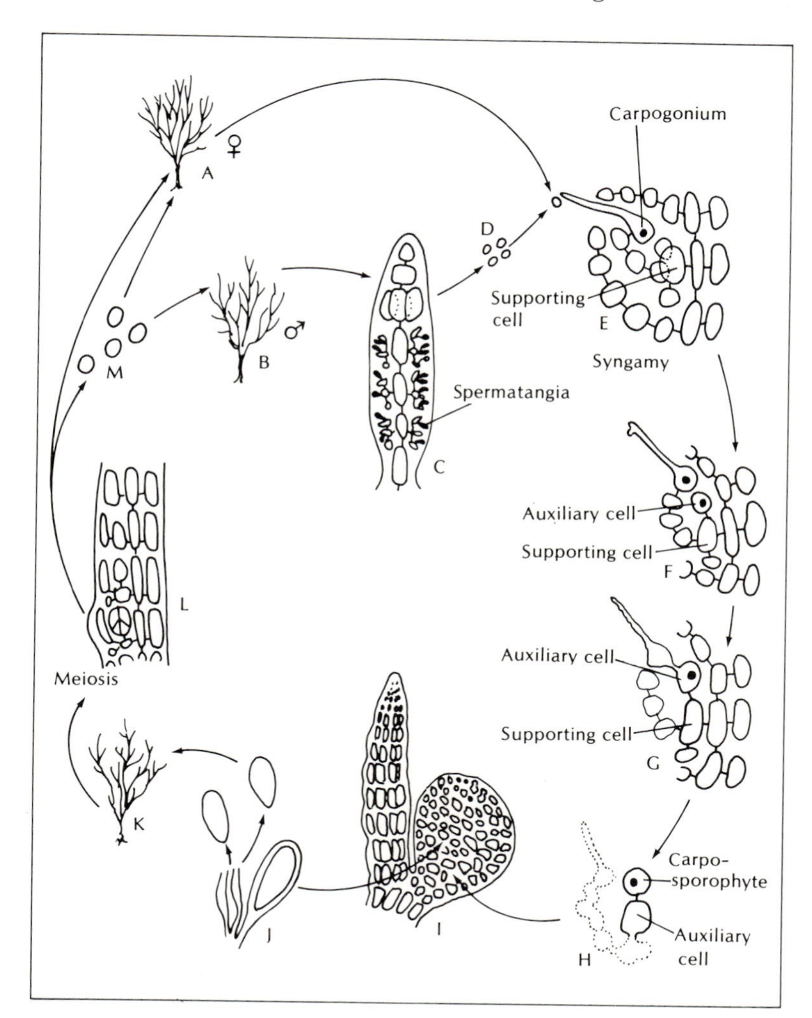

Fig. 12-2: Alternation of generations in a species of red alga of the genus *Polysiphonia*. A and B. Male and female haploid plants. C. In the male plants, gametes are formed. D. These gametes or spermatia are nonmotile cells released into the water surrounding these plants. E. A spermatium comes into contact with the carpogonium of a female plant. F. Fertilization, followed by cell division and nuclear migration, occurs. G. This results in a diploid nucleus in an auxiliary cell. H and I. Meiosis follows: many cell divisions produce a carposporophyte-containing haploid pericarp. J. Carpospores are released from the carposporophyte. K and L. Each carpospore produces a plant, within which another type of spore is formed. M. These latter spores are released and develop into male and female plants to complete the life cycle.

Table 12–2. Selected comparisons between the three algal phyla (divisions), which contain large numbers of multicellular species, and terrestrial plants (a continuation of Table 11–2).

Phylum	Photosynthetic pigments	Major storage products of photosynthesis	Cell wall
Rhodophycophyta	chlorophyll a and d, phycocyanin, and phycoerythrin	floridean starch	cellulose
Phaeophycophyta	chlorophyll a and c	mannitol and laminarin	cellulose plus algin
Chlorophycophyta	chlorophyll a and b	starch	cellulose
terrestrial plants	chlorophyll a and b	starch	cellulose

thallus in the red and brown algae and the tetrasporalian line of evolution in the green algae. The result is an impressive array of multicellular aquatic plants. Quite naturally, the question arises as to which of them might be ancestral to the land plants. Or perhaps, the land plants are polyphyletic and have ancestors in all three algal phyla. Answers become clearer when we make certain other comparisons between these algae and the terrestrial plants (see Table 12–2).

All these plants contain chlorophyll a among their photosynthetic pigments, but differ in terms of the other pigments, except for the Chlorophycophyta and the terrestrial plants. These latter two groups have the same photosynthetic pigments, food reserves, and cell wall components. This strongly suggests that the green algae were ancestral to these terrestrial plants. The red algae are the most different in terms of the characters cited. Note their photosynthetic pigments: these include phycoerythrin and phycocyanin, which occur only here and among the blue-green algae. This too has important evolutionary implications.

LAND PLANTS

The transition from an aquatic environment to a terrestrial one involves many adaptive changes. To set the stage for a more careful examination of that transition, we can start by looking at those features common to land plants.

1. There must be protection against evaporation. This, probably the most obvious adaptation, has been achieved by the development of epidermal tissue and, for some structures,

a special waxy layer or cuticle external to the epidermis or outermost plant tissue. Delicate tissues, such as leaves, most need the epidermis and its cuticle. Other structures, such as roots and stems and their counterparts in the less complex land plants, are effectively protected by the epidermis. The bark of trees is an obvious example.

2. Gaseous exchange with the environment must be possible. The leafy tissues must not be so completely covered that the release of oxygen and the uptake of carbon dioxide is prevented. Such a two-way flow of gases is essential to life. In the leaves of higher plants this flow is facilitated by special structures called stomata (Fig. 12–3). Under conditions of low humidity, cells on either side of the stomatal opening expand and effectively close the opening. Under other conditions, and depending on the needs of the nearby tissues, the stomata are opened to varying degrees.

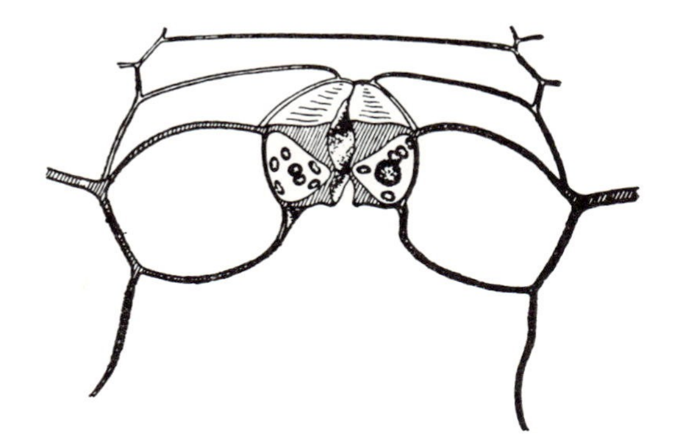

Fig. 12-3: A three-dimensional view of a *Zebrina pendula* stoma. The cell walls near the pore are quite thick, those farthest from the walls relatively thin. When the humidity is low, turgor increases, and the pressure bends the cell walls, which opens the stoma. A decrease in pressure closes the stoma by straightening the walls out. The principle can be illustrated by attaching adhesive tape (which represents a thick cell wall) along one side of a partially blown up balloon. The balloon will curve (the thin wall stretches more than the thick wall) if it is blown up further. Release of air (decrease in turgor or pressure) straightens out the balloon.

3. Water must be absorbed. Roots or root-like structures called rhizoids perform this function. The actual absorption by roots is carried out by root hairs. These microscopic cells extend at right angles from the root surface and provide the cell surface needed to absorb water and salt and other dissolved nutrients that may be present in the water.

4. Materials must be transported throughout the plant body. Photosynthetic products from leafy parts must be available to stems and roots and materials absorbed by roots must be available to stems and leaves. In the more complex land plants, special cells, which constitute the vascular tissue, conduct nutrients throughout the plant. Many land plants lack vascular tissue, but nonetheless transport is achieved by more generalized tissues. But since aquatic plants are surrounded by water, the plant tissues can exchange materials directly with the water, and there is no need for vascular tissue in the algae.

5. Land plants need support to keep them upright. Water is a buoyant medium and in its absence algae collapse into rather pathetic heaps. The special development of tough cell walls is used by plants for support. Additionally, some stem cells are specialized; there, the fiber cells carry the weight of plants. Consider, in particular, a giant redwood; it holds up thousands of tons, for centuries.

6. The gametes and especially, the early stages of new generations, must be protected. The terrestrial environment can be a relatively hostile place in which to germinate and survive. Spores and in particular seeds, which carry embryonic

plants, are adapted to survive dry conditions and to respond to wet conditions by germination and subsequent growth. This is achieved not only by outer protective coverings, e.g., the seed coat, but also by nutrients stored within the reproductive structure. Seeds carry a food supply in the endosperm, which, incidentally, humans and other animls have learned to eat as in cereal grains and various nuts.

In addition to these adaptions, we find certain other features common to all land plants. As was shown in Table 12–2, these plants all contain chlorphyll a and b, and starch is a major product of their photosynthesis. Also cellulose, the most common constituent of their cell walls, is the major component of their supportive tissues. Also important is the fact that their life cycles always alternate between sporophyte and gametophyte stages. As we shall see, such an alteration of generations can emphasize one or the other type of plant, but nonetheless, alternation always occurs.

It will be easiest to separate the land plants, first, in terms of whether or not vascular tissue is present and then to subdivide vascular plants by whether or not seeds are present. The vascular plants include the flowering plants, about two-thirds of all the metaphytan species (Table 12–1).

Non-vascular plants. There are two divisions or phyla here: the Bryophyta, or mosses, and the Hepaticae, or liverworts. They are often combined into the one phylum, the bryophytes. The essential common feature of mosses and liverworts is that the sporophyte generation is a kind of symbiont with the gametophyte. The latter grows out of the ground in moist places; the former grows out of the gametophyte. The life cycle of a moss is shown in Fig. 12–4. As is characteristic of gametophytes and sporophytes, the one producing the gametes is haploid, the other is diploid. Meiosis occurs in the diploid sporophyte, and haploid spores are formed and then released to germinate and thus start a new gametophyte generation.

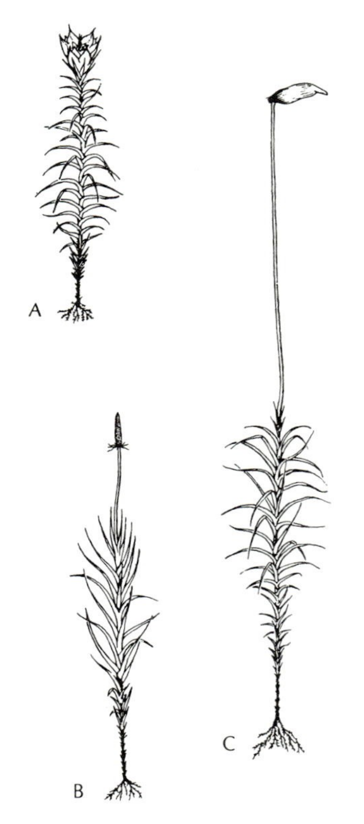

Fig. 12-4: Life cycle of a species of the moss *Funaria.* A. The male sperm of the gametophyte are carried in the antheridia at the terminal end. B. In the archegonium of the female gametophyte, a fertilized egg has produced a young sporophyte. C. A gametophyte bearing a mature sporophyte. The diploid sporophyte undergoes meiosis in the spore capsule (sporangium) at its tip. The haploid spores are released and germinate into haploid male or female gametophytes.

In mosses, water is absorbed through rhizoids. These are filamentous chains of cells that extend underground, like roots. The stem is a vertical growth of cells from the rhizoids. Never over a few inches long, it brings water and dissolved minerals up to the small leaflets that grow directly out of the stem. Active photosynthesis occurs in the leaflets. Because the cells are small, water, nutrients, and products of photosynthesis can be distributed by diffusion throughout the plant body. This is thought to be the reason why cells and tissue specialized for a transport function are notably absent.

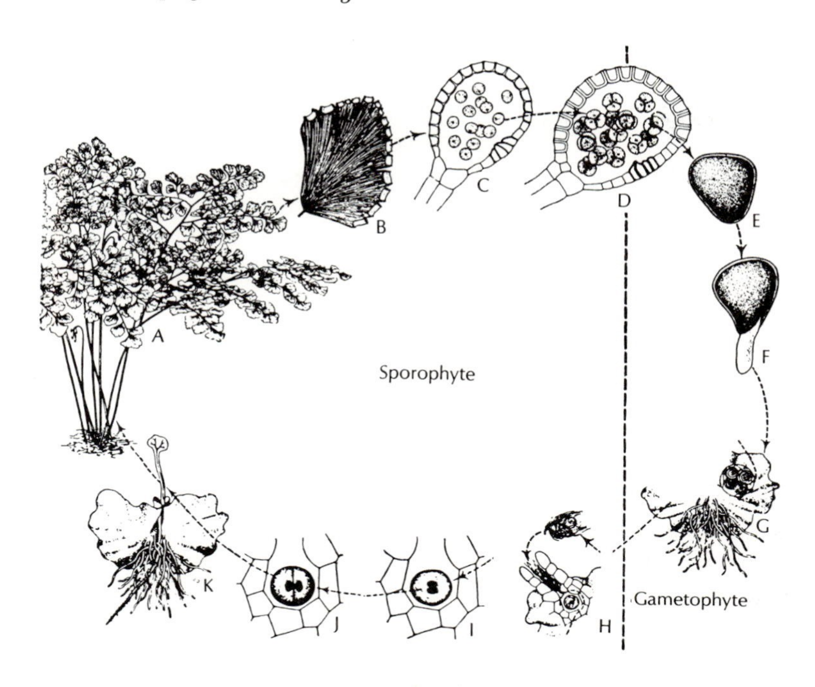

Fig. 12-5: The life cycle of a species of maidenhair fern (*Adiantum*). A. The diploid sporophyte carries sporangia with spores on the margins of the leaf-like pinnules, which make up the frond. B. A pinnule with marginal sporangia. C and D. Sporangium with spores. E and F. These germinating spores are haploid. G. The haploid gametophyte with antheridia and archegonium. H. An archegonium with entering sperm. I, J, and K. Fertilization of the egg and growth of a young sporophyte.

Vascular plants without seeds. Not only do all the remaining land plants show vascular tissues, their sporophyte and gametophyte generations differ from those seen in the mosses and liverworts. From among the various seedless vascular plants (Fig. 12–5), let us look at the ferns as typifying the essential features of this group.

A fern, as most commonly encountered, is a sporophyte, or diploid plant. Its most visible part is the frond (leaf), which extends gracefully upward from a root stock which in most cases has roots extending into the soil. On the underside of the frond, which can be undivided or subdivided into a delicate array of subparts, there are quite often sporangia containing spores. After meiosis, the spores formed here are released; upon germination, a spore produces a tiny, haploid gametophyte. This tiny plantlet develops both sperm-forming and egg-forming parts. A sperm fertilizes an egg and initiates a new diploid sporophyte. This life cycle is shown in Fig. 12–5.

Note, especially, the relative roles of the sporophyte and the gametophyte. Both are necessary for a complete life cycle, but here they draw independently from their environment to exist autonomously. There is no dependence of sporophyte on gametophyte, such as the symbiosis seen in liverworts and mosses. Furthermore, the sporophyte is the more conspicuous of the two plant stages or generations.

As regards the details of sporophyte structure, we find the already-mentioned root structures and frond. Vascular tissues are quite well developed in the stem of the frond (Fig. 12–6). Here we can see the xylem, which functions principally to transport water and dissolved minerals, and the phloem, which distributes dissolved foods (mostly carbohydrates)

In the ferns the vascular tissue branches off and extends into the branches of the frond. As we shall see when we discuss evolutionary trends in vascular tissue, this branching is correlated with the appearance of side branches or leaves along the stem. In the simpler vascular plants these *leaf traces* simply separate off from the central cylinder or *stele* of vascular tissues and extend toward the leaf bases. In the ferns and more complex vascular plants, they extend right into the frond branches and into the twigs and the leaves. This branching of a main stem is the functional basis for frond formation. It involves three related processes: overtopping, planation, and webbing. *Overtopping* refers to the occurrence of lateral extensions, but with further growth of the higher part of the plant.

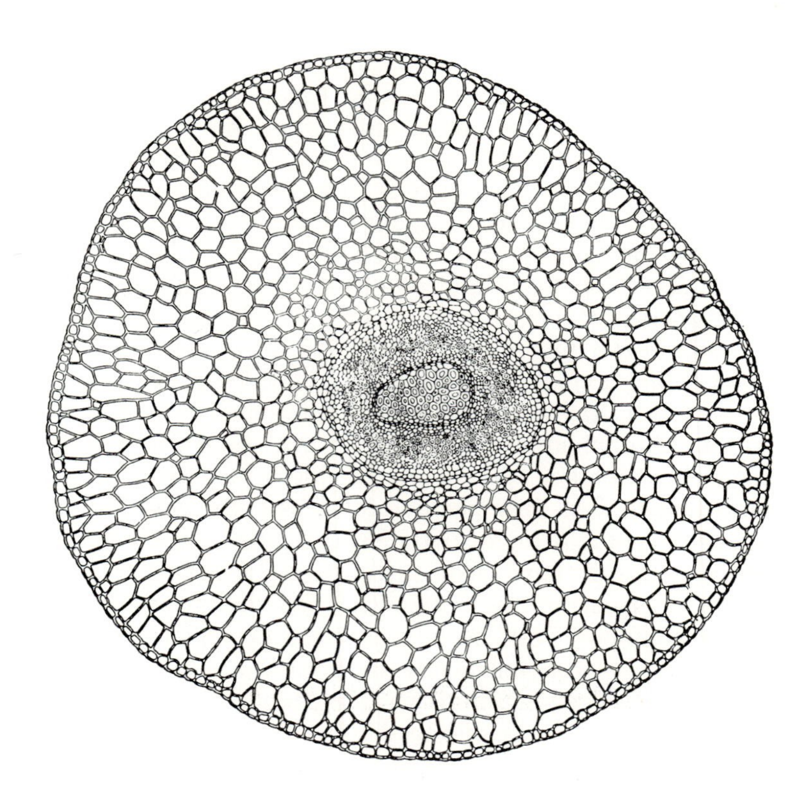

Fig. 12-6: Stem of a maidenhair fern (*Adiantum*) (cross section) showing the central stele with both xylem and phloem.

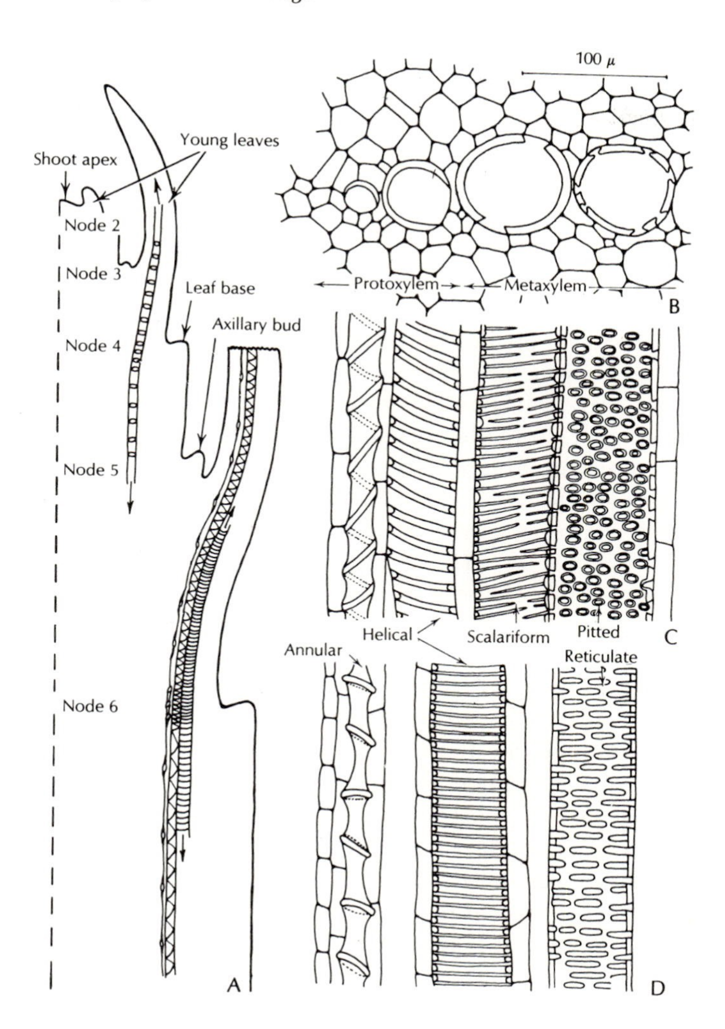

Fig. 12-7: The position and structure of tracheid cells. A. A stem (longitudinal section) with a growing tip at its upper end. Note the appearance of the new tracheid cells. B. The stem (cross section). C and D. Various kinds of tracheid cells (longitudinal sections).

The result is an elongation of the plant because one branch is the topmost. *Planation* describes the planar distribution of branches; that is, they tend to extend in one plane. The presumed advantage of this is to present a broad surface for the absorption of sunlight. *Webbing* is the appearance of leaf tissue between the branching stem tissue. This, too, aids in presenting a broad surface for absorbing light. Fern fronds formed from overtopping, planation, and webbing are called megaphylls; ferns are megaphyllous plants.

It is worth looking more closely at the tracheid cells of the xylem. These are the transport cells and, within the vascular plants as a whole, we find four kinds (Fig. 12–7). In ferns the most common kind of tracheid is the scalariform one, but ones with bordered pits are found in certain fern species.

Fern root structures are best considered prostrate stems, lying along the ground, from which fine filaments of cells extend into the earth or other substratum, such as rocks or even the bark of trees.

Ferns are world wide in distribution, but most typically are restricted to moist areas. One important reason for this restriction is that water—from rain, dew, or other sources—is necessary for fertilization to occur. The male gametes of ferns are flagellated and behave like actively swimming sperm cells. There must be water on the gametophyte surface to allow the sperm to swim from the structures where they are formed to the structures in which the nonmotile eggs are situated.

Most species have sporophytic fronds under 1 m tall, but there are still larger ferns growing in special tropical or semitropical habitats. In the late Paleozoic, there were more tree ferns, but evergreens and especially modern flowering shrubs and trees appear to have competed successfully with the tree ferns and replaced them as our dominant modern large plants.

Vascular plants with seeds. Pine trees and fruit trees are examples of seed-bearing vascular plants, and fruit trees are also flowering plants. Let us start with an examination of their sporophyte and gametophyte stages, and then discuss the major features of their stems, roots, and leaves, ending up with a special consideration of flowers.

Pine trees have both pollen-bearing cones and seed-bearing cones. Because of these cones the trees have their common name of conifers, or cone-bearers. The cones can occur separately, on different trees (a dioecious or "two-house" condition), or together on the same tree (a monoecius condition). In any case, the two kinds of cones are separate structures (Fig. 12–8). In each there occurs a haploid stage of development in the production of pollen and eggs and this is all that now remains of what we found as a gametophyte generation in ferns, mosses, or green algae. In the male, or pollen-producing cones, there are microsporangia. These are special structures located on the highly modified leaves that make up a cone. Within the microsporangia there are special cells called microgametocytes. (Their larger counterpart in the seed-producing cones are called, for obvious reasons, megagametocytes.) Meiosis in the gametocyte, followed by further special development, results in special pollen grains, each with two hollow sacs. These sacs help make the pollen airborne. Huge

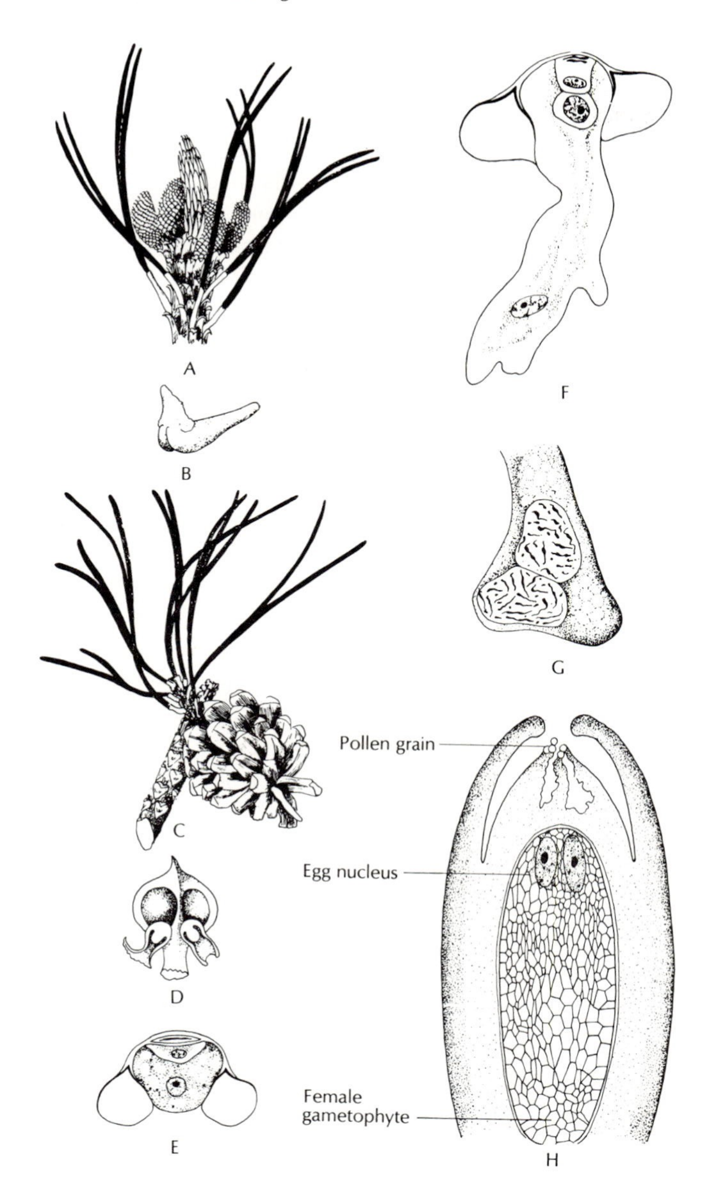

Fig. 12-8: Life cycle of a pine (*Pinus*). A. Pollen cones are clustered at the end of a shoot, which also bears a few needles (leaves). B. The microsporophyll, the male gametophyte, with pollen developing in sacs on its underside. C. The seed cone, which is the familiar pine cone, attached to a branch with needles. The megasporophyll, or female gametophyte, at the base of each cone scale, is formed inside the megasporangium. D. Two megasporangia, with an ovule in each, lie on the base of a scale. E and F. A germinating pollen grain. G. The two sperm nuclei of the pollen tube grow toward the female gamete. H. An early stage of seed development (cross section). Pollen grains stuck at the micropyle of the ovule will send out tubes to the female gametophyte. There, after the sperm nuclei of one tube has fertilized the female nuclei, the seed mature.

numbers of pollen are produced and, when ripe, are wafted by air currents. When pollen land on seed cones the ovules of which are ready to be fertilized, the pollen stick to the ovular openings and germinate. They proceed with development of a pollen tube and its special nuclei. This haploid structure is the male gametophyte, or microgametophyte. The female gametophyte, or megametophyte, lies within the megasporangium

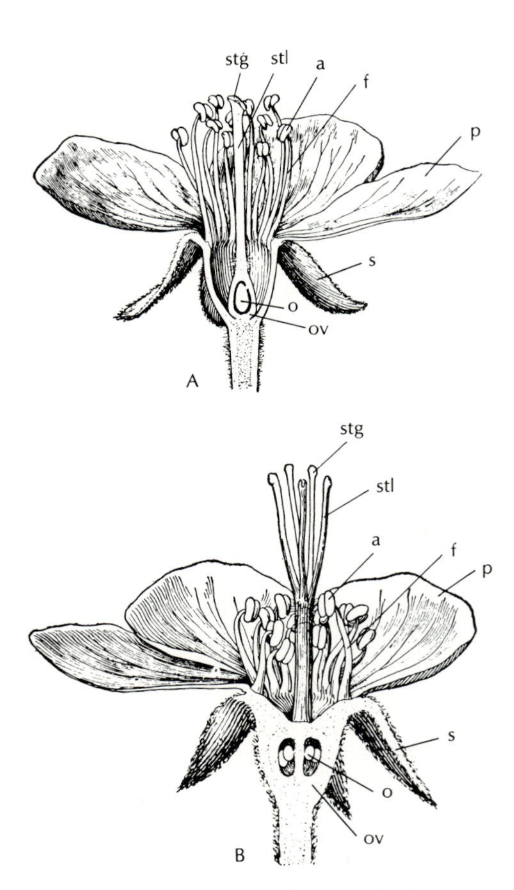

Fig. 12-9: Drawings of longitudinal sections through a plum flower (A) and an apple flower (B). Note the basic parts of a flower (s, sepals; p, petals; o, ovule; ov, ovary; stl, style; stg, stigma; f, filament; and a, anther). The ovary, style, and stigma make up the pistil, the filament and anther, a stamen.

of the seed cone. This seed cone is the familiar pine cone composed of woody scales set in a helical pattern around a central stem (Fig. 12–8). On the upper surface of each scale there will develop two seeds, which contain the new sporophyte; these are also edible pine nuts. Each developing seed comes from the megasporangium and its megasporocyte. The latter undergoes meiosis, and of the cells so produced one from each four becomes the haploid megagametophyte.

The pollen tube of the microgametophyte grows toward and finally reaches the megagametophyte. One cell from the pollen tube fuses with one egg cell, and this fertilization initiates the new sporophyte. This young sporophyte, the surrounding megagametophytic tissue, and some of the ovular tissue surrounding the megagametophyte (part of the parent sporophyte) together make up the pine seed. The megagametophyte provides stored food for the embryonic sporophyte. It is sometimes called an endosperm, but as we shall see it is not homologous to the endosperm of flowering plants. Both endosperms supply nutrients to the young sporophyte.

In the flowering plants (angiosperms), development is comparable. It differs from that just described largely because flowers rather than cones contain the gametophyte stage. A typical flower (Fig. 12–9) is dioecius, containing both male and female parts—the haploid microgametophyte and the haploid *embryo sac,* which is the female gametophyte (Fig. 12–10). The anther at the tip of a stamen contains microsporocytes. Each microsporocyte, when it undergoes meiosis, produces a tetrad of microspores, which germinate to form pollen grains. The pollen grain divides to form two haploid cells that together comprise the microgametophyte. When pollen grains are transferred—usually by insects or other animals, but also by the wind, depending on the species—to the stigma of a flower of the same species, the pollen can germinate and produce a pollen tube. One of the two initial cells in the pollen divides to form two *generative cells*. The total cell count in the microgametophyte is now three; the two generative cells enter the embryo sac.

The embryo sac is derived from megasoporocytes. Each of these undergoes meiosis to form four haploid products, three of which disappear leaving one megaspore. The megspore nucleus divides to produce eight nuclei. Three gather at each end of the developing embryo sac and form separate cell walls. The other two remain in the middle of the sac, which sac, as a result of the other six cells being formed, remains as a seventh

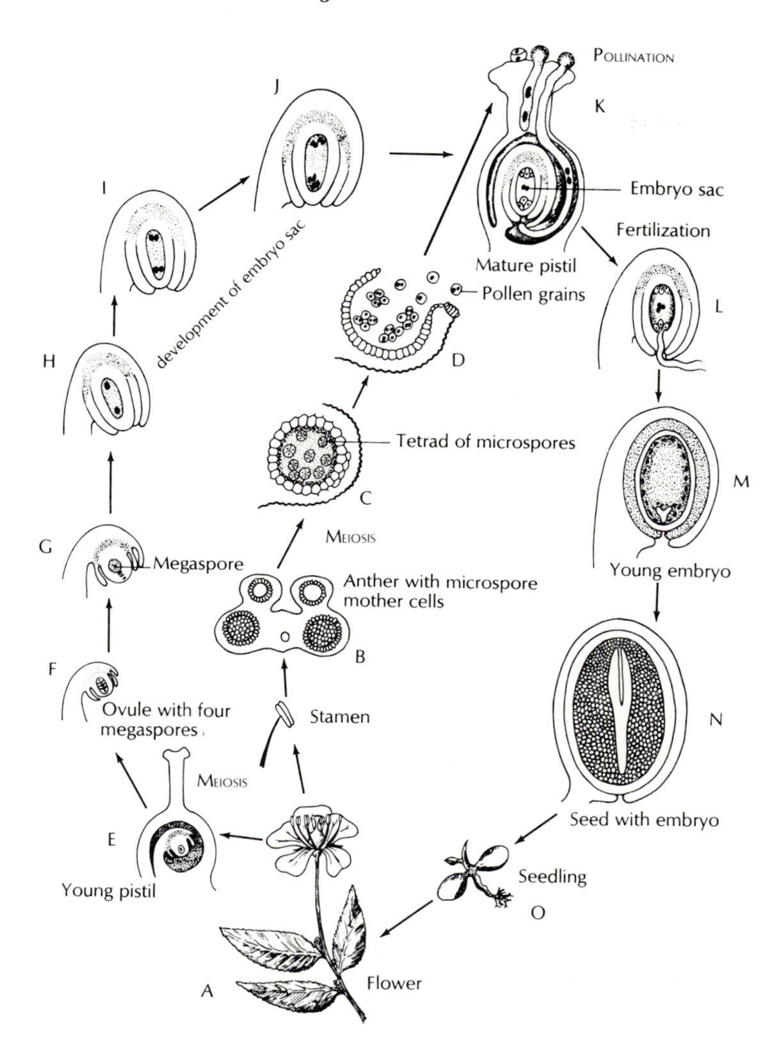

Fig. 12-10: Life cycle of a flowering plant. Both male (B through D) and female (E through J) gametophytes develop within the flower (A). The male gametophyte, which develops from the pollen, lies within the anthers and the microspore mother cell (B). Microspores (C) are formed and eventually release the pollen grains (D). An ovule forms within the young pistil (E). And within the ovule, a single megaspore develops (F and G). This megaspore develops further into the embryo sac (H, I, and J). When pollen lands on the stigma and extends a pollen tube toward the embryo sac (K), fertilization is imminent. It occurs (L), and growth of the embryo is initiated (M). The seed coat forms around the embryo to become a seed (N). When the seed germinates (O), the growth of the new plant has begun.

binucleate cell. All nuclei are haploid, and the embryo sac is the female gametophyte.

At fertilization, one of the two cells at the end of the embryo sac, the one closest to the pollen tube, functions as an egg. It allows entry by one of the generative cells of the pollen tube and its nucleus fuses with the generative nucleus. The other generative cell goes to the center of the embryo sac and fuses with the two haploid nuclei resting there. A diploid cell, which will form a new sporophyte, and a triploid cell, which divides rapidly to form the endosperm, result from this double fertilization. The seed, then, contains a diploid embryonic sporophyte, a triploid endosperm (which displaces the rest of the female gametophyte), and a seed coat derived from the diploid tissues of the parent sporophyte.

There are clear parallels between the gametophyte and sporophyte stages of cone-bearing and flowering seed plants. Both produce small, haploid, male and female gametophytes that depend on the parent sporophyte for survival. The seed contains an outer layer of parental tissue around endosperm and embryo. The diploid embryo is the start of a new sporophyte. The endosperm is derived from the haploid megagametophyte in conifers, but from new triploid tissue in flowering plants. Thus, the endosperms are not homologous in conifers and flowering plants.

Now we will consider stems, roots, and leaves before turning to a more detailed consideration of flowers. The stem supports a plant and connects roots and leaves. Hence, it has two principal functions: (1) support and (2) transport of metabolites. Transport is carried out by the vascular tissues of xylem and phloem; support depends largely on highly developed cell walls. Figure 12–7 illustrates various kinds of tracheid cells present in the xylem of vascular plants. In the confiers, tracheids with bordered pits are most common. They also occur in the flowering plants, but most commonly there is another cell type called *vessels*. These are tubular structures aligned end to end. Originally, these were single cells, but their boundaries disappeared where the adjacent cells touched, and the final result is a continuous transport tube.

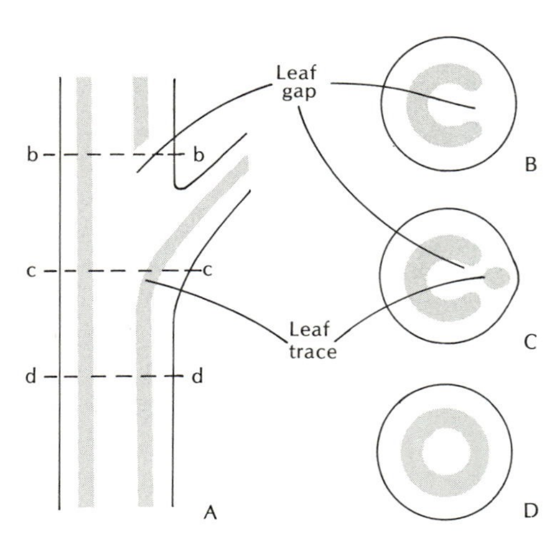

Fig. 12-11: Diagrams of traces and gaps in a branch with a leaf. A. A node where a leaf is extending to the right of the stem. B, C, and D. Sections through the stem shown in A at levels b-b, c-c, and d-d, respectively. The vascular tissue, shown as a shaded ring in D, is interrupted in B and C.

The many arrangements possible for xylem and phloem will be omitted here, as will further details on their development in young plants. Both are intriguing topics, but of more relevance to the anatomy and detailed phylogeny of various seed plant taxa then to the larger overall trends being surveyed here.

Worth mentioning in regard to stem patterns and vascular tissues is the presence of nodes and internodes. Nodes are the sites of branching of stems and internodes are the intervals between these sites. In the fern, leaf traces in the xylem are correlated with the presence of side branches in the stem. In the seed-bearing plants, the association of the central cylinder of conducting tissues is often interrupted by traces connecting with branches or leaves (Fig. 12–11). The patterns formed by these connections vary systematically with different groups of plants. For example, ferns often show only one trace to one of the lateral extensions of the fronds; in conifers, there are one or two; and in flowering plants, there are one, three, five, or more. In the flowering plants, the occurrence of one trace is thought to be reduction, by fusion, of at least three traces or loss of two out of three traces.

Roots function to anchor plants firmly in the ground and to absorb water and nutrients dissolved in the ground water. They are stemlike in their construction, but they lack nodes. The arrangement of conducting tissues in the root is like that found in the stem of the plant. In the flowering plants, or angiosperms, the arrangement of xylem and phloem varies considerably, from species to species. This variation, as we have just said, is useful in understanding the details of phylogeny within the flowering plants. As can be seen from our comments thus far, the xylem, with its different tracheids and vessels and its arrangement and relation to the phloem, represents a very informative aposemic or changeable character. Such characters are essential to deciphering evolutionary history (see Chapter 9). Roots transport absorbed nutrients to the rest of the plant and bring the metabolites formed in leaves to the root tissues. Absorption of water and its dissolved material is carried out by root hairs. These are lateral extensions (up to 1,500 μm) of single cells in the outer or epidermal layer of the root (Fig. 12–12). Root hairs persist for only a few days, and then are lost and replaced by new hairs from other epidermal cells. Roots do not have a cuticle or a stomata, structures typical of leaves.

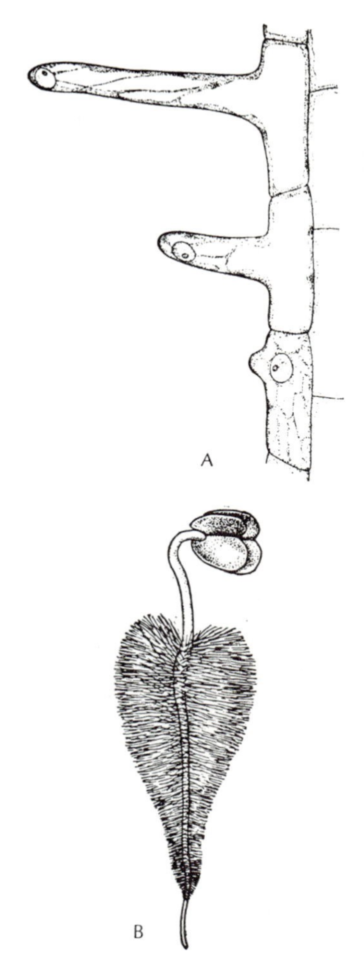

Fig. 12-12: Root hairs. A. A section of root showing the growth of single epidermal cells into root hairs. B. Extensive lateral growth of root hairs from a radish seedling growing in moist air.

Leaves function as the primary site of photosynthesis. In the conifers, leaves are seen as needles—long needles in pines, short needles in spruce trees, and so on—and in the flowering plants, they are of various sizes and shapes, but usually flattened to provide a broad surface for light absorption. Leaves in seed plants are termed microphylls, in ferns, megaphylls. Microphylls differ from megaphylls in the way the leaf arises—here as a short outgrowth from the surface tissues of the stem. The outgrowth is not always planar, since it can arise on any side of the stem. (The pattern of this outgrowth varies among different species.) And it seems that leaf traces, that is, extensions of vascular tissues, grow out into these outgrowths. In this way incipient leaves precede vascular tissue, rather than the other way around, as seen in the webbing of megaphylls. The prefixes *micro* and *mega* are misleading. They have essentially nothing to do with size (megaphyllous leaves and microphyllous leaves can be the same size), but refer to the origin of the leaf. Despite separate origins, similar patterns of broad flat expanses of photosynthetic tissues develop in the fern frond and in leaves on twigs. This is a clear case of convergent evolution (Fig. 12–13).

Flowers are now widely interpreted as being a stem with variously modified leaves. Actually, a better interpretation is to compare a flower to a young shoot with leaves in various stages of development. The Russian botanist Takhtajan, from the Botanical Institute in Leningrad, makes a convincing case for flowers as neotenous structures. By that he means they are stages of early development that persist into the adult. Or, conversely, that (sexual) maturity has arrived early. In the case of flowers, Takhtajan and others before him have concluded that flowers (Fig. 12–9) are derived from leaves. The green sepals—obvious at the base of roses, where the flower arises from its stem—are especially leaf-like. The venation (arrangement of veins) of petals is often reminiscent of leaf venation. And even such specialized structures as stamens and the pistil have leaf traces extending into them. Commonly, flowers are dioecius, since they contain both male and female reproductive parts—they produce the male gametophyte from the pollen and the female gametophyte or embryo sac from the megasporocyte. However, some flowers are monoecius, with flowers of each sex occurring on different plants.

Fig. 12-13: Convergence in the leaf of a flowering plant and the fronds of a fern. A leaf from ragweed (*Ambrosia artemisiifolia*) (A); a frond of a rattlesnake fern (*Botrychium virginianum*) (B).

EVOLUTIONARY TRENDS AND PHYLOGENY IN THE METAPHYTA

Aposemic traits are most informative when we come to tracing the course of evolution. Neosemic traits are also useful, but more so as markers for the initiation of an innovation than for tracing lines of historical development.

In the foregoing summaries of structures and life cycles, it became apparent that the Metaphyta show a number of aposemic or variable traits as well as certain neosemic ones. And among neosemes, such as vascular tissues, seed, and flowers, these also show aposemy. According to the methodology for phylogenetic analysis given earlier (Chapter 9), once homologies have been identified by positional and compositional relationships (direct similarities that identify plesiosemes) and by serial relationship (indirect similarities that identify aposemes), we must proceed to the designation of a plesiomorph. (Neosemes show no homologies.) The plesiomorph is the actual form (fossil or living, embryonic or adult) that most closely resembles the ancestral form or forms of the group or groups in question. Let us now look for one or more metaphytan plesiomorphs, and then, from that starting point, see what can be said about evolutionary trends, in general, and phylogenetic relations, in particular, within the Metaphyta.

Metaphytan plesiomorphs. The ancestor of the metaphytans must have been eukaryotic, multicellular, photosynthetic, and aquatic. The first three characters are the most conservative or plesiosemic ones we can find in this kingdom. And the last character, though aposemic, since the metaphytans occur in marine and freshwaters as well as on land, is plesiosemic among those forms arguably close to ancestral forms. The multicellular algae are the plants that come most quickly to mind in response to these four requirements for metaphytan ancestry. What can we say about the red, the brown, and certain green algae in this regard?

The characters given in Table 12–2 and our previous discussion of these algae suggest that they are not especially closely related. First, regarding the red algae, their lack of any motile cells and unique photosynthetic pigments place them closer to the prokaryotic blue-green algae than to the eukaryotes. It is thought that the red algae arose from the blue-green algae; either through endosymbiosis or through transformation of a prokaryotic alga into a eukaryotic red algae. The problem here is much the same as for the origin of the protistans, in particular, the origin of the green algae. The plesiomorph of the red algae would have a relatively simple thallus and a simple life cycle. A possible candidate might be *Porphyra perforator*. It must be emphasized, however, that there is a large gap—the gap separating prokaryotes and eukaryotes—between this red alga and the blue-green algae.

Turning next to the brown algae, we have a very difficult problem. These organisms suggest some similarity to the protistan golden algae (Chrysophycophyta). (Compare data from Table 12–2 and Table 11–3.) But there is no series of homologous semes that allows a convincing ancestor-descendant relationship to be established between the brown and the golden algae. There is a gap here; not as large as that between prokaryotes and eukaryotes, but nonetheless large. The first aspect of the problem is that there is no good candidate for a plesiomorph for the brown algae. All the brown algae are multicellular and have fairly complex to very complex life cycles. It is as if all the simpler ones have lost out by competition to the highly evolved ones. The result is that forms suggestive of the ancestral form are missing. The second problem is that we do not find any multicellular protists in the golden or other algae

Fig. 12-14: An early vascular plant, the rhyniophyte *Rhynia gwynne-vaughani,* a possible plesiomorph of vascular plants.

that show convincing homologies with the brown algae. The photosynthetic pigments of the Phaeophycophyta, Chrysophycophyta, and Pyrrophycophyta are somewhat alike, but only in a general way. Photosynthetic products, cell wall chemistry, occurrence of flagella, and patterns of multicellular organization are not arguably homologous. We do not really know yet what to say about the ancestor and origin of the brown algae except that the origin was probably from the Protista and that the ancestor and its related forms are extinct.

Thus far, we see the two metaphytan algal phyla as having separate origins. What about the rest of the Metaphyta, the multicellular terrestrial plants? Again, from information given earlier (Table 12–2), we see good evidence for homologies in the pigments, photosynthetic products, and cell walls of these plants and the green algae (Chlorophycophyta). The problem now comes down to locating a plesiomorph for the land plants and then seeing if it can be homologized to any of the multicellular, aquatic green algae.

One outstanding candidate for the plesiomorph of the vascular land plants is *Rhynia gywnne-vaughani* (Fig. 12–14). This is a fossil plant of the Silurian and Devonian periods of the Paleozoic (about 400×10^6 years ago). This plant is described by paleobotanists as a leafless, rootless, branching stem. Part of the stem was prostrate on the ground and from it there extended tufts of slender filaments—*not* roots—into the ground. Presumably they took up water and minerals. The aerial part of the plant had only one specialized structure at the end of some of the stems, interpreted as sporangia, for the formation of spores. Within the sporangia are the expected tetrads of cells, a characteristic of spores. These structures identify the plant as a sporophyte. Stomata are seen on the stems and this suggests that the stems were photosynthetic. Furthermore, the fossils are so well preserved that a very simple vascular bundle, or stele, can be identified in the center of the stem. Tracheids of the stele are even identifiable as annular (Fig. 12–7). They make up the xylem, and surrounding it is the recognizable phloem.

Except for such plants as the duckweed *Lemna,* no other vascular plant is as simply organized as *Rhynia*. In the case of *Lemna,* there is every reason to think of its simplicity as being due to a reduction of parts from much more complex plants.

Now *Rhynia* as a plesiomorph poses at least three problems: (1) How did more complex vascular plants, with roots and leaves and gametophyte as well as sporophyte stages, arise from it? (2) What was ancestral to *Rhynia*? (3) *Rhynia* may well be the plesiomorph for the vascular plants, but what about a plesiomorph for the nonvascular mosses and liverworts?

Rhynia *and the subsequent evolution of vascular plants.* Ideally, in a careful phylogenetic analysis, we want to find the actual plants that serve as intermediates between *Rhynia* and present-day vascular plants. This would fulfill the serial relationship and indirectly establish homologies between *Rhynia* and such plants as the ferns, conifers, and flowering plants described above. More specifically, we need to show the evolutionary development of roots and leaves. These would be neosemic traits, since both are absent in *Rhynia.* When both arise, they would then develop aposemically to reach the stages seen in the ferns and the seed plants. Since a stem with a stele is already present in *Rhynia,* we must look for its aposemic development into forms with more complex tracheids and leaf traces before we reach the level of organization manifested by ferns and seed plants. Similarly, we want to see how sporophytes and gametophytes evolved, aposemically.

Starting with roots and leaves, we find a truly useful series of forms intermediate to *Rhynia* and other vascular plants (Fig. 12–15). In this series, it should be emphasized, we are in all probability not looking at a series of direct descendants of one plant from the other, but at plants specialized in their own right and representing side branches from some main line of evolution that produced today's vascular plants. These intermediates mark the general direction of evolutionary change, rather than being actual forms that arose sequentially one from the other. For reasons discussed in Chapter 10, it is most unlikely that we shall find many missing links like *Archaeopteryx* (Fig. 10–15A), which convincingly displays both reptilian and avian characters. Admittedly, though, fossil forms are the most likely place to find them. This series going from *Rhynia* through *Sawdonia, Asteroxylon,* to the club mosses or lycopods, is a remarkably complete documentation of neosemic innovation and aposemic changes subsequent to it.

In terms of leaves, *Sawdonia* has pointed extrusions on its stem, but they are not connected to vascular tissues; hence, they are not true leaves. In *Asteroxylon,* the lateral extensions of the vascular strands look like leaf traces. They extend from the central stele toward the leaflike scales. And then, in present-day club mosses of the genus *Lycopodium,* we see true vascularization of the scale-like growths on the stem; they are leaves.

Roots emerged gradually from prostrate stems into struc-

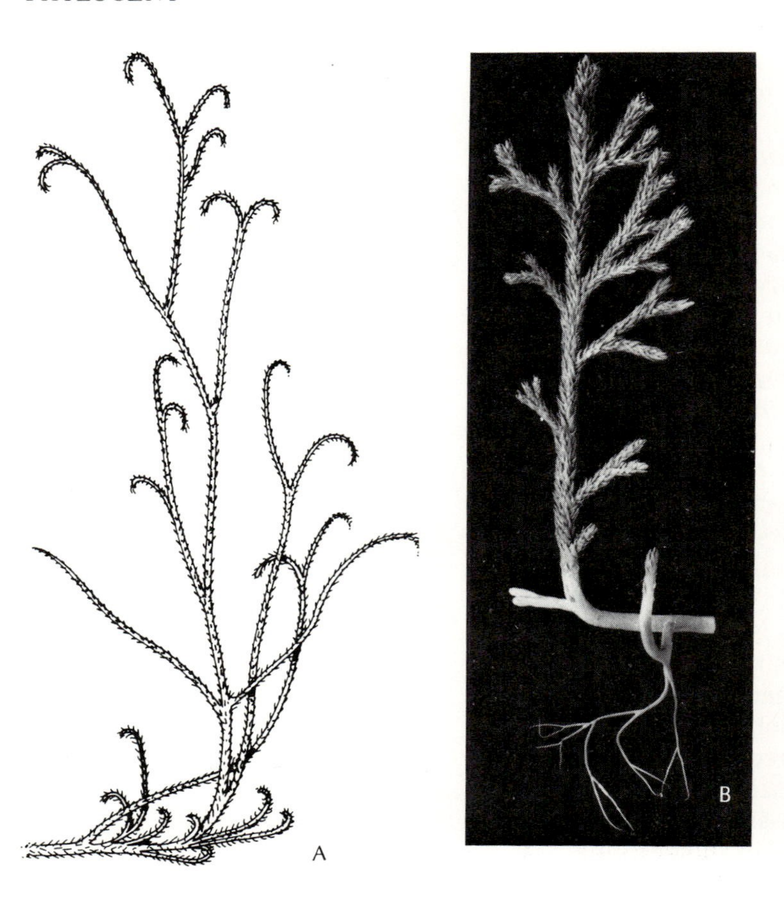

Fig. 12-15: A selection of fossil and living plants representing aposemic change from *Rhynia* to modern vascular plants. *Sawdonia* (A); *Asteroxylon mackiei* (B); *Lycopodium lucidulum* (C); and *Lycopodium obscurum* (D).

tures adapted to root functions, i.e., uptake of water and nutrients, and anchoring of plants. Though we referred to the root as a neosemic structure—it is absent in *Rhynia* and present in all seed plants—it apparently emerged by the gradual transformation of stem-like parts.

It is worth noting that no structure or function emerges suddenly. Evolution proceeds by the slow accumulation of advantageous mutational changes. Complex structures and their functions are the result of many such changes (thousands or more) and to expect that they would occur simultaneously and produce a new functional part of an organism that is adaptively significant, is so improbable as to be dismissed. Rather, a slow accumulation of changes, each one beneficial in terms of survival, is the tactic used by evolution. Sometimes complex structures having one function can change their function, and thus selection for one function becomes a preadaptation for another. This is precisely what we think happened in the evolution of the root. It evolved first as a stem and then, when prostrate, began to function as a root. That being selectively advantageous, it evolved further and specialized as a root.

Asteroxylon (Fig. 12–15) shows stem-like structures that clearly look as if they extended into the soil. In the club mosses we see genuine roots. They have the specialized structures and associated functions characteristic of roots. Some place in the line of evolution between *Asteroxylon* and *Lycopodium,* real root structures evolved. This appears neosemic because we look at the ends of the series, i.e., forms without true roots and then forms with roots.

The tracheids of the vascular tissue (Fig. 12–7) were evolving throughout this sequence, going from annular through helical and scalariform to ones with bordered pits and, eventually, to vessels. *Rhynia* has annular tracheids. *Asteroxylon* has helical ones; the lycopods and some ferns have scalariform ones; bordered pits are frequent in seed plants; and vessels characterize flowering plants. Furthermore, it is important to know that in the embryonic development of plants, annular, helical, and scalariform tracheids can all appear, and in that order. It is clearly a case of an evolutionary history persisting in the development of an individual. This recapitulation, in which ontogeny recapitulates phylogeny, was discussed in Chapter 9. In the nineteenth century, Ernst Haeckel pronounced this to be the Biogenetic Law. As we said earlier, further work has shown many exceptions to this "law," but here, in tracheid development, is one of its more obvious manifestations.

Also regarding vascular tissues, there is convincing aposemic evidence in the changes shown by the organization of the xylem and phloem, the significance of pith, etc. These details cannot be covered, here, but are amply documented in texts that cover plant anatomy and its evolution. (See references at the end of this chapter.) Finally, it can be noted that certain flowering plants, such as certain cacti, entirely lack vessels. This is another case of simplification through reduction and loss.

Last, we consider the evolution of gametophyte and sporophyte stages. The plants shown in Fig. 12–15 are all sporophytes. In the sporangia of the lycopods there is a gametophyte stage, but whether gametophytes occur in *Asteroxylon, Sawdonia,* and *Rhynia* is still conjectural. That is, although there is every reason to believe they occurred in those plants, they have not as yet been identified. A different problem is the two different patterns of occurrence of gametophytes and sporophytes in the ferns and club mosses. In the ferns, there is a separate plant for each stage; in the club mosses, the gametophyte develops within the sporophyte, as it does in the seed plants. In ferns the situation appears to be a dead end—it has gone no further than what we see. The ferns are an example of *homosporous plants,* i.e., plants that produce only one type of spore. This spore germinates into a gametophyte with male gametes in its antheridia and female gametes in its archegonia.

In contrast, *heterosporous plants* have two kinds of spores. Megaspores formed in a megasporangium produce a megagametophyte, or female gametophyte. Correspondingly, microspores formed in a microsporangium produce the male gametophyte, or microgametophyte. In the special club mosses of the genus *Selaginella,* one plant contains both microsporangia and megasporangia. Within these the two types of spores produce the microgametophytes and megagametophytes. The

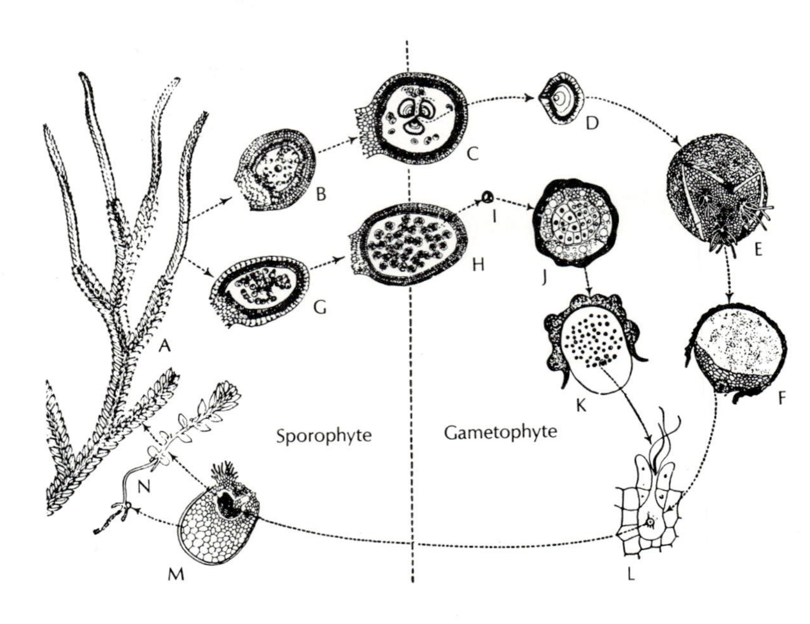

Fig. 12-16: Life cycle of the spike moss *Selaginella.* (Not a true moss, the spike moss is closely related to lycopods. See Fig. 12-15, C and D.) A. The terminal ends of the sporophyte bear microsporangia and macrosporangia. B through F. The megasporangium produces the megaspore (D) and goes on to develop into the female gametophyte (F). G through K. The microsporangium produces microspores (I) and, finally, the male gametophyte, which produces a sperm-like gamete. L. Fertilization restores the diploid condition and initiates a new sporophyte. M and N. Growth of the new sporophyte from the remnants of the old female gametophyte.

sperm from the former swim, when water is available, to the egg from the latter, and fertilization occurs (Fig. 12–16) Here clearly the plant is dioecius, and the gametophytes depend on the sporophyte for survival, as in all seed plants. It seems that this dependence emerged as a result of the sporophyte being better adapted to survival on land; the gametophyte, then, retained only its most essential function, that is, gamete formation. All other aspects of the gametophyte stage were reduced and lost.

This reconstruction of evolutionary trends places the seed plants at the end of a phylogenetic sequence going from *Rhynia* through the fossil plants (Fig. 12–15A and B) to the club mosses. Ferns and their relatives are left as a side branch of the vascular plants; they are not on the main phylogenetic line giving rise to the seed plants and their flowering members.

The flowering plants. The emergence of the flowering plants is still something of a mystery. Darwin, in 1879, wrote his botanist friend Hooker that "the rapid development so far as we can judge of all the higher plants within recent geological time is an abominable mystery. I would like to see the whole problem solved." Seeds emerged as a further development of sporangia, for example, appearing within the cones of the conifers. Then seeds appeared within the flower. But where did the flowers come from?

A possible candidate for the plesiomorph of flowering plants is a magnolia such as *Magnolia soulangeana* (Fig. 12–17). As with all plesiomorphs, it shows a series of characters that are conservative and therefore representative of the ancestral condition, although a magnolia plant as a whole is not consistently a conservative plant. The actual ancestor of the flowering plant has not been found in the fossil record, and efforts are still being made to reconstruct it (Box 12–1). In terms of the flower, which is the critical feature here, we see a structure still bearing parts highly suggestive of evolution from a twig.

A long fascinating story can be summarized by looking at some of the trends in floral evolution (Box 12–1). This story tells us that flowers underwent an adaptive radiation and in some cases, reverted to pollination by wind, which probably occurred in the ancestral plants from which they somehow evolved. We say somehow because Darwin's "abominable mystery" is still with us. Statistical analyses, such as that by Kenneth Sporne, of the University of Cambridge, England, can tell

Fig. 12-17: Flower (A) and flower receptacle (B) of *Magnolia soulangeana*.

Box 12–1. Trends in floral evolution and the search for the primitive angiosperm.

As expected, we see a tremendous variety in flowers, since a flower represents an important innovation in seed plants. It is a neosemic trait and has diversified enormously with respect to fertilization and seed formation. Flowers show an almost bewildering variety of aposemic changes. Their variation includes size, color, and shape of the different parts of the flower as well as different attractants to insects in terms of odors and nectars. Also, some flowers are pollinated by birds, bats, or other animals, others by windborne pollen. Starting with the idea that a flower is a specialized twig, we see several trends in the evolution of flowers.

1. *Flowers have tended to flatten out.* The floral axis has telescoped. As it shortened, it became less stem-like and more an attractive display of parts adapted to pollination and seed production.

2. *Flowers have shifted from a helical to a cyclical arrangement of parts.* This is related to point 1, since leaves often arise in a spiral pattern around a vertical stem. As the stem shortened and the flower flattened, of necessity, other parts arose symmetrically around the floral axis. The sepals formed a whorl of parts followed by whorls of petals and stamens (Fig. 12–9).

3. *Flower parts became reduced in number.* Petal numbers were often reduced to three, four, or five, and petals often fused so that their apparent number became as low as one (Fig. 8–2).

4. *Radial symmetry of the flower sometimes changed to bilateral symmetry.* This was often related to fusion of parts, to produce trumpets of the petals, or to the pollinating agent. Box 8–1 and Fig. 8–2 show this relation to such agents.

5. *Fusion and multiplication of parts occurred throughout the flower.* Fusion of petals occurred, but so did multiplication of petals. "Double" flowers are prized horticulturally, but often appear to be at a disadvantage in nature, e.g., most wild roses have relatively few petals, compared to florists' varieties. Stamens can fuse as can pistils, and sometimes the two are associated in one complex structure.

6. *The relative position of floral parts can change.* Usually the pistil lies above the point of attachment of petals and stamen, but it may lie below (Fig. 12–9). Fusion of parts (point 5) can make interpretation of the relative positions of flower components difficult.

7. *Flowers can aggregate.* Flower clusters are common, and in some cases, the clusters are differentiated into parts that only carry petals and that surround an

Hibiscus flower with fused pistil and stamen

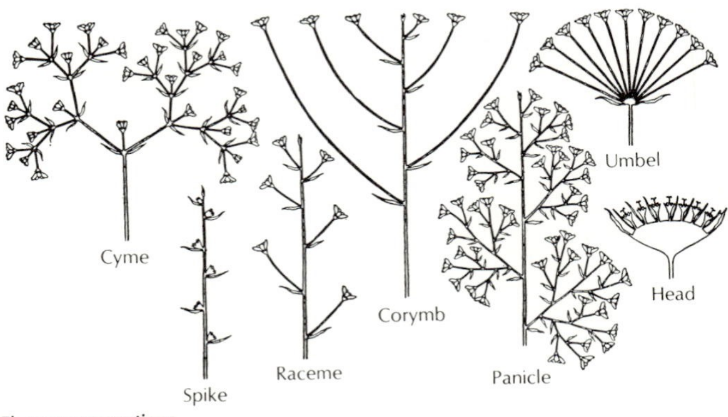

Flower aggregations

inner area of many separate flowers. The sun-flower, the daisy, and dogwood flowers are examples of these complex flower heads. Other patterns of aggregation are seen in the accompanying figure. A selective advantage of aggregates is that a single visit by a pollinator secures many different pollinations. Furthermore, flower clusters may be more visible to pollinators than single flowers.

From considerations such as the foregoing, we can deduce what is probably the ancestral type of flower. And primitive characters other than floral ones are being looked for. Kenneth Sporne, of Cambridge University, England, has been a leader in this type of study. He first determined that characters thought to be primitive occurred simultaneously more often than they would occur on the basis of chance alone. Upon statistical analysis (two characters present together, two absent together, and one or the other present while the other was absent) he found, indeed, that the frequencies at which they occurred together were significant. But this would also be true for advanced characters and, therefore, how

Character correlations in dicotyledons	1 Woody habit	2 Scalariform end-plates	3 Scalariform side-wall	4 Apotracheal parenchyma	5 Unstoreyed wood	6 Leaves alternate	7 Stipules present	8 Secretory cells	9 Leuco-anthocyanins	10 Flowers unisexual	11 Flowers actinomorphic	12 Petals free	13 Stamens pleiomerous	14 Carpels pleiomerous	15 Carpels free	16 Axile placentation	17 Pollen binucleate	18 Seeds arillate	19 Two integuments	20 Integument bundles	21 Ovules crassinucellate	22 Endosperm nuclear	23 Pollen pauciaperturate	24 Tapetum glandular
1 Woody habit		+		■	■	+		+	+	+	+	+	+	+		+	+		+		+	+	+	+
2 Scalariform end-plates	+		+	+	+				+		+						+						+	
3 Scalariform side-wall		+			+					+		+				+								
4 Apotracheal parenchyma	■	+			+	+			+		+		+				+			+		–	+	
5 Unstoreyed wood	■	+	+	+						+													+	
6 Leaves alternate	+			+									+					+				+		
7 Stipules present								+	+			+	+	+		+		+	+		+	+		
8 Secretory cells	+						+		+		+	+	+	+					+	+	+	+		
9 Leuco-anthocyanins	+	+		+			+	+		+	+	+	+				+		+	+	+	+		
10 Flowers unisexual	+		+		+				+		+										+			
11 Flowers actinomorphic	+	+		+				+	+	+		+	+	+					+		+		+	+
12 Petals free	+		+				+	+	+		+		+	+		+		+	+		+	+	+	+
13 Stamens pleiomerous	+			+		+	+	+	+		+	+		+	+			+	+	+	+	+		
14 Carpels pleiomerous	+						+	+			+	+	+		+	+		+	+		+		+	
15 Carpels free													+	+		■			+				+	
16 Axile placentation	+		+				+					+		+	■								+	
17 Pollen binucleate	+	+		+					+															+
18 Seeds arillate						+	+					+	+	+					+	+	+	+		
19 Two integuments	+						+	+	+		+	+	+	+	+			+			+	+		
20 Integument bundles				+				+	+				+					+			+	+		
21 Ovules crassinucellate	+						+	+	+	+	+	+	+	+				+	+	+		+		
22 Endosperm nuclear	+			–		+	+	+	+			+	+					+	+	+	+			
23 Pollen pauciaperturate	+	+		+	+						+	+		+	+	+								
24 Tapetum glandular											+	+					+							
Pre-Oligocene (Muller)	+	+		+			+	+					+						+	+		+	–	
Pre-Tertiary (CGH)	+			+		+	+	+	+	+							+			+	+	+		

Box 12–1 (continued)

could one be sure these characters were primitive? Here Sporne turned to fossil plants to see how often his chosen characters occurred in the fossil record. Correlations between the traits in question and their presence in the fossil record were significant. It now seems apparent that Sporne has found a complex of primitive traits. Their occurring together is given in the preceding, where + is a positive correlation, −, a negative correlation; blank square, no correlation; a black square, a meaningless correlation. [Redrawn by permission from K. R. Sporne (1977) in Beck (1977), Fig. 1.]

Note the bottom two lines of this chart. They show whether or not there is a correlation (positive, negative, or none) between the 24 characters and their occurrence in pre-Oligocene material (over 38,000,000 years old) and in pre-Tertiary material (over 65,000,000 years ago. As the plus signs show, many of the characters are correlated.

Sporne goes on to derive an "advancement index," which is a measure of evolutionary advance in terms of these 24 characters. It assesses the primitive nature of a taxon such that on a scale of 100, a low number designates a relatively primitive group and a high number an advanced group. For example, the family containing the magnolias (Magnoliaceae) is the lowest of the angiosperms, with a score of 20. Certain mangroves are also very low. This kind of analysis has allowed Sporne to draw a novel phylogenetic diagram (p. 423).

Although the details are not clear in this diagram because Sporne used standard but unfamiliar taxonomic terms, the concept of evolutionary advancement is quite clear. First, we see that each taxon demarcated by a continuous solid line is an order and that the orders contained within a dotted line comprise a subclass of flowering plants. The family Magnoliaceae is contained within the order Magnoliales, which is contained within the subclass Magnoliidae. Now note that, in addition to the Magnoliales, only three other orders come close, but do not touch the advancement index line of 20. These three also contain rather primitive forms. In fact, each class contains orders showing varying degrees of evolutionary advancement. This gives a very clear impression of the mosaic nature of evolutionary change in terms of different groups having different degrees of innovation.

Finally, note the absence of forms at the center of the diagram. Truly primitive or ancestral forms are missing. This is another way of pointing to the gap between the magnolias as plesiomorphic angiosperms and their ancestors. It also leaves unanswered the question whether there was one or more origins of flowering plants from ancestral forms. Some botanists argue strongly for a polyphyletic origin of the angiosperms. Sporne believes the non-flowering cycads are "the most promising candidates" for the forerunners of the flowering plants, largely because of the organization of their sporangia relative to carpel structure.

Chemical evidence, such as the presence of flavonoids, is being used phy-

us that a plausible ancestor for the flowering plants is the curious group of plants known as the cycads. They are not conifers, but members of a group that seems to have had a parallel, but somewhat independent evolution. But that still begs the question of how they arose. That must be answered in selectionist terms. Something about the flowering plants gave them a survival advantage over other plants.

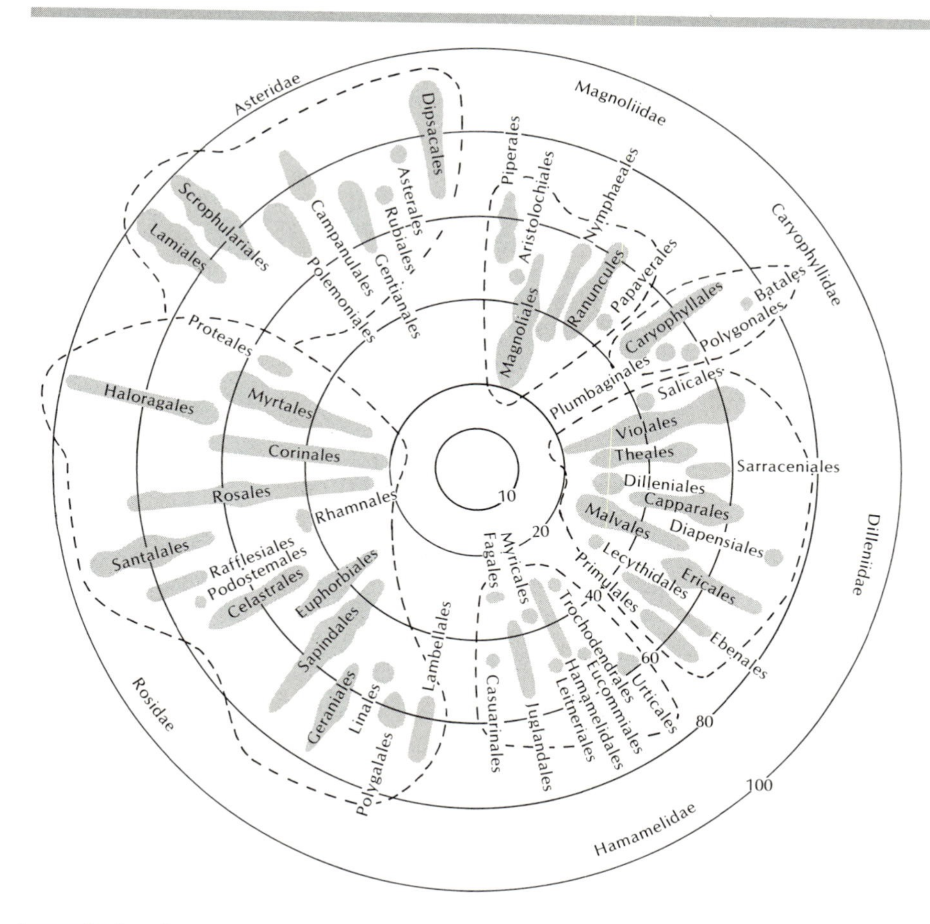

logenetically. These organic compounds occur in the vast majority of vascular plants and show great variation (an aposemic character); they are under genetic control regarding their synthesis, and they are easy to handle chemically. The flavonoids are already used to help define species and genera; how much they will add to the evolutionary story remains to be seen.

A recent approach to this question is that of Philip Regal, of the University of Minnesota. He starts at an obvious point, i.e., the flower. This feature of these plants is obviously unique and neosemic, and it would not exist unless it had a selective advantage. Other angiosperm features can be found to one degree or another in other seed plants. These include broad leaves, substances that repel predators, and vessels instead of the simpler tracheids. The special success of the angiosperms must be connected with their flowers, for why else do they and no other plants have flowers? Regal's answer depends on both pollination and seed dispersal. He argues cogently that insect pollination is a much more precise mechanism for ensuring pollination than wind pollination. Many insects are species specific for the plants they pollinate (Box 8–1 and Fig. 8–2). But this is not the whole story, since flying insects were present in the Paleozoic and flowering plants only emerged in the early Mesozoic. That brings seed dispersal in the picture. Birds and mammals are primary agents in spreading seeds, through eating colored fruits and eliminating undigested seeds and in carrying seeds as they stick to feathers and fur. Birds and mammals emerged in the late Mesozoic as dominant animals. Actually, Regal feels that the two features of animal pollination and seed dispersal, important as they are, have to be augmented with still other features. These include a feedback between new flowers, with their nectars and fruits and new nectivorous and frugivorous animals. This accelerated the evolution of both the plants and the animals. As new angiosperms evolved, the preexisting conifers and their relatives were reduced in number through competition and suffered further loss through the inadequacy of wind pollination, whereas insect pollination ensured adequate gene flow and genetic diversity.

All in all, Regal's view of angiosperm evolution is a complex of selective advantages arising from floral evolution that allowed initially herbaceous plants to exploit a variety of scattered niches. They could, however, effectively cross-breed, thanks to pollinating insects, and continue to have seed spread widely, thanks to birds and mammals. Thus this nexus of different selective advantages gave them an edge in survival, which they have not relinquished to this day. In fact, they have exploited it inexorably to take over a position of dominance among land plants.

This has been a long answer to the question of the connection between *Rhynia* as a plesiomorph and the rest of the vascular plants. The length was unavoidable, given the richness of the evolution of land plants. The remaining questions can be treated more briefly.

What was ancestral to Rhynia? The best answer available today is that a branching, multicellular green alga was ancestral to *Rhynia*. The French botanist Lignier advanced a useful theory that is consistent with what was just described regarding the evolution of the vascular plants (Fig. 12–18). Quite clearly, Lignier is looking for aposemic changes connecting an aquatic green alga to a plant that is adaptive to the land. The land plant he envisages is not unlike *Rhynia* with its prostrate and upright stems, the latter branching and leafless. Lignier's land plant seems to be evolving by overtopping and some webbing of delicate terminal branches. And there is development of a prostrate stem into a root-like structure. This green alga only vaguely resembles known examples because the regular branching he describes, called dichotomous, is common in the red and brown algae, but not in the green forms. Lignier's response might have been that the dichotomously branched green algae have evolved into land plants—they are today's examples of those forms. But this is not convincing evolutionary biology, since it says nothing about the selection pressures for the proposed evolutionary changes.

If we look at those pressures now, we have to consider the question of the invasion of land. Why and how would a well-adapted aquatic plant invade the land? The answer seems to lie in the realization that for freshwater algae—the green algae are the most common freshwater forms—their habitat, when it consists of shallow ponds or streams, often, and even annually, dries out. Adaptations for surviving desiccation would be advantageous, and such adaptations would be preadaptations for living on land. When we consider the evolution of green algae in such terms, it is not improbable that terrestrial forms evolved. But we have no fossil evidence or other data to document that evolutionary breakthrough. There is a gap. It could well be another case of tachytelic evolution, wherein a form adapted to one adaptive zone invades a new zone, evolutionary changes are rapid, and no fossils are found. Furthermore the intermediates are not really successful aquatic plants nor are they successful land plants. They lose out in competition to both. Hence, no intermediates survive. But we cannot, from present information, document Lignier's hypothesis.

This again illustrates the frustrations of phylogenetic research: the concept of evolution encourages us to look for phyletic series, but the action of natural selection tells us we must both expect and accept gaps. Perhaps the most disconcerting aspect of such a gap is our lack of insight into how the primitive transport tissue—the stele—of *Rhynia* arose. It is disconcerting for two reasons. First, despite the relative simplicity of the rhyniophyte stele (it is a thin strand of long, slender cells in the middle of the stem), it makes a rather sud-

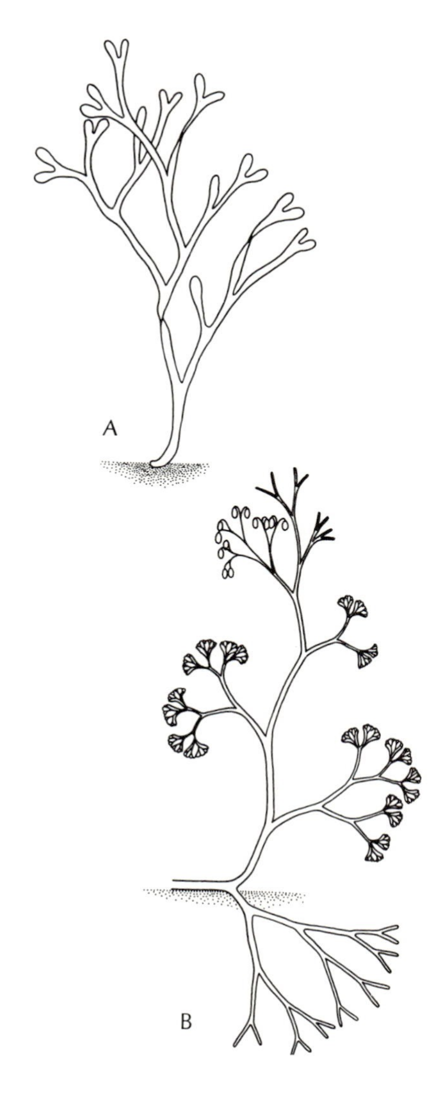

Fig. 12-18: Lignier's hypothesis on the evolution of land plants. A. Diagram of a branching green alga, a hypothetical algal ancestor of the land plants. Presumably this freshwater form lived in waters that showed an annual recession and necessitated, therefore, adaptations that anticipated terrestrial living. B. Another hypothetical plant derived from the alga (A). Here the upright part of the plant is growing by overtopping and webbing of the ends of the branches. Possible reproductive bodies are indicated on the upper branches. And a major branch, presumably the forerunner of a root system, extends into the soil.

den or genuinely neosemic appearance. It cannot be homologized with any green algae cells. Such apparently sudden and discontinuous changes, as we have emphasized, are inconsistent with the known process of evolutionary change. We can only hope that new data from living or fossil plants that represent a useful missing link here will become available. The second reason is the origin of the nonvascular plants.

The origin of mosses and liverworts. Phylogenetically, these known nonvascular plants are off on an evolutionary side branch, quite distinct from the other metaphytan species. The reasons for this are their lack of vascular tissue and the dependence of their sporophyte generation on their gametophytic one. Thus, although there are useful homologies between the green algae and these bryophytes as regards pigments, photosynthetic products, and cell walls, the course of evolution from the ancestral algae was probably independent of that taken by *Rhynia* and other vascular plants. This could even mean—and it is not unlikely—that land habitats were successfully invaded at least twice by descendents of the green algae. And these two invasions were based on two different kinds of adaptations, notably descendents with and without vascular tissues and their associated patterns of alternation of generations. But again there is a gap. The transitional forms between the green algae and ancestral mosses and liverworts are not in evidence today any more than are those forms between the green algae and vascular plants.

It may be that these gaps can be narrowed, if not closed, by demonstrating conservative or plesiosemic molecular characters. Molecular evolution might reveal changes that indicate important homologies between multicellular green algae and plesiomorphic vascular and nonvascular plants. Of course, a fossil plant like *Rhynia* will have few proteins in which amino acids can be sequenced, but the lycopods may turn out to be quite informative.

The Fungi

The second major group of multicellular organisms, the kingdom Fungi, is usefully discussed now, since it too arose in all probability from the protistan protophyta. But whereas the Metaphyta arose from photosynthetic forms, the Fungi arose from non-synthetic ancestors.

The Fungi are eukaryotic and predominantly multicellular decomposers, with diverse types of organization.

FUNGAL DIVERSITY

The bodies of these organisms are sometimes unicellular, but more often organized into filaments. Each filament is called a *hypha;* collectively, they make up a mycelium (Fig. 1–20). The mycelium, which may be highly branching, extends into the environment that nourishes the fungus. Typically, as decomposers, the fungi live in moist areas where there is an abundance of organic material. Hence their occurrence in dead wood or the litter on forest floors. But they also invade animal tissues—athlete's foot is caused by a fungus. They are severe problems in terms of food spoilage, since they cause molding of breads and vegetables, but they are also helpful in food production, especially in beer brewing and wine making, which depend on the fermentation of yeasts, and in certain cheeses. The colored patches in blue cheese are the result of fungal growth. In brief, fungi grow just about anywhere there is organic material that can be decomposed. Decomposition occurs through the release of enzymes that degrade substrates in the immediate vicinity of the hypha. These substrates are macromolecular constituents produced by other organisms because such molecules contain necessary building blocks for further growth of the fungi. Proteins supply amino acids, nucleic acids nucleotides, and carbohydrates and lipids sugars and other carbon compounds (Box 1–1). These smaller molecules are assimilated into the fungal cells and used there for vegetative growth.

Reproductive functions involve sexual reproduction as well as asexual spore-formation. Although the reproductive structures of the familiar puffballs, mushrooms, toadstools, and brackets of shelf fungi (Fig. 12–19) are rather complex structures, they constitute the lesser part of fungal growth. The usually invisible mycelium comprises the mass of the organism and is quite simply organized. Only in the reproductive structures does complexity approach that seen in the thalli of the red or the brown algae, for example.

Fig. 12-19: Reproductive structures in fungi. A. A cluster of spore-bearing brackets of the multi-zoned polystictus (*Polystictus versicolor*) grows on a dead black birch. B. The curious triplex earthstar (*Geaster triplex*), with its whitish central spore sac, grows on leaf litter.

FUNGAL PHYLOGENY

The fungi still present many unsolved phylogenetic problems. There is wide agreement that they are polyphyletic; that is, that there are as many plesiomorphs as there are independently originated groups of fungi. It also means that the evolutionary trends within this kingdom are complex.

Polyphylesis of the fungi. In a phenetic approach, the

fungi were classified into four divisions, i.e., Phycomycetes, Ascomycetes, Basidiomycetes, and Fungi Imperfecti. These divisions largely depended on reproductive structures, with the last being a catch-all for forms whose reproductive structures (and, therefore, life cycles) were unknown. This approach did not satisfy those who felt that classification should be more than just a way to store and retrieve information, i.e., that it should also reflect evolutionary history.

A phylogenetic classification is given in Table 12–3. We see here that the Phycomycetes are missing, and in their place are three other divisions. These are thought to be close to monophyletic groupings. The Fungi Imperfecti, now designated Deutoromycota, are accepted as probably having no sexual reproduction; they reproduce asexually by spore-formation and elongation of hyphae. Then three unusual groups are placed here, but there is no real conviction that they are genuine fungi (Fig. 12–20). The first of these are the lichens. There are 15,000 to 20,000 species of these organisms, which represent symbiotic relation between a fungus and an alga and show identifying features from the blue-green and green algae and from the Basidiomycota, Ascomycota, and Deuteromycota. Obviously, classification and evolutionary relations are difficult beyond the recognition that their status is special and that they are arbitrarily put in with the fungi. The other two groups are the cellular slime molds and the plasmodial slime molds, which seem to have independent origins. Both have a spore-forming stage that is fungal in appearance, which is now believed to represent convergent evolution. Both have an ameboid stage in their life cycle, during which feeding is by phagocytosis. In this stage their character is clearly that of

Table 12–3. Major divisions of the fungi and of forms often classified with them.

Division	Common name
Myxomycota	plasmodial slime molds
Acrasiomycota	cellular slime molds
Chytridiomycota	chytrids
Oomycota	many known as water molds
Zygomycota	(no common name)
Ascomycota	sac fungi (the molds)
Basidiomycota	puffballs, mushrooms, etc.
Deuteromycota	(no common name)
Mycophycophyta*	lichens

*This name was suggested by Bold, but not actually incorporated into his classification scheme. From H. C. Bold (1973).

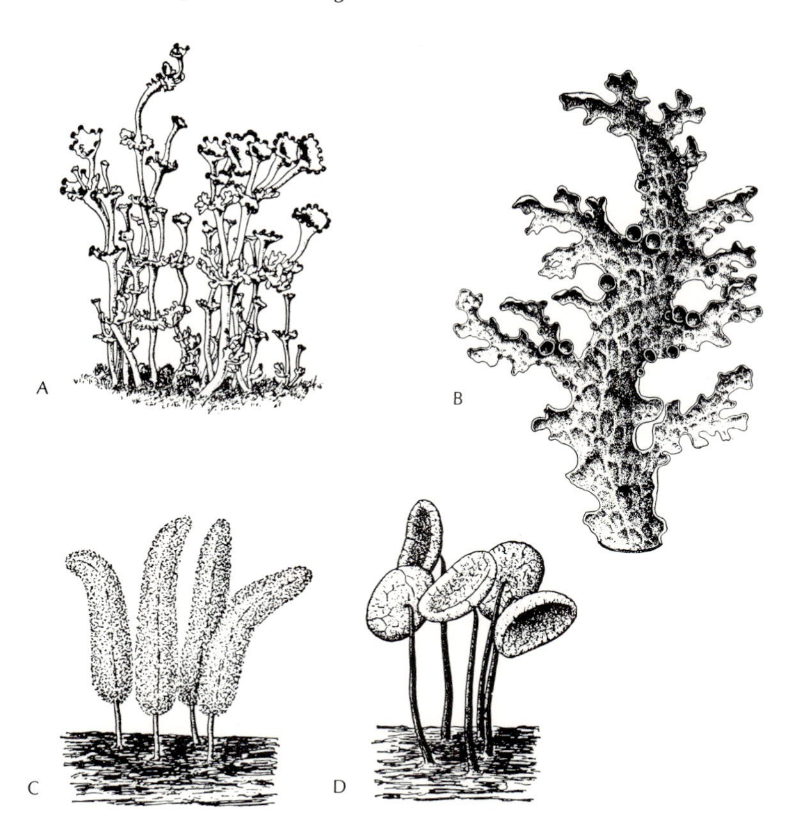

Fig. 12-20: Representative lichens and slime molds. The lichens *Cladonia verticillata* (A) and *Laboria pulmonaria* (B); the slime molds *Comatricha sp.* (C) and *Trichamophora sp.* (D).

unicellular ingestor or animal, which strongly indicates a protozoan nature. In fact, many schemes of protozoan classification include the slime molds. They are included here only because they are commonly classified among the fungi; it is an Adansonian solution to their taxonomic status, not an evolutionary one.

There remains the two major divisions of the Ascomycota and Basidiomycota. These, plus the former Phycomycophyta, are often combined into one division as the Eumycophyta or true fungi. It is within this grouping that phylogenetic trends, if they can be studied anywhere in the fungi, are most apparent.

TRENDS IN FUNGAL EVOLUTION

The so-called true fungi generally have a mycelial organization. This is probably best interpreted as parallel evolution, rather than any sort of trend. In the Oomycota and Zygomycota, the mycelial hyphae are continuous tubes without cross-

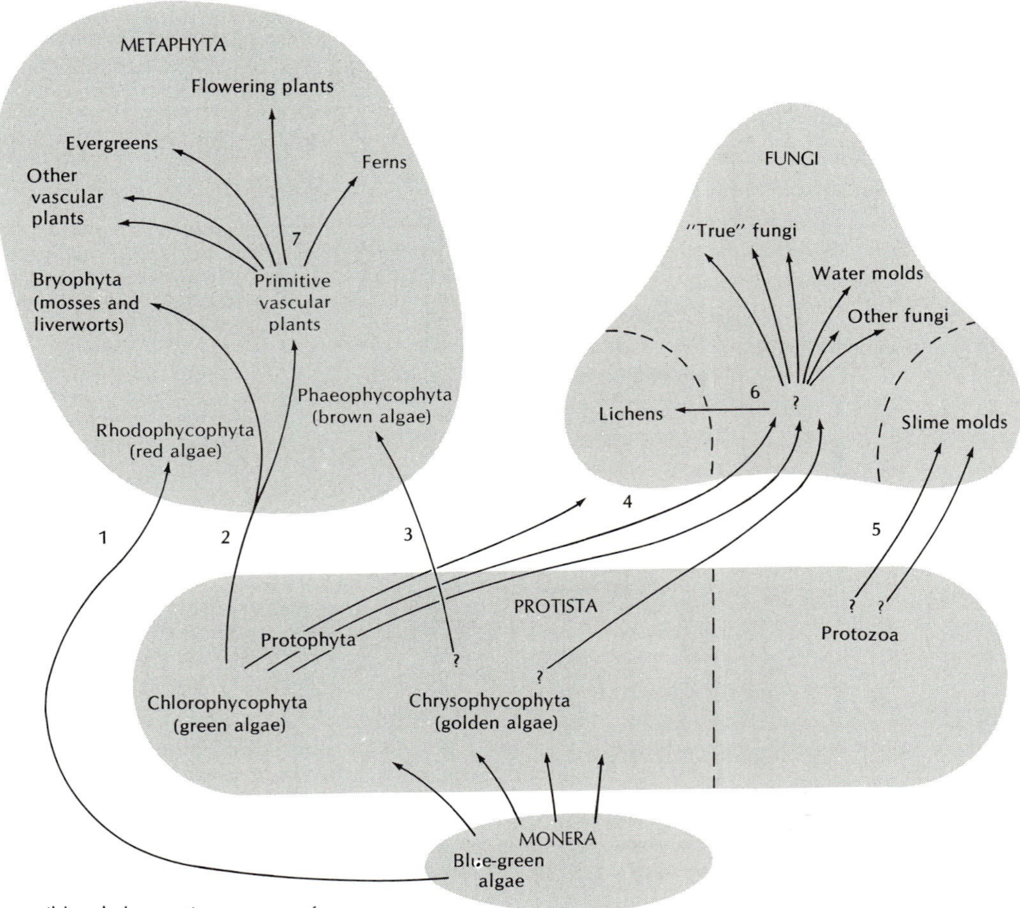

Fig. 12-21: A possible phylogenetic summary from the Metaphyta and Fungi. The following problem areas are (1) the origin of the red algae from the blue-green algae is based almost exclusively on the common occurrence of certain photosynthetic pigments; (2) there is a large gap between the aquatic green algae and the primitive terrestrial plants; (3) we have only meager evidence for believing that the brown algae arose from the golden algae; (4) fungi are clearly polyphyletic in origin; (5) slime molds may have originated from the protozoa; (6) phyletic relations among the various divisions of the fungi are obscure; and (7) the adaptive radiation of the vascular plants is only now being worked out through a combination of anatomical studies on fossil and living plants and through statistical and chemical analyses in many areas.

walls. They may have originated from the coenocytic green algae (Box 11–3) following a loss of plastids. The similarity of flagellated cells in certain green algae and in the Chytridiomycota and Oomycota suggest an origin from the Chlorophycophyta. Also, similarities in reproduction of the red algae and certain Ascomycota suggest that as a line of evolutionary change. And arguments are brought forward to homologize the reproductive structures of the Ascomycota and the Basidiomycota.

What does all this add up to?

For those willing to consider evolutionary trends as working hypotheses for phylogenetic work, the following statements can be made:

1. There were separate algal origins of the Oomycota, the Zygomycota, and the Chytridiomycota.
2. The Ascomycota arose from red algae and gave rise to the Basidiomycota.

A more conservative treatment is to withhold evolutionary pronouncements for the present and simply treat each of these groups as a possibly monophyletic division for which phyletic relations are as yet unknown.

The broad outlines of phylogenetic development considered in this chapter are summarized in Fig. 12–21.

References

Beck, C. B. (ed.), 1977. *Origin and Early Evolution of Angiosperms*. Columbia University Press, New York.

Bold, H. C., 1973. *The Morphology of Plants*. 3rd ed. Harper & Row, New York.

Crawford, D. J., 1978. Flavonoid chemistry and angiosperm evolution. *Botanical Review 44:* 431–456.

Esau, K., 1960. *Anatomy of Seed Plants*. New York.

Delevoryas, T., 1962. *Morphology and Evolution of Fossil Plants*. Holt, Rinehart and Winston, New York.

Delevoryas, T., 1977. *Plant Diversification*. 2nd ed. Holt, Rinehart and Winston, New York.

Dobzhansky, T., 1970. *Genetics of the Evolutionary Process*. Columbia University Press, New York.

Regal, P. J., 1977. Ecology and evolution of flowering plant dominance. *Science 196:* 622–629.

Sporne, K. R., 1971. *The Mysterious Origin of Flowering Plants*. Oxford Biology Reader. Oxford University Press, London.

Sporne, K. R., 1977. Character correlations among angiosperms and the importance of fossil evidence in assessing their significance. In Beck (see above). pp. 312–329.

Takhtajan, A., 1977. Neoteny and the origin of flowering plants. In Beck (see above). pp. 207–219.

THIRTEEN

The Metazoa: Invertebrate Animals

OUR DISCUSSIONS of animal evolution will focus on the central fact that animals are consumers; they ingest their food.

The origins of consumers, as unicellular animals or protozoa, were considered in Chapter 11. We return to that story now and continue it to include multicellular consumers, or the kingdom Metazoa. The diversity of the Metazoa is so great—they contain well over one million described species (Table 13–1 and Fig. 13–1)—that it will be best to divide their treatment into two parts. This chapter will consider the metazoans that lack a backbone, i.e., the invertebrates, and the next chapter will consider the metazoans that have backbones, i.e., the vertebrates and their evolutionary relatives (the phylum Chordata).

Invertebrate animals range in size from microscopic worms, free-living and symbiotic, to huge forms such as huge giant squids, which, including their longest tentacles, can be as much as 20 m in length. Within this size range, the variety of forms derives from a combination of various appendages attached to bodies that may or may not be segmented, but that are usually bilaterally symmetrical. Colors range throughout the possible spectrum and are combined in sometimes extraordinary ways. The invertebrates include insects and snails, star-

Table 13–1. The major animal phyla and the estimated numbers of their species.

Phylum	Living species	Fossil species	Total species	Described species
	From Muller and Campbell (1954)			From Dobzhansky (1970)
Protozoa	16,250	8,750	25,000*	28,350
Porifera	2,240	1,760	4,000*	4,800
Coelenterata**	4,500	4,500	9,000	5,300
Platyhelminthes	6,000	negligible	6,000	12,700
Nematoda†	6,000	negligible	6,000	11,000
Annelida	6,000	negligible	6,000	8,500
Arthropoda				838,000
Insecta	753,920	6,080	760,000*	
other arthropods	58,600	4,400	63,000	
Mollusca	41,610	31,390	73,000*	107,000
Echinodermata	950	4,050	5,000	6,000
Chordata	33,640	24,360	58,000*	43,000
Mammalia††	3,552	3,698	7,250*	
Minor phyla	3,000	3,000	6,000	6,850
total			1,028,250	1,071,500

*These are "phyla for which reasonable estimates were available," according to Muller and Campbell. Nonetheless, note the discrepancies with Dobzhansky's estimates and note, also, that Dobzhansky's figures come from sources published more than ten years after the appearance of Muller and Campbell's paper, during which thousands of new descriptions were published. This indicates the problems inherent in comparing such figures.
**Cnidaria.
†Roundworms, often included in the Aschelminthes.
††A class within the Chordata, but given here independently of that phylum.

fish and corals, and worms of all kinds—in all, about a million different species. And all of them are consumers.

Active consumers

The term predator is usually reserved for animals that hunt their prey. A lion preys on zebras and antelopes, an octopus often preys on crabs, many spiders prey on insects, and so on. We do not think of zebras or antelopes as preying on grass; nor do crabs that browse on algae or butterflies that obtain nectar from flowers strike us as predators. Nonetheless, in all these cases, a similar function is carried out; this includes locating, and orienting to food, attaining it by capture or simply moving up to it (to graze, for example), and then ingesting it. Predators are simply the most dramatic example of this set of common coordinated behaviors. And underlying these behaviors is a set of common structural features. Notably, there is a nervous system for sensing and coordinating; a muscular system for moving, which also demands some sort of skeleton for muscle attachment and for transforming muscular action into mechanical action. And these are all integrated with a functional mouth for ingestion.

EXTERNAL ANATOMY

A typical organizational plan that integrates functions and their underlying structures is bilateral symmetry and cephali-

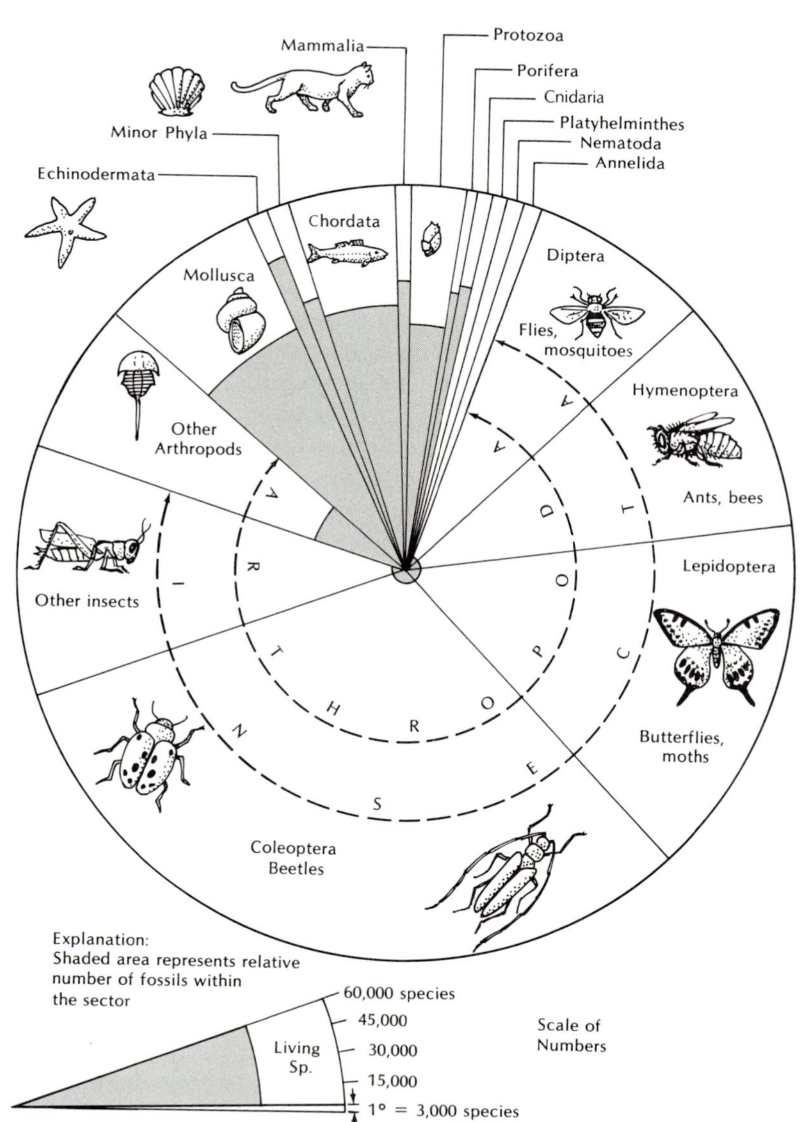

Fig. 13-1: The relative abundance of animal species, with the number of fossil and living species compared.

zation. Bilateral symmetry refers to a body that can be divided through only one plane into mirror halves (Fig. 1–6). Cephalization refers to the tendency to collect specific structures toward one end of an animal body, which structures, collectively, are called a head. These include various sense organs—eyes, ears, nose, or another olfactory organ—and the mouth. Let us now look at how both bilateral symmetry and cephalization are adapted to the life a consumer, and especially, a predator, leads.

A head is selectively advantageous for an organism that first searches out its food and then eats it. In most animals, it is that part of the body that first encounters the environment; thus, the head contains sense organs to inform the body where it is going and where the food is. Such organs are coordinated with the organs of locomotion through the nervous system. And obviously, for the above reasons, it will also include the organ of ingestion, the mouth and its associated parts. A head is adaptive over the range of habits shown by consumers, from herbivores to active carnivorous predators.

Bilateral symmetry makes adaptive good sense in much the same way. A freely moving consumer that knows up from down and right from left has an advantage in being oriented to its environment. The location of prey, orientation towards it, and controlled approach or pursuit is enhanced by bilateral symmetry. Appendages are seen as a further refinement of such a body. They add to the maneuverability and the feeding capabilities of the animal.

Thus far our discussion has viewed the metazoan externally. How is it internally organized?

INTERNAL ANATOMY

The major phyla of the invertebrates and the phylum Chordata are coelemate. That is these animals have an internal cavity or *coelom*. Strictly speaking a coelom is a cavity lined by mesoderm, the middle germ layer (Box 13–1). The other invertebrate phyla are either pseudo-coelemate (the cavity has mesoderm only partially surrounding it) or acoelemate (lacking a body cavity). The presence of a coelom then raises at least two questions: How did it arise? What is its adaptive significance?

As to the origin of the coelom, there have been many theories. It arose from the cavities present in an excretory organ and it is therefore a nephrocoel; or, it is a gonocoel, arising from a cavity in a gonad; or it might be a hemocoel, having originated from the blood sinuses. And so it goes: it seems that someone has proposed any internal cavity as a source of the coelom. Clearly, then, one task in our discussion is to investigate the merit of these proposals. To do so we will need to turn to the adaptive significance of the coelom. Complex structures and organization arise and are maintained through selection pressures. Hence we must ask, What need is served by the coelom? To answer that we must consider the organization and function of the whole invertebrate. This includes at least two more major anatomical features, namely, appendages and segmentation.

We have already mentioned *appendages* as being important to consumers for locomotion and food capture. (Some appendages, such as the antennae of insects, function as sense organs and other appendages function as gills. We will, for the most part, limit our attention to the locomotory and grasping appendages. This means we shall exclude shells, horns, feathers, and the like.) Externally, appendages are obvious extensions of the animal body. Internally they are intimately connected with the body musculature and the skeleton (if there is one) and supplied by nerves and blood vessels. Such complex structures must also have a significant evolutionary history. What is it? Is it connected with coelomic evolution? The answer appears to be a definite yes, as we shall see after looking at segmentation.

Segmentation is the transverse partitioning of a body into similar parts; it results in a repetition of body parts along the body's length. An earthworm is often cited as a good example of segmentation, displaying, as it does, dozens of segments both externally and internally. Internally these worms do show a formidable repetition of parts (Fig. 13–2), since most segments are each furnished with a part of the digestive and circulatory systems, a ganglion on the ventral nerve cord, an excretory organ, and muscles for extension and retraction of the locomotory bristles called setae. However, this repetition of parts is not complete. The anterior and posterior ends are specialized, and the reproductive structures are limited to certain segments. Also certain segments have specialized portions of the circulatory system, or hearts. But overall segmentation is an obvious part of the organization of these animals. Segmentation is also apparent in insects and other arthropod groups (see Chapter 1 and Appendix I). Hints of segmentation are also seen in chordates (the backbone, which consists of the vertebrae, and the attached ribs). So again, questions arise: How did segmentation originate? What is its adaptive value?

Box 13–1. Germ layer origins in animal development.

We have already commented on the importance of development in the expression of the genotype as the adult phenotype and the difficulties and complexities involved in understanding the expression of the epigenotype as a functionally adapted phenotype. The concept of *germ layers* is helpful here. Germ layers are specific layers of the early embryo that develop into tissues and organs. Germ layers, therefore, are intermediate to the molecular dimensions of the genotype, at one extreme, and the organismic dimensions of the adult phenotype, at the other.

Typically there are three germ layers: the outer *ectoderm,* the inner *endoderm,* and the middle *mesoderm.* These three layers are not apparent until the gastrula stage of development.

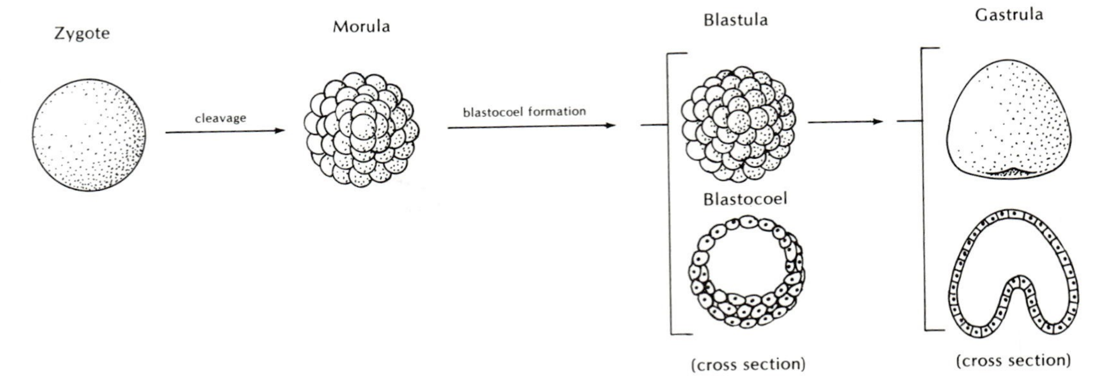

Early in development, the zygote undergoes cleavage to form, first, a solid ball of cells called the morula. Cleavage continues and a hollow space, the blastocoel appears among the cells. The embryo is now a blastula. Then one area of the blastula folds in or invaginates, which initiates the formation of the gastrula. The inner cells form the endoderm, which becomes the lining of the digestive system of the adult, and the outer cells form the ectoderm, which becomes the skin, the nervous system, and the sense organs; the middle mesoderm has two general patterns of formation, schizocoelous and enterocoelous, respectively, to produce muscles, supportive tissue, and linings of body cavities. As can be seen, the schizocoelous mode gives rise to mesoderm from special cells, which divide to form daughter cells that spread over the surfaces of the ectoderm and endoderm. The cavity they eventually enclose is the coelom.

In the enterocoelous mode, the mesoderm arises as an out-pocketing of the endoderm. Two pockets are formed, which pinch off and, by cell division, grow and expand to line the surfaces of the ectoderm and endoderm. The space enclosed by this mesoderm is also a coelom.

And can these questions be answered in conjunction with parallel questions relating to appendages and the coelom?

Before we grapple with these interrelated questions, we will consider one more idea for background information. This is the idea that the coelom, the appendages, and segmentation all apparently evolved within the Metazoa. In all three cases

The adult organs and organ systems arising from the germ layers are shown in the table below.

In Box 7–1, the development of a frog is shown; it differs from the diagrams given here, which apply to invertebrate animals. However, certain parallels do exist. Also in Box 7–1, figures 1–3 show early cleavage, 4 and 5, the blastula, and 6–10, the gastrula. In figure 7, invagination starts at the blastopore and the archenteron is the primitive cavity of the digestive system. In figure 10, at the top of the figure, the outermost ectoderm is clearly demarcated from the inner endoderm. Eventually the space between them will contain the neural tube (derived from ectoderm, see figures 11–13) and the mesoderm and its derivatives.

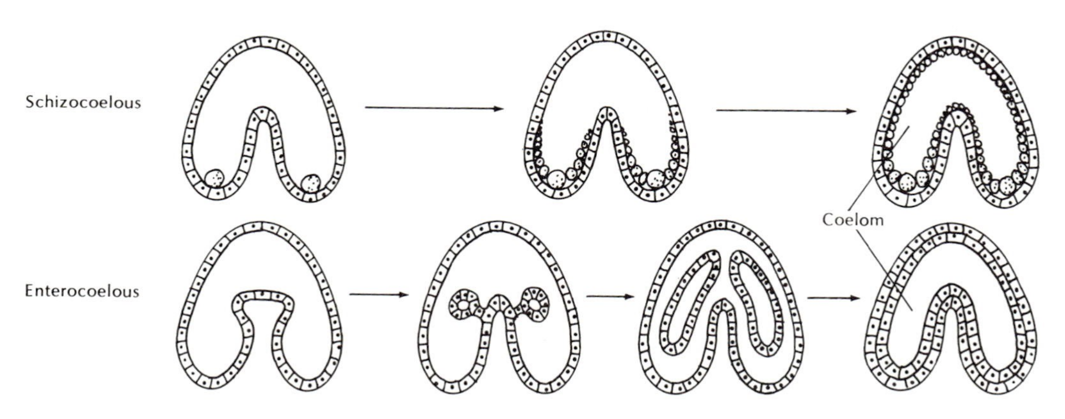

Germ layer	Adult structure	Position in the adult
ectoderm	epidermis and associated structures	outermost layers of the skin, linings of the oral and anal ends of the digestive system
endoderm	gastrodermis	lining of the digestive system (except at the oral and anal ends) and the lining of the lungs, liver, and certain other organs
mesoderm	(a) mesenchyme—loosely connected mass of cells	between epidermis and gastrodermis
	(b) well-differentiated mesoderm—various organs and organ systems	between epidermis and gastrodermis

we have metazoans without the feature in question: there are acoelomate invertebrates; invertebrates with no apparent appendages; and non-segmented invertebrates. We have, in effect, a neosemic change followed by an aposemic change of all three characters. Furthermore, the occurrence of these characters can be correlated—the more complex invertebrates have

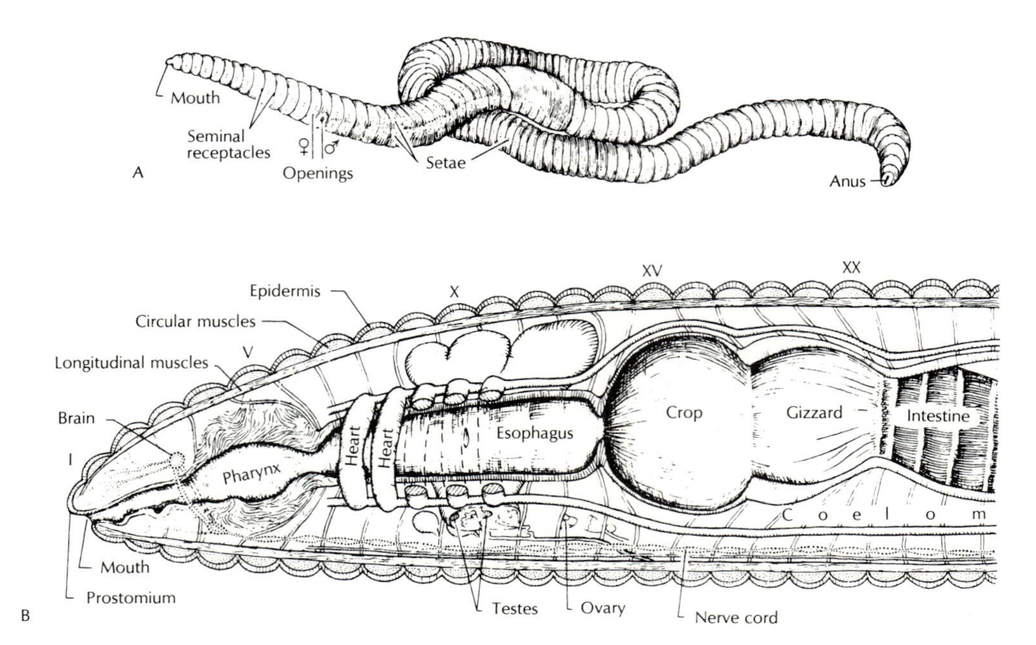

Fig. 13-2: Segmentation in the earthworm *Lumbricus terrestris*. A. This drawing of the whole worm shows segments along the length of the body. B. The distribution of various organs (note the two hearts) are shown in the cutaway drawing of the anterior segments I through XXIII.

variously developed coelomic cavities, appendages, and segmentation, whereas the apparently less complex invertebrates often lack all three of these important characters. These correlations lead to an obvious question: Is it possible that acoelomate, unsegmented forms lacking appendages evolved into coelomate, segmented forms with appendages?

That is precisely the question we shall try to answer. But in doing so, we must keep in mind that evolution does not always go from simpler to more complex forms. Recall the many cases of reduction found in the Metaphyta. Consider the possibility that symbiotic and sedentary animals, like corals and starfish (acoelomate and coelomate, respectively), are obviously not active predators so how then to explain their bilateral symmetry and cephalization? Recognize, also, that there may be many kinds of coelomic cavities (nephrocoels, gonocoels, etc.), and we know there are many kinds of appendages (jointed and unjointed appendages, fins, appendages with pincers or claws, tentacles, etc.) Therefore, although we have managed to reduce the problem to the correlated evolution of the coelom, the appendages, and the segments, a marvelously diverse array of possibilities still exists. In other words, although we can phrase our approach to the problem in a few words, the reality covered by the problem is still complex, covering as it does, the main evolutionary trends of the invertebrate animals. And we must always remember that evolution proceeds from one adaptively advantageous phenotype to the next (Box 13–2).

Box 13–2. Why have organisms not evolved wheels?

Answering this question is a useful exercise in understanding the limits to evolution. Though the innovative potential of natural selection is impressive, it cannot do just anything. The real limitation or constraint on evolutionary innovation is the process of natural selection.

A wheel is a circular structure able to rotate around its center on an axle, or if motive power is supplied through the axle, the axle and wheel are fixed together and rotate together. The latter is more complicated than the former, so let us set aside the latter and consider a simple wheel consisting of a fixed axle, and a movable rim that spins about the axle.

The first concern would be to ask if a wheel could be functional in an organism. It could be, given the appropriate terrain. Wheeled vehicles, produced through human cultural evolution, are extremely advantageous not only on roads, but on a variety of flatlands, e.g., grassy fields, sand, and snow. Animals also live in such areas. Perhaps wheels might be advantageous on sandy ocean floors, too. At least initially one can think that wheels, if they existed, could be a useful means of locomotion. (Perhaps even more useful would be moveable treads, such as those on bulldozers. But that is really a complication of wheel design. Let us stick to the simple wheel.) Again the question, Why are there no truly wheeled animals?

The best answer we can give is that not all the steps leading to a wheel will be of selective advantage. Put another way, we can say that natural selection and evolution have no foresight. What that means is that even though a certain structure, such as a wheel, might be functional as a fully developed structure, it will not evolve unless each step in its development is also of selective value. There is no conceivable advantage, say, to a disc-like fleshy part, with an endoskeleton, before they develop into a functional wheel. In fact, they are conceivably disadvantageous. Hence, the early stages in wheel evolution would be selected against, and therefore, a wheel would never evolve.

The essential lesson regarding the limits to evolution is this: For any new structure or function to evolve, *each* stage or step in its evolution must be selectively advantageous. Darwin faced this problem squarely when critics challenged him to explain the origin of the human eye through natural selection. The critics claimed that the image-forming capacity of an eye is only the product of a fully evolved eye. What then was the selective advantage of an intermediate stage in which images were not formed? Darwin's answer was, of course, that eyes evolved from light-sensitive organs in which the initial function was not image formation, but simply light detection, i.e., the primitive structure was simply a photoreceptor. To increase the efficiency of the photoreceptor, a collecting and focusing device would be of selective advantage and that would be a preadaptation for a lens. With a lens, the transition to an image-forming eye became a functional and selectively advantageous option. Hence, all stages in the evolution of a true eye are conceivably of selective advantage and the eye is, therefore, an evolvable structure.

Metazoan origins

According to the phylogenetic methodology being followed here (see Chapter 9), the way to read aposemic trends or determine whether a character is a neosemic emergence, or its absence is reductive loss, is to determine the plesiomorph of the group in question. Given the plesiomorph, we can refer the questions of loss or emergence and of direction of change to this representative of the primitive or ancestral form. This is what we will do now, so that a metazoan plesiomorph will tell us how to read the changes present in acoelomate or coelomate forms and in animals with or without appendages and segments.

THE METAZOAN PLESIOMORPH

Four lines of evidence are used, singly or in combination, to identify a plesiomorph. These are the fossil record, embryonic data, distributional information, and phenoclines. Here, the fossil record and distributional data are of no use. The fossil record does not help because all the major invertebrate phyla appeared at about the same time, all well evolved (Fig. 10–17). We cannot use that evidence to decide which group came first, except to eliminate terrestrial forms and the chordates. Their fossils appear much later than those of other metazoans. Distributional data are meaningless when we deal with groups as large as phyla; they are really useful at the levels of families, genera, and species.

Embryonic evidence is of some use, especially in terms of trends within the Metazoa. In Box 13–1 the origins of germ layers and different mechanisms of mesoderm development and coelom formation were described. The general conclusion, *if we take the law of recapitulation at face value*, is that acoelomate organisms preceded coelomate forms because the blastula, which is first a solid mass of cells, precedes the gastrula, which has a cavity. However, this law has to be interpreted carefully because development or ontogeny is notoriously variable. At best, conclusions drawn from embryonic data are merely suggestive, perhaps especially so when we consider as complex a problem as the metazoan plesiomorph.

This leaves only the phenocline approach, which, as we described it earlier (Fig. 9–13), involves an indirect, probabilistic argument. Basically, what we are doing here is looking at various aposemic changes and deciding which of them all center—if they do—in one species or in a closely related group of species. In particular, we ask, are there acoelomate, unsegmented, invertebrates without appendages? And, if the answer is yes, is there any other evidence, especially fossil, embryonic, or other phenoclinic, that does or does not support that invertebrate as a metazoan plesiomorph?

Three invertebrate phyla have traditionally been viewed as among the most primitive of the invertebrates. They are the Porifera (the sponges), the Cnidaria (the jellyfish, sea anemones, and related forms), and the Platyhelminthes (the flatworms, especially the free-living ones called Turbellaria). Using the more rigorous concept of plesiomorph rather than simply primitive form, we shall now see that a strong case can be made for the turbellarian flatworms as the metazoan plesiomorph.

Sponges as plesiomorphs. Sponges are filter feeders. They pass a continuous flow of water through their bodies and remove from that flow various food particles. The flow is maintained by the flagellated collar cells, which are located in internal chambers (Fig. 13–3). Sponges are acoelomate and they

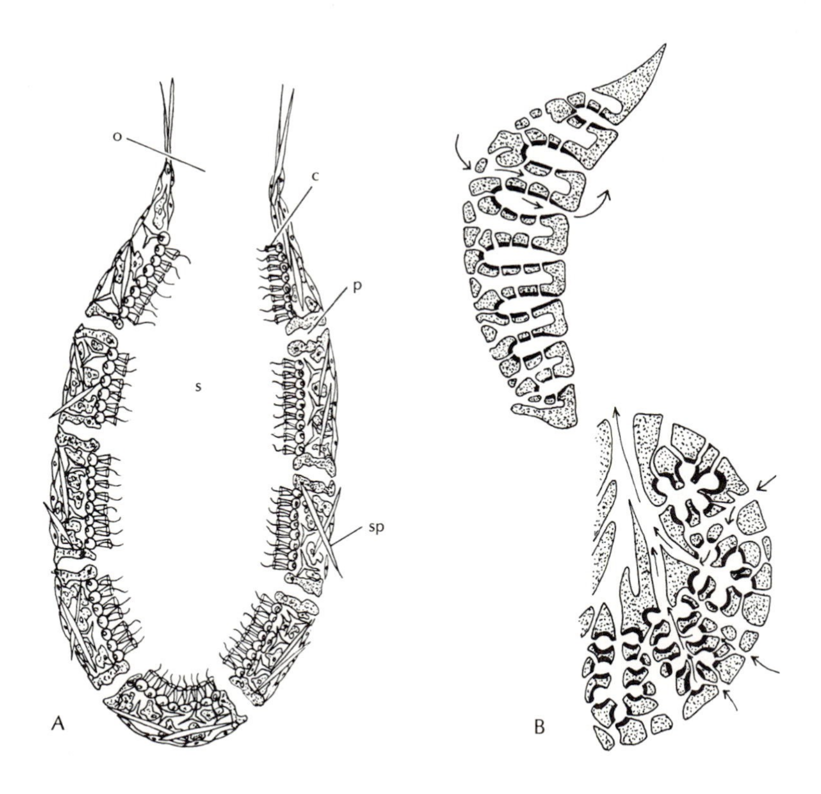

Fig. 13-3: Flagellated chamber of sponges. A. A simple, ascon-type sponge is shown in cross section (c, choanocyte or collared cell with flagellum; o, osculum or excurrent opening of the chamber; p, porocyte with an incurrent channel through it; s, spongocoel or cavity of the chamber; sp, spicule). B. Two complex sponge structures shown in cross section. A syconoid type (above) has flagellated chambers incorporated into the wall of the spongocoel. In a leuconoid type of organization (below), clusters of flagellated chambers feed into a common excurrent channel.

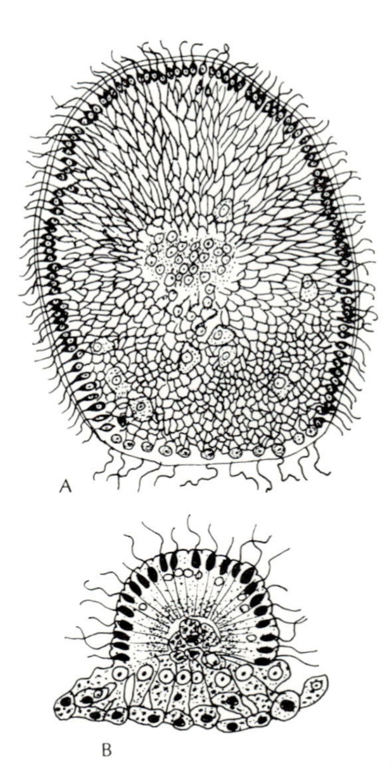

Fig. 13-4: Sponge larvae showing, especially, cell-type distribution. A. The parenchymella of *Halisarca* is totally surrounded by flagellated cells. B. The amphiblastula of *Leurosolenia* has flagellated cells surrounding one pole.

lack appendages and segmentation. Furthermore, they are either asymmetrical or radially symmetrical, and show no evidence of cephalization. It is true that the flow of water through their bodies gives them a polarization of inflow pores and one or a few outflow pores, but this is not what we mean by cephalization. The occurrence of a nervous system in the sponges is still open to question. The fact they are sedentary filter feeders means their locomotory and sensory capabilities are very poorly developed. They close their pores when stimulated either mechanically or chemically. Their dispersion is carried out by free-swimming larvae. Their embryology is complicated in a unique way. Their two major larval forms (Fig. 13–4) are found nowhere else among living things.

Sponges combine some of the features we expect in a metazoan plesiomorph, but beyond that they have features that are related to their sedentary habit as filter feeders that set them quite apart from other animals. Their uniqueness is emphasized by their special larval development. Thus they are not convincing candidates for the role of metazoan plesiomorph.

Cnidaria as plesiomorphs. There are two body forms here, the medusa and the polyp (Fig. 1–13). Both are radially symmetrical, with tentaculated appendages, and they are acoelomate and unsegmented. Food is captured by the tentacles and their stinging cells, furnished with nematocysts. The immobolized prey is drawn to the mouth, in the center of the body, and ingested. Digestion occurs in the gastrovascular cavity and undigested material is eliminated through the mouth. Development is from the fertilized egg, through a larval stage, called a planula, before the adult stage of polyp or medusa (Fig. 13–5). The zygote first forms a hollow blastula, with cleavage patterns differing among different cnidarians. This coeloblastula then becomes a solid gastrula (stereogastrula), which is the planula. The outer epidermal layer of the planula is made up of a single layer of cells, each cell having one cilium. These cilia allow the planula to swim about before it is transformed into a sedentary polyp or a slow-swimming medusa (Fig. 13–5). In this transformation, the planula settles down on its anterior pole, the solid internal mass of cells splits to form a cavity lined with endoderm (the future gastrodermis) (Box 13–1), and a mouth is formed at the former posterior end of the planula.

In summary, about the only thing that even vaguely recommends a cnidarian as a metazoan plesiomorph is that it is acoelomate and unsegmented. Beyond that it does have appen-

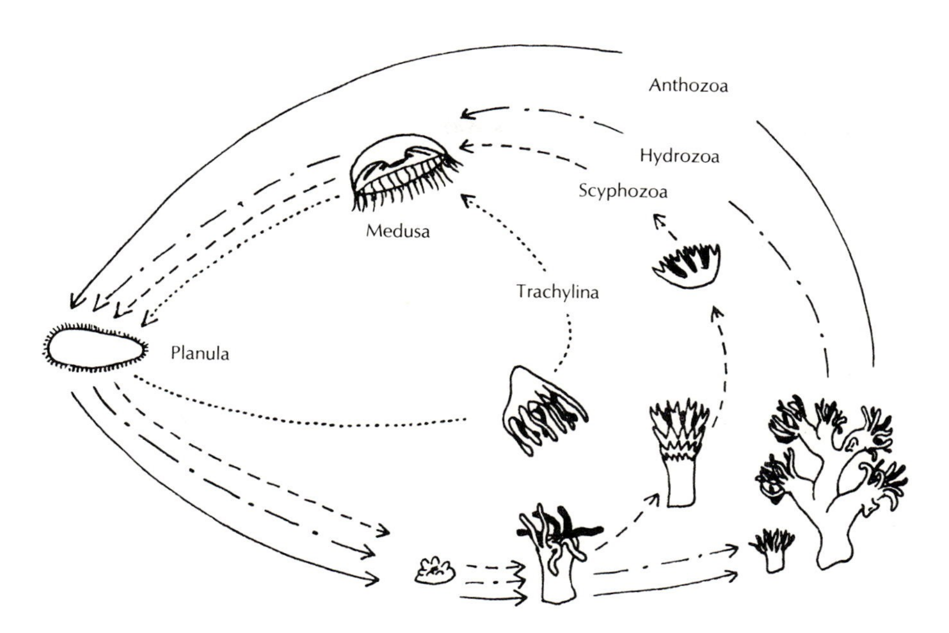

Fig. 13-5: Various cnidarian developmental or life cycles. Note how all cycles include the planula larva, but the fate of the larva differs, depending on the group being considered.

dages (tentacles), it is radially symmetrical, it is not really cephalized (the anterior end of the planula becomes the posterior end of the adult), and its development goes from a cavitated ball of cells to a solid gastrula, which, if it recapitulates anything, suggests a very complex phylogeny.

Turbellaria as plesiomorphs. These free-living flatworms are bilaterally symmetrical, clearly cephalized, acoelomate, unsegmented, and essentially lacking in appendages. They capture food by a mouth located either anteriorly or somewhere between the anterior end and a mid-ventral position. The epidermis is ciliated, with many cilia per cell that aid in locomotion. Muscles are present throughout the body and there is a digestive cavity, except in the order Acoela. This order lacks such a cavity (Fig. 13–6). Food is ingested through the mouth

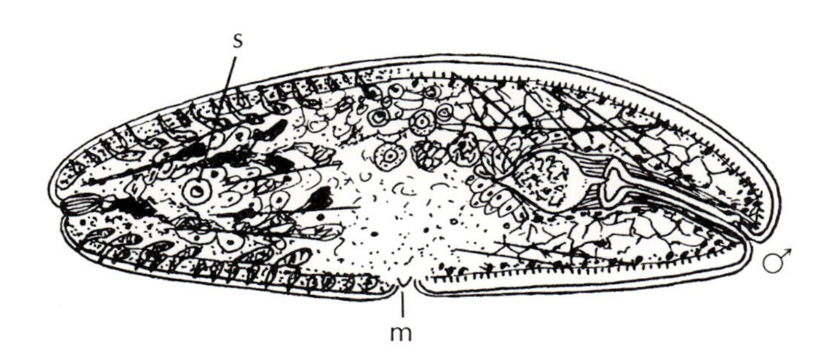

Fig. 13-6: The flatworm plesiomorph *Haploposthia rubra* (frontal section). The mouth (m) is on the ventral surface with a statocyst (s) embedded centrally toward the anterior end. The male genital opening lies at the posterior end (♂).

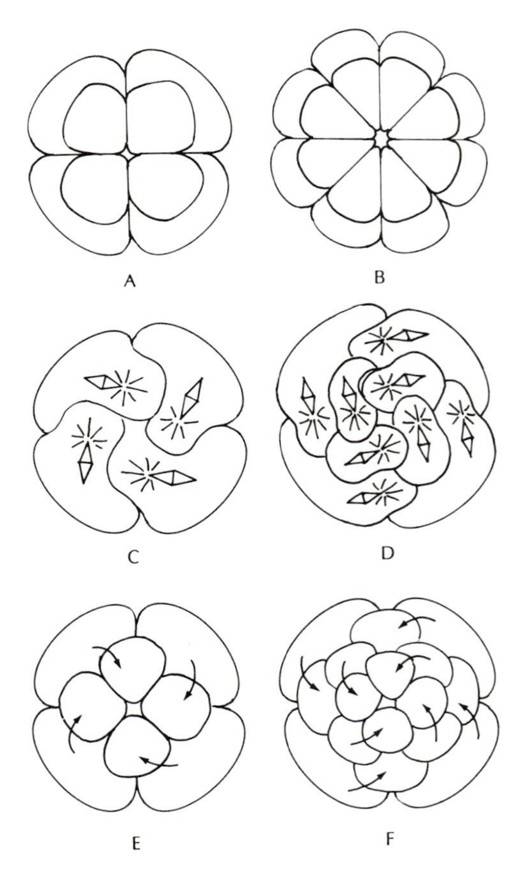

Fig. 13-7: Regular or radial cleavage (A and B) contrasted with spiral cleavage (C through F). Cleavage patterns are viewed from above, looking down on the animal pole of the zygote. The difference in the two cleavage patterns becomes apparent at the third cleavage. A and E. In spiral cleavage, there is a slight rotation of blastomeres (arrow). C and D. The spindles of dividing cells are seen here.

and is taken into food vacuoles, which are surrounded by a loose mass of cells (the inner parenchyma). The food of these worms ranges from unicellular algae to protozoa to other micrometazoa. Their development starts from a zygote, which undergoes spiral cleavage (Fig. 13–7), going on to form a solid blastula and then a solid stereogastrula. The stereogastrula develops into a young worm that leaves the gelatinous egg capsule and then starts feeding and matures, with internal differentiation of its male and female reproductive structures. These worms are hermaphroditic and when two of them come together to mate, they exchange sperm to cross-fertilize each other. The fertilized eggs are laid in clusters and develop as described above.

These turbellarian flatworms, and especially the acoel worms, thus, are good candidates for the metazoan plesiomorph, although the occurrence of spiral cleavage is puzzling.

THE EMERGENCE OF MULTICELLULAR ANIMALS

We discuss here our search for the origin of the metazoan plesiomorph before returning to our major problem, which is to determine how the coelom, segments, and appendages arose. As might be expected, there have been various suggestions as to how the acoel flatworms originated. First, we should ask what homologies, if any, can be found between these worms and other organisms. A rigorous search for homologies, in the terms discussed earlier (Chapter 9), suggests that a ciliated protozoa has some homologies with an acoel worm and there are also three important absences of homology (Table 13–2).

How can this combination of apparent homologies and the absence of homologies be explained? Our knowledge of how organisms diverge evolutionarily tells us there will be a mosaic of changes. Direct homologies relate plesiosemic characters, showing us that selection pressures have changed little regarding them. Indirect homologies relate aposemic characters where change in response to selection has occurred. And, finally, the absence of homologies identifies neosemic traits where new features have emerged in response to selection. If we recall that ciliates function as bilaterally symmetrical, anteriorly-posteriorly polarized micropredators, we see how cellularization, once it occurred, could go on to sexual reproduction and embryogenesis. In other words all absences of homology can be understood in terms of one innovative breakthrough. The argument goes like this:

1. The ciliates respond to the selection pressures exerted

Table 13–2. Homologies between the ciliate plesiomorph *Stephanopogon mesnili* and the flatworm plesiomorph *Haploposthia rubra*.

Semes	*Stephanopogon mesnili*	*Haploposthia rubra*
Probable Homologies:		
general body organization	bilateral symmetry; anterior-posterior polarity; surface ciliation; ectoplasm and inner and outer endoplasm	bilateral symmetry; anterior-posterior polarity; surface ciliation; epidermis and inner and outer parenchyma
nuclei	all diploid except for meiotic products	all diploid except for meiotic products
digestion	surface mouth producing food vacuoles	surface mouth producing food vacuoles
secretary and extrusible structures	trichocysts in ectoplasm	sagittocysts in epidermis
feeding	ingestion of algae	ingestion of algae
sexual processes	conjugation and cross-fertilization	copulation and cross-fertilization
Absence of Homologies:		
cellular organization	one complex cell	internal cellular organization
sensory and coordinatory structures	not identified	nerve net and "brain"
reproductive structures	haploid nuclei	haploid gametes and female genitalia
reproduction	asexual	sexual and embryogenesis

These probable homologies reflect plesiosemes and aposemes to differing degrees in the six semes listed. The absence of homologies in the remaining semes indicates neosemic innovations, all arguably derived from the one fundamental innovation of internal cellularization.

on any predator, i.e., there is a premium on locomotory-sensory coordination and on increase in size, so as to exploit larger organisms for food.

2. Increases in complexity and size would be facilitated by cellularization of the body. This can occur in a multinucleate form through the appearance of membranes around the separate nuclei.

3. A further consequence of internal cellularization of a formerly multinucleate cell body is that gametes, also, become cellularized, so that sexual reproduction is possible. Since male and female gametes have the potential to become separate cells, and to produce separate or independent zygote; thus

sex and reproduction can now combine as sexual reproduction.

4. The undifferentiated zygote must undergo embryogenesis to achieve its adult form.

In this way, a multinucleate ciliate can evolve into a multicellular acoel flatworm. All the changes are selectively advantageous to a further exploitation of the predaceous mode of life. Note that *Stephanopogon mesnili* (Fig. 11–7F) is our best candidate for the ciliate plesiomorph. This suggests a branching of evolutionary ways in the primitive ciliates. One line, giving rise to the modern ciliated protozoa, responded to selection pressures on these micropredators by evolving a large polyploid macronucleus as well as a diploid micronucleus to meet the needs of increased size. The kinetide (cilium, kinetosome, and related structures) was fully exploited to produce a highly coordinated, agile hunter, which also continued to take advantage of the high reproductive potential of asexual reproduction.

The other line, giving rise to the flatworms, became internally cellularized, thus organizing for internal specialization and, through sexual reproduction and embryogenesis, recombining genotypes to specialize further.

Ciliates are the most highly evolved unicellular consumers; acoel flatworms are primordial to bilaterally symmetrical multicellular consumers.

This view of the origin of the Metazoa is not widely accepted today. It is an outgrowth of a rigorous concept of homology as the theoretical basis for all phylogenetic analysis. It is presented here as the point of view most consistent with that type of phylogenetic analysis, an analysis that also recognizes the species as the unit of evolution by natural selection. Other views on the origin of the metazoans are summarized in Box 13–3.

Evolutionary trends in the invertebrates

Now we are ready to return to the key task we set ourselves—to explain the correlated, but apparently diverse origins of coelomic cavities, segmentation, and appendages. As we have been at pains to make clear, that task can only be approached by determining the starting point—the plesiomorph—of metazoan evolution and then looking for adaptive changes that would account for the occurrence of the characters in question (Box 13–2). The most convincing study of this problem is that by R. B. Clark of Bristol University, England. His central concern is

Box 13–3. Alternative views on the origin of the Metazoa.

Ernst Haeckel, the nineteenth-century German biologist, was the father of phylogenetic speculation. In particular he proposed in his gastraea theory that multicellular animals arose from an evolutionary sequence that started with the freshwater green alga *Volvox* (Box 11–3) and proceeded via sponges to the cnidarians. Today, this sequence makes no sense. How can a colony of photosynthetic freshwater plant cells become the filter-feeding chemoheterotrophic marine sponges and, then, turn into polyps and jellyfish with stinging tentacles? What lured Haeckel into this speculation was an exercise in morphology, not evolution by natural selection. *Volvox,* as a hollow ball of cells, suggested a sponge blastula, some of which turn inside out to place flagellated cells internally in the flagellated chamber. Then, somehow the incurrent pores of the sponge were plugged up, the excurrent pore or osculum somehow became a mouth, and tentacles were pulled out—again, somehow—around it, thus transforming a sponge into a cnidarian polyp or jellyfish. Homologies were not critically analyzed and natural selection (see Box 13–2) was ignored.

One point of view put forward by a sponge specialist, the late Madame Odette Tuzet, of France, ignores a volvocine ancestor and starts from protozoan collared cells, to derive the sponges, and then goes on to the Cnidaria. There is considerable support for the first step—protozoan to sponge—but little for the second one. Another authority in this area, the late Dr. Libbie Hyman, of the American Museum of Natural History, cautiously suggested that a colony of flagellates (presumably zooflagellates) give rise to a bottom-crawling multicellular form that became the cnidarian planula. That then gave rise to the modern Cnidaria. Dr. Hyman relegated the sponges to a dead end in evolution.

More recently the Swedish invertebrate zoologist, Professor Jösta Jägersten, of Uppsala University, has continued the Haeckelian viewpoint by proposing a series of intermediate forms extending from a colonial flagellate to produce, as separate evolutionary branches, the Cnidaria, the sponges, and the bilaterally symmetrical Metazoa. This is simply a more detailed version of Haeckel's theory. Again, there is no rigorous analysis of homologies, and natural selection is ignored.

The fundamental problem here is always to recognize characters descended from a common ancestor and then to explain changes in them and in the appearance of new characters in terms of selection.

the way animals get around, their locomotion. As we have seen, locomotion is an integral part of the consumer's way of life. Clark bases his analysis on a consideration of the basic modes of locomotion available to animals, and from this he obtains very convincing answers to the questions of the origins and adaptive significance of coelomic cavities, segmentation, and appendages.

TYPES OF LOCOMOTION

Modifying Clark's point of view somewhat, we can say there are really four ways animals get around. (1) They use cilia; (2) they use muscles in the part of the body that is in contact with the substratum (there can also be combinations of these first

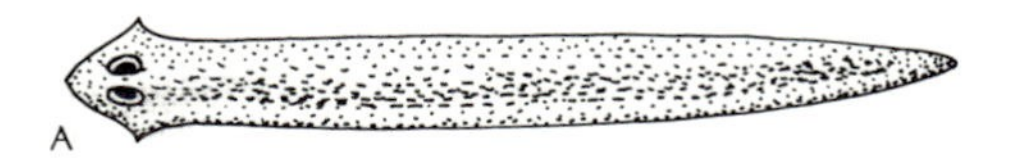

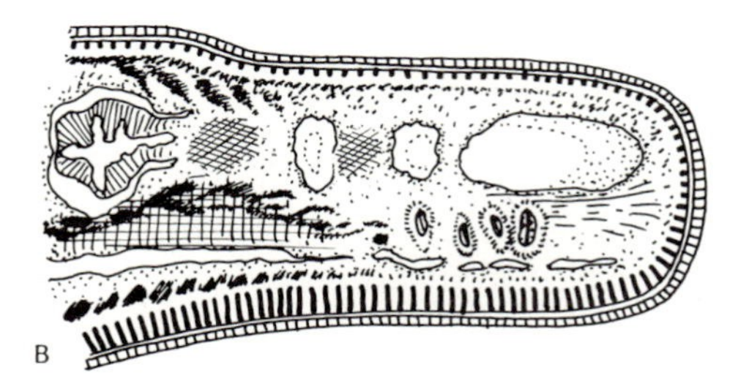

Fig. 13-8: A land planarian and its adaptations. Drawings of a whole animal (A). Cross section showing ciliated ventral surface and musculature (B).

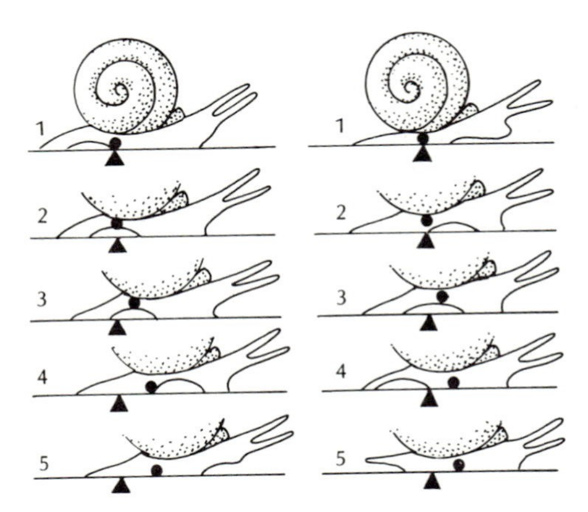

Fig. 13-9: Pedal locomotion in a snail. An anteriorly directed contraction wave (left); a posteriorly directed wave (right). Both are known to result in forward motion. Compare the positions of the reference dot on the snail's foot to the reference triangle on the substratum.

two modes, as in the larger flatworms); (3) they use peristaltic motions of their bodies; and (4) they use appendages (these last two modes can also be used in combination).

Ciliary locomotion. This is limited to small animals from ciliated protozoa to flatworms. (Ciliary locomotion is also used by small larvae throughout the metazoan kingdom.) But there is a limit to the bulk that can be moved by microscopic cilia. Apparently, the size of a cilium also has a limit. Cilia are exceedingly uniform in structure, wherever they occur, which implies that there is one optimal way to build them. So when the needs of locomotion surpass the capacity of cilia, other means of locomotion evolve, if the organism is to survive. In the ciliated protozoa, contractile fibrils are widespread, that is, cilia and these fibrils occur together. The fibrils control body form, but seem never to be used for locomotion. But the story is different for flatworms.

Pedal locomotion. This type of locomotion results from waves of muscular contractions passing along the body wall which is in contact with the substratum. Cellularized flatworms retain ciliation, but many of them are so large that the cilia are relatively ineffective in locomotion. Especially in land planarians (Fig. 13–8), where the buoyancy supplied by water is absent, pedal locomotion is essential. Here, cilia still help by beating against the slime secreted by the epidermis of the planarian, but muscles are largely responsible for moving the body. There are also muscles that bend the body various ways and thus aid in directional movement.

The phylum that has most exploited pedal locomotion is the Mollusca. A snail (Fig. 13–9) is a good example. Here the muscular foot, aided by a slime secretion that helps with traction, efficiently moves the animal over sand, rocks, or vegetation and across surfaces as smooth as the glass walls of aquaria. The mollusc foot is a fascinating aposemic development: in clams and oysters, it can be extended into the sand for burrowing purposes, pulling the shelled body into the sand behind it; and in octopuses, the foot has evolved into a tentaculated structure. In these organisms, pedal locomotion is, of course, replaced by these other means of getting around.

Peristaltic locomotion. Peristalsis depends on the presence of longitudinal and circular muscles in the body wall (Fig. 13–10). The coordinated, but antagonistic action of these muscles produce the waves of contractions and expansions of the body that is called peristalsis. Such peristaltic action is seen most clearly in a burrowing animal (e.g., the earth-

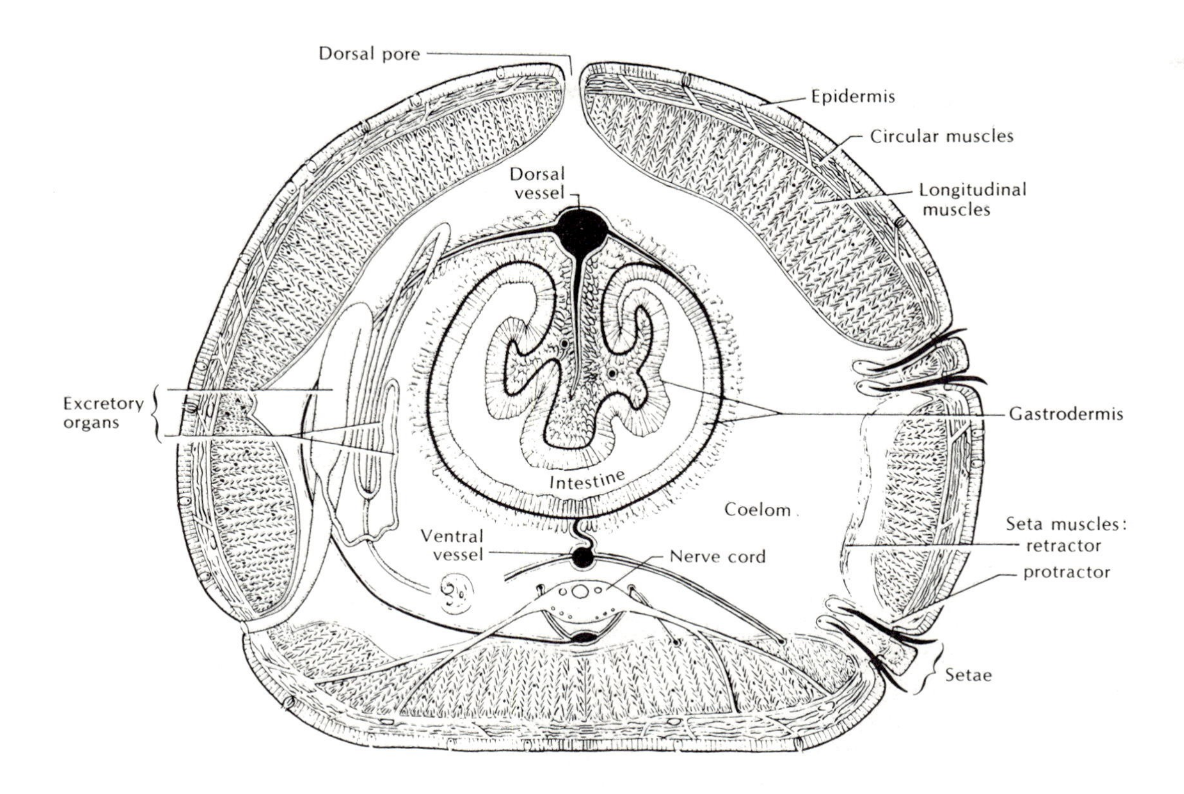

Fig. 13-10: An earthworm (cross section). Note especially the placement of longitudinal and circular muscles. The left half shows excretory organs, but no setae, the right half setae, but no excretory organs.

worm). It is here that we see the importance of the simultaneous presence of the coelom and of segmentation.

The coelom here is a fluid-filled cavity against which muscles can exert their force. It thus functions as a hydrostatic skeleton. To illustrate this, compare pinching an empty balloon and an inflated one, especially one filled with a fluid. In an empty balloon, contraction at one site has no effect on the rest of the balloon. But squeezing a filled balloon forces an expansion elsewhere. Similarly, if circular muscles squeeze a fluid-filled body, it must bulge out on either side of the squeezed area. Shape is restored by contraction of the longitudinal muscles and relaxation of the previously contracted circular muscles. By coordinating these muscular efforts, a body can be propelled through a semisolid medium. It can burrow (Fig. 13–11).

But, why segmentation? This is a further refinement of peristaltic action. Segments allow localization of the muscle action on the fluids in the coelom. If the coelomic cavity were

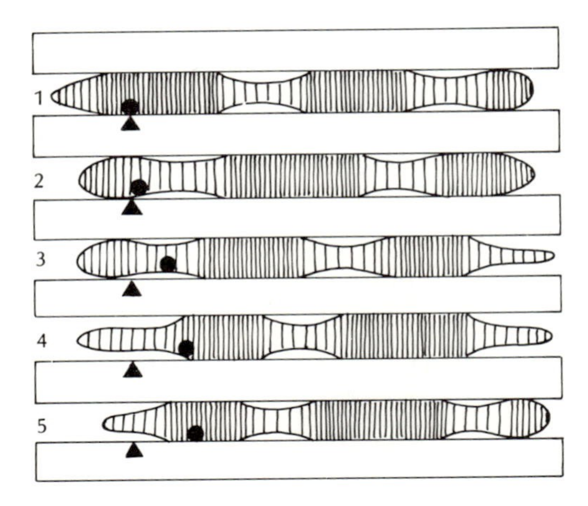

Fig. 13-11: Peristaltic burrowing by an earthworm. The alternating waves of contraction of longitudinal muscles (thickened body) and circular muscles (thinned body) are obvious. Locomotion is to the right as shown by the relative positions of the dot and the triangle.

a long one, the effect of localized squeezing would be damped and lost by a spreading throughout the length of the cavity. If, however, this cavity were divided into compartments or segments, by cross-walls, local squeezing would have a strong local effect. Also, if these compartments were innervated separately—but coordinated, too, with neighboring compartments—then control of muscle action would be more precise. All of this controls peristalsis and burrowing activity (Fig. 13–11).

Locomotion by appendages. The locomotory appendages on the earthworm are the bristles called setae (Fig. 13–10). These extend laterally from the body and are obvious aids in burrowing. When longitudinal muscles contract and the body thickens, the setae are extruded and stick into the walls of the burrow. This gives the worm's body a strong anchor, and as the body elongates (through contraction of nearby circular muscles) it is thrust forward. (See the middle diagram, right-hand end, of Fig. 13–11.) Incidentally, when a robin tries, or you try, to pull a worm from its burrow, it anchors itself—or attempts to—by the combined action of the setae and the longitudinal muscles.

Other members of the phylum Annelida (the segmented worms) have more elaborate appendages (see Appendix I). Some of these have multiple functions; they aid in burrowing or in swimming, and often act like gills.

Appendages are most complex and diversified in the arthropods—jointed-leg invertebrates. The appendages of a lobster are a good example (Fig. 1–7); in addition to locomotory and ingestatory appendages, they also include sensory (antennae and eyestalks) and respiratory ones (gills). A grasshopper also has obvious and powerful locomotory appendages, including wings, in the adult. These appendages are such obvious aids to the life of a consumer that no further commentary is needed on them, except to mention some functional interrelations between them and the coelomic cavity and segmentation.

In segmentation there will be many appendages, if each segment has its own pair. Classic examples of this are centipedes and millipedes (Fig. 13–12). However, elaborate appendages are not compatible with a burrowing mode of life, and therefore, we can expect changes in the relation of coeloms, segments, and appendages as the habits of consumers and the selection pressures on them change.

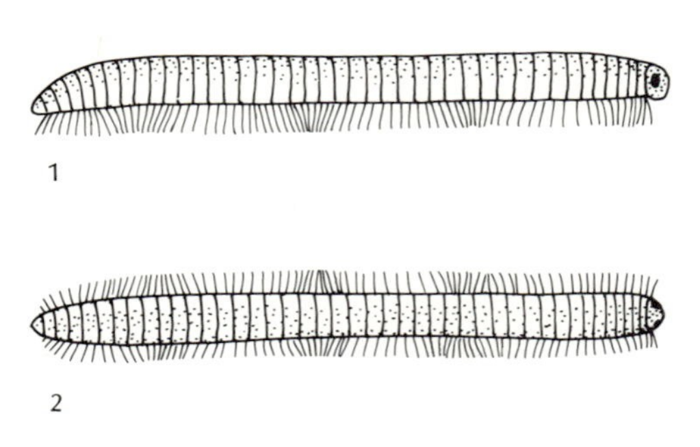

Fig. 13-12: Locomotion of a many-legged millipede. Note the coordinated waves of leg movement.

Consider again an agile insect like a grasshopper or a cockroach. What has happened to segmentation here? More details are given in Fig. 13–13, but the overall point is this: segments have fused and have also become specialized. The head is made up of several segments so highly integrated as to appear as a functional and structural entity. The thorax gives more evidence of segmentation, especially in terms of paired legs. And the abdomen is still clearly segmented. What can be argued here is that with the transition to appendages as the basis for locomotion, segmentation lost its selective advantages. Legs were reduced in number to those few optimally adapted to the locomotory needs of an insect, and in general, segments fused as the body met the needs of an agile consumer that ran around for its food rather than burrowing or even crawling around for it.

What happened to the coelom? Its role as a fluid-filled cavity on which muscles of the body wall act for locomotion no longer exists. The coelom is now simply a space within which lie specialized body parts. The body wall functions to anchor muscles needed to activate legs and wings. For that purpose, a skeleton is also useful and insects have evolved a tough chitinous exoskeleton, which also provides a protective outer coat, but at the same time limits growth. The limitation to growth is solved by molting, a process whereby the exoskelton is shed at regular intervals and growth occurs before the new exoskeleton hardens.

All this contributes to the viewpoint that the ancestors of the arthropods probably were segmented worms. Appendages and the exoskeleton were the neosemic innovations that were exploited in a variety of aposemic ways. As a consequence, earlier needs for an internal body cavity and for segments were greatly modified and even reduced. Both of these aspects of the body are now somewhat vestigial; they only remain as evidence of the phylogenetic history of these animals.

Fig. 13-13: Segmentation in a cockroach. The abdominal segments are clear on this immature roach whose wings have not yet covered its abdomen. Segmentation of the thorax is indicated by the three pairs of thoracic legs; segmentation of the head is not apparent.

INVERTEBRATE PHYLETIC HISTORY

We are now in a position to draw together the major lines of the foregoing discussion so as to see how invertebrate diversity has arisen within these multicellular consumers. Let us return again to Clark's point of view (Box 13–4).

The first bilaterally symmetrical multicellular consumers were probably flatworms. They lived in the Precambrian seas, preying upon algae and protozoa. Technically, they are called benthic predators; that is, they are bottom-crawling consumers. Ciliary locomotion was at first sufficient to propel

Box 13–4. Clark on metazoan evolutionary theories.

By quoting directly from Clark's important study, *Dynamics in Metazoan Evolution,* we get a very good feel for both the approach of that author and the strengths and limitations of his analysis.

". . . three general criticisms can be made of the whole trend of discussion of the subject [i.e., evolution of the coelom and segments]. . . .

"First, most of the theories deal particularly with the structures from which the coelom might have been derived and the intermediate stages through which they passed in the course of their evolution, but hardly at all with selective advantage gained from the possession of either a coelomate or a metameric [segmented] type of organization. The advantages must have been considerable to judge from the success of the coelomates and, particularly, of the segmented coelomates. This failure to consider the adaptive significance of the coelom and of segments resulted in much speculation degenerating into a mere juggling with morphological types in order to arrange them in a tidy sequence. . . .

"A second confusing aspect . . . has been the failure to distinguish the evolution of the coelom from the evolution of metamerism. . . .

"A third reason for a great deal of confusion [arose because] . . . it was still not clear what in fact constituted a coelom [p. 27–28].

"If, as I have attempted to show in previous chapters, the coelom serves primarily as a hydrostatic organ [or fluid skeleton] it is difficult to conceive of any function that it might have subserved in small animals moving by ciliary activity; rather, we should expect the coelom to make its first appearance in relatively large animals and not during the early stages of metazoan evolution. This view of the function of the coelom implies a rejection of the gastraea theory in its modern form. Whatever other criticisms may be advanced against the gastraea theory [see Box 13–3], its most disturbing feature is the absence of any functional justification of the evolution of the coelomic compartments at so early a stage in metazoan phylogeny. . . .

"If the gastraea theory is rejected, we are left with a planuloid or an organism comparable in its fundamental structure to the modern Acoela as the most primitive bilaterian [i.e., bilaterally symmetrical animal] from which the remaining Metazoa evolved. . . .

"Whether the Protobilateria [or ancestors of the bilaterally symmetrical Metazoa] evolved from colonial or solitary protistans, they are likely to have been ciliated externally. They are likely to have had, or to have evolved modest powers, comparable to those of ciliates, of changing their shape by means of peripheral [or cortical] contractile elements. Composed of relatively few cells, they are likely to have fallen in the size range of the larger ciliates and smaller Acoela, that is of the order of 1 mm in length. Modern organisms that answer to this description are able to swim by ciliary activity and generally do so [pp. 212–213].

"The coelomic activity may have been derived from enlarged and cavitated gonads in some animals, as postulated in the gonocoel theory. A gastric pouch or pouches may have become separated from the main digestive cavity in others, as claimed by supporters of the enterocoel theory. . . . It is conceivable, too, that in some animals a cavity appeared *de novo* within the mesoderm and that this is reflected in modern animals by the schizocoelous method of coelom formation. There is evidence, chiefly embryological, that might be quoted in support of all these possibilities; none of it is conclusive and it is most unlikely that more reliable evidence will ever be forthcoming, so that the precise method of evolution of the coelomic cavities in modern phyla must always remain in doubt. We are left, however, with a strong belief that conditions favoring radiative evolution existed at the time of appearance of the coelom, a variety of theories to account for its evolution in different groups of animals and conflicting evidence as to its origin. The conclusion that it is polyphyletic seems inescapable" (pp. 216–217).

them along in their unending search for food, but this was later augmented by muscular activity, and pedal locomotion, in particular. These early seas, especially shallow sunlit waters were highly productive of life forms, with various monerans and protistans and probably multicellular algae and fungi. Sponges and flatworms were probably the most highly evolved consumers. The bottom or benthic, community, a rich deposit of decaying material, was a rich source of food for microherbivores and microcarnivores, like the flatworms, and for filter feeders, like the sponges. As plants and animals became larger (and, particularly, multicellular), the herbivores and carnivores also increased in size. This was still part of the Precambrian evolution of the Metaphyta, Fungi, and Metazoa.

At some time in this era, the predators became so large and so active that protection against them was a real selective advantage. Three responses were possible: become distasteful, burrow, or develop armor. All three apparently occurred. Both plants and animals could become distasteful. But armor and burrowing were more animal specialities. Armor could be simply tough outer layers of the body or even shells. Burrowing could be digging into the subtratum, or secreting tubes, which are also a kind of armor. And some predators burrowed or otherwise covered themselves while in the pursuit of prey. Therefore, there was selection pressure for a coelom and segmentation, for those who burrowed and for strong muscular locomotion, for those who carried around protective shells. In other words, the first acoelomate flatworms probably underwent an adaptive radiation as they became specialized, becoming various types of multicellular consumers. Eventually, some became coelomate and segmented, giving rise to the Annelida, for example; others retained pedal locomotion, the Mollusca, for example. The arthropods were a further adaptive development of forms, with segmented bodies and appendages.

The Cambrian period began about 600 million years ago, and as animals developed hard parts (shells and exoskeletons) and bodies large enough to leave tracks in the mud (including worm tracks), the fossil record became abundant. In the early Paleozoic period (Fig. 10–9), we find good fossil evidence of all the major invertebrate groups, including some now extinct, and of the major trends of invertebrate evolution according to Clark's unifying theme of locomotion among predaceous consumers. We will return to the diverse array of opinions on the origin of the coelom—gonocoel, nephrocoel, or what?—and then look more closely at the origins and evolutionary history of specific invertebrate phyla in order to pinpoint specific problems in invertebrate phylogeny.

Origin of the coelom. Clark implied that all theories of the origin of the coelom might be correct. Thus, the coelom originated in different ways in different animal phyla, and therefore, it is not a truly homologous structure. The multiple origin reflects different selection pressures on the early unsegmented worms—benthic animals in whom muscles were evolving as an aid to locomotion. Those worms, plowing through the detritus of the bottom of shallow seas in search of food, probably used any internal, fluid-filled body cavity to enhance the action of their musculature. (The digestive cavity could not be used, since its fluid contents were not constant.) Hence, a gonocoel, nephrocoel, hemocoel, or other cavity could have been selected as a coelom. If this is so, we must accept the coelom as usually being analogous, not homologous, among different groups. But how can we find the homologous ones? Here embryological evidence is of the most use.

Earlier (Box 13–1), we saw that there were two major modes of coelom formation, the schizocoelous and the enterocoelous. The arthropods, annelids, and moluscs are schizocoelous; the chordates, echinoderms (starfish and their relatives), and the Brachiopods (or lamp-shells) are enterocoelous. But this major distinction in modes of coelom formation is not entirely helpful for within the schizocoelous forms we must determine whether coeloms are genuinely homologous. For example, in the molluscs the blood sinuses are about all that represent the coelom. These arise within the embryonic mesoderm, apparently from the blastocoel. The fact that they are surrounded by mesoderm during development of course defines these sinuses as coelomic cavities. Blood reaches them from arteries and is returned to the heart by veins.

In the annelids, the well-developed coelom could be a derived opening of the excretory structures, since that organ system connects directly with the coelomic fluid. The phyletic history of the annelid coelom is not clear. It is such a well-developed structure and so intimately associated with the various organ systems in the surrounding tissues that to decide whether it originated from this or that organ is difficult. It may be that, evolutionarily, it arose simply from a space in the loosely organized mesenchyme (mesoderm) of an acoelomate ancestor. That would make it a true mesodermal cavity or coelom.

In the arthropods, the embryonic origin of the coelom is

especially difficult to resolve because of the complexity of development. To begin with, the presence of yolk changes the cleavage pattern to what is called superficial cleavage. As this term implies, only the outer part, or superficial layer, of the egg cleaves. The zygote nucleus divides many times and the division products come to lie near the egg cortex. This area cleaves to produce a layer of cells surrounding the inner yolk mass. The next steps, in insects, lead to a worm-like larva. When this larva enters a relatively immobile pupal stage, special masses of tissue, the imaginal discs, transform the pupa into a legged and often winged insect. The development of the coelom in such cases is very hard to follow in the larval, pupal, and the adult stages. At least it has no straightforward homology with the coelom of annelids and molluscs, and it may not even be homologous.

These difficulties in embryological interpretations of coelomic origin are summarized most cogently in the third quotation from R. B. Clark, in Box 13–4.

Cleavage patterns and larval types. While on the topic of development and its phylogenetic information, it will be worth looking at spiral cleavage and larval types in somewhat more detail.

Spiral cleavage (Fig. 13–7) is so widespread in the invertebrates that the term Spiralia is often used for animals ranging from acoelomate flatworms through pseudocoelomate roundworms (Nemathelminthes) to the annelids and molluscs. In fact, it also includes the arthropods because their superficial cleavage is thought to be a modification of spiral cleavage. It thus appears to be a fundamental homology implying a phyletic relationship among all these animals. The presence of acoelomate, pseudocoelomate, and coelomate groups may suggest an evolutionary trend. Another aspect of Clark's work, which was not covered here but which integrates the work of many other workers, leads to the conclusion that the pseudocoel has had a parallel, but independent evolution from the coelom. Both arose from the acoelomate condition, but beyond that not much can be said with assurance regarding the evolution of the pseudocoel. In most cases, it appears to develop embryonically from the blastocoel, and its function is often, but not always related to that of a fluid skeleton.

The pseudocoelomates are a peculiarly fascinating group, best interpreted as a collection of evolutionary experiments occurring at the time of the great adaptive radiation of the coelomate worms. Most pseudocoelomates are grouped together in the phylum Aschelminthes (perhaps better a superphylum) for taxonomic convenience. A phenetic classification, it does not reflect phylogenetic history accurately.

The general phyletic relationship implied by spiral cleavage can be sorted out by looking at later stages of development, namely the larva. The Platyhelminthes have no real larval forms. The non-symbiotic flatworms, after cleavage, directly develop into an immature worm, which completes its maturation as a free-swimming predator. In the different phylum of the nemertine worms, there is a ciliated pilidium (Fig. 13–14A), a larval form not seen elsewhere in the invertebrates. The nemertines are important phylogenetically because, although they are still acoelomate, they show important evolutionary innovations in their digestive and circulatory systems, as well as a specialized feeding structure (Fig. 13–15).

The annelids and molluscs both have trochophore larvae (Figs. 13–14B) and, additionally, the molluscs have some further development into larval forms—a veliger, in marine forms, and a glochidium, in freshwater forms. Curiously, the cephalopods (squids and octopus) have no well-developed larval forms.

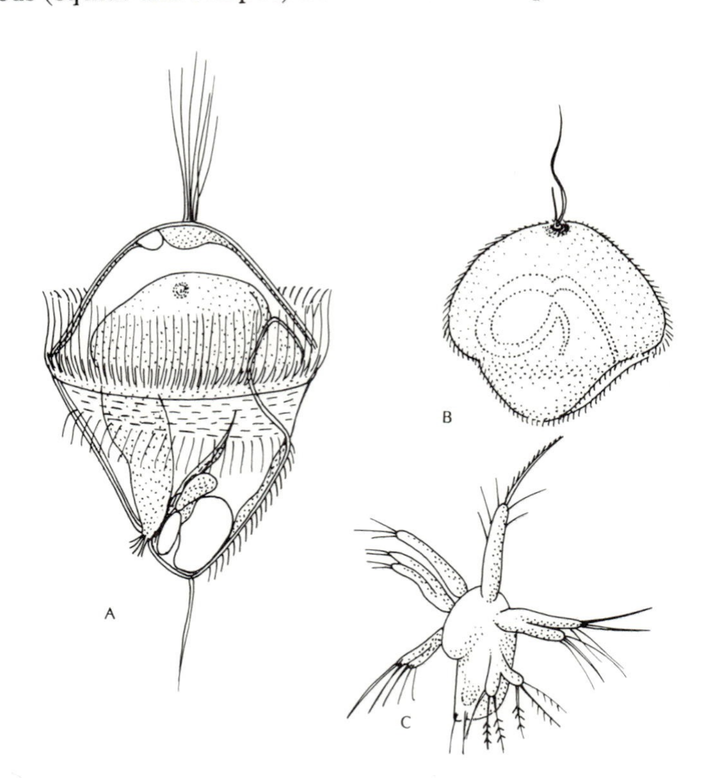

Fig. 13-14: Selected invertebrate larvae. A six-day-old trochophore larva of an annelid (A); a pilidium larva of a nemertine worm (B); and a nauplius larva of a crustacean (Arthropoda) (C).

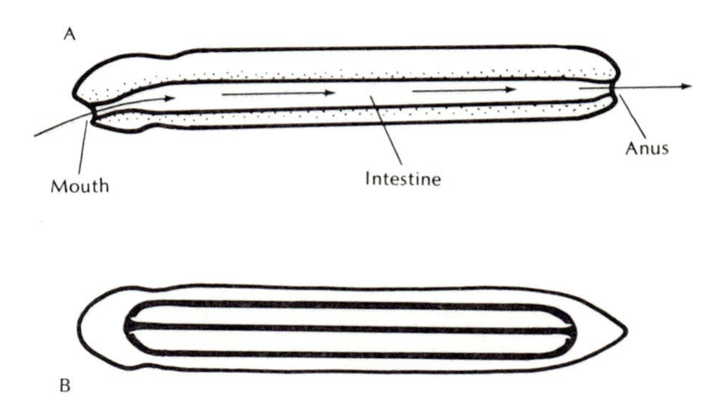

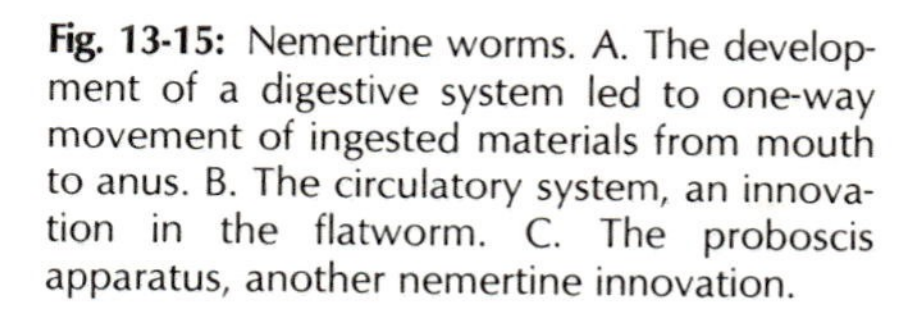

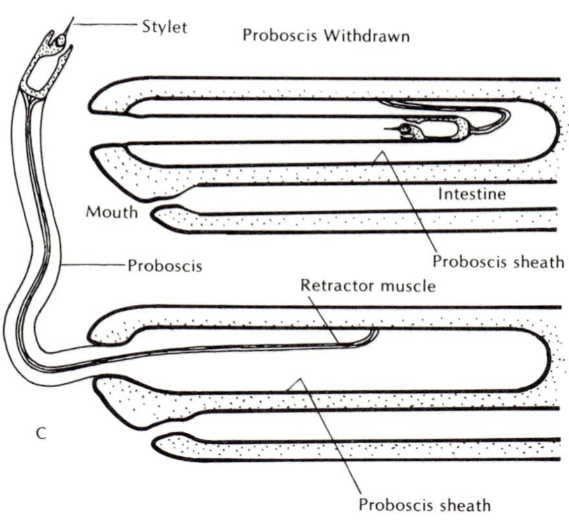

Fig. 13-15: Nemertine worms. A. The development of a digestive system led to one-way movement of ingested materials from mouth to anus. B. The circulatory system, an innovation in the flatworm. C. The proboscis apparatus, another nemertine innovation.

The terrestrial arthropods develop directly into young adults or, as in insects, they may go through a series of larval molts. The aquatic forms, crustaceans such as crabs, shrimps, and lobsters, all have a characteristic nauplius larva (Fig. 13–14C).

The differences among these larvae—if we again cautiously invoke the biogenetic law or idea that ontogeny or embryonic development is a kind of review of phylogenetic history—suggest that spiral cleavage preceded various major lines of development. The fact that larvae show certain profound differences—pilidium versus trochophore—suggests that there was an early divergence among these phyla. On the other hand, the similarity in the trochophore larva of annelids and many molluscs suggests a common, though still very ancient ancestry for these two important phyla.

There still remains the problem as to how this spiral pattern of cleavage arose. About the only possible solution today is that when cleavage first arose in the ancestral flatworms it had to occur in some fashion and spiral cleavage was that pattern.

Intuitively, many workers object. They feel that spiral cleavage is a highly ordered pattern, and therefore it had to be a product of considerable evolution. However, the fact of the matter is that we do not know whether or not spiral cleavage is a highly complex or a very simple pattern. We simply do not know how much effort—bioenergetically, in terms of cellular biosynthesis and organization of subcellular organelles—is involved in cleavage patterns. Hence, speculations based on intuitive guesses regarding the complexity of cleavage are meaningless. We, thus, can say no more than what was said above: the first animal zygotes had to cleave somehow and apparently spiral cleavage is that first pattern. Therefore, spiral cleavage recapitulates nothing and is the primitive pattern of cleavage in the bilaterally symmetrical metazoans.

PHYLOGENETIC PROBLEMS

One view of the evolution of the invertebrate animals is given in Fig. 13–16. A first look at it reveals at least two significant points: a fan-like adaptive radiation arising from the flatworms and many question marks regarding the origins of various groups. Let us comment on each of these points to draw to a conclusion this review of invertebrate phylogeny.

Radiata and Bilateria. The adaptive radiation just referred to is a product of the evolution of bilaterally symmetrical animals, the Bilateria. At one time, animals were separated, in phenetic fashion, on the basis of their symmetry into the Radiata and the Bilateria. The former contained the Coelenterata and the Echinodermata. The last phylum contains starfish, sea urchins, sea cucumbers, and the like (Appendix I). Very importantly, however, as we shall see in Chapter 14, the echinoderm larvae are bilaterally symmetrical. The radial symmetry of the adults reflects their special embryonic development; hence, these are not fundamentally radially symmetrical forms. Furthermore, a careful look at the Coelenterata show them to contain the Spongiara, Cnidaria, and Ctenophora. The Spongiara or sponges (now called the Porifera) were only occasionally radially symmetrical or were closer to radial symmetry than to any other type of symmetry. A single flagellated chamber could be conceived of as radially symmetrical (Fig. 13–3), but no adult living sponge is composed of one such chamber. Sponges really are asymmetrical arrangements of flagellated chambers. That left only the Cnidaria (the medusae and the polyps) and the Ctenophores, or comb-jellies, as radiate animals. When these two groups were placed in separate

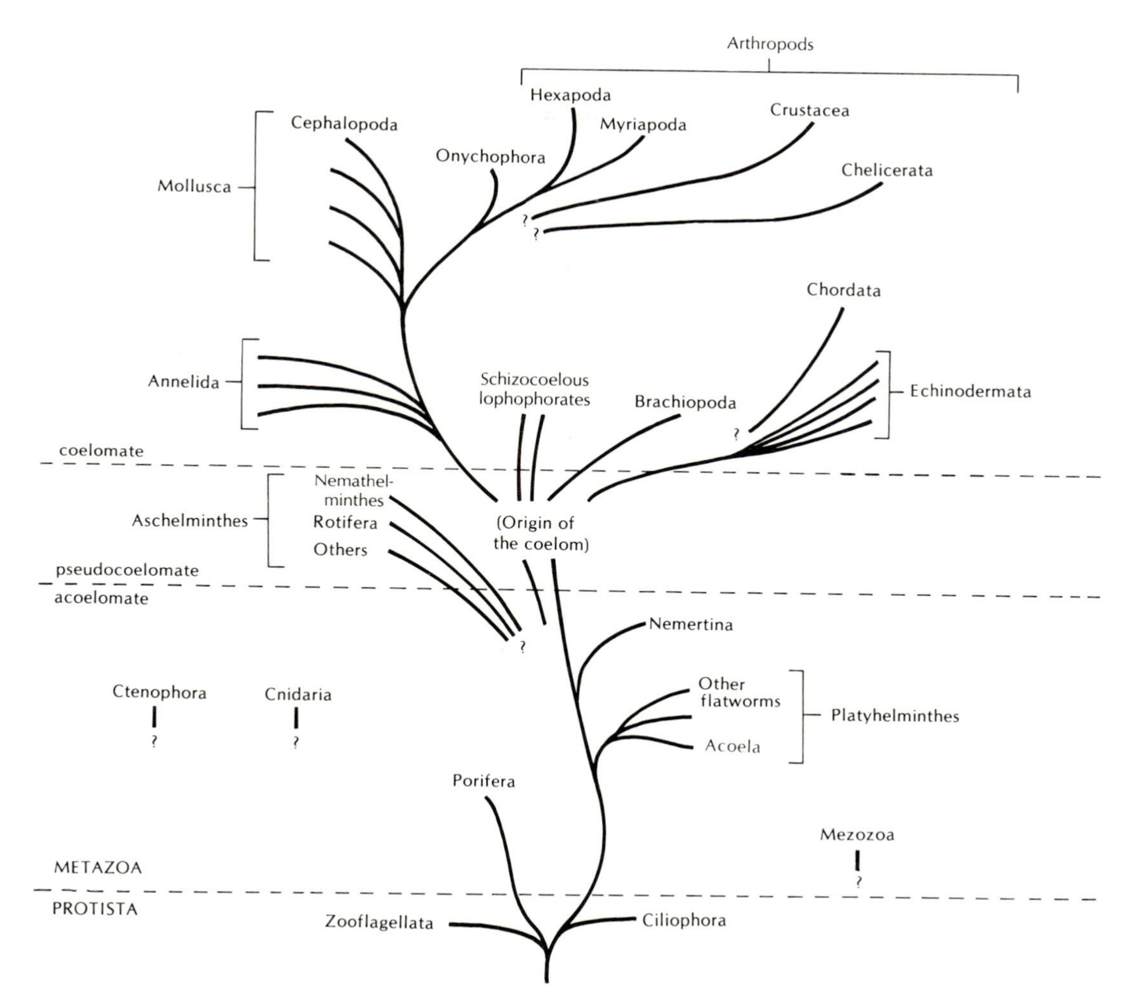

Fig. 13-16: A dendrogram view of invertebrate evolution.

phyla, not only was the concept of the phylum Radiata utterly demolished, but the phylum Coelenterata was also. The term is still used erroneously for the phylum Cnidaria. At best it is a common name, i.e., the coelenterates, for the cnidarians.

More important than this historical discussion of the so-called radiate phyla is the present-day realization that they are all dead ends, evolutionarily. Earlier views of these phyla (Box 13–3) have become less and less convincing with the refinement of our knowledge of these animals and with the application of rigorous phylogenetic analysis.

One also sees the term Eumetazoa used to include all

metazoans, except the sponges. This is an attempt to emphasize the importance of tissues in the multicellular organization found in the Cnidaria and the bilaterian animals, but it is not an especially useful category, since the Cnidaria, once thought to have only two tissue layers (diploblastic), are now known to possess three. They are triploblastic as are the Bilateria. The most useful approach is to view the Bilateria as a subkingdom of the Metazoa—if a taxonomic designation is needed—and to characterize the Cnidaria, Ctenophora, and Porifera as a phenetic grouping having a symmetry that approaches radial to varying degrees.

The Bilateria are probably all derived from flatworm ancestors, with the acoel flatworms being plesiomorphic. The acoels, in turn, could have had a common ancestor with present-day, ciliated protozoa. This ancestor would have been a multinucleate ciliate, something like the ciliate plesiomorph *Stephanopogon mesnili*. The sponges seem likely to have homologies between their collared cells and the collar cells of certain colonial zooflagellated protozoa, called choanoflagellates. The origins of the Cnidaria and the comb-jellies (Ctenophora) are still, frankly, a mystery. Hence the question mark in Fig. 13–16. Equally mysterious is the origin of another group, called the Mesozoa, that we have not mentioned previously. They are kidney symbionts in certain fishes. Their minimal multicellular structure is, in all probability, more the result of reduction in adapting to their hosts than of being in any real sense primitive.

As we look back over the origins of the kingdom Metazoa we see that it, like the other kingdoms of multicellular forms, must be polyphyletic. The flatworms probably arose from ciliated protozoa; the sponges from flagellated protozoa; and the Cnidaria, Ctenophora, and Mesozoa from we know not what. If theirs were a protozoan ancestry, it was in all probability quite different from that of the flatworms and sponges. Other comparative comments on the origins of multicellularity are pursued further in Box 13–5.

Phylogeny within the Bilateria. A variety of phyla within the bilaterian invertebrates can be discussed only briefly here. We have already mentioned the psudocoelomate phyla (Appendix I), especially those of the nemertine worms (Fig. 13–15) and the roundworms (Fig. 1–11). Also deserving mention is a group we can call the schizocoelous lophophorates. They are filter feeders, using ciliated tentacles called lophophores to extract food from the water around them with a coelom formed by the schizocoelous mode (Box 13–1). Their embryology teases us with some interesting features about what may be the early evolution of the coelom. However, their sedentary habit has changed the selection pressures on maintaining a coelom and so now they also confuse the picture. This is emphasized by another lophophorate phylum, the Brachiopoda, or lamp-shells, mentioned earlier. They show an enterocoelous form of coelom development. Hence, something, still unclear, was going on in the evolution of these lophophorate groups that may yet be important in understanding the early evolution of the coelom. Why did the lophophorate forms evolve both enterocoelus and schizocoelous coelomic cavities? It is at this crucial point that the greatest problems in invertebrate evolution are still unresolved. Clark's analysis defines in substantial detail the selection pressures leading to coelom, segment, and appendage formation. But the actual course of events is still obscure.

Last we return to "The Big Three" among the invertebrates—the Annelida, Mollusca, and the arthropods. (We have been at pains not to refer to the Arthropoda, but only to the arthropods. The reason for that was mentioned in Chapter 1 and we will address it directly after a few final comments on the Annelida and Mollusca.)

That the segmented worms, or annelids, and the molluscs share a common ancestor can be argued from certain homologies. In addition to sharing the basic characteristics of the bilaterian morphology, they also have in common spiral cleavage, schizocoelous formation of the coelom, and a trochophore larva. But whereas the annelids are extensively segmented, the only suggestion of segmentation in the molluscs is in the superficial repetition of certain features in the chitons (Fig. 13–17A). Basically, the molluscs are not segmented. There is, however, a living fossil called *Neopilina galathea,* which is a relative of the modern chitons (Fig. 13–17B); this deep-sea animal (collected by Danish scientists in 1953 while they were dredging at about 3,000 m in the Pacific) shows clear evidence of internal segmentation. This finding suggests that certain of the molluscs might at one time have been segmented. Some experts have urged that *N. galathea* is a fine candidate for the mollusc plesiomorph, but others believe it is a side branch in molluscan evolution. Perhaps the evolution of the foot (starting from pedal locomotion) and a protective shell removed pressures for any further evolution of a coelom or segments. The hemocoel (a blood sinus), it will be recalled, is about all that

Box 13–5. The origins of multicellularity.

There are, in principle, two basic ways to go from unicellular to multicellular organisms. One is *colonial integration,* the other *internal cellularization*. This can be shown diagrammatically:

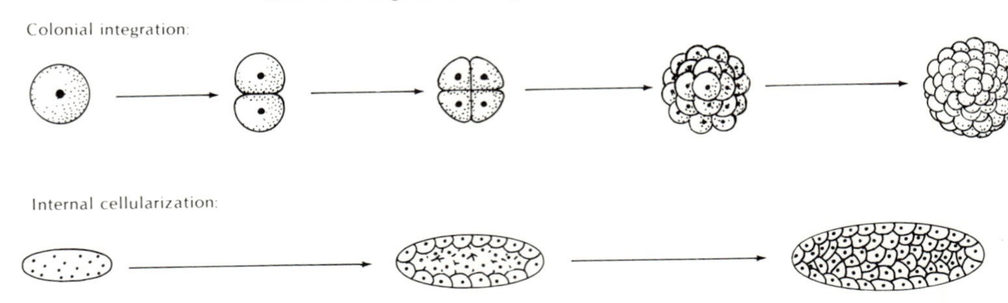

As the diagram suggests, colonial integration results from the products of cell division remaining together, first as a colony and then, in an integration of their activities, as a superorganism. Such an evolution results in the subordination of individual organisms to the needs of the group. A variety of superorganisms, not shown here, can be produced by this path of evolutionary development. The variety of multicellular algae, fungi, land plants, and sponges document that point.

In internal cellularization, a multinucleate cell is subdivided internally by cell membranes forming around each nucleus. It may be that the endoplasmic reticulum was a source of such new cellularization. In the final drawing, the multicellular organism is shown with an outer epidermal layer of cells and an inner mass of less differentiated cells. Here the original cell is subdivided; the integrity of the original organism is maintained and its subparts become specialized as a result of cellularization. This is not a superorganism in the sense that that term was used for colonial integration. It is an organism with enhanced capabilities, due to an increased specialization of its parts.

Both modes of evolution achieve much the same final product, namely, an organism composed of differentiated cells carrying out specialized functions. Of course, that is precisely the advantage of multicellularity, for it offers greater functional complexity and, therefore, more adaptability than can unicellular forms. Unicellular forms are highly specialized—that is why they survive—but they do not have the array of functions multicellular forms have.

At this point it is worth reiterating a maxim formulated by the great American geneticist T. M. Sonneborn, who has studied the genetics of *Paramecium,* in particular. He says that the major tactic of evolution is "Repeat, then vary." This tactic is seen to operate at the molecular, subcellular, cellular, tissue, and organ

remains of the mollusc coelom. Thus, we can accept both the annelids and molluscs as highly evolved invertebrate phyla. The occurrence, though, of such active and highly specialized molluscs as the octopuses and squids (Class Cephalopoda) in the Mollusca suggests a special degree of evolution in the direction of complex behavior—including memory and learned behavior. In this sense they are more complex or "higher" than the annelids, hence their position in Fig. 13–16.

levels and at the population level, too. What Sonneborn is saying is that biological complexity and its resulting functional versatility comes about by taking a unitary structure and compounding it, and then when the units vary, they form a complex structure.

For example,

1. The first protein macromolecules were probably rather monotonous arrays of similar amino acids. As the amino acid composition varied, the proteins became more diverse in function. Compare the diversity of proteins and nucleic acids and their variable subunits with the relative uniformity of carbohydrates and lipids (Box 1–1).

2. Among cell organelles, consider the potential of the kinetide (Fig. 11–9). Initially, there was probably only one per cell. Then it became repeated, as in the more complex zooflagellates and ciliates. And within the ciliates we see some variation as certain kinetides served ingestatory and locomotory functions and some, even sensory functions. They also combined to form such new structures as membranelles or special locomotory tufts called cirri.

3. Cells in a simple algal colony were at first all alike—they repeated themselves. Then they varied in shape and function to form holdfasts, vegetative cells, and reproductive cells. Recall, in particular, the beautiful sequence that illustrates this in the volvocine evolution of the green algae (Box 11–3). Presumably the same thing occurred in the evolution of the Acoela. Their ancestor is thought to be derived from a multinucleate cell, which cellularized. Initially these cells were probably all very much alike, but became varied, due to a changing embryological development, to form epidermal cells and inner digestive cells. Further change gave rise to different tissue cells.

4. Ontogeny, of course, uses "repeat, then vary" as the basis for differentiating versatile adults from relatively undifferentiated zygotes and their cleavage products.

5. A homogeneous gene pool can evolve into a polymorphic one, a polytypic one, and then, possibly, more than one gene pool (Fig. 7–9).

Our last comment here is to point out that only the Bilateria are thought to have become multicellular by internal cellularization. All other multicellular plants, fungi, and animals (namely, sponges, because we do not know the origins of the Cnidaria, Ctenophora, and Mesozoa) became multicellular through colonial integration. Or, put another way, only the active predators became multicellular by way of internal cellularization. It may be that continuity of coordinated or integrated functions in a unicellular organism could only be maintained by internal cellularization. Colonial integration, on the other hand, takes individuals and tries to reintegrate them as a colony. For producers, decomposers, and filter-feeding consumers, this is obviously adequate. But for active consumers, and the evolution of that coordination that eventually is served by a nervous system, internal cellularization was perhaps not only a preadaptation, but perhaps also a functional necessity.

Finally, now, the arthropods. An incisive analysis of arthropod functions and structures, with a view toward evolutionary questions, has recently been completed by the distinguished British scientist S. M. Manton. Her conclusion is that the arthropods are, in all probability, polyphyletic. A significant amount of parallel evolution occurred, once three different groups, more or less simultaneously and independently, evolved an exoskeleton and jointed legs. Dr. Manton separates

Fig. 13-17: Early stages in molluscan evolution. A. *Neopilina galathea,* unsegmented shell, dorsal view (left), ventral view (right). Note the pairs of gills, among other structures. B. The chiton *Lepidochitona cinereus.* These shell plates appear, at least superficially, to be segmented, dorsal view (left), ventral view (right). Note, especially, the series of gills along the right side, again suggesting segmentation.

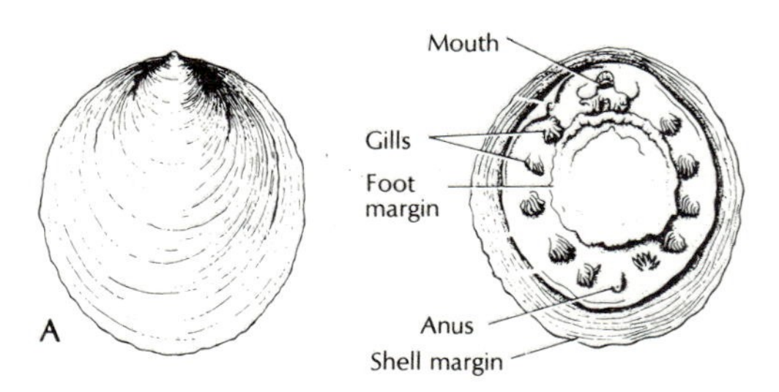

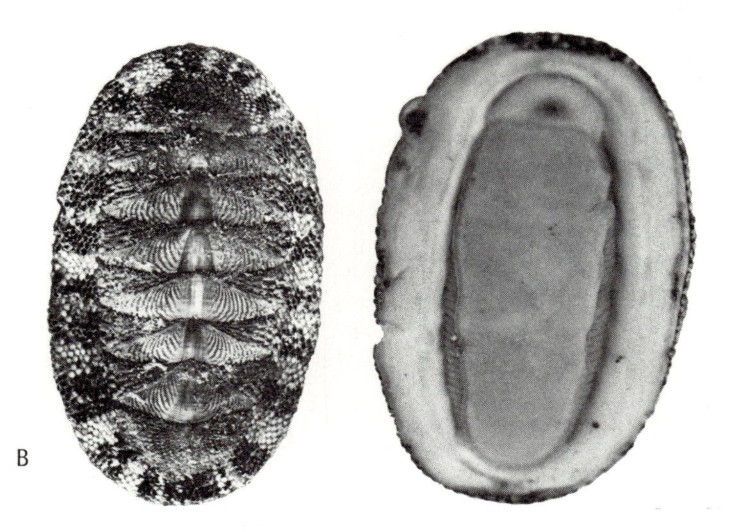

the arthropods into three phyla; the Chelicerata, which includes spiders and such relatives as the improbable horseshoe crab; the Crustacea, which includes crabs, shrimps, and lobsters; and the Unirama, which includes insects, millipedes and centipedes and a special group called the Onychophora. It is the Unirama that we will comment on briefly here.

Let us start with the Onychophora. *Macroperipatus* (Fig. 13–18) is an unusual mixture of worm-like and uniramate features. The body is segmented and each segment carries a pair of legs that are clawed the way insect legs often are. The body wall has longitudinal and circular muscles. For breathing, external openings called spiracles allow air to pass into the body by means of tracheae. Spiracles and tracheae are also found in insects. Finally, the eggs of *Peripatus* develop via spiral cleavage, another feature shared with other uniramate forms. *Peripatus,* for reasons such as the foregoing, can be considered a plesiomorph of the phylum Unirama. If we ask where such animals came from, a first guess would indicate a terrestrial annelid-like ancestor. But there are such differences in the development of *Peripatus* and the annelids that it seems safer to guess that the segmented worm-like ancestor of *Peripatus* was not the ancestor of present-day annelids.

For the Crustacea and Chelicerata there are no missing links of the *Peripatus* sort to show us their origins. These groups are separate from each other and from the Unirama for a variety of reasons, i.e., absence of homologies, such as differences in their respiratory organs. The crustacean respiratory organs are typically gills; the Chelicerata have special respiratory organs called gill books and lungs, with a few having special tracheae. The tracheae appear to be a convergence with those in the Unirama rather than an homology. The most obvious difference between these three phyla is in the general organization of the body. The Chelicerata have a fused head and thorax—a cephalothorax—with four pairs of legs. The Crustacea still retain a pair of appendages for each segment, with the exception of some head segments, and their appendages are variously specialized (Fig. 1–7). The Unirama are more complicated. In the Onychophora and the Myriapoda—the many-legged centipedes and millipedes—we see a head and a long segmented body with many pairs of legs. In the Hexapoda—the six-legged insects—there is a head, a thorax, and an abdomen, with all the legs (and wings, when present) arising from the thorax.

Details of homologies are omitted here. The interested reader, willing to learn the necessary technical details of arthropod structure and function, will find Dr. Manton's analysis of evolution in these animals fascinating and rewarding.

The foregoing summaries of invertebrate groups, seen through the unifying concept of locomotion and the ways of life of mobile consumers, provides some understanding of invertebrate diversity. The story is far from complete. Perhaps the best perspective on this aspect of phyletic history is supplied by once more going to R. B. Clark and the concluding paragraph of his work on the dynamics of metazoan evolution. He says,

> It is an indispensable principle that structure must be considered in relation to function; in isolation it is meaningless. Reconstructing the evolutionary history of animals is one of the most fruitful unifying disciplines in zoology, but it can progress and unify only if some attempt is made to view the whole of zoological knowledge. Perhaps the volume of information at our disposal is already far too great for this to be pos-

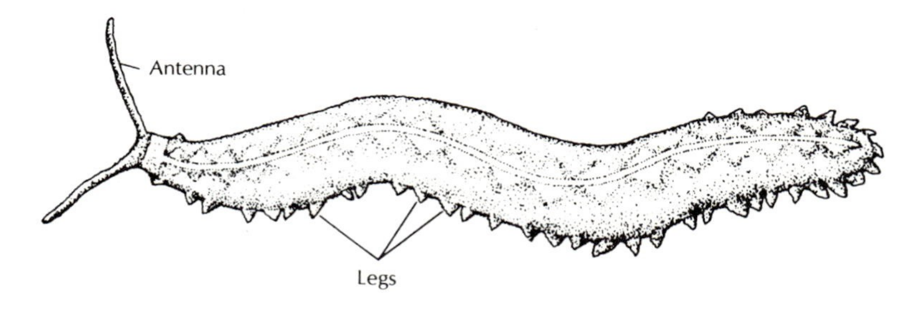

Fig. 13-18: *Macroperipatus geayi.* One of the curious onychophorans occurring in the leaf mold of tropical forests.

sible, but occasionally by raising our eyes from the immediate problems that confront us, we achieve a partial synthesis which gives fresh direction to our enquiries (reprinted, by permission from R. B. Clark 1964 *Dynamics of Metazoan Evolution*. pp. 260–261. Clarendon Press, Oxford).

Nowhere is the diversity of life greater and, hence, more challenging to the student of phylogeny than among the invertebrate animals. Slowly, and even painfully, the earlier guesses regarding the course of invertebrate history are being replaced by more rigorously informed insights—fresh directions—into understanding that phylogeny.

References

Barrington, E.J.W., 1967. *Invertebrate Structure and Function.* Houghton Mifflin, Boston.

Clark, R. B., 1976. *Dynamics in Metazoan Evolution.* Oxford University Press, New York.

de Beer, G. R., 1940. *Embryos and Ancestors.* Clarendon Press, Oxford.

Dobzhansky, T., 1970. *Genetics of the Evolutionary Process.* Columbia University Press, New York.

Dougherty, E. C., Z. A. Brown, E. D. Hanson, and W. Hartman (eds.), 1963. *The Lower Metazoa: Comparative Biology and Phylogeny.* University of California Press, Berkeley, Calif.

Grassé, P. P., 1952–1977. *Traité de Zoologie.* Masson et Cie, Paris.

Hanson, E. D., 1977. *The Origin and Early Evolution of Animals.* Wesleyan University Press, Middletown, Conn.

House, M. R. (ed.), 1979. *The Origin of Major Invertebrate Groups.* Academic Press, London.

Hyman, L. H., 1940–1961. *The Invertebrates,* vols. 1–6. McGraw-Hill, New York.

Jägersten, J., 1972. *Evolution of the Metazoan Life Cycle: A Comprehensive Theory.* Academic Press, New York.

Kerkut, G. A., 1960. *Implications of Evolution.* Pergamon, New York.

Manton, S. M., 1977. *The Arthropoda: Habits, Functional Morphology, and Evolution.* Oxford University Press, Oxford.

Muller, S. M., and A. Campbell, 1954. The relative number of living and fossil animal species. *Systematic Zoology 3:*168–170.

Valentine, J. W., 1970. How many marine invertebrate fossils? A new approximation. *Journal of Paleontology 44:*410–415.

Yonge, C. M., and T. E. Thompson, 1976. *Living Marine Molluscs.* Collins, London.

FOURTEEN

The Metazoa: Chordates and Related Animals

TO UNDERSTAND the evolution of the chordates, it is useful to start by examining fishes. These aquatic animals display all the main features of a chordate—notochord, dorsal nerve cord, and gills—and, additionally, they are an excellent reference from which to pose two sets of problems: Where did fishes come from? What did they give rise to?

Structure and function in chordates

In Chapter 13, our study of invertebrate evolution centered on three morphological features—coelomic cavities, body segmentation, and appendages—as they served the needs of mobile predators. The same approach is useful in examining fishes.

Fishes and other chordates possess a well-defined coelom, their appendages are obvious and varied, but their segmentation poses a special problem. Concentrating on what we see in a bony fish, such as a perch (Fig. 14–1), segmentation is apparently restricted to the skeleton and the body muscles and to the nervous system as it innervates blocks of muscles. Why should this be so? The key lies with locomotion, and again Clark has brought together a convincing analysis of the situation.

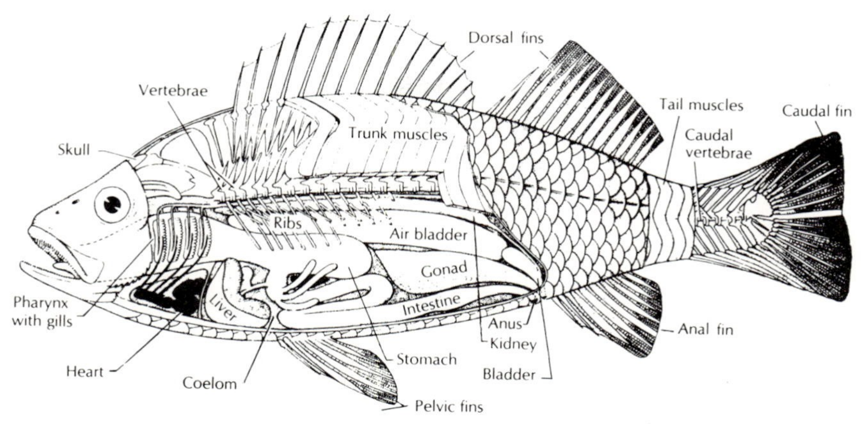

Fig. 14-1: The bony fish *Perca,* a yellow perch (cut-away longitudinal section) The pectoral fins (not shown) are posterior and ventral to the gills.

Fish swim by using their fins and by right-left or lateral undulations of the whole body. As we will see below, fins (except for the caudal or tail fin) were absent in early fishes. Hence, undulations were the primitive basis for fish locomotion (Fig. 14–2). Clark and others view the beginning of fish locomotion in terms such as the following. We must start with the notochord. This is a cartilaginous rod that runs almost the length of the animal along its dorsal side. The notochord is present in fish embryos and is replaced in adult bony fishes by the vertebrae of the so-called backbone. Considering this as an

Fig. 14-2: Undulatory swimming movements in a fish. Arrows, top diagram, indicate the forces exerted, by undulations, on the watery medium. The sum of those forces, a posterior push, propels the fish forward.

example of ontogenetic recapitulation of phylogeny—the biogenetic law, again—we then conclude that the notochord was present as a kind of flexible skeletal rod in very early fishes. It was used in swimming as a device to straighten out bodies bent by asymmetric muscular contractions. To generate directed motion, contractions must be coordinated. Waves of contractions and relaxations would do it. Look again at Fig. 14–2. On the inside of the curve in a fish body, the muscles are contracted; they are relaxed on the opposite side. Further posteriorly, where the body curve goes the other way, the muscles are contracted on the opposite side from the first curve and the relaxed muscles are, also, comparably reversed. As contraction waves pass alternately down one side of the body and then the other, the body undergoes the undulations or wriggling or "fish-tailing" that results in swimming.

Segmentation of muscles arose in this system as a means of achieving localized contractions of the body. If the muscles along one whole side contracted simultaneously, it would result in a C-shaped fish. And contractions on the other side would simply reverse the C-shape. Localized contractions could result in a S-shaped fish, and when waves of contractions, moving posteriorly, acted against a stiff, but flexible notochord, undulatory swimming became possible. The segmentation of muscle allowed localized muscle contractions to occur, and when nerves innervated each segment separately, coordinated waves of contraction became possible.

Segmentation of the skeleton became a necessity if fish were to continue to swim by right-left undulations. Such a skeleton, of course, also allowed localized attachment of muscles and so enhanced their effectiveness. As the skeleton and muscles evolved, the notochord was replaced as the stiffening and supporting system of the body, but the segmentation of bones and muscle was retained to preserve the flexibility needed for swimming.

None of these selection pressures for segmentation affected any other organ systems. Hence, we see no suggestion of segmentation in the digestive, excretory, reproductive, endocrine, or integumentary systems. What slight suggestion we see in the circulatory and respiratory systems comes from running blood vessels through the gills. But, gills are not, by any stretch of the imagination, an example of segmentation. They are a regional repetition of respiratory organs and affect only a minor portion of the body. Hence segmentation in fishes appears in only three organ systems. (Even the nervous system is only partially segmented, since the brain is not subdivided into comparable subparts.)

The coelom, seen in these terms, is an unsegmented cavity wherein unsegmented organ systems are located. Their relatively unrestricted location there does not interfere with the undulations of the muscles and skeleton, nor do the latter interfere with coelomic organs. Hence, the chordate coelom is presently adapted to act as a compartment for unsegmented organ systems of the chordate body. That, however, does not explain its origin. We will return to that shortly.

Appendages in fish, in all probability, arose as adjuncts to swimming behavior. The dorsal fin acts as a hydrodynamic stabilizer; the pectoral and pelvic fins (Fig. 14–1) aid in controlling speed of swimming and in maneuvering. All this is necessary for the predaceous mode of life pursued by fishes. Any structures that improved swimming abilities would be selected for. And once present, fins could be preadaptations for other functions as we will see later.

ADAPTATIONS IN FISH

Before turning to the problem of the origin of fishes and of chordates, in general, let us look at the subsequent evolution of fishes by first commenting on fish evolution itself and then by surveying what evolved from fishes.

Fish evolution. The earliest known fishes are Ordovician fossils, some 500 million years old. These fossil remains are from a kind of fish called an ostracoderm (Fig. 14–3). Os-

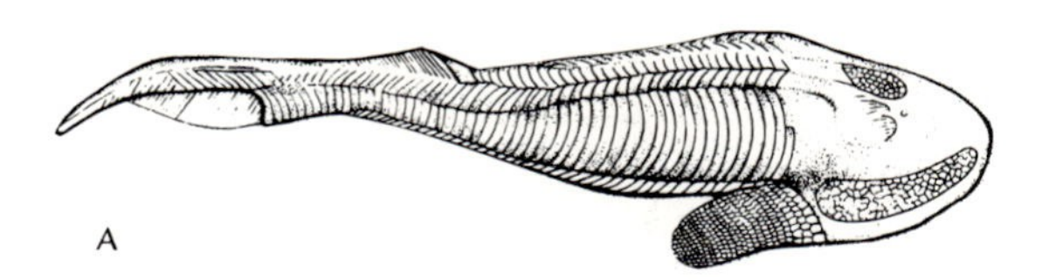

Fig. 14-3: The ostracoderm fish *Hemicyclaspis*. Note the single anterior lobe, an extension of the skull structure. It is probably not homologous to the pectoral fins of modern fishes. Bony plates set in the skin are a kind of armor protecting the body. A reconstruction (A); fossil remains (B).

tracoderms are notable for their being jawless and for having slim, tapering bodies with only a tail fin, for propulsion in swimming, and two pectoral lobes, of doubtful function. Such lobes may have functioned as stabilizers during swimming, and to hold the fish upright while it was resting on the bottom. From a form such as this, there evolved more than 30,000 different species of fish. These can be divided into the two great modern groups of Chondrichtyes and Osteichthyes. The

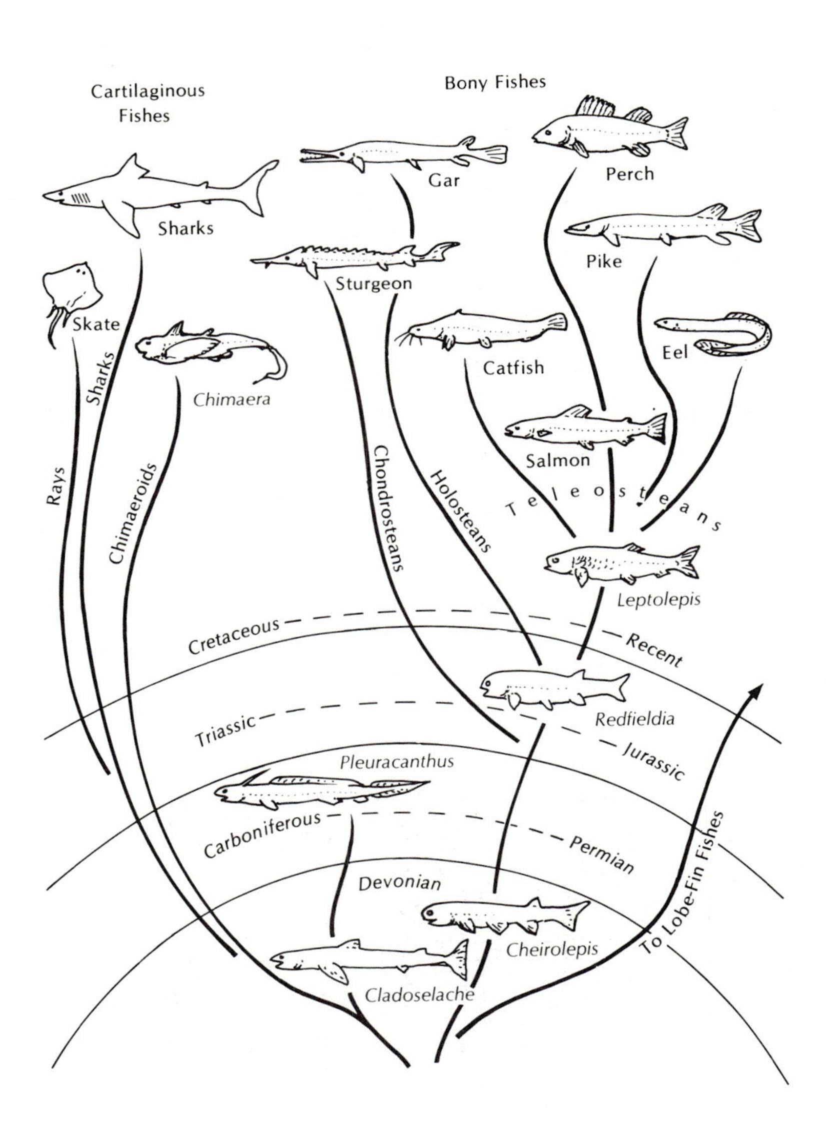

Fig. 14-4: Fish evolution.

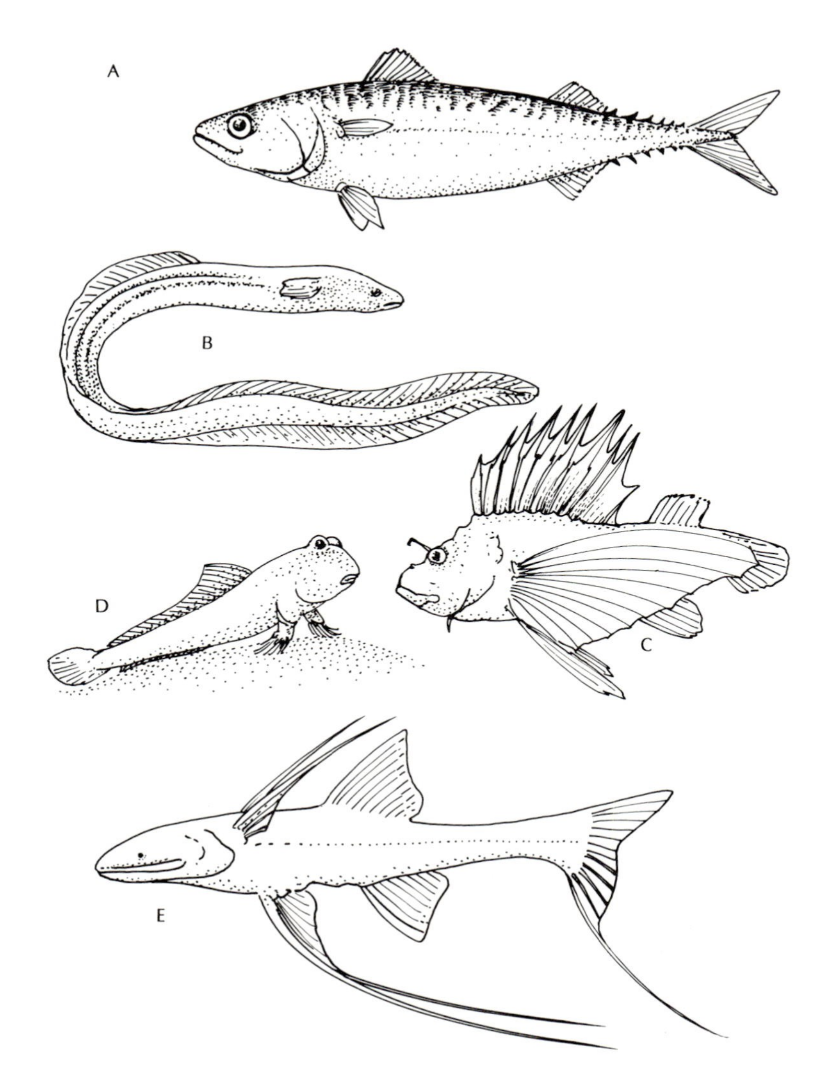

Fig. 14-5: Some modern bony fishes. A mackerel (A); an eel (B); a scorpion fish (C); a mud-skipper (D); and a deep-sea fish (E).

former have a cartilaginous skeleton, the latter a bony skeleton. Sharks belong to the former group and tuna, trout, and perches (Fig. 14–1) to the latter group. A dendrogram summarizing the diversity of fishes is given in Fig. 14–4. This dendrogram, because of the great evidence of branching, suggests that once fishes appeared they underwent a great adaptive radiation. The most successful fishes, judging from all modern fishes, are those with jaws and with dorsal, pectoral, and pelvic fins. Some have additional fins, such as anal fins, and some (the eels) have lost certain fins (Fig. 14–5A,B).

Some fish have adapted their fins to functions in addition to swimming—such as the brilliant display of the scorpion fish (Fig. 14–5C) and for crawling across mud flats (Fig. 14–5D). Starting from the small, jawless ostracoderms, fishes have adapted themselves to a wide range of aquatic habitats. These include the invasion of freshwater from the original marine habitat and of waters of all latitudes and depths including those where no light penetrates. The deep-water fish show some of the most bizarre morphology imaginable, but careful study reveals them to be adaptations to life at these dark depths. Here the predator generates its own lights and artificial lures to bring prey within reach of the fearfully toothed jaws (Fig. 14–5E).

Perhaps the most lasting, overall impression of fish evolution is conservatism in the cartilaginous fishes and continual change in the bony fishes. The shark-like shape of *Cladoselache,* an ancestral cartilaginous fish, reappears in modern sharks (Fig. 14–4). In between those two extremes, the body flattened, as seen in skates, rays, and their relatives. The adaptations behind these rather minimal changes are thought to relate to food-getting. *Cladoselache* is believed to have been an actively swimming predator of Devonian seas. Then, for some reason, its food decreased and the cartilaginous fishes adapted themselves to the capture and the ingestion of hardbodied molluscs and crustaceans. (Their teeth also flattened for their new role of crushing and chewing.) Then, as bony fishes became more plentiful, the active predators with sharp, cutting teeth reevolved as our modern sharks.

In the bony fishes the evidence for continual change is quite real. No Devonian fish is alive today—all have been replaced. It is hard to pinpoint a particular cause here. In fact, some authorities consider it to be a result of a successful group of organisms becoming progressively more efficient in their way of life. As the fish air bladder allowed fish to maintain their position at any depth, and as their bodies changed to produce faster and more agile swimmers, they presumably became better able to live in water with less and less expenditure of energy. In this sense, they progressively became more efficient fish. As one of the great authorities on vertebrate evolution, J. Z. Young, of University College, London, said, ". . . in a sense fishes have not found a new environment. But they have found endless new ways of living in water" (From *The Life of Vertebrates,* p. 243).

FURTHER VERTEBRATE EVOLUTION

The emergence of Amphibia from the fishes, reptiles from amphibians, and birds and mammals from reptiles is one of the greatest passages we have in the history of life. It is great for two reasons; it documents the inventiveness of natural selection, and the documentation is remarkably convincing. This is not to say that all phylogenetic problems are solved; they are not. For example, we are still filling in many details on the emergence of various groups, none the least of them being that of humans. But, taking the case of humans, the wealth of data from comparative anatomy, the fossil record, developmental biology, molecular and chromosomal evolution all argue convincingly for a primate ancestor for humans. Major evolutionary trends proceeding from fishes to all other major vertebrate groups are clear. The actual working out, case by case, of phylogenetic details is the ongoing work in this great area of evolutionary study. Here we will comment on the highlights of what we know today and what remains to be learned.

Amphibian evolution. In contrast to the fishes, the 4,000 or more species of amphibians show a more limited evolutionary picture. True, they did invade the land, but given that important innovation they appear as a group to be stuck there at the border, hovering between water and land. They have remained amphibians for hundreds of millions of years.

There is a close resemblance between the skulls of the earliest amphibians and the bony fishes from which they obviously evolved (Fig. 14–6). Further evolution of the amphibians produced *Eryops*, a large amphibian, about 2 m long, and from it modern toads and frogs evolved. Another somewhat parallel line of evolution produced modern salamanders and newts (Fig. 14–7). Between the frog line and the salamander line there has been considerable parallel evolution. In

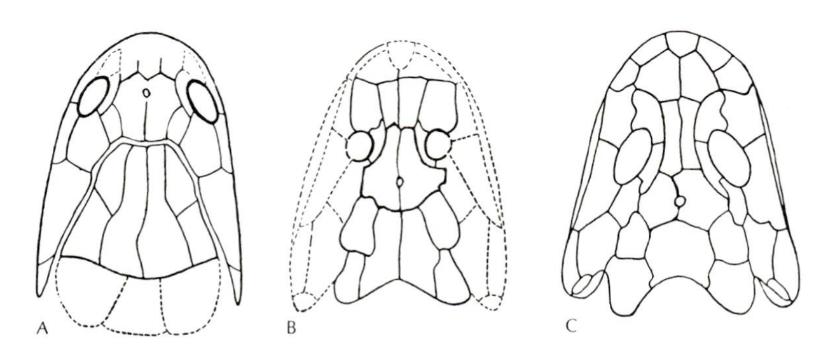

Fig. 14-6: Aposemic changes in skulls of early bony fishes and an early amphibian. A. *Osteolepis,* an ancestor of the lobe-finned fish. The lobe-finned fishes include the present-day lung fish that negotiate dry land as they move from one freshwater pool to another. This ability is thought to be the essential preadaptation for invading land. Hence, it seems very plausible that lobed fins evolved into amphibian limbs. B. *Elpistostege,* a mid-Devonian form transitional to the amphibians, lived about 375 million years ago. C. *Ichthyostega* is clearly a four-limbed vertebrate (a tetrapod). It would be a good plesiomorph for the amphibians. Its fossils, found in upper Devonian rocks, are about 350 million years old.

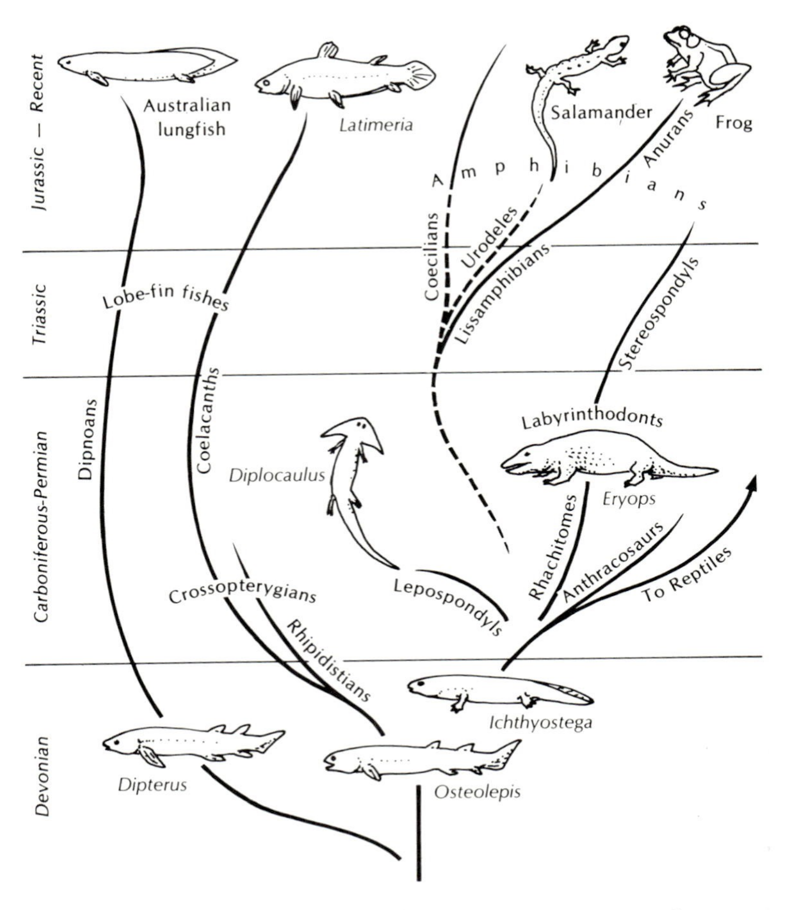

Fig. 14-7: Phylogeny of the lobe-finned fishes and amphibians. The lobe-finned forms arose from the plesiomorph of all bony fishes *Cheirolepis* (see Fig. 14-4). Some authorities derive all lobe-finned fish, including the lung fish, from *Osteolepis*. A different derivation is given here. But there is agreement that *Ichthyostege*, the amphibian plesiomorph, is derivable from *Osteolepis*.

both, there are species that have reduced their dependence on water and others that have become more aquatic; there are species that are burrowers and others that have even become largely tree dwellers or aboreal. In all cases, the adaptations to land demanded specializations not present in the ancestral fishes. The fish air-bladder has evolved into a lung for the terrestrial amphibians. In the sense that these specializations allowed a successful invasion of land, the amphibians represent an "advance" over fishes. The first vertebrates truly at home on land are, however, the reptiles.

Reptilian evolution. The reptiles started their evolution from an organism like *Seymouria,* their plesiomorph (Fig. 14–8). From a modest vertebrate such as this, not much more than one-half a meter in length, the magnificient adaptive radiation of the reptiles took place (Fig. 14–9). Present-day turtles and tortoises, lizards in a large variety, snakes, and crocodiles and alligators all arose from a *Seymouria*-like ancestor, as did the extinct reptiles like aquatic ichthyosaurs (Fig. 9–10) and the plesiosaurs and that rich array of our favorite fossil animals, the dinosaurs.

In the reptiles, as in the bony fishes, the fossil record tells us there has been continual change. It seems that in these terrestrial forms and in their descendants, the birds and mammals, no species remains unchanged for more than a few million years. There is a continual transformation of one species into another, with extinction and diversification, too. Among the reptiles, there was a repeated reinvasion of water from terrestrial ancestors. This is evidenced by turtles, crocodiles, various aquatic lizards, and water snakes. (The sea snakes are especially well adapted to water.) Among the fossil forms are aquatic ichthyosaurs and plesiosaurs. But on the other hand, reptiles also opened a new adaptive zone by becoming airborne. Certain of the winged forms, pterosaurs, were huge, having a wing-span greater than 8 m. And among those reptiles that remained terrestrial, there were many adaptations to various kinds of terrestrial life. Some remained small, others became enormous. Some were vegetarians, others meat eaters. There were many successful experiments at bipedal locomotion, *Tyrannosaurus* being an outstanding example (Fig. 10–15).

As with the fishes, there seems to be a continual improvement in efficiency in their mode of life. It is hard to say just how the later reptiles were better adapted than their Mesozoic ancestors, but the fact is many ancient reptilian forms have persisted. And they survive along with, and often in competition with, their yet more highly evolved descendants, the birds and the mammals.

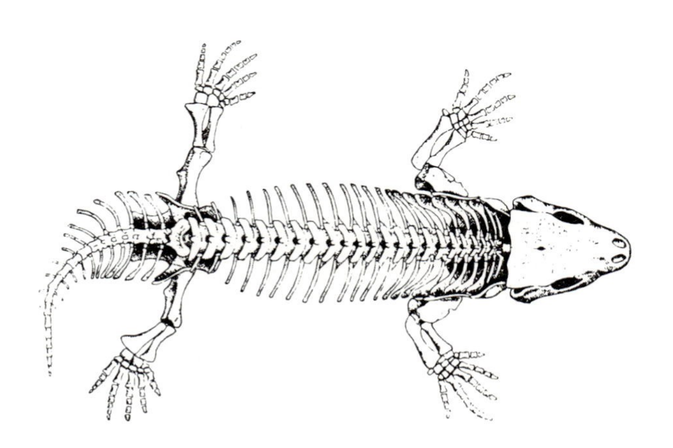

Fig. 14-8: *Seymouria* skeleton. This Permian reptile lived about 250 million years ago.

Avian evolution. The plesiomorph of the order Aves, or the birds, is, of course, *Archeopteryx* (Fig. 10–14). A comparison of the reconstructed skeleton of *Archeopteryx* and that of a modern pigeon is instructive (Fig. 14–10). The differences reflect the recent reptilian ancestry of *Archeopteryx* and the 150 million years of evolution that extend from *Archeopteryx* to the modern bird. Flight, which is the key adaptation of birds, is responsible for the most striking difference between ancestor and descendant in this figure. That is seen especially in the breastbone to which the flight muscles of the pigeon breast are attached.

Birds have gone through a great many adaptations, and rather than present their phylogeny, which can be followed in great detail, it is informative to look simply at their adaptive radiations (Fig. 14–11).

Bird evolution seems to have started in the Jurassic, about 150 million years ago. There are few good fossils for reasons that are not entirely clear. It may be that as flying vertebrates chances for fossilization were very poor. Most phyletic rela-

Fig. 14-9: Phylogeny of the reptiles. Note the key position of *Seymouria,* as plesiomorph, and the adaptive radiation that follows. There are four large groups of reptiles. Modern turtles and tortoises come from Anapsida, modern lizards and snakes from the lepidosaurian branch of the Diapsida. Birds arose from the archosaurian branch of the Diapsida, mammals from the Synapsida. The Parapsida was dominated by the now extinct aquatic reptiles.

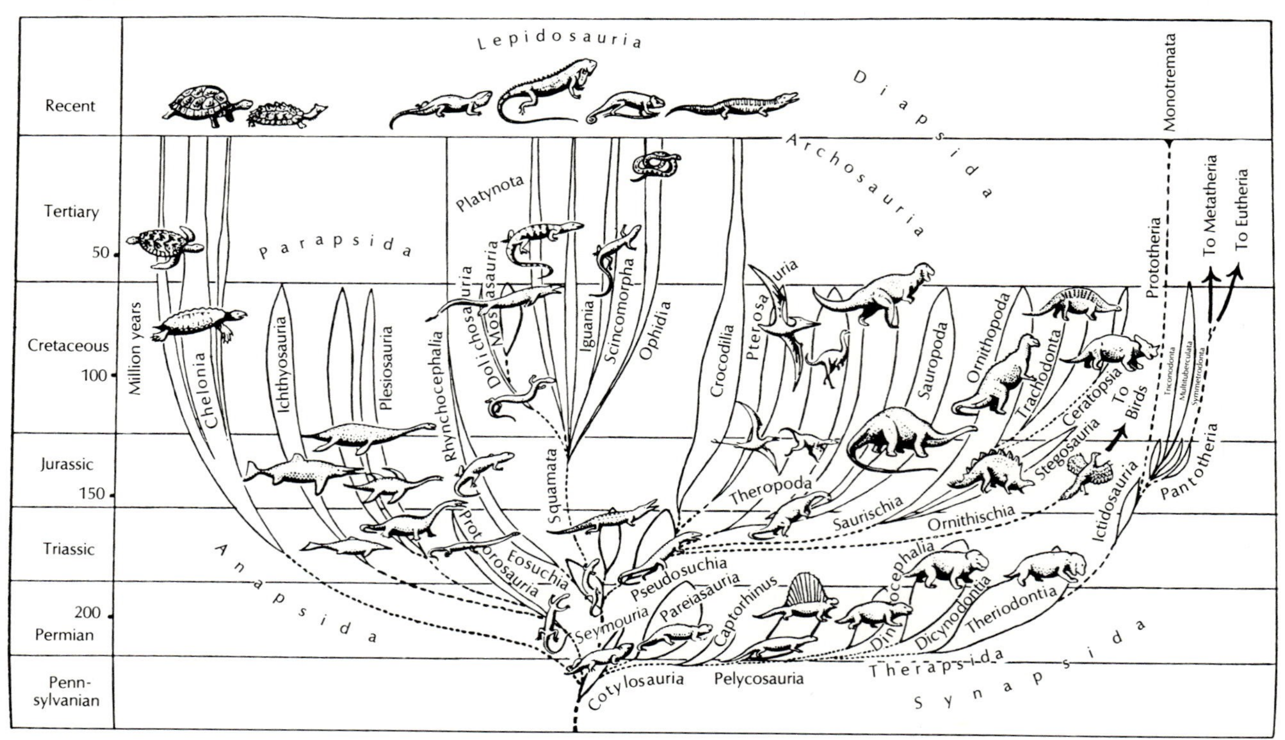

tions have to be inferred, therefore, from homologies of extant forms—adult and embryonic.

In addition to flight, another notable innovation is the maintenance of a constant body temperature (homiothermy), popularly known as being warm-blooded. This neosemic trait opened up new habitats for birds. They carry a livable temperature around with them and can, therefore, survive the frigid climates of the Arctic and Antarctic. Feathers are thought to have originally evolved from reptilian scales as a device for conserving body heat and possibly food capture (Fig. 10–15). Their later use in gliding and then flying seems to have been a genuine fringe benefit from earlier adaptations. But warm-bloodedness has a cost, too; it takes significant fuel to maintain a constant body temperature. Birds have put a premium on being agile, active animals. Their large brain and superb motor coordination have resulted in a body that is continually active, and successful in adapting to a great variety of niches. This activity deriving from homiothermy is an advantage also shared by the mammals.

Fig. 14-10: Comparison of the skeletons of *Archaeopteryx* (A), the plesiomorph of birds, and a modern pigeon (*Columba*) (B). Certain homologous bones are shaded in both skeletons. Note the larger brain case in the pigeon, but also note that the terminal digits (homologous with our fingers) are reduced in the wings. The breastbone and ribs of the pigeon are more highly developed than their *Archaeopteryx* homologs, as are the pelvic bones, where the legs join the body. Note, finally, differences in the tail.

Mammalian evolution. J. Z. Young has said that "Mammals might be defined as highly percipient [i.e., with keen sensory perceptions] and mobile animals, with large brains, warm blood, and a waterproofed, usually hairy skin, whose young are born alive" (*The Life of Vertebrates,* p. 535). These vertebrates are reliably believed to have arisen from a distinct reptilian line called the synapsid. [The term synapsid refers to a type of skull (Fig. 14–12). Birds, for example, arose from one group within the diapsids—reptiles of another skull type. Both derive from *Seymouria,* the reptilian plesiomorph.] By the Carboniferous period, mammal-like forms are seen in the fossil record, and fully evolved mammalian forms appear in the mid-Jurassic, about 150 million years ago.

If we are to name a mammalian plesiomorph, the early Triassic organism called *Cynognathus* would be a good candidate (Fig. 14–13). This dog-like and dog-sized reptile carried a variety of mammalian characters. Its teeth indicated that it chewed its prey into small pieces rather than swallowing it whole, as reptiles usually do. Details of the spine and rib cage are mammalian, and the limbs, especially, were mammalian. These appendages are attached under the body rather than to the sides, as in lizards, and the knees point forward and the elbows backward. This suggests a land vertebrate that could really run, rather than scuttle rapidly. All in all, we see evidence of a very active, carnivorous predator. From this beginning there started an adaptive radiation of mammals. This gave rise to some egg-laying mammals—the only exceptions to live birth in mammals—of which, today, we have two representatives, the Australian spiny anteater and duck-billed platypus. Then there are marsupials whose young are born alive, but very immature and who move at birth to a ventral pouch on the mother, as in the kangaroo, and grow there until they are able to move about by themselves. And finally there are the placentals. These are the dominant mammals. There are some thousands of placental species, whereas the marsupials are represented by a couple of hundred species at the most.

The placentals are believed to have arisen as small insectivorous mammals. Here the adaptive radiation of mammals becomes a veritable evolutionary explosion (Fig. 14–14). Several evolutionary trends can be mentioned.

1. There is an increase in size in many mammals, starting

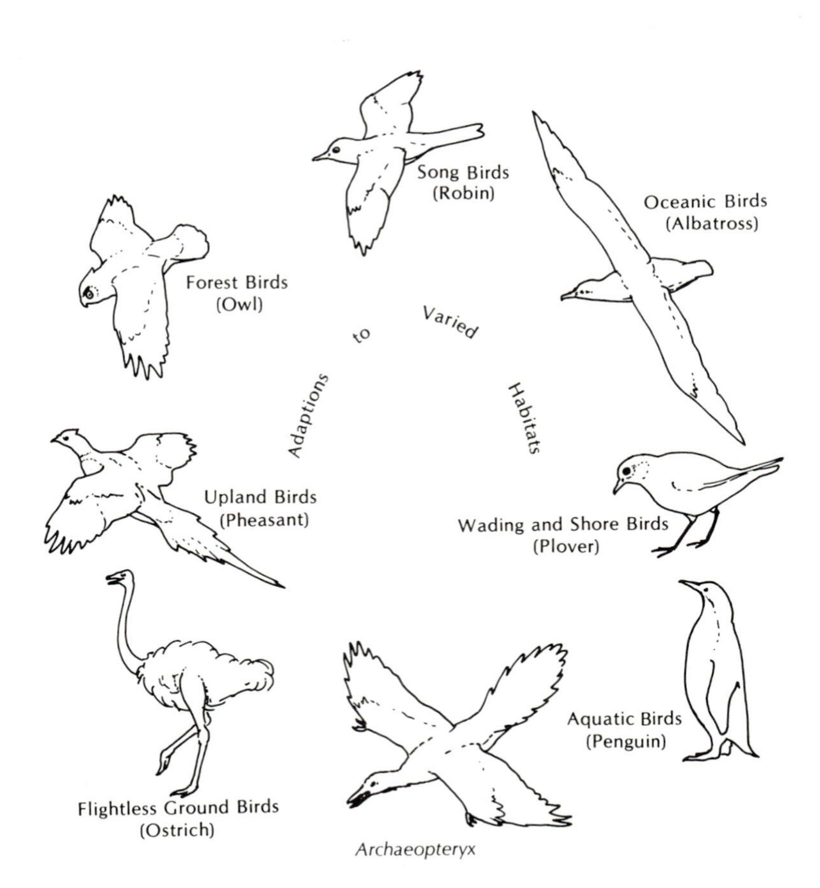

Fig. 14-11: Bird adaptations. Note, especially, the differences in beaks, legs, and feet.

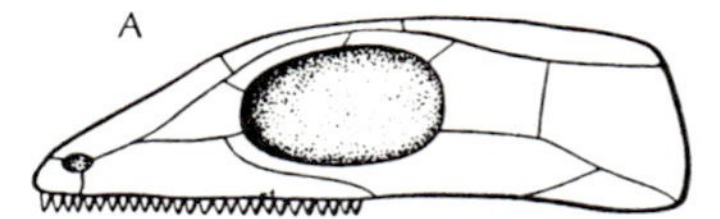

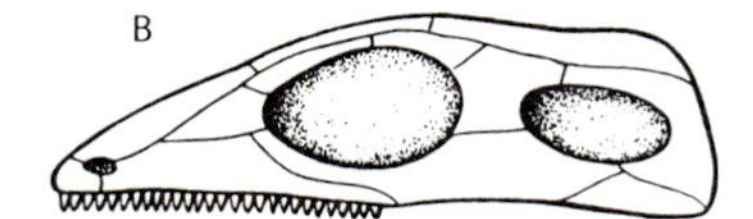

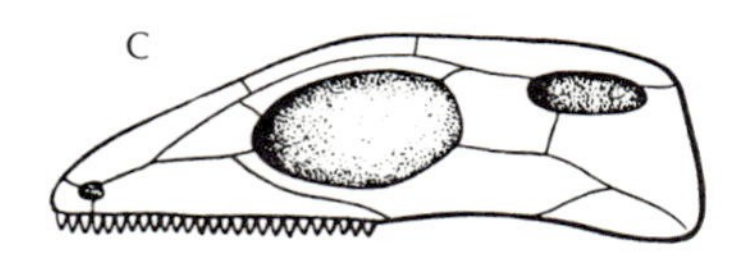

Fig. 14-12: Types of reptilian skulls. The key differences relate to the temporal openings, that is, the openings behind the eye socket. Anapsid skull (primitive reptiles and turtles) (A); synapsid skull (reptiles leading to mammals (B); parapsid skull (many reptiles, including most living ones; reptiles leading to birds) (C). (See Fig. 14-9 for the reptilian distribution of these skull types.)

from the insectivorous ancestor. Larger body size is an aid in heat conservation, for it decreases the surface to volume ratio of the animal's body. (See the discussion of this topic in Chapter 11.)

2. Limbs often get longer and become specialized for running, holding prey, grasping tree limbs, etc.

3. The teeth become fewer in number and more specialized. Whereas reptilian jaws are typically equipped with many similar, small, sharp teeth, mammals have the incisors, canines, premolars, and molars familiar in our dog and cat pets and in ourselves.

4. The brain increases in size, and in specialization, too, since certain parts, in particular, are larger. These parts include the non-olfactory parts of the cortex and larger frontal lobes. Other changes presumably associated with complex behavior and memory, also occur. We will return to this topic when we consider human evolution at the end of the chapter.

Again, turning to the insights of J. Z. Young we come across this surprising summary of mammalian evolution: "Warmth, enterprise, ingenuity, and care of the young have been the basis of mammalian success throughout their history" (*The Life of Vertebrates*, p. 535). Warmth refers to the vital activity engendered by homiothermy. Enterprise and ingenuity depend on a highly developed nervous system that not only coordinates the active mammalian body, but is also continually feeding in sensory information, which is efficiently processed by and stored in the complex mammalian brain. Such a mechanism for inquiring into and analyzing the world around us reaches its apogee in the Primates, which includes monkeys, apes, and humans. And care of young implies a system of learning that allows transmission of experiences in a way that by-passes genes and development. It is a cultural transmission. Not all mammals can be said to have a culture—that view is usually applied only to humans and their hominid relatives. But rearing of the young, which occurs in all mammals, is a beginning.

Fig. 14-13: Reconstruction of *Cynognathus,* the mammalian plesiomorph. The fossil record on which this is based is about 200 million years old.

THE PROTOCHORDATES

The foregoing account is exclusively that of vertebrate evolution. The vertebrates are the largest subgroup (usually a subphylum, see Appendix I) within the chordates, hence the attention paid to them in the preceding pages. However, when we turn to the problem of chordate origins we must consider other groups along with the vertebrates. These include the cephalochordates and urochordates. Sometimes the hemichordates are included in the phylum Chordata (Fig. 14–15). Collectively, we can refer to these three groups as the protochordates. It is a catch-all name for organisms that show chordate features, but that are not included in the Vertebrata. In considering the whole of chordate evolution, we have to explain the phyletic relations of the protochordates and the vertebrates. As the term protochordate suggests, those forms may yield useful information regarding chordate origins.

Chordate origins

The first step in analyzing the origin or origins of a group of organisms is to locate a plesiomorph or plesiomorphs. In the case of the chordates, that analysis can start from the knowledge that today there are two chief candidates for the Chordate plesiomorph. One is *Amphioxus* (Fig. 14–15A) and the other is the so-called tadpole larva of a urochordate or tunicate (Fig. 14–15B). The argument in support of these candidates largely depends on phenoclines and embryology. In terms of embryology, we find two characters or semes that are often argued to be homologous.

1. Cleavage in the cephalochordates, urochordates, and vertebrates is radial as compared to the spiral cleavage seen in most invertebrates (Fig. 13–7).

2. Similarly, coelomic development is enterocoelous,

namely, by an out-pocketing of the primitive gut (Box 13–1).

In terms of phenoclines, the vertebrates show a more complex development of organ systems than amphioxus or the tadpole larva. The most important exception is the organization of the gills and associated structures. In amphioxus these are especially well developed. In the tadpole they are prominent,

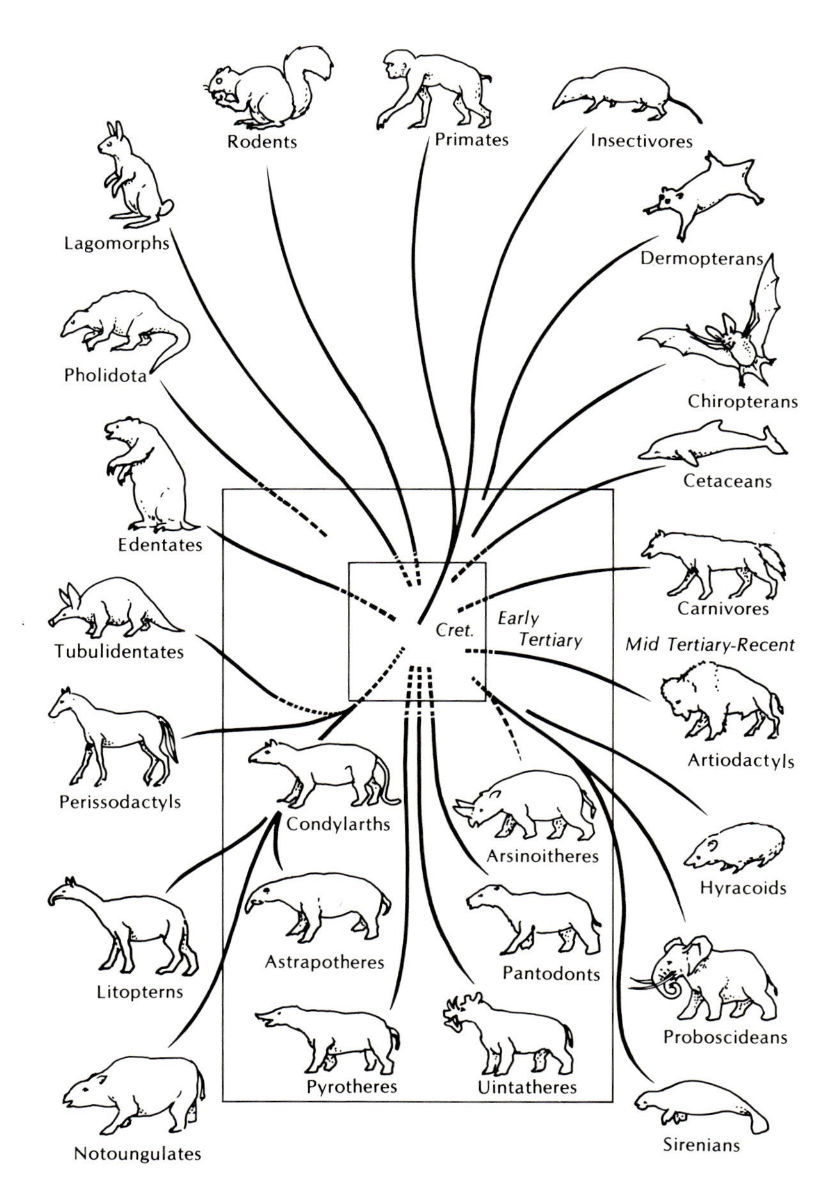

Fig. 14-14: Adaptive radiation among the orders of placental mammals. These Eutheria are shown (far right) in Fig. 14-9.

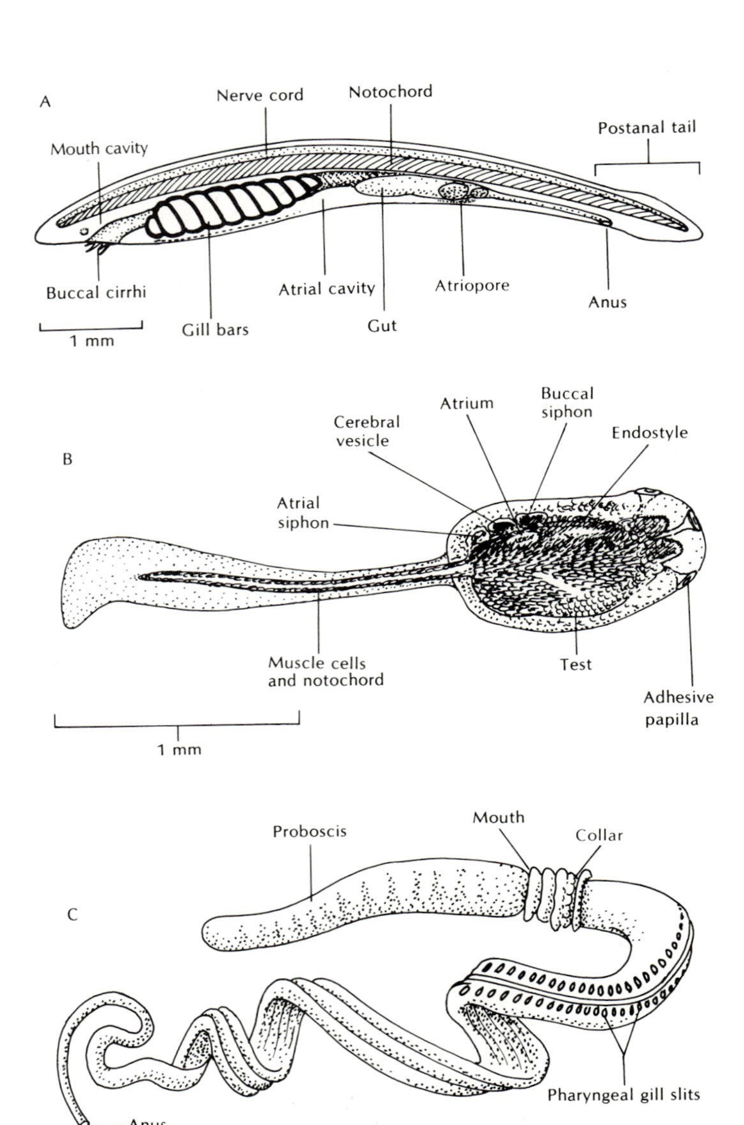

Fig. 14-15: The protochordates. A young *Amphioxus* (A), member of the cephalochordates; tadpole larva of the tunicate or urochordate *Amaroucium* (B); *Balanoglossus* (C), an acorn worm, belonging to the hemichordates.

but come into functional prominence only in the adult. And in the fishes, gills are respiratory organs; fish are not filter feeders. In other vertebrates, gills are successively reduced in importance as one moves on to the birds and mammals. But this reduction of gills in the more complex vertebrates, and their prominence in protochordates, is another phenocline that correlates inversely with those from other organ systems. These arguments, then, provide evidence for a common ancestry for cephalochordates, urochordates, and vertebrates, and allows us to conclude that the former are closer to the ancestral form (as yet unspecified here) than are the vertebrates.

Our discussion now can be divided into two parts: (1) The transition from the chordate plesiomorph to the vertebrates and (2) the origin of the chordate plesiomorph.

EVOLUTION FROM CHORDATE PLESIOMORPH TO VERTEBRATES

Within the vertebrates, the ostracoderm fish *Hemicyclaspis* (Fig. 14–3) is an excellent plesiomorph for that group. Thus, we can ask how *Hemicyclaspis* evolved from *Amphioxus* or from a tunicate tadpole larva. Which is the more convincing ancestor of ostracoderm fishes?

Amphioxus as a plesiomorph. The cephalochordate amphioxus (Fig. 14–15A) shares many homologies with a fish like *Hemicyclaspis*. They both have a notochord, a dorsal nerve cord, and gills. (They both lack jaws and well-developed fins, other than the tail fin.) It is hard to compare internal anatomy because the internal organs of the fossil *Hemicyclaspis*, were not preserved. (The notochord and dorsal nerve cord are inferred to be present.) No embryological comparisons are possible; there are no fossil *Hemicyclaspis* embryos. Overall, we can see that amphioxus is a burrower and a swimmer. It filters its food out of its watery environment. We can only guess at the mode of life of *Hemicyclaspis*. It most probably did not burrow, but was protected from larger predators (clawed arthropods?) by the bony plates in its skin (Fig. 14–3). What it fed on as it rested on the ocean floor or swam about is not known at this time.

The Cambrian is the period in which we think the ostracoderms first emerged. The seas then were probably abundantly furnished with algae and invertebrate animals. There was sufficient oxygen in the atmosphere to dissolve into at least shallow waters. The calcium and phosphates of these waters could have been the source of skeletal materials. (It has been suggested that deposits of calcium phosphate in animals were first a device for storing relatively scarce phosphorus. They might thus have been a preadaptation for protective shells, which, in turn, might have been a preadaptation for the evolution of skeletons.) Out of this combination of environmental factors it is plausible to see selection acting so as to transform a small burrower-swimmer like amphioxus into an ostracoderm fish. The essential point is to move from the exploitation of filter feeding in bottom muds to active ingestion of other plants and animals.

This scheme, however, does not answer the following puzzling question: Why does a filter feeder like amphioxus have the heavily muscled body of an active swimmer? One answer is that amphioxus originally might have been an active predator that evolved as an efficient filter feeder when the ostracoderms went on to become, as jawed fishes, the more efficient active predators. Rather than competing with these forms, amphioxus found its own niche, by exploiting filter feeding, where it has survived up to today. It retains the body of an active swimmer so as to move rapidly from one area to another when it leaves the mud. It is not competing with burrowing worms or with jawed fishes, since its niche combines something from both the niches of those other groups of animals. In this view, amphioxus is a specialized relict of a chordate ancestor that was a free-swimming ingestor. It represents one surviving branch from that ancestral form, whereas ostracoderms represent another, vastly more successful branch. It is plesiomorphic only in that it is closer to the ancestral form than is a fish like *Hemicyclaspis*.

The tunicate larva as a plesiomorph. The plausibility of the tadpole larva as a plesiomorph depends on neotenous development. Clearly the adult tunicate (Fig. 14–15B), as a well-adapted sessile filter feeder, is not likely to have evolved into a free-swimming ostracoderm. Furthermore, the adult tunicate nervous system is minimal, and there is no evidence of a notochord. In other words, its specializations are antagonistic to two of the characters diagnostic of chordates. But the tunicate larva contains all three chordate features, although the dorsal nervous system is only minimally apparent. Through neotenous development, the free-swimming larva could become sexually mature and then evolve into a free-swimming, jawless fish. Such is the argument here.

The importance of neoteny, according to the British biologist A. C. Hardy, is that it provides an escape from specialization. How else could a sessile, filter-feeding tunicate evolve into a free-swimming, fish-like form? There is no other known way, but that does not mean fish did evolve this way. If the tadpole larva were the only plausible plesiomorph, its neotenous evolution would be forced on us until a better alternative became apparent. *Amphioxus* is such an alternative, as we have seen. Biologists readily admit that neoteny *can* occur, but that it *must* have occurred in the evolutionary emergence of the vertebrates from protochordates is not necessarily admitted. The real point here is to find the selective advantage for such neotenous evolutionary development and for that we must consider the Cambrian environment again.

As noted in the preceding discussion of *Amphioxus,* the Cambrian seas were well stocked with marine algae and invertebrate animals. The invertebrates surely contained an array of forms, ranging from burrowing and sessile worms through crawling molluscs to crawling and swimming arthropods. Where would the emerging chordates find a niche? Looking at the urochordates or tunicates, two answers are possible. Considering an adult tunicate, the answer is a sessile filter feeder, attached to the bottom in shallow waters. Considering the tadpole larva, one might argue for a small, actively swimming predator, but not very convincingly. The tadpole larva is supplied with yolk that lasts until there has been settling onto the substrate and metamorphosis into a recognizable adult form. In other words, the tadpole does not feed. Its real function is simply dispersal. Although relatively short-lived as a larva, there is sufficient time for the tadpole to travel away from its parent and find its own attachment site. Hence, the motile larva is not a biogenetic recapitulation of a free-swimming ancestral form, but an adaptation for dispersal, as is true of many aquatic animal larvae.

So we turn to the adult, the sessile filter feeder. Since it would be in competition with sponges and the species in the lophophorate phyla, it would somehow have to be more efficient than these forms, or utilize different food. The latter is probably the case, although there is no good evidence on which to judge. Nonetheless, the urochordates have survived as filter feeders.

But none of this makes it clear why there should be a selective advantage for neotenous development of the tadpole larva. In fact, the tenor of the preceding discussion suggests there is none. So although neoteny *can* happen, it cannot be convincingly argued that it happened here.

We come to the tenuous conclusion that, from our present knowledge, amphioxus is a more reasonable chordate plesiomorph than is a urochordate tadpole larva. The next problem, then, is to try to find out where amphioxus came from. In the process, we also want to explain the origin of the urochordates and the remaining protochordaaes, the Hemichordata.

THE ORIGIN OF THE CHORDATE PLESIOMORPH

To solve this problem, we must, of course, search for homologies between amphioxus and other non-chordate or inver-

Box 14–1. Homologies among a vertebrate, selected protochordates, and selected echinoderms.

Semes	Occurrence of semes in different groups					
					Echinodermata	
	Vertebrata (e.g., *Hemicyclaspis*)	Cephalochordata (e.g., *Amphioxus*)	Urochordata (e.g., *Amarouciam*)	Hemichordata (e.g., *Balanoglossus*)	Stylophora (e.g., *Reticulocarpos*)	Asteroidea (e.g., *Luidia*)
A. *Adult organism*						
1. Body form	jawless fish with bony skin plates	fish-like, poorly developed head	rounded body with large pharyngeal pouch	worm-like, anterior proboscis	modified bilateral symmetry, skin plates and tail	pentaradiate symmetry, rough skin
2. Coelom	present	present	much reduced	much reduced	present	present
3. Notochord	absent in adult	present	absent in adult	absent in adult	present (?)	absent
4. Gill slits	present	present	present	present	present	absent
5. Dorsal nerve cord	[presumably present]	present	much reduced	much reduced	present (?)	absent
6. Nerve-muscle connections	[nerve fibers extend to muscle]	muscle fibers extend to nerves	?	?	?	muscle fibers extend to nerves
7. Acetylcholine neurotransmitter	[probably present]	present	present	present	?	present
8. Binding of iodine	present (+ thyroxine)	present (+ thyroxine)	present (+ thyroxine)	present (no thyroxine)	?	?
9. Phosphagen	[phosphocreatine]	phosphocreatine	phosphocreatine or phosphoarginine	phosphocreatine and phosphoarginine	? ?	phosphocreatine and phosphoarginine
B. *Development*						
10. Cleavage	[presumably radial]	radial	radial	radial	[presumably radial]	radial
11. Coelom formation	[presumably enterocoelous]	enterocoelous	enterocoelous	enterocoelous	[presumably enterocoelous]	enterocoelous
12. Parts to coelom	?	tripartite	?	tripartite	?	tripartite
13. Coelomic derivatives	?	pore and pulsatile vesicle	none	pore and pulsatile vesicle	?	pore and pulsatile vesicle
14. Nature of larva	?	amphioxides	tadpole	tornaria	?	auricularia
C. *Ecology*						
15. Niche	free-swimming ingestor	semi-sessile filter feeder	sessile filter feeder	burrowing filter feeder	crawling filter feeder	crawling ingestor

The table compares fifteen semes in selected organisms representing certain chordates and echinoderms. Obviously, much information is lacking, especially when fossil forms and contemporary forms are compared. (When informed guesses are made regarding fossil forms, the data are bracketed to indicate their uncertainty.) From this, some preliminary phylogenetic speculations can be attempted.

First, we will discuss the fifteen semes in the table.

1. Body form: symmetry varies significantly here, from the bilateral symmetry of the fish-like forms to the radial symmetry of a starfish. Bilaterally symmetrical larvae represent intermediate forms that join the extreme forms and allow homologizing of this seme.

2. Coelom: Again there is significant variation, and again embryonic forms allow us to homologize this seme among the six different organisms. (See also semes 11, 12, and 13, below.)

3. Notochord: This structure is definitely absent in the starfish, but reportedly present in their relative, *Reticulocarpus.* It is present in all the other forms, though only in the embryos (larvae) in some.

4. Gill slits: Only the starfish lack these. They are variously evolved in the others.

5. Dorsal nerve cord: This seme is much like seme 4 in its distribution.

6. Nerve-muscle connections: These are relatively new findings and data are not available for the three groups of protochordates. One wonders if the conditions in amphioxus and the starfish represent homologous or convergent conditions. And how did the amphioxus condition change to that seen in the vertebrates?

7. Acetylcholine neurotransmitter: Acetylcholine transfers nerve impulses from one nerve cell to another, chemically. It is not the only neurotransmitter known, hence its almost uniform occurrence in these groups may be a significant homology, as some authorities have argued.

8. Binding of iodine: iodine is an essential element in the hormone thyroxine, which, in higher vertebrates, is produced by the thyroid gland. Here we see the evolution of the production of this hormone.

9. Phosphagen: This refers to those compounds used for high-energy phosphate bonds that function in energy transfer. Why some urochordates use phosphocreatine and some phosphoarginine is not known. (These data do not apply strictly to *Amaroucium,* but they are reported this way to provide the necessary perspective.)

10. Cleavage: Radial cleavage is illustrated in Fig. 13–7.

11. Coelom formation: See Box 13–1 for the enterocoelous mode of forming a coelomic cavity.

12. Parts to coelom: Curiously, three pairs of coelomic pouches are formed by the enterocoelous mode in the forms so indicated in the table. Their fates are seen in seme 13.

13. Coelomic derivatives: the occurrence of an asymmetripore from one coelemic pouch and of a pulsating vesicle from another is a highly complex and precise aspect of larval development. To have such evolution occur in three different groups has led a British expert, Dr. Q. Bone, to remark that this "is perhaps the one sure foothold in these shifting sands of phylogenetic speculation, for it is inconceivable that such peculiarities could have arisen independently." That is, they must be homologous, not convergent semes.

14. Nature of larva: The tornaria (A) and auricularia larva (B) are so much alike as to be considered homologous larval forms. The other larvae are quite different, although they share such traits as listed in semes 2, 3, 4, 10, and 11. (Amphioxides is a large larval form known to feed, as a free-swimmer, on smaller planktonic organisms.)

15. Niche: It is hard to compare these general characterizations of complex

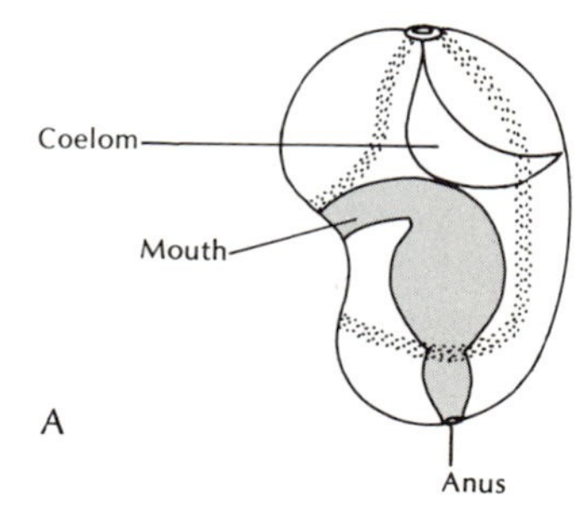

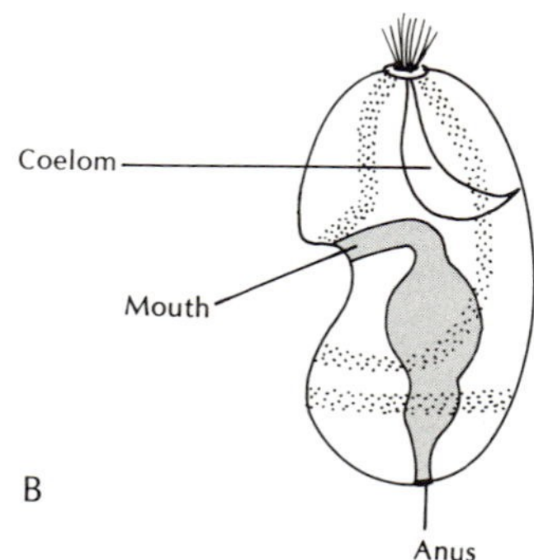

Box 14–1 (continued)

modes of life from the point of view of homology. If they are to be homologized, it will be through the feeding habits of the larval forms within each niche, providing they are homologizable (see seme 14). The larval forms show how very different adult forms can arise from a common, though distant ancestral stock.

What conclusions can we draw from such comparisons?

First, we note that *Hemicyclaspis* and *Amphioxus* have the most semes in common, if we accept our informed guesses (in brackets) regarding the former.

Second, the echinoderms *Reticulocarpos* and *Luidia* share certain semes, as might well be expected. The absence of information from the fossil form limits the conclusions we can draw.

Third, *Amaroucium* and *Balanoglossus* have their own specializations as well as sharing some with the other four organisms.

Overall, the real argument for a distant common ancestor comes down to supportable homologies in notochord, gill slits, dorsal nerve cord, radial cleavage, and enterocoelous tripartite coeloms (semes 3, 4, 5, 10, 11 and 12). These relate the vertebrates to the protochordates and *Reticulocarpos,* and the latter to the remaining echinoderms. The actual phyletic relations among these organisms is discussed further in the text. As stated there, the problem is not yet resolved.

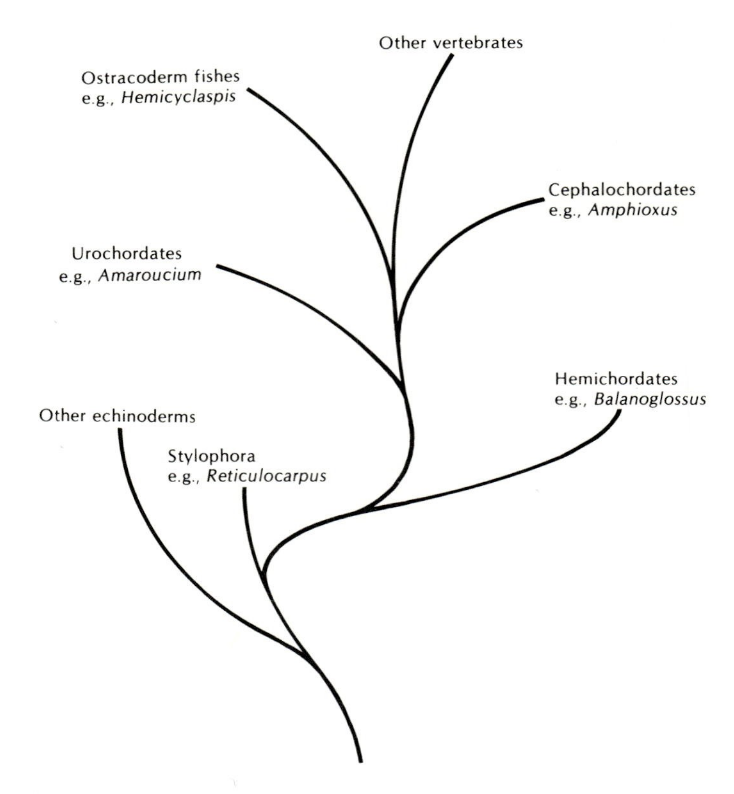

Fig. 14-16: Dendogram of a possible interpretation of phyletic relations among the organisms characterized in Box 14-1.

tebrate animals. In Box 14–1 are tabulated various semes, most of which show homologous relationships between two or more of the six organisms examined there. These semes, in some degree, characterize these groups of organisms. Many other traits shared by these and other animals are ignored. But these latter traits are largely plesiosemes, conservative traits that tell us little about evolutionary change. The semes in Box 14–1 are mostly aposemes, i.e., they are variable characters, and homologies can be established by finding intermediate forms to bridge the differences between the semes. Here larval forms are very important. It is, of course, the aposemes that are most informative regarding the path of evolutionary history.

The conclusions that can be drawn from a study such as that seem in Box 14–1 can be summarized, as in Fig. 14–16. In words, the dendrogram says echinoderms and hemichordates share a common ancestor. Although the similarities between the auricularia and tornaria larvae had long been known, it was the discovery by R.P.S. Jeffries, a British paleontologist, of *Reticulocarpos* (Fig. 14–17) and related forms that makes a common ancestor of the echinoderms and chordates much more plausible.

Whereas biologists such as Q. Bone classify these new finds as members of the echinoderm class Stylophora, Jeffries sees them as a chordate subphylum, the Calcichordata. Jeffries has provided us with a different phylogeny than that given in Fig. 14–16. Jeffries' dendrogram, slightly modified, is seen in Fig. 14–18. It differs from that in Fig. 14–16 in two

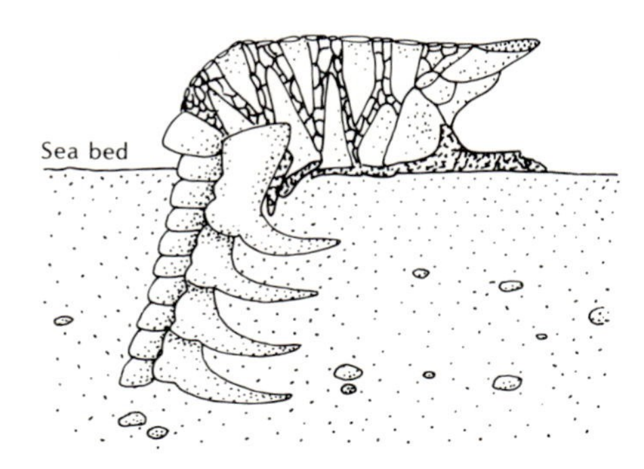

Fig. 14-17: Reconstruction of the fossil relative of the echinoderm *Reticulocarpos* (side view).

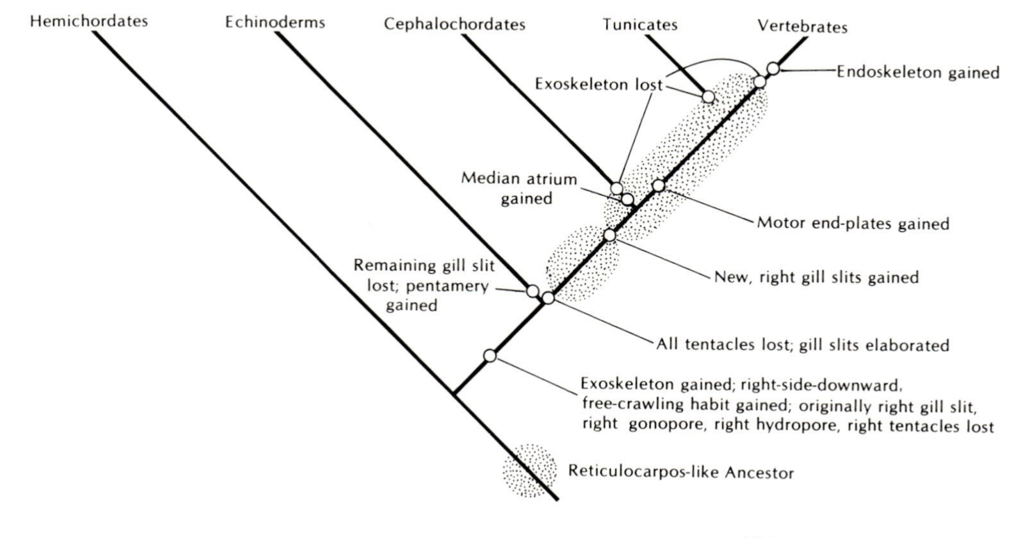

Fig. 14-18: Phyletic relations among echinoderms, protochordates, and vertebrates, as interpreted by Jeffries. Note that a form like *Reticulocarpos* (Fig. 14-17) is plesiomorphic to this line of evolutionary change. Various anatomical changes (aposemic) in that plesiomorph, listed along the main line of evolution, are proposed by Jeffries. These changes account for the evolutionary divergences he proposes and are supported to some degree by fossil forms (shaded areas).

important aspects : (1) the hemichordates precede the echinoderms in evolutionary development and (2) the cephalochordates precede the urochordates or tunicates. It is differences of opinion such as these that are most baffling to the non-expert. How can experts differ so greatly? The answer is that different experts use different data and different methods to evaluate those data. The dendrogram in Fig. 14–16 is based on a consideration of homologies. (It does not go into the details that a rigorous analysis, culminating in calculations of phyletic distance or R values, demands, as in Chapter 9.) The one in Fig. 14–18 emphasizes morphological evidence almost exclusively and makes little explicit reference to homologies and to the selective advantages of the postulated changes. Eventually such differences are resolved; we are now at a stage that is, we hope, a prelude to such a resolution.

Other views. In fact we should take a moment here to report something like the real range of opinions offered on this subject of chordate origins. Two more opinions are worth noting.

The most extreme one is that of the Canadian biologist J. R. Nursall (Fig. 14–19A). His view is that the organisms in many of the major invertebrate phyla and the Chordata each had a separate origin from protozoan ancestors. True, he would admit, the echinoderms and chordates probably shared a protozoan ancestor different from those giving rise to other phyla. Such a viewpoint tries to avoid the problems arising from deriving one invertebrate phylum from another. It says, in effect, one specialized group of multicellular animals did not (could not?) give rise to another specialized group of multicellular animals. Rather, the argument continues, the groups all arose from relatively unspecialized protozoan ancestors. Unfortunately, apart from the basic common sense of not deriving one highly specialized group from another, there is no evidence—no analysis of homologies, we could say—to support this radical view.

The other notable view is that of the American behaviorist Donald D. Jensen, who, in following some ideas on the origins of behavior, presented a provocative study of homologies starting from nemertine worms (Fig. 14–19B). In brief, Jensen is saying that the chordates orginated from relatively unspecialized worms. The nemertines are readily derived from flatworms, as many biologists would agree. Then from the nemertines, with their extrusible proboscis (Fig. 13–15c), we have the start of specializations leading to a dorsal nerve cord, a no-

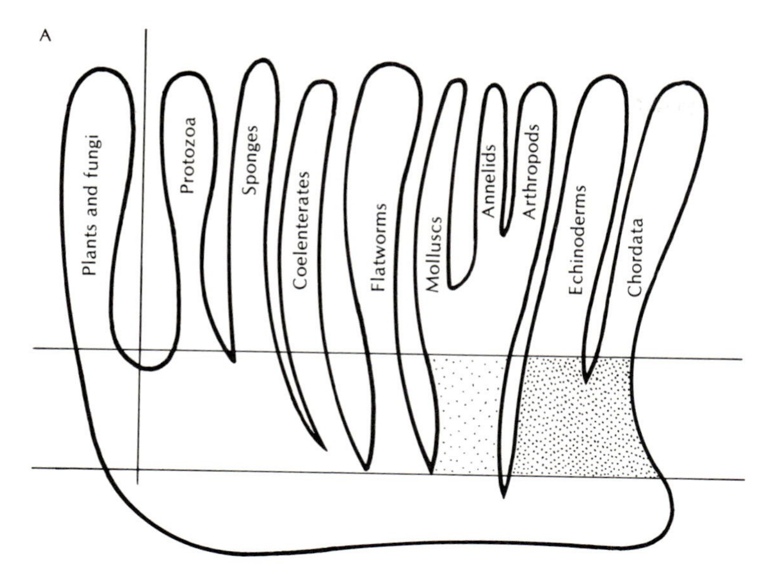

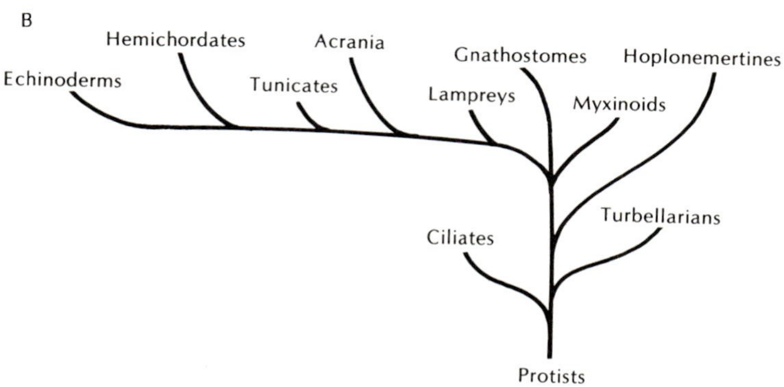

Fig. 14-19: Two phylogenies (compare Fig. 14-18). A. Nursall's phylogeny derives all major multicellular phyla directly from different unicellular forms. B. Jensen's phylogeny derives the primitive fishes from certain nemertine worms (hoplonemertines).

tochord, and gill slits. The protochordates, with their semi-sessile and sessile tendencies, are all off-shoots of the line going from the hoplonemertines, in particular, to myxinoids or hagfishes. In his line of thinking, the free-living worms became free-swimming and evolved lateral undulatory swimming (how?) and went on to form the primitive fishes. Not all biologists agree with Jensen's attempts to homologize the hoplonemertine proboscis and related structures to the three basic characters of chordates. And certainly, Jensen has not considered the evolution of the rest of the chordate body—segments and coelom. But what remains is, as we said, a provocative view that uses again the early worms as the source of adaptive radiation of the bilaterally symmetrical animals.

Note that, in Figs. 14–16 and 14–18, the question remains

as to where the echinoderms and hemichordate ancestors came from. From the worms? Possibly, but enough has been said to show that, although phylogenists are hard at work on these questions, as yet there are no widely accepted answers. At this stage of knowledge we are still sorting out various plausible answers.

Human evolution

Let us complete this chapter and book by looking at human evolutionary history. We leave the puzzle of chordate origins and turn to the other end, so to speak, of chordate evolution. We consider now the evolution of that one species on this earth that is aware of its evolutionary history. That very awareness is one of the most fascinating and complex problems in the whole area of evolutionary studies. How can we explain the emergence of intellectual curiosity and the knowledge it produces? How can we explain those other activities and their products that we rightly cherish so deeply—music, poetry, and the arts in general? In brief, can we explain not only the emergence of ourselves as another complex chordate, but as a human chordate, and in particular, a *humane* one? Comparable questions are not asked of other species probably because we really cannot get inside another species to explore the possibilities of its emotional, creative, and psychic urges. Hence the questions we pose here regarding not only our biology, but also our humanity, are unique questions. We can therefore anticipate that they will demand unique answers.

Our approach will be to start from our biological heritage, since our biology is definitely shared with other mammals. Then we will try to trace our special history through the apes, culminating in the appearance of humans. But, as we have just said, we will try to understand ourselves not just as specialized chordates or specialized mammals or even specialized apes, but we will also try to see how far evolutionary principles can go in helping us understand our human nature—our humaneness. We build on biology and transcend it as we go on to explore our social and finally our emotional and psychic dimensions.

HUMAN ORIGINS

That humans are also mammals is attested to by our anatomy (Fig. 14–20E), our development (Fig. 1–5), and our warm-blooded physiology, to mention three major lines of evidence.

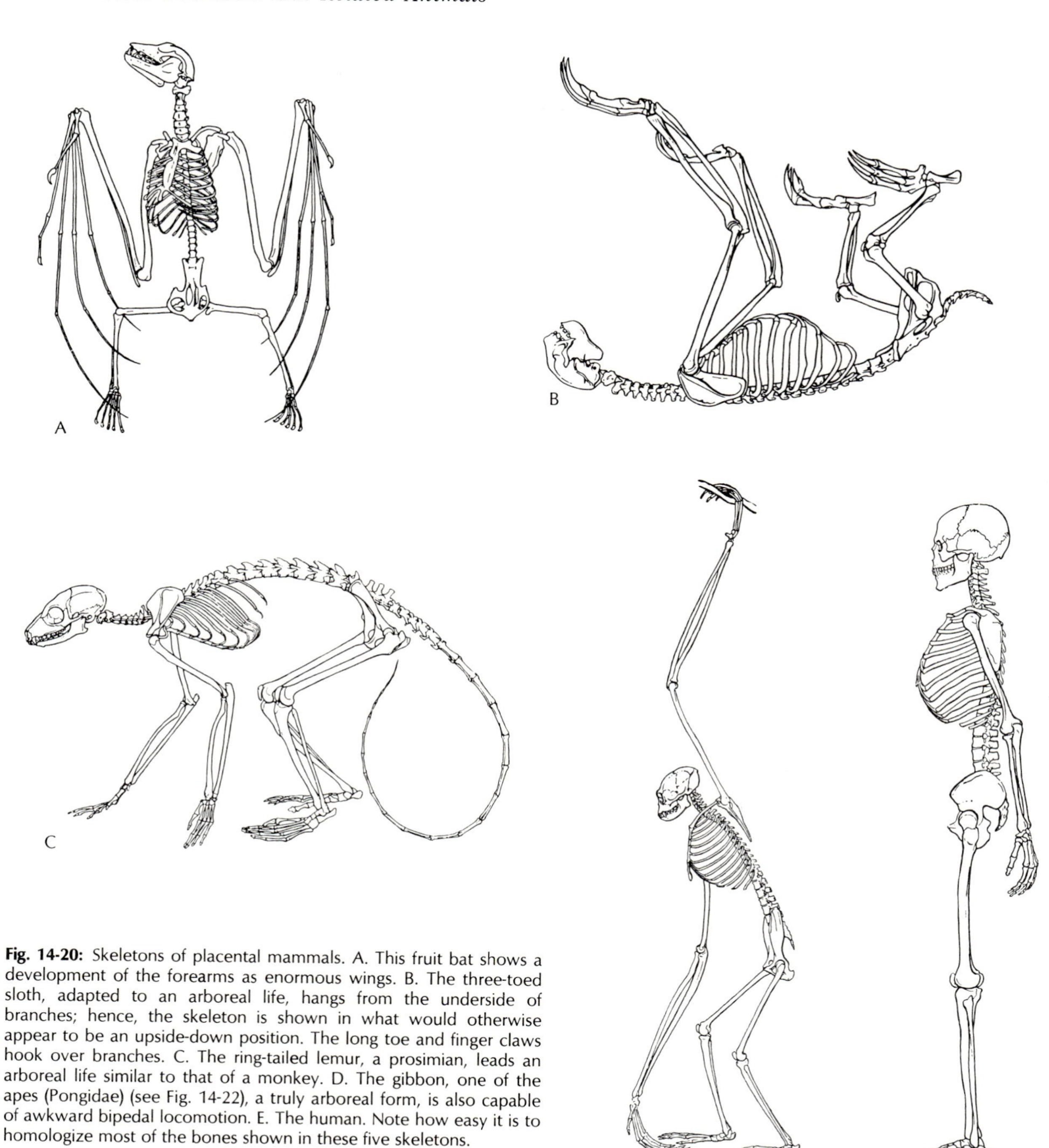

Fig. 14-20: Skeletons of placental mammals. A. This fruit bat shows a development of the forearms as enormous wings. B. The three-toed sloth, adapted to an arboreal life, hangs from the underside of branches; hence, the skeleton is shown in what would otherwise appear to be an upside-down position. The long toe and finger claws hook over branches. C. The ring-tailed lemur, a prosimian, leads an arboreal life similar to that of a monkey. D. The gibbon, one of the apes (Pongidae) (see Fig. 14-22), a truly arboreal form, is also capable of awkward bipedal locomotion. E. The human. Note how easy it is to homologize most of the bones shown in these five skeletons.

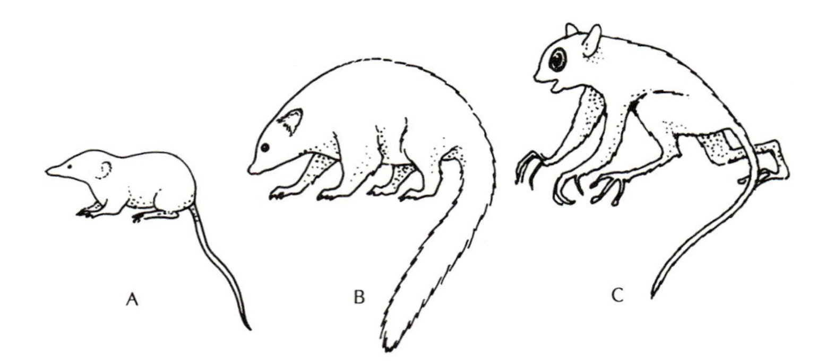

Fig. 14-21: Selected prosimians. Tree shrew (A); lemur (B); and tarsier (C).

We can resume the evolutionary story where we left it, with the emergence of mammals earlier in this chapter, by recalling the view of mammalian evolution given in Fig. 14–13. Now we will concentrate on the evolution of the order Primates, which includes humans, apes, monkeys, and a variety of other tree-dwelling forms called the prosimians or antecedents of monkeys. These latter include tarsiers, lemurs, and tree-shrews (Fig. 14–21). Their phylogenetic relations can be diagrammed in a rather straightforward manner (Fig. 14–22). These relations are all based on a convincing analysis of homologies taken largely from anatomical data. There is general agreement on these phyletic relations; details of actual transitional forms are still being pursued. We will focus on such details now.

Fossils linking apes and humans. New evidence is emerging every year. The result is that we can draw a fairly complete picture of how humans emerged from non-human ancestors.

The divergence of hominid evolution from pongid or ape evolution is often placed in the ape-like organism called *Ramapithecus*. This animal, known only from a few teeth and jaw fragments found in East Africa and the Siwalik Hills of India and Pakistan, is usually thought to have lived about 9 to 14 million years ago. However, recent reevaluation by Lisa Tauxe, of Columbia University, of the Siwalik rocks that contains *Ramapithecus* fossils shows that those rocks are now reliably estimated to be 8 million years old. Since we know so little about the general body structure of *Ramapithecus*, we cannot say for sure whether or not it walked erect. This is not to say that erect posture is the sole criterion of being a hominid. But it is an important trait, since it is associated with freeing the forelegs from a locomotory habit and allowing hands to evolve.

Also an upright posture has anatomical effects on the musculature of the back, legs, and neck, all of which further differentiates the hominids from the apes. But, first, how is *Ramapithecus* connected to the clearly ape-like forms, that is, to forms clearly ancestral to modern apes? Here the evidence is fragmentary, but such fossils as *Proconsul* and *Aegyptopithecus* are promising transitional forms. These lived in the Miocene epoch (about 15 to 25 million years ago). It was in the next more recent epoch, the Pliocene, that *Ramapithecus* appeared, and later in that period there emerged the first hominids. And finally humans of the genus *Homo* emerged toward the end of the Pliocene, and its members were firmly established on earth

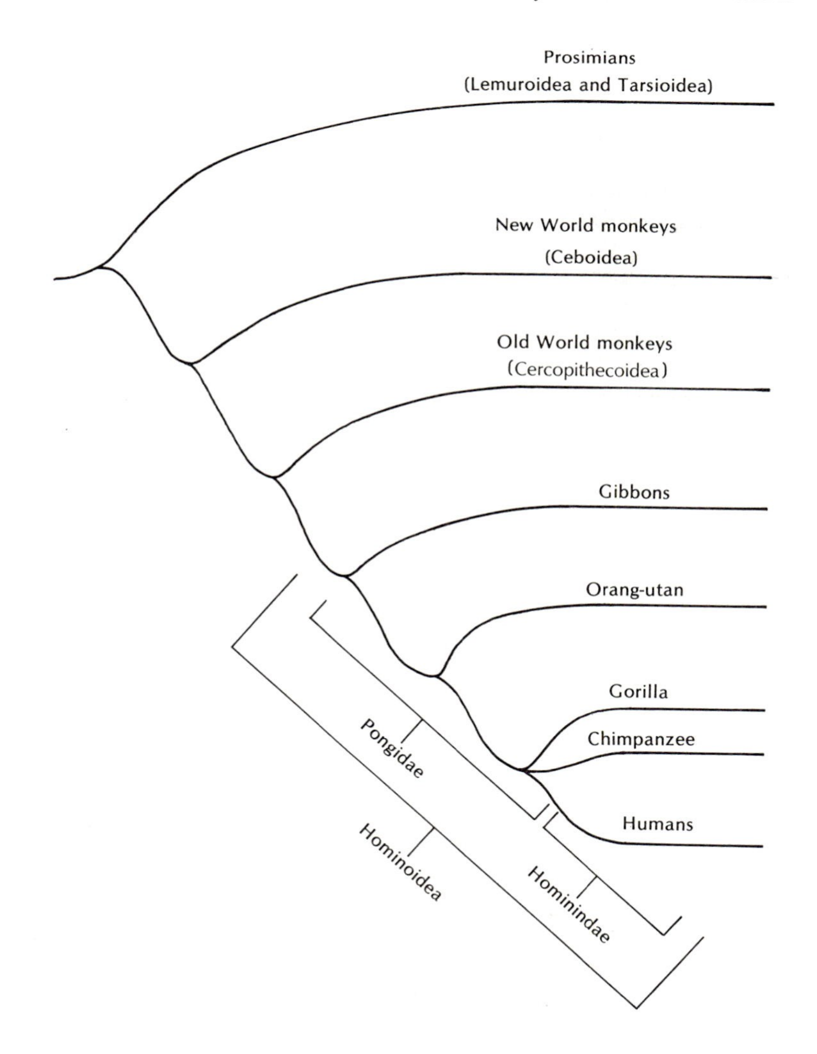

Fig. 14-22: Phylogeny of the major groups within the primates. The hominoids (superfamily Hominoidea) includes the apes (family Pongidae) and the hominids (family Hominidae).

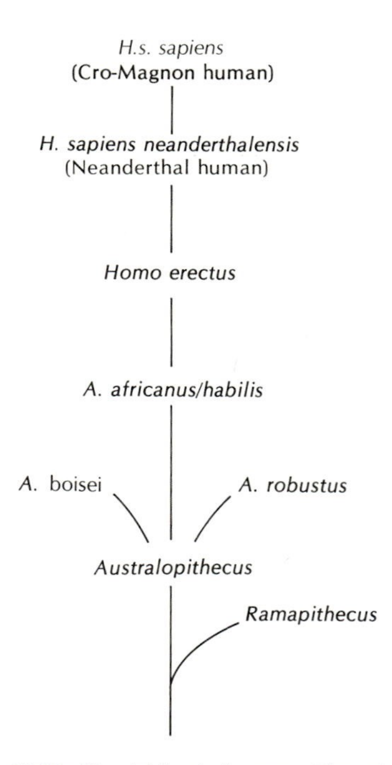

Fig. 14-23: Hominid phylogeny. The time scale shrinks rapidly as one progresses up this diagram. There are millions of years between *Ramapithecus* and *Australopithecus,* only 60,000 years between Neanderthal humans and Cro-Magnon humans.

during the most recent epoch the Pleistocene, which started about one million years ago.

Recent work has inserted three hominid forms between *Ramapithecus* and the genus *Homo,* within which we are located (Fig. 14–23). These three forms are included in the genus *Australopithecus,* and their fossils have all been found in South and East Africa, where, it is now concluded, hominids must have originated. *Australopithecus africanus* or *A. habilis*—there is no clear demarcation between these two—are considered to be on the main line of evolution toward the genus *Homo. Australopithecus boisei* died out about one million years ago and is thought to be a side branch in hominid evolution. Similarly, *A. robustus* is another side branch. This latter hominid is not as robust as *A. boisei,* but is definitely larger than *A. habilis.* It, too, apparently died out about a million years ago. More recently, yet another australopithecine has been reported. This is *A. afarensis,* discovered in East Africa and thought to have been alive about 2.9 to 3.6 million years ago. Since this is such a new find, its man-like and ape-like characteristics are still being debated.

Now let us look more closely at the main line of hominid evolution represented by *A. africanus/habilis.* The first remains, about 2 million years old, were called *A. africanus* by Raymond Dart, who found them in South Africa. Then in East Africa, very similar remains, about 1.75 million years old were found, and associated with them were stone tools. The tools are the reason for the name *A. habilis* or tool-user. And the presumption is that *A. africanus* evolved into *A. habilis;* hence, there is a certain interchangeability of names. The findings pertaining to *A. habilis* are largely the work of Louis Leakey and his wife Mary, excavating Olduvai Gorge in East Africa. At that remarkable site they have shown that *A. habilis* lived for over half a million years and a slow evolution occurred with a transition to *Homo erectus.* The remains of *H. erectus,* also found at Olduvai, span a period from about one million to half a million years ago.

Not all workers agree with this treatment of the australopithecines and their connection with humans in the genus *Homo.* The problem lies in the interpretation of data. Look at Fig. 14–24, where the material used for the foregoing study, as it existed a few years ago, is summarized. Quite clearly, a great deal of reconstruction was necessary to infer the completed skeleton from the parts on hand. Hence there is room for various interpretations. But in all three cases, the skulls are complete. They support the interpretation just outlined. Notice that *A. africanus/habilis* has no skull crest, which suggests that jaw muscles and their attachment were more like *Homo* than in the case of the other two. Also, when all the *A. habilis* skulls are examined together, and the average taken of their brain volume, the value obtained is 642 cubic centimeters (cc). This lies between the value of 1,200–1,600 cc for modern humans and 450 cc for Dart's South African australopithecines, the original *A. africanus*.

In addition to the work at Olduvai Gorge, the work at two other sites must be described. One site is a remote spot in southern Ethiopia called Omo. It is part of that geological formation called the Rift Valley, which is a gigantic north-south crack in East Africa. Olduvai Gorge is also part of the Rift Valley. The geological advantages of Omo are that it can be dated more precisely than can Olduvai because its sedimentary layers are over 2,000 feet thick, whereas at Olduvai they are compressed into only about 100 feet of sediments. Also, whereas the oldest sediments at Olduvai are about 2 million years old, those at Omo go back about 4 million years. What was found at Omo? Between a group of French paleontologists and a similar group of Americans, hominid remains going back 3.7 million years were found. These belong to *A. boisei*-type hominids. All told, *A. boisei* seems to have lived in the Omo area for about 2 million years. Also, teeth and a thigh bone that resembled those of *A. africanus* were found. These are about 3 million years old. Additionally at Omo, there has been found a great richness in other mammal fossils: 80 different species of mammals have been identified and their evolutionary changes, in parallel with the hominid changes, have been examined. All of this provides us with a paleoecological view that is uniquely informative. We now have a chance to reconstruct the environment in which hominid evolution occurred. Work is continuing at Omo, and those curious about human evolution are eagerly following reports from there.

The last site to be mentioned is one at the eastern shore of Lake Rudolf. This is not far from the Omo site and was discovered on a brilliant hunch of Richard Leakey, the son of Mary and Louis Leakey. His finds can be grouped into two parts. First, he located enough material of the *A. boisei* type to provide the equivalent of population data, i.e., means and the associated range of variation, on these hominids. They tell us that *A. boisei* differs from *A. habilis* type, which makes up the other part of Richard Leakey's finds. This material seems to

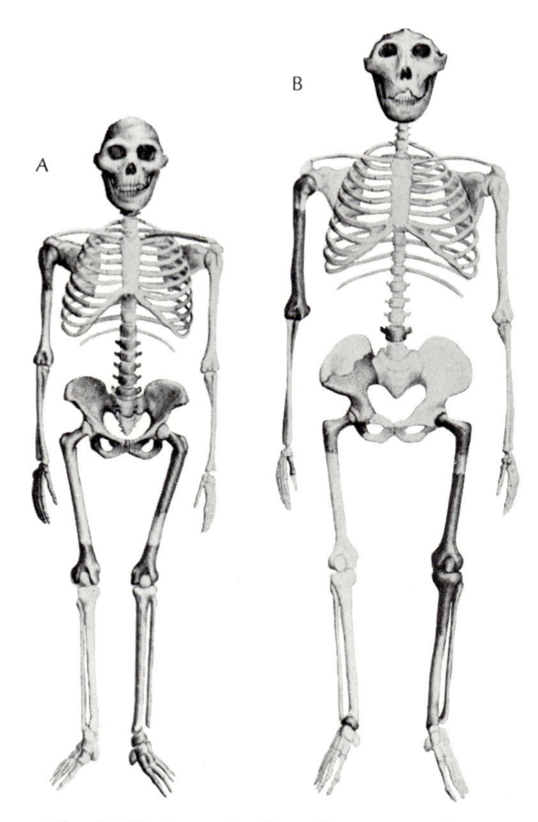

Fig. 14-24: Reconstruction of two australopithecine skeletons. Known fossil remains (as of a few years ago) are shaded. Unshaded parts are conjectural, but highly probable. A. *Australopithecus africanus/habilis*. This male hominid was 4-1/2 to 5 feet tall and weighed 80 to 100 pounds. The female was smaller. B. *Australopithecus africanus/boisei*. This large hominid weighed as much as 200 pounds.

represent the possible transition from *A. habilis* to *H. erectus*. Apparently both *A. boisei* and *H. erectus* coexisted in this region near Lake Rudolf for over one million years. Eventually, *A. boisei* disappeared. Careful speculation suggests that the smaller, tool-using type may have been more successful than the larger vegetarian type, who may never have used tools. Further finds by Richard Leakey have been a jaw and skull, about 2.5 million years old, that are *Homo*-like. It appears that members of *Homo* and *Australopithecus* coexisted, as would be expected if one merged into the other. *Homo,* with its larger brain, seems to have outcompeted *A. habilis* and finally emerged as the dominant hominid in East Africa.

From East Africa *H. erectus* spread to Asia and Europe. In Europe, the remains of the Neanderthal humans are dated as occurring about 1,100,000 years ago. This *H. sapiens neanderthalensis* was then succeeded by Cro-Magnon humans, which mark the start of modern humans (*H.s. sapiens*), about 40,000 years ago (Fig. 14–23). That outlines the fossil links between apes and humans. Now we can leave what has been an almost purely biological account and turn to the social and cultural factors, which are the context for the expression of those capabilities that especially distinguish modern humans from their ancestors.

SOCIAL AND CULTURAL EVOLUTION

In the grasslands of East Africa where *Homo erectus* emerged from australopithecine ancestors, hominids were hunter-foragers. Their food included berries, nuts, roots, grubs, birds and their eggs, rodents, and game. In their hunting, it is plausible that they killed mammals as large as wild pigs and antelopes. This is attested to by their weapons and the bones of antelopes and pigs found at what are identified as their "occupation sites," as Mary Leakey calls them. Hers is the special authority that has transformed a welter of stone artifacts, bone chips, and other assorted rubble and remnants into a theory from which the life of the early hominids could be understood. The substance of her decades of painstaking work is too extensive to be covered here. A partial summary of her findings is given in the description in Box 14–2. The central point of Mary Leakey's extraordinary work is that it establishes, beyond doubt, that the australopithecines had advanced further into the early stages of culture than most of us could imagine by looking only at their biological remains.

This can be brought into clearer perspective by looking

Box 14–2. The Leakeys at work.

The following extensive quotation, which catches the flavor and substance of Mary Leakey's important work, is from Maitland Edey's book *The Missing Link*. This thoughtful popular account of the work on human origins is to be contrasted with other accounts of the same subject, such as Robert Ardrey's *African Genesis* and Desmond Morris's *The Naked Ape*. The latter writings, lively, extremely well-written interpretations of the work of scientists, lack the careful, skeptical turn of mind that marks first-rate scientific work. Edey's book is as readable as *African Genesis* or *The Naked Ape* and is considerably more accurate in its interpretations.

Herewith, then, is the passage regarding Mary Leakeys's work on hominid artifacts:

"Ten occupation sites have been identified at Olduvai, out of about 70 that contain fossils or tools, scattered along a 12-mile stretch of the gorge. One floor has its cultural debris arranged in a most peculiar manner. There is a dense concentration of chips and flakes from tool manufacture, mixed in with a great number of small smashed-up animal-bone fragments—all of it crowded into a roughly rectangular area some 15 feet wide and 30 feet long. Surrounding this rectangle is a space three or four feet wide where there is hardly any of this cultural junk; the ground is nearly bare. But outside that space, the material becomes relatively abundant again. How can this extraordinary arrangement be explained?

"The most obvious explanation is that the densely littered central section was a living site, that it was surrounded by a protective thorn hedge, that the hominids who lived there relaxed safely inside that hedge while they made their tools and ate their food, and that whatever they did not simply drop right there on the floor they tossed out over the hedge.

"At another site is a roughly circular formation of stones about 14 feet across. This is as attention-getting as a trumpet blast on a summer night. Not only are there very few other stones on that occupation floor, but what stones there are are widely and haphazardly scattered. By contrast, the circle is a dense concentration of several hundred stones carefully arranged in a ring by somebody—somebody who also took the trouble to make higher piles of stones every two or three feet around the circle.

"That this configuration should survive after nearly two million years is

Box 14–2 (continued)

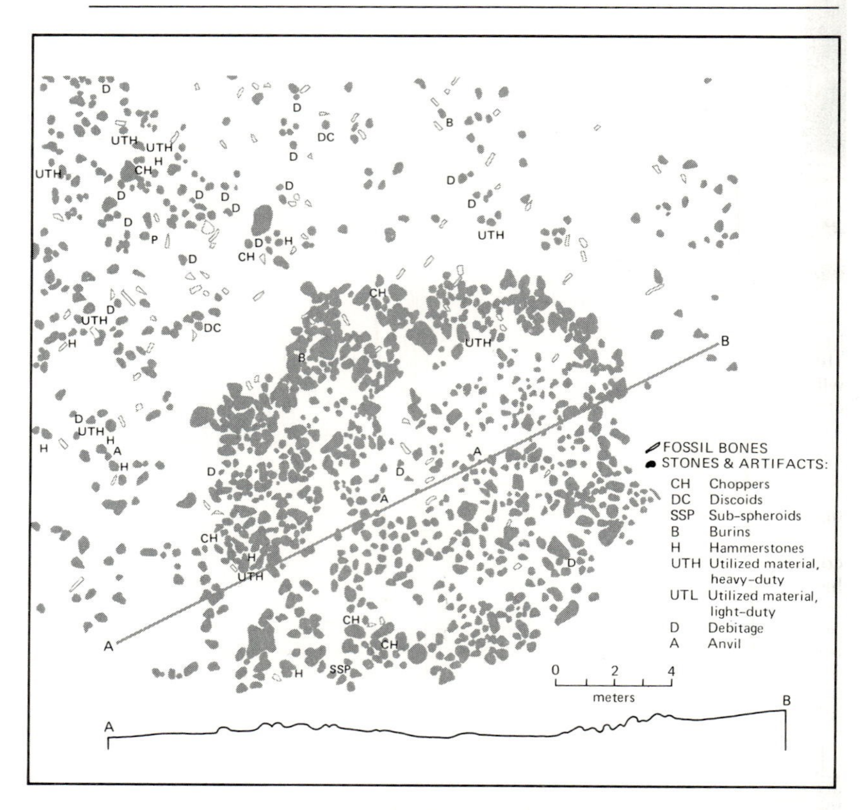

staggering. It suggests a shelter, of a kind that is being made today by the Okombambi tribe of South West Africa. They, too, make low rings of stones, with higher piles at intervals to support upright poles or branches, over which skins or grasses are spread to keep out the wind.

"Although the predictable debitage of flaked debris is found inside the stone circle, indicating that some activity took place there, there is more evidence of a wider variety of activities having taken place outside. This makes sense. The interior dimensions of this somewhat irregular circle are only about eight feet by ten or twelve feet, which would have made things rather crowded in there if it had been the home of several people. Furthermore, that group had in it some extremely good hunters or scavengers. The surrounding area contains the fossilized remains of giraffes, hippos, many antelopes and the tooth of a deinotherium, an extinct elephant. Those people were eating a lot, and may have found it more convenient to do their eating out in the open rather than in the confines of the shelter.

"Whether they actually killed those large animals, whether they chased them into swamps and helped them die, whether they brought home meat from found carcasses, whether they pre-empted the kills of other carnivores, the chronicle of Olduvai does not say. But it makes clear that when an extremely large carcass became available they did cut it up and eat it. There are two sites in Olduvai Gorge known to have been butchering sites. One contains the skeleton of an elephant, the other that of a deinotherium. Since those animals weighed several tons each, it was obviously impossible to move them; the thing to do was to settle down at the carcass and chop and chew away at it until its meat was gone. On the evidence at these butchering sites, that is exactly what happened. At each, there is an almost complete skeleton of a huge animal, its bones disarranged as if they had been tugged and hacked apart. And, lying among the bones are the discarded choppers and other stone tools that did the hacking.

"The Olduvai hominids were very catholic eaters. Certain sites are rich in antelope bones, some with their skulls cracked open at the precise point on the front where the bone was thinnest. Others are crammed with the shells of large tortoises. One is littered with snail shells. Another contains a giraffe head, but nothing else belonging to that giraffe—clearly it was lugged in to be eaten at home. A site higher up in Bed II reveals an increasing dependence on horses and zebras, which means that the climate had become drier by that time and was encouraging the spread of open grassland. There is also a marked increase of scrapers in Bed II, which suggests the beginning of an effort to work hides and leather.

"The clues are many, and enthralling. What do we make of little concentrations here and there of very small bones, most of them broken into tiny pieces? Was any hominid so quixotic as to collect handfuls of skeletal fragments of mice, shrews, small birds and lizards and then carefully place them in piles? It seems wildly unlikely, and Mary Leakey has concluded that these strange little heaps are probably what is left of hominid feces. This means that our ancestors were eating those small animals whole, bones and all, much as modern man might eat a sardine. The bones were ground up into very small fragments by chewing, then passed through the intestines and deposited right where they were found.

"Mrs. Leakey's attention to these minute details is extraordinary. At one site she has collected more than 14,000 bone fragments, so small that all together they weigh only 15 pounds. Her measurement and classification of tools are equally painstaking. She can tell exactly the mix of 14 different kinds of tools at any one of the important sites in which she has done extensive work. . . .

"With this wealth of fantastic information coming from Olduvai, there can be no question that by two million years ago hominids were living in a remarkably advanced state of culture—at a level that no one in his wildest dreams would have believed possible a few decades ago. Since progress in the early Stone Age was abysmally slow, this means that the beginnings of the Oldowan tool industry are far older than Olduvai—how much older no one yet has the slightest idea. But in 1969 word began to trickle back from East Rudolf and Omo that there were tools there too. The first official news of this came in 1970, when Mary Leakey published a paper describing some implements from her son Richard's East Rudolph dig at Koobi Fora. The next year two experts went to Koobi Fora to help Richard Leakey analyze his site. They were Glynn Isaac, a prehistorian from the University of California at Berkeley, and Kay Behrensmeyer, a geology student from Harvard. Their analysis confirmed another stunner: an ocupation floor containing animal bones and Oldowan choppers and flakes that is probably three-quarters of a million years older than Olduvai.

"What is particularly promising about Koobi Fora along with several other nearby sites in East Rudolf, is that when the geology and dating of the whole area are worked out, and one place properly linked to another in time, it will then be possible to connect directly the area's exceptionally rich hominid fossil remains with its tool industry finds—and, to learn more about Australopithecus as a maker and user of tools 2.6 million years ago" (pp. 125–127).

first at the revolution that has occurred in the last three decades of our understanding of the social life of apes. This received its greatest impetus from the pioneering work of Jane Goodall, the British primate biologist, who spent years getting to know intimately the private lives of wild chimpanzees in the Gombe Stream forests near Lake Tanganyika in East Africa. If we humans used to think of chimpanzees as overgrown monkeys or as big monkeys that could be absurdly human when appearing on television dressed in our clothes, we must now see them totally differently, thanks to Jane Goodall. She showed that they possess a flexible social organization; they communicate by gestures, touch, facial expressions, and vocalization; they are tool-makers and tool-users; on occasion, they are bipedal; and they hunt and they relish the eating of flesh. Comparable work on the other apes—gibbons, orangutans, and especially, the gorilla—have also revolutionized our appreciation of their lives.

This gives us a new perspective on the australopithecines. No longer can we consider these "ape-men" as aberrantly bipedal escapees from forests who somehow managed to romp around the grasslands of East Africa and survive. Rather, their invasion of the grasslands as hunter-foragers was associated with various preadaptations that made the transition from a largely arboreal life to that of a ground-dwelling hunter-forager very successful. The preadaptations include upright stance, tool-using, vocalization, social organization, and hunting skills. These all occur in some form in chimpanzees and gorillas. It is this combination and the availability of grasslands that combined to let a *Ramapithecus*-like form evolve into *Australopithecus* and then *Homo*.

The key to understanding the evolutionary development of the hominids is the connection between their biology and their social habits and their environment. Older theories about the one key step from ape to human, e.g., tool-using or bipedal stance or language, etc., are now seen to be meaningless. There was no one step; but many steps, all reinforcing one another. For example, bipedal posture freed the hands for the use of weapons and tools. The use of weapons meant better hunting, of tools, the ability to deal with the prey better. Hunting put a premium on cooperative behavior—between hunter and hunter and between hunters and those who handled the prey as food. This social activity encouraged vocalization and other forms of communication. This led to better hunting skill, which also reinforced improvement of tools. And that encouraged more handicrafts and preserved the tendency to an upright posture. And so on. The result was a life-style increasingly better able to compete on the grasslands. There was competition with other large predators, such as wild-dogs, hyenas, cheetah, leopards and even lions. And, also, since the hominids varied among themselves (recall the differences between *Australopithecus africanus/habilis, A. robustus,* and *A. boisei*), they competed within their own genus for survival.

It is, however, worth noting that recent finds by Mary Leakey and her associates at the Laetoli Beds in northern Tanzania have settled the argument over whether an upright posture preceded the evolution of a human-like skull and cranium. There are now unequivocal fossil footprints of bipedal hominids from 3.6 million years ago and the skull fragments associated with hominid remains of this age are more ape-like than human. In the article describing these finds, Leakey and Hay say, "With the hands free and available for purposes not connected with locomotion it is perhaps surprising that no form of artifact has been found [in the Laetoli Beds]. But the concept of tool-making may well have been beyond the mental ability of these small-brained creatures" (Leakey and Hay, 1979, p. 323).

The emergence of intelligence. Thus far, we have said very little about that special commodity called intelligence. It is hard to discuss because it is an ambiguous term. Let us first discuss what we do not mean by intelligence in our present discussion and then turn to what we may mean. By intelligence we *do not* mean whatever it is that I.Q. tests measure. Nor do we refer solely to mathematical genius or other special computational abilities. Also excluded is simple quickness or alertness of mind (mind is also an ambiguous term). Now, then what *do* we mean by intelligence? Our context for using the term is survival in East African grasslands 1 to 3 million years ago. That may confuse rather than help the situation. But essentially it means we want to see intelligence as being adaptively advantageous. In that context, intelligence refers to abilities that help a hunter-forager society of australopithecines to survive. Thus, we refer to abilities to plan successful hunts, design better tools, communicate more effectively, handle food resources more efficiently and safely, train offspring more carefully, and store essential information; all of these involve intelligence in some degree. Associated with that is one key factor: the size of the brain. We have only referred to that factor previously in terms of brain volume and noted

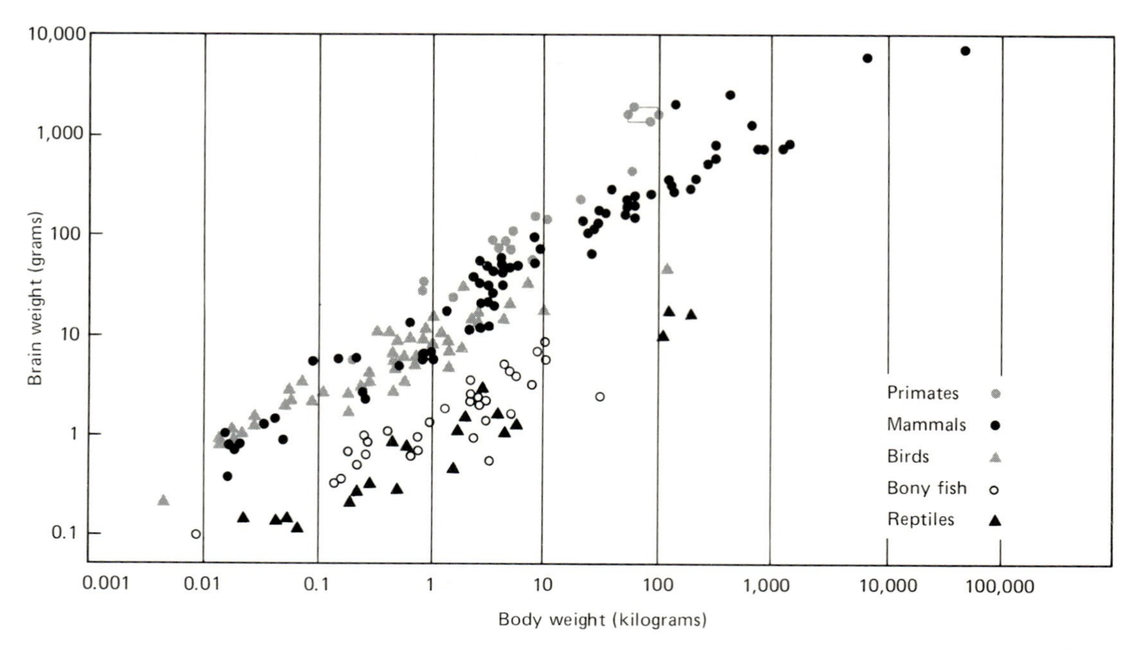

Fig. 14-25: Brain weight in vertebrates, changes in different groups.

that it increases dramatically as we proceed from *Australopithecus africanus* (450 cc) to *A. habilis* (642 cc) to modern humans (1,200–1,600 cc). In fact if we take a longer-range look at brain size in the vertebrates as a whole (Fig. 14–25), we see that there is a dramatic increase as we leave the fishes and another rapid increase as we reach the primates. Modern humans are at the crest of what might be another dramatic increase.

We know enough today about brain function to say that much of our human ability to think abstractly is correlated with functions of the forebrain, especially with functions of the left half of the cerebrum and its cortex. Thinking abstractly involves the ability to solve problems apart from actually experiencing those problems in everyday form. If an australopithecine could think abstractly, a hunt could be planned before it was actually underway; a better tool could be envisaged; a better way to express feelings and ideas could be found; food usage problems could be easily solved; offspring could be better and more quickly informed about the total environment; and knowledge could be stored as a cultural heritage to be used as needed rather than having to be discovered anew by each individual.

When we view intelligence in that light, it is indeed a complex phenomenon and difficult to describe in detail. But its essential function in survival becomes obvious. So much so that intelligence seen as problem-solving in a given environment becomes a genuine selective advantage. It has adaptive value. We still, today, do not know enough about the physiology of the brain to know exactly how it goes about generating solutions to problems. It is a fascinating mixture of symbol use and patterns, including mathematical ones. But it is clear that problem-solving is a capability that can be improved by practice. It has been suggested that *A. habilis,* as a tool-using hunter-forager, made more use of problem-solving capabilities than did *A. boisei,* who was, it seems, a stolid vegetarian (perhaps like today's gorillas) and seemed not to have used tools. That may explain how the former gained a competitive advantage over the latter, why brain size continued to increase through selection, and why *A. habilis* survived to become *Homo erectus.*

The most important consequence of the emergence of intelligence is that the rate of change in hominid life increased enormously. Before the emergence of intelligence—problem-solving ability is not purely hominid, primates also show it in varying degrees—evolutionary change depended on genetic (mutational) change. With the emergence of intelligence, information was stored and revised and improved extragenetically. Change was freed, in part, from purely biological constraints and could, to some extent, be controlled culturally. By culture we mean, loosely, the accumulated social habits and knowledge of a group of organisms. In this sense, culture depends on intelligence. And in this sense, intelligence introduced a whole new dimension—the extragenetic one—into organic evolution. That dimension has reached its most elaborate and sophisticated expression in the intellectual life of humans today.

As Carl Sagan, an American scientist, said in his lucid exploration of the evolution of intelligence, *The Dragons of Eden,* "Only an extragenetic learning system can possibly cope with the swiftly changing circumstances that our species faces" (p. 4). We add to that statement that it is the extragenetic system itself that is causing much of the rapid change we are experiencing. Ability to change rapidly is the major advantage our species has in competing with other species. The effect we humans have on our environment is greatly increasing the rate of extinction of other species. Why? They still depend on a genetic system for coping with environmental changes. We deal with those changes extragenetically.

Extragenetic evolution, superimposed on genetic evolution, adds a unique further dimension to the statement made earlier of our being the only species aware of evolution. That further dimension is one of many-faceted responsibility. We are responsible for most of the rapid change in the world today, and we also bear the responsibility for its effects. It is also our moral responsibility to monitor and control those effects.

The capability to solve problems—we have called it intelligence here—carries with it such correlated abilities as the creation of poems, music, great literature, and philosophical and ethical systems. All of that can be reduced to nothing if we cannot survive on this earth. The changes we humans have introduced on this earth threaten not only the existence of other species, but ourselves, too, as we press on toward overpopulation and its concomitant dangers of overconsumption and pollution. This is a new phenomenon in the history of this earth. No other species has so dominated and ravaged the earth as has *Homo sapiens* today. But we are aware of that domination, of the forces that have lead to that position, and of the dangers inherent in it. Understanding evolution can increase our understanding of ourselves and of our precarious position today. If we are to survive, and survive with dignity and fulfillment of our emotional and psychic capabilities, we must, along with all other pertinent knowledge, use responsibly our understanding of evolution.

References

Barrington, E.J.N., and R.P.S. Jeffries (eds.), 1973. *Protochordates. Symposium of the Zoological Society of London, No. 36.* Academic Press, London.

Bone, Q., 1979. *The Origin of Chordates.* 2nd ed. Oxford Biology Readers. Oxford University Press, London.

Berrill, N. J., 1955. *The Origin of Vertebrates.* Clarendon Press, Oxford.

Colbert, E. H., 1965. *Evolution of the Vertebrates.* 2nd ed. Wiley, New York.

Dimond, S. J., and D. A. Blizard (eds.), 1977. *Evolution and Lateralization of the Brain. Annals of the New York Academy of Sciences. Vol. 299.*

Edey, Maitland, and the Editors of Time-Life Books, 1973. *The Missing Link.* Time-Life International (Nederland) B.V.

Goodall, J., 1971. *In the Shadow of Man.* Houghton Mifflin, Boston.

Hardy, A. C., 1954. Escape from specialization. In *Evolution as a Process.* J. Huxley, R. C. Hardy, and E. B. Ford (eds.), pp. 122–142. Allen and Unwin, London.

Hecht, M. K., P. C. Goody, and B. M. Hecht (eds.), 1976. *Major Patterns in Vertebrate Evolution.* Plenum Press, New York.

Leakey, M., 1979. *Olduvai Gorge: My Search for Early Man.* Collins, London.

Leakey, M. D., and R. L. Hay, 1979. Pliocene footprints in the Laetoli Beds at Laetoli, northern Tanzania. *Nature 278:* 317–323.

Leakey, R. E., and R. Lewin, 1977. *Origins.* Dutton, New York.

Lewontin, R. C., 1977. Biological determination as a social weapon. In *Biology as a Social Weapon.* The Ann Arbor Science for the People Editorial Collective (eds.). Burgess, Minneapolis, Minn.

Romer, A. S., 1959. *The Vertebrate Story.* 4th rev. ed. University of Chicago Press, Chicago.

Sagan, C., 1977. *The Dragons of Eden.* Random House, New York.

Sarich, V. M., and A. C. Wilson, 1973. Generation time and genomic evolution in primates. *Science 171:* 1144–1147.

Schaller, G. B., 1973. *Golden Shadows, Flying Horses.* Knopf, New York.

Symons, D., 1979. *The Evolution of Human Sexuality.* Oxford University Press, New York.

Tauxe, L., 1979. A new date for *Ramapithecus. Nature 282:* 399–401.

Young, J. Z., 1962. *The Life of Vertebrates.* 2nd ed. Oxford University Press, Oxford.

Wilson, E. O., 1975. *Sociobiology: The New Synthesis.* Harvard University Press, Cambridge, Mass.

Appendices

I. THE CLASSIFICATION OF ORGANISMS

In the first, generally accepted classification of living things in the Western world, Linnaeus (see Chapter 1) recognized two kingdoms, Plantae and Animalia. Since Linnaeus' time, other schemes have been proposed and one widely used today, and the one used in this volume, is that proposed by R. H. Whittaker (Fig. 1–23). Whittaker retained Linnaeus' names of Plantae and Animalia, but since the content of those kingdoms is considerably changed, it seemed more useful to use the terms Metaphyta and Metazoa for multicellular plants and animals, respectively. (Note that the common names of plant and animal are still used to refer, generally, to producers and consumers, whether they are unicellular or multicellular.)

There have been occasions in this text to refer to different philosophies of classification, e.g., phenetic, Adansonian, or numerical taxonomy, systems devoted simply to efficient storage of information and systems that also attempt to illustrate phyletic relations. This is not the place to sort out and criticially evaluate those philosophies. However, since this text is devoted to evolutionary concerns, we will first present a diagram using common names, for the most part, which summarize the phylogenetic conclusions reached in Part 4 of this volume. But it must be recalled that such a diagram does not specify the many problems still remaining in phylogenetic analyses and that alternate views are current today in the various problem areas. After the phylogenetic diagram there will be various illustrations of the major taxa and more detailed taxonomic listings.

A phylogenetic summary

The diagram (Fig. I–1) that follows is recognizably derived from Whittaker's five-kingdom scheme of classification. In many places its details differ from Whittaker's point of view. The reasons behind the interpretations given here are discussed in Chapters 11 through 14.

Recall here that the Monera are all prokaryotic and the remaining four kingdoms are all eukaryotic. (Note how this is treated in the new classification, given below, developed by Whittaker and Margulis.) And also recall that the Monera and Protista are predominantly unicellular and the other three kingdoms are multicellular. Finally, in terms of these larger trends, note the occurrence of haploidy and diploidy in these kingdoms. For further details see, of course, the appropriate chapters in

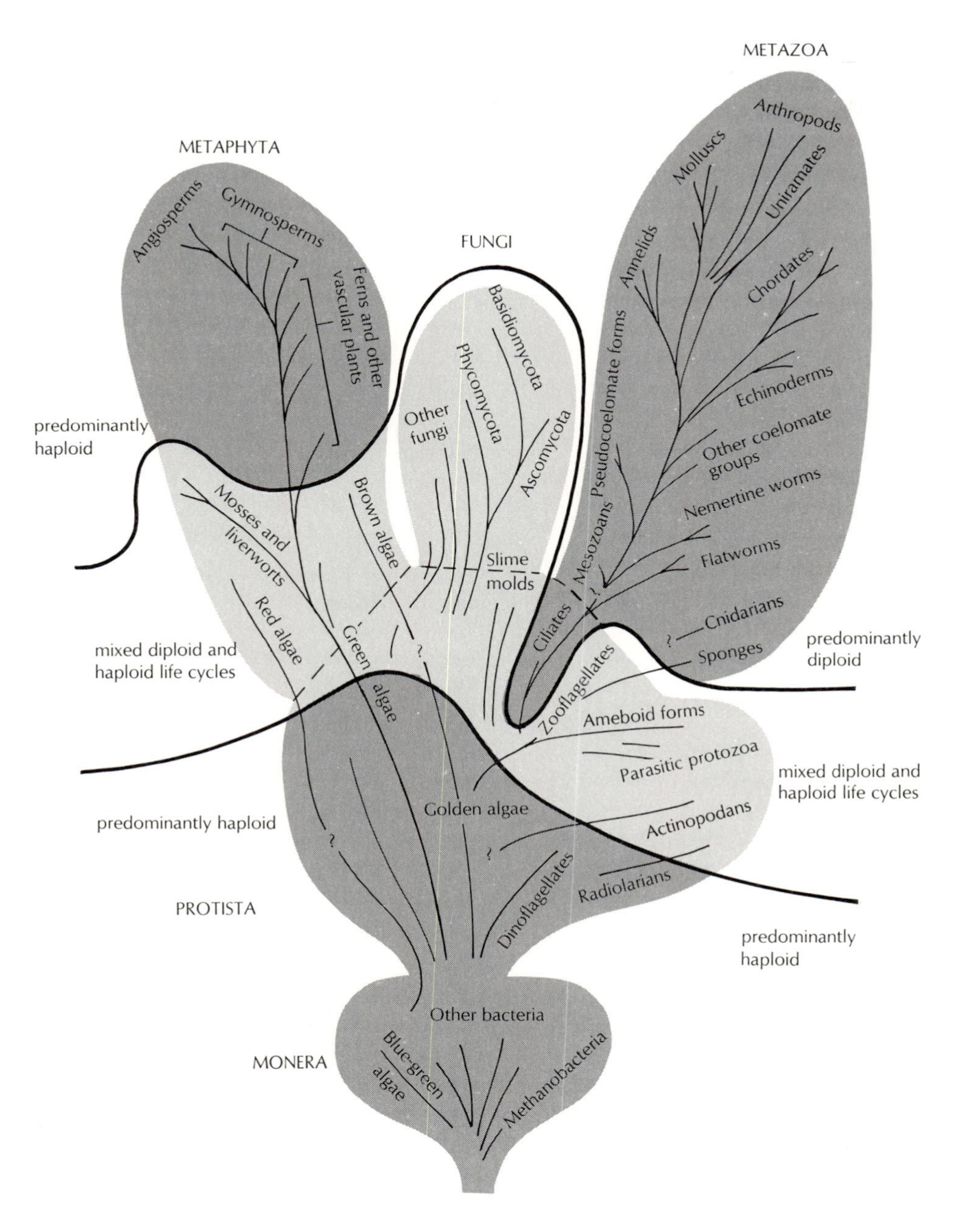

Fig. I-1: A phylogenetic summary of the five kingdoms, using common names.

Part 4, which also contain more detailed dendrograms of the various taxa discussed there.

The photographs following Table I–1 can be used to make more real the major taxa referred to in Fig. I–1. Other illustrations are also given in Chapters 1 and 11 through 14. (All photographs taken by Earl D. Hanson.)

Systems of classification

First let us look at a taxonomic treatment found in a very popular introductory biology text, that provided in W. T. Keeton's *Biological Science*, 3rd ed. (Table I–1). The numbers in parentheses refer to estimated numbers of species.

Table I–1. A five-kingdom classification (Keeton, 1980).

KINGDOM MONERA

DIVISION SCHIZOMYCETES. Bacteria (1,400)
- CLASS MYCOPLASMATA. *Mycoplasma, Acholeplasma*
- CLASS RICKETTSIAE. *Rickettsia, Coxiella*
- CLASS CHLAMYDIAE. *Chlamydia*
- CLASS ACTINOMYCETES. *Streptococcus, Staphylococcus, Arthrobacter, Actinomyces, Streptomyces*
- CLASS EUBACTERIA. *Escherichia, Rhizobium, Spirillum, Salmonella, Nitrosomonas, Serratia, Thiocystis*
- CLASS MYXOBACTERIA. *Myxococcus, Chondromyces*
- CLASS SPIROCHETES. *Leptospira, Spirocheta, Treponema*

DIVISION CYANOPHYTA. Blue-green algae (1,800). *Gloeocapsa, Microcystis, Oscillatoria, Nostoc, Scytonema*

DIVISION PROCHLOROPHYTA. *Prochloron*

KINGDOM PROTISTA

SECTION PROTOPHYTA: Algal protists

DIVISION EUGLENOPHYTA. Euglenoids (800). *Euglena, Eutreptia, Phacus, Colacium*

DIVISION CHRYSOPHYTA
- CLASS CHRYSOPHYCEAE. Golden-brown algae (650). *Chrysamoeba, Chromulina, Synura, Mallomonas*
- CLASS HAPTOPHYCEAE (or Prymnesiophyceae). Haptophytes and coccolithophores. *Isochrysis, Prymnesium, Phaeocystis, Coccolithus, Hymenomonas*
- CLASS XANTHOPHYCEAE. Yellow-green algae (360). *Botrydiopsis, Halosphaera, Tribonema, Botrydium*
- CLASS EUSTIGMATOPHYCEAE. Eustigmatophytes. *Pleurochloris, Visheria, Pseudocharaciopsis*
- CLASS CHLOROMONADOPHYCEAE. Chloromonads. *Gonyostomum, Reckertia*
- CLASS BACILLARIOPHYCEAE. Diatoms (10,000). *Pinnularia, Arachnoidiscus, Triceratium, Pleurosigma*

DIVISION PYRROPHYTA. Dinoflagellates (1,000). *Gonyaulax, Gymnodinium, Ceratium, Gloeodinium*

DIVISION CRYPTOPHYTA. Cryptomonads. *Cryptomonas, Chroomonas, Chilomonas, Hemiselmis*

SECTION PROTOMYCOTA: Fungal protists

DIVISION HYPHOCHYTRIDIOMYCOTA. Hyphochytrids (25). *Rhizidiomyces*

DIVISION CHYTRIDIOMYCOTA. Chytrids (1,000). *Olpidium, Rhizophydium, Diplophlyctis, Cladochytrium*

SECTION GYMNOMYCOTA: Slime molds

DIVISION PLASMODIOPHOROMYCOTA. Plasmodiophores (or endoparasitic slime molds). *Plasmodiophora, Spongospora, Woronina*

DIVISION LABYRINTHULOMYCOTA. Net slime molds. *Labyrinthula*

DIVISION ACRASIOMYCOTA. Cellular slime molds (26). *Dictyostelium, Polysphondylium*

DIVISION MYXOMYCOTA. True slime molds (400). *Physarum, Hemitrichia, Stemonitis*

SECTION PROTOZOA: Animal-like protists

PHYLUM MASTIGOPHORA
- CLASS ZOOFLAGELLATA (or Zoomastigina). "Animal" flagellates (5,000). *Trypanosoma, Calonympha, Chilomonas, Trichonympha*
- CLASS OPALINATA. Opalinids (200). *Opalina, Zelleriella*

PHYLUM SARCODINA. Pseudopodal protozoans (11,500)
- CLASS RHIZOPODEA. Naked and shelled amoebae, foraminiferans. *Amoeba, Pelomyxa, Entamoeba, Arcella, Globigerina, Textularia*
- CLASS ACTINOPODEA. Radiolarians, heliozoans, acantharians. *Aulacantha, Acanthometron, Actinosphaerium, Actinophrys*

PHYLUM SPOROZOA. Sporulation protozoans (3,600)
- CLASS TELOSPOREA. *Monocystis, Gregarina, Eimeria, Toxoplasma, Plasmodium*
- CLASS PIROPLASMEA. *Babesia, Theileria*

PHYLUM CNIDOSPORA. Cnidosporans (1,100)
- CLASS MYXOSPOREA. *Myxobolus, Myxidium, Ceratomyxa*
- CLASS MICROSPOREA. *Nosema, Haplosporidium, Mrazekia*

PHYLUM CILIATA. Ciliates (6,000). *Paramecium, Stentor, Vorticella, Spirostomum*

KINGDOM PLANTAE

DIVISION CHLOROPHYTA. Green algae (7,000). *Chlamydomonas, Volvox, Ulothrix, Spirogyra, Oedogonium, Ulva*

DIVISION CHAROPHYTA. Stoneworts (300). *Chara, Nitella, Tolypella*

DIVISION PHAEOPHYTA. Brown algae (1,500). *Sargassum, Ectocarpus, Fucus, Laminaria*

DIVISION RHODOPHYTA. Red algae (4,000). *Nemalion, Polysiphonia, Dasya, Chondrus, Batrachospermum*

DIVISION BRYOPHYTA (23,600)
- CLASS HEPATICAE. Liverworts. *Marchantia, Conocephalum, Riccia, Porella*
- CLASS ANTHOCEROTAE. Hornworts. *Anthoceros*
- CLASS MUSCI. Mosses. *Polytrichum, Sphagnum, Mnium*

DIVISION TRACHEOPHYTA. Vascular plants
- **Subdivision Psilopsida.** *Psilotum, Tmesipteris*
- **Subdivision Lycopsida.** Club mosses (1,500). *Lycopodium, Phylloglossum, Selaginella, Isoetes, Stylites*
- **Subdivision Sphenopsida.** Horsetails (25). *Equisetum*
- **Subdivision Pteropsida.** Ferns (10,000). *Polypodium, Osmunda, Dryopteris, Botrychium, Pteridium*
- **Subdivision Spermopsida.** Seed plants
  - CLASS PTERIDOSPERMAE. Seed ferns. No living representatives
  - CLASS CYCADAE. Cycads (100). *Zamia*
  - CLASS GINKGOAE (1). *Gingko*
  - CLASS CONIFERAE. Conifers (500). *Pinus, Tsuga, Taxus, Sequoia*
  - CLASS GNETEAE (70). *Gnetum, Ephedra, Welwitschia*
  - CLASS ANGIOSPERMAE. Flowering plants
    - SUBCLASS DICOTYLEDONEAE. Dicots (225,000). *Magnolia, Quercus, Acer, Pisum, Taraxacum, Rosa, Chrysanthemum, Aster, Primula, Ligustrum, Ranunculus*
    - SUBCLASS MONOCOTYLEDONEAE. Monocots (50,000). *Lilium, Tulipa, Poa, Elymus, Triticum, Zea, Ophyrys, Yucca, Sabal*

KINGDOM FUNGI

DIVISION OOMYCOTA. Water molds, white rusts, downy mildews (400). *Saprolegnia, Phytophthora, Albugo*

DIVISION ZYGOMYCOTA. Conjugation fungi (250)
- CLASS ZYGOMYCETTES. *Rhizopus, Mucor, Phycomyces, Choanephora, Entomophthora*
- CLASS TRICHOMYCETES. *Stachylina*

DIVISION ASCOMYCOTA. Sac fungi (12,000)
- CLASS HEMIASCOMYCETES. Yeasts and their relatives. *Saccharomyces, Schizosaccharomyces, Endomyces, Eremascus, Taphrina*
- CLASS PLECTOMYCETES. Powdery mildews, fruit molds, etc. *Erysiphe, Podosphaera, Aspergillus, Penicillium, Ceratocystis*
- CLASS PYRENOMYCETES. *Sordaria, Neurospora, Chaetomium, Xylaria, Hypoxylon*
- CLASS DISCOMYCETES. *Sclerotinia, Trichoscyphella, Rhytisma, Xanthoria, Pyronema*
- CLASS LABOULBENIOMYCETES. *Herpomyces, Laboulbenia*
- CLASS LOCULOASCOMYCETES. *Cochliobolus, Pyrenophora, Leptosphaeria, Pleospora*

DIVISION BASIDIOMYCOTA. Club fungi (15,000)
- CLASS HETEROBASIDIOMYCETES. Rusts and smuts. *Ustilago, Urocystis, Puccinia, Phragmidium, Melampsora*
- CLASS HOMOBASIDIOMYCETES. Toadstools, bracket fungi, mushrooms, puffballs, stinkhorns, etc. *Coprinus, Marasmius, Amanita, Agaricus, Lycoperdon, Phallus*

KINGDOM ANIMALIA

SUBKINGDOM PARAZOA

PHYLUM PORIFERA. Sponges (5,000)
- CLASS CALCAREA. Calcareous (chalky) sponges. *Scypha, Leucosolenia, Sycon, Grantia*
- CLASS HEXACTINELLIDA. Glass sponges. *Euplectella, Hyalonema, Monoraphis*
- CLASS DEMOSPONGIAE. *Spongilla, Euspongia, Axinella*
- CLASS SCLEROSPONGIAE. Coralline sponges. *Ceratoporella, Stromatospongia*

SUBKINGDOM AGNOTOZOA

PHYLUM PLACOZOA (1). *Trichoplax*

PHYLUM MESOZOA (50)
- CLASS DICYEMIDA. *Dicyema, Pseudicyema, Conocyema*
- CLASS ORTHONECTIDA. *Rhopalura*

SUBKINGDOM METAZOA

SECTION RADIATA

PHYLUM COELENTERATA (or Cnidaria)
- CLASS HYDROZOA. Hydrozoans (3,700). *Hydra, Obelia, Gonionemus, Physalia*
- CLASS SCYPHOZOA. Jellyfishes (200). *Aurelia, Pelagia, Cyanea*
- CLASS ANTHOZOA. Sea anemones and corals (6,100). *Metridium, Pennatula, Gorgonia, Astrangia*

PHYLUM CTENOPHORA. Comb jellies (90)
- CLASS TENTACULATA. *Pleurobrachia, Mnemiopsis, Cestum, Velamen*
- CLASS NUDA. *Beroe*

SECTION PROTOSTOMIA

PHYLUM PLATYHELMINTHES. Flatworms (10,000)
- CLASS TURBELLARIA. Free-living flatworms. *Planaria, Dugesia, Leptoplana*
- CLASS TREMATODA. Flukes. *Fasciola, Schistosoma, Prosthogonimus*
- CLASS CESTODA. Tapeworms. *Taenia, Dipylidium, Mesocestoides*

PHYLUM GNATHOSTOMULIDA (100). *Gnathostomula, Haplognathia*

PHYLUM NEMERTINA (or Rhynchocoela). Proboscis worms (650)
- CLASS ANOPLA. *Tubulanus, Cerebratulus*
- CLASS ENOPLA. *Amphiporus, Prostoma, Malacobdella*

PHYLUM ACANTHOCEPHALA. Spiny-headed worms (500). *Echinorhynchus, Gigantorhynchus*

PHYLUM ASCHELMINTHES
- CLASS ROTIFERA. Rotifers (1,700). *Asplanchna, Hydatina, Rotaria*
- CLASS GASTROTRICHA (200). *Chaetonotus, Macrodasys*
- CLASS KINORHYNCHA (or Echinodera) (100). *Echinoderes, Semnoderes*
- CLASS NEMATODA. Round worms (12,000). *Ascaris, Trichinella, Necator, Enterobius, Ancylostoma, Heterodera*
- CLASS NEMATOMORPHA. Horsehair worms (230). *Gordius, Paragordius, Nectonema*

PHYLUM ENTOPROCTA (60). *Urnatella, Loxosoma, Pedicellina*

PHYLUM PRIAPULIDA (8). *Priapulus, Halicryptus*

PHYLUM ECTOPROCTA (or Bryozoa). Bryozoans, moss animals (4,000)
- CLASS GYMNOLAEMATA. *Paludicella, Bugula*
- CLASS PHYLACTOLAEMATA. *Plumatella, Pectinatella*

PHYLUM PHORONIDA (15). *Phoronis, Phoronopsis*

PHYLUM BRACHIOPODA. Lamp shells (300)
- CLASS INARTICULATA. *Lingula, Glottidia, Discina*
- CLASS ARTICULATA. *Magellania, Neothyris, Terebratula*

PHYLUM MOLLUSCA. Molluscs
- CLASS AMPHINEURA
  - SUBCLASS APLACOPHORA. Solenogasters (250). *Chaetoderma, Neomenia, Proneomenia*
  - SUBCLASS POLYPLACOPHORA. Chitons (600). *Chaetopleura, Ischnochiton, Lepidochiton, Amicula*
- CLASS MONOPLACOPHORA (6). *Neopilina*
- CLASS GASTROPODA. Snails and their allies (univalve molluscs) (40,000). *Helix, Busycon, Crepidula, Haliotis, Littorina, Doris, Limax*
- CLASS SCAPHOPODA. Tusk shells (350). *Dentalium, Cadulus*
- CLASS PELECYPODA. Bivalve molluscs (7,500). *Mytilus, Ostrea, Pecten, Mercenaria, Teredo, Tagelus, Unio, Anodonta*
- CLASS CEPHALOPODA. Squids, octopuses, etc. (600). *Loligo, Octopus, Nautilus*

PHYLUM POGONORPHORA. Beard worms (100). *Siboglinum, Lamellisabella, Oligobrachia, Polybrachia*

PHYLUM SIPUNCULIDA (250). *Sipunculus, Phascolosoma, Dendrostomum*

PHYLUM ECHIUROIDA (80)
- CLASS ECHIURIDA. *Echiurus, Urechis, Ikeda*
- CLASS SACTOSOMATIDA. *Sactosoma*

PHYLUM ANNELIDA. Segmented worms
- CLASS POLYCHAETA (including Archiannelida). Sandworms, tubeworms, etc. (5,400). *Nereis, Chaetopterus, Aphrodite, Diopatra, Arenicola, Hydroides, Sabella*
- CLASS OLIGOCHAETA. Earthworms and many freshwater annelids (3,100). *Tubifex, Enchytraeus, Lumbricus, Dendrobaena*
- CLASS HIRUDINEA. Leeches (300). *Trachelobdella, Hirudo, Macrobdella, Haemadipsa*

PHYLUM ONYCHOPHORA (65). *Peripatus. Peripatopsis*

PHYLUM TARDIGRADA. Water bears (300). *Echiniscus, Macrobiotus*

PHYLUM PENTASTOMIDA. Tongue worms (60). *Cephalobaena, Linguatula*

PHYLUM ARTHROPODA

Subphylum Trilobita. No living representatives

Subphylum Chelicerata
- CLASS EURYPTERIDA. No living representatives
- CLASS XIPHOSURA. Horseshoe crabs (4). *Limulus*
- CLASS ARACHNIDA. Spiders, ticks, mites, scorpions, whipscorpions, daddy longlegs, etc. (55,000). *Archaearanea, Latrodectus, Argiope, Centruroides, Chelifer, Mastigoproctus, Phalangium, Ixodes*
- CLASS PYCNOGONIDA. Sea spiders (500). *Nymphon, Ascorphynchus*

Subphylum Mandibulata
- CLASS CRUSTACEA (26,000). *Homarus, Cancer, Daphnia, Artemia, Cyclops, Balanus, Porcellio*
- CLASS CHILOPODA. Centipeds (3,000). *Scolopendra, Lithobius, Scutigera*
- CLASS DIPLOPODA. Millipeds (8,000). *Narceus, Apheloria, Polydesmus, Julus, Glomeris*
- CLASS PAUROPODA (300). *Pauropus*
- CLASS SYMPHYLA (130). *Scutigerella*
- CLASS INSECTA. Insects (900,000)
  - ORDER COLLEMBOLA. Springtails. *Isotoma, Achorutes, Neosminthurus, Sminthurus*
  - ORDER PROTURA. *Acerentulus, Eosentomon*
  - ORDER DIPLURA. *Campodea, Japyx*
  - ORDER THYSANURA. Bristletails, silverfish, firebrats. *Machilis, Lepisma, Thermobia*
  - ORDER EPHEMERIDA. Mayflies. *Hexagenia, Callibaetis, Ephemerella*
  - ORDER ODONATA. Dragonflies, damselflies. *Archilestes, Lestes, Aeshna, Gomphus*
  - ORDER ORTHOPTERA. Grasshoppers, crickets, walking sticks, mantids, cockroaches, etc. *Schistocerca, Romalea, Nemobius, Megaphasma, Mantis, Blatta, Periplaneta*

ORDER ISOPTERA. Termites, *Reticulitermes, Kalotermes, Zootermopsis, Nasutitermes*
ORDER DERMAPTERA. Earwigs. *Labia, Forficula, Prolabia*
ORDER EMBIARIA (or Embiidina or Embioptera). *Oligotoma, Anisembia, Gynembia*
ORDER PLECOPTERA. Stoneflies. *Isoperla, Taeniopteryx, Capnia, Perla*
ORDER ZORAPTERA. *Zorotypus*
ORDER CORRODENTIA. Book lice. *Ectopsocus, Liposcelis, Trogium*
ORDER MALLOPHAGA. Chewing lice. *Cuclotogaster, Menacanthus, Menopon, Trichodectes*
ORDER ANOPLURA. Sucking lice. *Pediculus, Phthirius, Haematopinus*
ORDER THYSANOPTERA. Thrips. *Heliothrips, Frankliniella, Hercothrips*
ORDER HEMIPTERA. True bugs. *Belostoma, Lygaeus, Notonecta, Cimex, Lygus, Oncopeltus*
ORDER HOMOPTERA. Cicadas, aphids, leaf hoppers, scale insects, etc. *Magicicada, Circulifer, Psylla, Aphis, Saissetia*
ORDER NEUROPTERA. Dobsonflies, alderflies, lacewings, mantispids, snakeflies, etc. *Corydalus, Hemerobius, Chrysopa, Mantispa, Agulla*
ORDER COLEOPTERA. Beetles, weevils. *Copris, Phyllophaga, Harpalus, Scolytus, Melanotus, Cicindela, Dermestes, Photinus, Coccinella, Tenebrio, Anthonomus, Conotrachelus*
ORDER HYMENOPTERA. Wasps, bees, ants, sawflies. *Cimbex, Vespa, Glypta, Scolia, Bembix, Formica, Bombus, Apis*
ORDER MECOPTERA. Scorpionflies. *Panorpa, Boreus, Bittacus*
ORDER SIPHONAPTERA. Fleas. *Pulex, Nosopsyllus, Xenopsylla, Ctenocephalides*
ORDER DIPTERA. True flies, mosquitoes. *Aedes, Asilus, Sarcophaga, Anthomyia, Musca, Chironomus, Tabanus, Tipula, Drosophila*
ORDER TRICHOPTERA. Caddisflies. *Limnephilus, Rhyacophila, Hydropsyche*
ORDER LEPIDOPTERA. Moths, butterflies. *Tinea, Pyrausta, Malacosoma, Sphinx, Samia, Bombyx, Heliothis, Papilio, Lycaena*

SECTION DEUTEROSTOMIA

PHYLUM CHAETOGNATHA. Arrow worms (60). *Sagitta, Spadella*

PHYLUM ECHINODERMATA
CLASS CRINOIDEA. Crinoids, sea lilies (630). *Antedon, Ptilocrinus, Comactinia*
CLASS ASTEROIDEA. Sea stars (1,600). *Asterias, Ctenodiscus, Luidia, Oreaster*
CLASS OPHIUROIDEA. Brittle stars, serpent stars, basket stars, etc. (2,000). *Asteronyx, Amphioplus, Ophiothrix, Ophioderma, Ophiura*
CLASS ECHINOIDEA. Sea urchins, sand dollars, heart urchins (860). *Cidaris, Arbacia, Strongylocentrotus, Echinanthus, Echinarachinus, Moira*
CLASS HOLOTHUROIDEA. Sea cucumbers (900). *Cucumaria, Thyone, Caudina, Synapta*

PHYLUM HEMICHORDATA (90)
CLASS ENTEROPNEUSTA. Acorn worms. *Saccoglossus, Balanoglossus, Glossobalanus*
CLASS PTEROBRANCHIA. *Rhabdopleura, Cephalodiscus*

PHYLUM CHORDATA. Chordates
Subphylum Urochordata (or Tunicata). Tunicates (2,000)
CLASS ASCIDIACEA. Ascidians or sea squirts. Ciona, Clavelina, Molgula, Perophora
CLASS THALIACEA. *Pyrosoma, Salpa, Doliolum*
CLASS LARVACEA. *Appendicularia, Oikopleura, Fritillaria*
Subphylum Cephalochordata. Lancelets, amphioxus (30). *Branchiostoma, Asymmetron*
Subphylum Vertebrata. Vertebrates
CLASS AGNATHA. Jawless fishes (50). *Cephalaspis, Pteraspis, Petromyzon, Entosphenus, Myxine, Eptatretus*
CLASS PLACODERMI. No living representatives
CLASS CHONDRICHTHYES. Cartilaginous fishes (625). *Squalus, Hyporion, Raja, Chimaera*
CLASS OSTEICHTHYES. Bony fishes (30,000)
SUBCLASS SARCOPTERYGII
ORDER CROSSOPTERYGII (or Coelacanthiformes). Lobe-fins. *Latimeria*
ORDER DIPNOI (or Dipteriformes). Lungfishes. *Neoceratodus, Protopterus, Lepidosiren*
SUBCLASS BRACHIOPTERYGII. Bichirs. *Polypterus*
SUBCLASS ACTINOPTERYGII. Higher bony fishes. *Amia, Cyprinus, Gadus, Perca, Salmo*
CLASS AMPHIBIA (2,600)
ORDER ANURA. Frogs and toads. *Rana, Hyla, Bufo*
ORDER URODELA. Salamanders. *Necturus, Triturus, Plethodon, Ambystoma*
ORDER APODA. *Ichthyophis, Typhlonectes*
CLASS REPTILIA (6,500)
ORDER CHELONIA. Turtles. *Chelydra, Kinosternon, Clemmys, Terrapene*
ORDER RHYNCHOCEPHALIA. Tuatara. *Sphenodon*
ORDER CROCODYLIA. Crocodiles and alligators. *Crocodylus, Alligator*
ORDER SQUAMATA. Snakes and lizards. *Iguana, Anolis, Sceloporus, Phrynosoma, Natrix, Elaphe, Coluber, Thamnophis, Crotalus*
CLASS AVES. Birds (8,600). *Anas, Larus, Columba, Gallus, Turdus, Dendroica, Sturnus, Passer, Melospiza*
CLASS MAMMALIA. Mammals (4,100)
SUBCLASS PROTOTHERIA
ORDER MONOTREMATA. Egg-laying mammals. *Ornithorhynchus, Tachyglossus*
SUBCLASS THERIA. Marsupial and placental mammals
ORDER MARSUPIALIA. Marsupials. *Didelphis, Sarcophilus, Notoryctes, Marcopus*
ORDER INSECTIVORA. Insectivores (moles, shrews, etc.). *Scalopus, Sorex, Erinaceus*

ORDER DERMOPTERA. Flying lemurs. *Galeopithecus*
ORDER CHIROPTERA. Bats. *Myotis, Eptesicus, Desmodus*
ORDER PRIMATES. Lemurs, monkeys, apes, humans. *Lemur, Tarsius, Cebus, Macacus, Cynocephalus, Pongo, Pan, Homo*
ORDER EDENTATA. Sloths, anteaters, armadillos. *Bradypus, Myrmecophagus, Dasypus*
ORDER PHOLIDOTA. Pangolin. *Manis*
ORDER LAGOMORPHA. Rabbits, hares, pikas. *Ochotona, Lepus, Sylvilagus, Oryctolagus*
ORDER RODENTIA. Rodents. *Sciurus, Marmota, Dipodomys, Microtus, Peromyscus, Rattus, Mus, Erethizon, Castor*
ORDER CETACEA. Whales, dolphins, porpoises. *Delphinus, Phocaena, Monodon, Balaena*
ORDER CARNIVORA. Carnivores. *Canis, Procyon, Ursus, Mustela, Mephitis, Felis, Hyaena, Eumetopias*
ORDER TUBULIDENTATA. Aardvark. *Orycteropus*
ORDER PROBOSCIDEA. Elephants. *Elephas, Loxodonta*
ORDER HYRACOIDEA. Coneys. *Procavia*
ORDER SIRENIA. Manatees. *Trichechus, Halicore*
ORDER PERISSODACTYLA. Odd-toed ungulates. *Equus, Tapirella, Tapirus, Rhinoceros*
ORDER ARTIODACTYLA. Even-toed ungulates. *Pecari, Sus, Hippopotamus, Camelus, Cervus, Odocoileus, Giraffa, Bison, Ovis, Bos*

Skeleton of the Irish elk (extinct)

Plate 1. The moss *Polytrichum commune,* lying diagonally across the photograph, shows both gametophytic and sporophytic portions. At the lower right, rhizoids connect with the portion bearing slender, radiating leaflets; both constitute the haploid gametophyte. The diploid sporophyte is the slender stem that terminates, upper left, in a capsule contain spores.

Plate 2. A rock covered with lichens. The darkish, leaf-like plants are toad-skin lichens (*Umbilicaria papillosa*). Among them is a smaller, light gray lichen.

Plate 3. The club-moss (*Lycopodium complanatum*) seen here as a sporophyte. The vertical stems bear special structures within which are spores; the structures, below, bear small leaves.

Plate 4. These fronds of the fern *Polypodium virginianum* bear conspicuous circular structures (sori) on their undersides. These contain spores; hence the fronds are haploid sporophytes.

Plate 5. Pollen from the male cones of the Japanese black pine (*Pinus thunbergii*), air-borne in a breeze, appears like a fine dust. The male cones are the clusters of small projections lying typically at the base of a twig without needles. The one large female cone, to the right, looks like a typical, woody pine cone.

Plate 6. The small bracket fungus (*Polysticticus versicolor*) is growing on the fallen trunk of a beech tree. In the upper left corner are a larger bracket fungus and, to its right, a small toadstool (both unidentified).

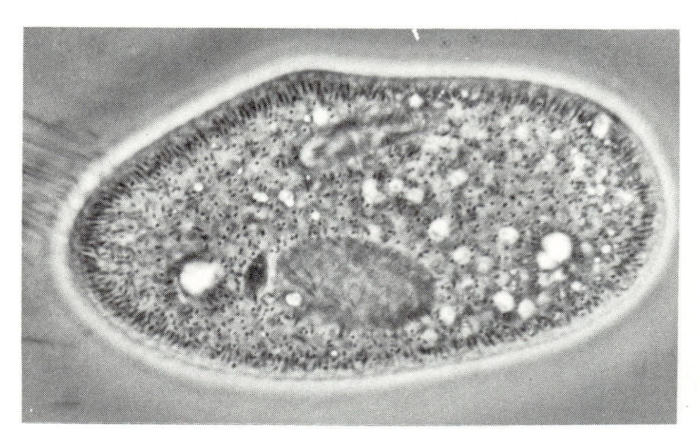

Plate 7. A ciliated protozoan (*Paramecium trichium*) seen under a special optical microscope, which brings out details of its internal structure. The large gray, oval macronucleus is just below the center of this unicellular animal. To the left of the macronucleus is a smaller, dark, spindle-shaped micronucleus; the clear area to the left of that is a contractile vesicle (another lies toward the other end of the cell), which regulates cellular water content; and the permanent mouth is the gray, ovoid structure above the macronucleus. This cell is about 120 μm long.

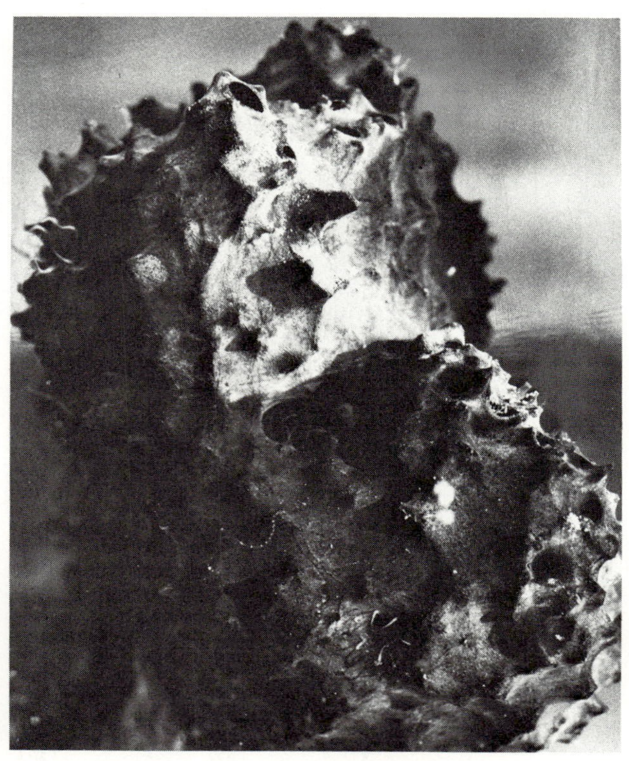

Plate 8. This marine sponge, of the genus *Hircinia,* appears as a brownish lump of material. The dark openings are the pores for water excretion. The many tiny openings, all over the sponge, take in water and food particles suspended in it. Compare Fig. 1-14, which diagrams water movements into and out of sponges of varying complexity.

Plate 9. Members of the Cnidaria, these rose corals (*Manicina areolata*), a few inches across, lie in shallow marine waters. Here the feeding polyps are withdrawn into the protective stony structure they secrete.

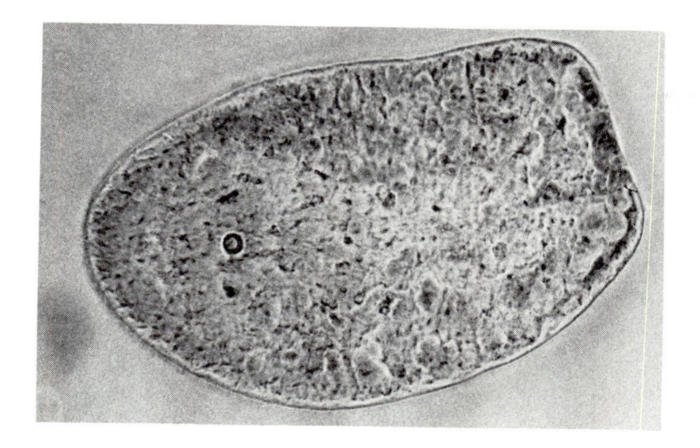

Plate 10. To the left, center, of this freshly hatched acoel flatworm (*Amphiscolops langerhansi*) (head region) is a round statocyst, presumably some kind of orienting organ. On either side of the statocyst are two dark blobs of light-sensitive pigments, the eyespots. This little worm, photographed through a microscope, is about 0.25 mm (250 μm) long.

Plate 11. This marine annelid (*Sabellastarte magnifica*) is known as the magnificent feather duster worm. The feathery looking tentacles collect food particles from the surrounding water and pass them to the mouth situated at their base. This segmented worm lives in a tough leathery tube, into which it can withdraw its tentacles in a flash.

Plate 12. This mollusc is a snail, *Oliva reticularis.* The siphon at its anterior end circulates water through its lungs. On either side of the siphon are eye-stalks, each tipped with a small, dark eye.

Plate 13. The two pincers of this box crab (*Calappa gallus*) cover the front of the animal; its two knob-like eyes protrude above, its slender feet below.

Plate 14. This spider (*Argiope argentata*) typically joins two legs together into four pairs as it waits in its web for food to arrive in the form of some flying insect.

Plate 15. The sea-urchin (*Arbacia punctulata*) illustrates the spiny exterior that gives the echinoderms their name, i.e., spiny skinned. The delicate, light-colored structures that terminate in light discs are the tube feet attached to the glass of the aquarium. See also the tube feet in the starfish in Fig. 1-10.

Plate 16. These marine tunicates are members of the phylum Chordata. They typically attach themselves to mangrove roots in tropical waters. Each animal in the colony of this species (*Ecteinascidia turbinata*) is tinged soft blue. They are filter feeders.

Plate 17. The lizard fish (*Synodus foetens*) usually rests on the bottom among assorted algae, but it will dart suddenly upward to catch other fish for food.

Plate 18. This Cuban rock iguana (*Cyclura nubila*), about four feet long, is a vegetarian.

Plate 19. This young rhesus monkey (*Macacus rhesus*) is from a colony imported to roam on a Caribbean island until needed for experimental medical research.

It is worth noting that, in the first edition of Keeton's book, a Linnaean two-kingdom scheme of classification was used. The second edition contained a three-kingdom scheme, i.e., Monera, Plantae, and Animalia, and the third edition a five-kingdom classification, given here. Note especially his treatment of the fungi, since he has separated them into the kingdom Protista and the kingdom Fungi. This is one response to the problems encountered in Chapter 12. Note also the subkingdoms in the kingdom Animalia.

A quite different approach to classification is given in Table I–2. This is the most recent attempt by R. H. Whittaker and Lynn Margulis to present their current phylogenetic thinking as a scheme of classification.

Table I–2. Kingdoms of Organisms (Whittaker and Margulis, 1978).

Superkingdom Prokaryota. *Anucleate organization*

I. Kingdom Monera. Prokaryotic cells. Nutrition absorptive, chemosynthetic, photoheterotrophic, or photoautotrophic, with anaerobic, facultative, microaerophilic or aerobic metabolism. Reproduction asexual and chromonemal; genetic recombination unidirectional or viral mediated. Nonmotile or motile by bacterial flagella composed of flagellin proteins or by gliding. Solitary unicellular, filamentous, colonial, or mycelial.

- **Form-Superphylum Protobacteria,** lacking cell walls
  - **Phylum Mycoplasmata,** mycoplasms
  - **Phylum Rickettsiae,** Rickettsias
- **Superphylum and Phylum Spirochaetae,** spirochetes
- **Form-Superphylum Eubacteria,** true bacteria with cell walls
  - **Phylum Fermenting bacteria;** unable to synthesize porphyrins
  - **Phylum Anaerobic sulfate reducers;** synthesis of heme proteins limited
  - **Phylum Methane bacteria;** anaerobic chemotrophs reducing CO_2 to CH_4
  - **Phylum Nitrogen-fixing aerobic bacteria**
  - **Phylum Pseudomonads**
  - **Phylum Aerobic endospore bacteria,** Gram-positive
  - **Phylum Micrococci,** Gram-positive aerobes with full Krebs cycle
  - **Phylum Aerobic Gram-negative heterotrophic eubacteria**
  - **Phylum Chemoautotrophic bacteria**
- **Superphylum Photomonera,** photosynthetic prokaryotes
  - **Phylum Photobacteria,** non-oxygen-eliminating photosynthetic bacteria
  - **Phylum Prochlorophyta,** green oxygen-eliminating prokaryotes
  - **Phylum Cyanophyta** or **Cyanobacteria,** blue-green algae
- **Superphylum** and **Phylum Actinomycota;** Gram-positive coryniform and mycelial bacteria
- **Superphylum** and **Phylum Myxobacteria,** aerobic gliding bacteria

Superkingdom Eukaryota. *Nucleate organization*

II. Kingdom Protista or **Protoctista.** Eukaryotic cells with solitary and colonial unicellular organization (Protista), or also including simpler multicellular forms (Protoctista). Nutrition ingestive, absorptive, or, if photoautotrophic, in photosynthetic plastids. Premitotic and eumitotic asexual reproduction; in eumitotic forms meiosis and karyogamy are present, but detailed cytology, life cycle, and ploidy levels vary from group to group. Flagella (or cilia) (≡undulapodia) life cycle, and ploidy levels vary from group to group. Undulapodia composed of tubulin microtubules in the 9 + 2 pattern. Lack embryos and complex cell junctions.

- **Branch Protophyta,** plant-like protists (or protoctists).
  - **Form-Superphylum Chromophyta** or **Chromobionta,** yellow and brown flagellate algae and allies
    - **Phylum Chrysophyta,** s.s., golden algae
    - **Phylum Bacillariophyta,** diatoms
    - **Phylum Xanthophyta,** yellow-green algae
    - **Phylum Haptophyta,** haptophytes or coccolithophores
    - **Phylum Eustigmatophyta,** eustigmatophytes
    - **Phylum Dinoflagellata** or **Pyrrophyta,** s.s., dinoflagellates
    - **Phylum Cryptophyta,** cryptomonads
    - **(Phylum Phaeophyta)**
  - **Form-Superphylum Chlorophyta,** s.l., or **Chlorobionta,** green algae
    - **Phylum Chlorophyta,** s.s., grass-green algae
    - **Phylum Siphonophyta,** siphonaceous syncytial green algae
    - **Phylum Prasinophyta,** prasinophytes
    - **Phylum Zygnematophyta** or **Gamophyta,** conjugating green algae
    - **Phylum Charophyta,** stoneworts
    - **Phylum Euglenophyta,** euglenoid flagellates
  - **(Superphylum Rhodophyta)**
- **Branch Protomycota,** fungus-like protists (or protoctists)
  - **Form-Superphylum Mastigomycota,** primitive fungi with chytrid or simple mycelial organization, flagellated stages
    - **Phylum Hyphochytridiomycota,** hyphochytrids with single, anterior, mastigonemate flagellum
    - **Phylum Chytridiomycota,** true chytrids with single, posterior flagellum
    - **Phylum Oomycota,** oosphere water molds, with biflagellate, heterokont zoospores
  - **Form-Superphylum Gymnomycota,** slime molds with life cycles including separate cells, aggregations, and sporulation structures
    - **Phylum Plasmodiophoromycota,** plasmodiophores
    - **Phylum Labyrinthulomycota,** cell-net slime molds

Phylum Acrasiomycota, cellular or pseudoplasmodial slime molds
Phylum Myxomycota, syncytial or plasmodial slime molds
Branch-Protozoa, animal-like protists (or protoctists)
Form-Superphylum Sporozoa, s.l., protozoans with sporulation stages
Phylum Apicomplexa or **Sporozoa,** s.s., sporozoan parasites
Phylum Cnidosporidia, cnidosporidian parasites with polar capsules
Form-Superphylum Sarcodina, s.l., pseudopodial protozoans
Phylum Caryoblastea or **Pelobiontida,** amitotic amoebae
Phylum Rhizopoda or **Sarcodina,** s.s., naked and shelled amoebae
Phylum Actinopoda, radiolarians, heliozoans, and acantharians
Phylum Foraminifera, foraminiferans
Form-Superphylum and **Phylum Zoomastigina,** animal flagellates
Superphylum and **Phylum Ciliophora,** ciliates and suctorians
(Superphylum Agnotozoa)
(Superphylum Parazoa)

III. Kingdom Fungi. Primarily multinucleate organisms with eukaryotic nuclei dispersed in a walled and often septate mycelial syncytium (some forms secondarily unicellular). Nutrition absorptive. Flagella lacking, no motility except protoplasmic streaming. Zygotic meiosis, haploid spores produced; mycelia haploid or dikaryotic.

(Superphylum Gymnomycota)
(Superphylum and **Phylum Oomycota)**
Superphylum Eumycota, amastigote fungi
Phylum Zygomycota, conjugation fungi
Phylum Ascomycota, sac fungi
Phylum Basidiomycota, club fungi
Form-Phylum Deuteromycota, fungi imperfecti
Form-Phylum Mycophycophyta, lichens, polyphyletic symbioses of fungi and algae into functional "plants"

IV. Kingdom Animalia. Multicellular animals with wall-less eukaryotic cells. Nutrition primarily ingestive with digestion in an internal cavity, but some forms are absorptive and some lack a digestive cavity; phagocytosis and pinocytosis present. Reproduction mainly sexual with anisogamous fertilization and gametic meiosis; haploid stages other than gametes almost lacking above the lowest phyla. Motility of the organism (or in sessile forms of its parts) based on contractile fibrils. Develop (except Agnotozoa) from blastulae; complex connections between cells in tissues

Subkingdom and **Form-Superphylum Agnotozoa**
Phylum Placozoa, placozoans
Phylum Mesozoa, mesozoan parasites
Subkingdom and **Form-Superphylum Parazoa**
Phylum Porifera, sponges
Phylum Archaeocyatha, extinct
Subkingdom Eumetazoa
Branch Radiata, radially symmetrical animals
Superphylum Coelenterata
Phylum Cnidaria, coelenterates
Phylum Ctenophora, comb jellies
Branch Bilateria, bilaterally symmetrical animals
Grade Acoelomata
Superphylum Prothelminthes or **Platyhelminthes,** s.l., acoelomate worms
(Phylum Mesozoa)
Phylum Platyhelminthes, s.s., flatworms
Phylum Nemertea or **Rhynchocoela,** ribbon worms
Phylum Gnathostomulida, gnathostome worms
Grade Pseudocoelomata
Superphylum Aschelminthes, pseudocoelomate worms
Phylum Gastrotricha, gastrotrichs
Phylum Rotifera, rotifers
Phylum Kinorhyncha, kinorhynch worms
Phylum Acanthocephala, spiny-headed worms
Phylum Nematoda, roundworms
Phylum Nematomorpha, horsehair worms
(Superphylum and **Phylum Entoprocta)**
Grade Coelomata
Subgrade Schizocoela or **Proterostoma**
Superphylum Tentaculata, lophophorate organisms
Phylum Entoprocta or **Kamptozoa,** pseudocoelomate polyzoans
Phylum Bryozoa or **Ectoprocta,** coelomate polyzoans
Phylum Phoronida, phoronid worms
Phylum Brachiopoda, lamp shells
Superphylum and **Phylum Mollusca,** molluscs
Superphylum Coelhelminthes or **Annelida,** s.l., schizocoel worms
Phylum Priapulida, priapulid worms
Phylum Sipuncula, peanut worms
Phylum Echiurida, echiuroid worms
Phylum Annelida, s.s., segmented worms
Superphylum Arthropoda, s.l., arthropods
Phylum Onychophora, Peripatus
Phylum Tardigrada, water bears
Phylum Pentastoma, linguatulids or tongueworms
Phylum Arthropoda, s.s., arthropods
Subgrade Entercoela or **Deuterostoma**
Superphylum and **Phylum Echinodermata,** echinoderms
Superphylum Metahelminthes, enterocoel worms
Phylum Pogonophora, beard worms
Phylum Chaetognatha, arrow worms
Phylum Hemichordata or **Enteropneusta,** acorn worms
Superphylum and **Phylum Chordata,** chordates

V. Kingdom Plantae. Primarily autotrophic plants, multicellular with walled and frequently vacuolate eukaryotic cells and photosynthetic plastids (together with related organisms which lack the latter). Simply multicellular-advanced tissue organization; development from solid embryos. Reproduction primarily sexual with haploid and diploid stages, the former progressively reduced toward higher members of the kingdom. Plasmodesmata.

Superphylum and **Phylum Rhodophyta,** red algae
Superphylum and **Phylum Phaeophyta,** brown algae
(Superphylum Chlorophyta)
Superphylum Embryophyta
  Phylum Bryophyta, liverworts, hornworts, and mosses
  Phylum Tracheophyta, vascular plants
    Class Psilopsida or **phylum Psilophyta,** whisk ferns
    Class Lycosida or **phylum Lycopodophyta,** club mosses, quillworts, lepidodendra
    Class Sphenopsida or **phylum Sphenophyta,** horsetails and calamites
    Class Pteropsida or **phylum Filicinophyta,** ferns and seed ferns
    Class Cycadopsida or **phylum Cycadophyta,** cycads
    Class Cnetopsida or **phylum Gnetophyta,** *Gnetum, Ephedra,* and allies
    Class Ginkgopsida or **phylum Ginkgophyta,** ginkgoes
    Class Conopsida or **phylum Coniferophyta,** conifers
    Class Angiospermae or **phylum Angiospermophyta** or **Anthophyta,** flowing plants

Note here the use of taxonomic categories that go far beyond those proposed by Linnaeus (Table 1–4 and Box 1–2), i.e., Superkingdom, Subkingdom, Branch, and Superphylum. Though this classification has real phylogenetic intentions, it contains sections that are also phenetic, i.e., systems for filing information in which no phyletic sense is retained in the taxon used. This can be seen, for example, in the Superphylum Sarcodina, which is a grabbag of protozoa that have pseudopodia. This is certainly a polyphyletic taxon and sheds little light on the evolutionary history of pseudopodia. This classification is a compromise between phylogenetic interpretations and taxonomic convenience and utility.

The last taxonomic scheme (Table I–3) is a rather detailed treatment of the phylum Chordata, and especially the subphylum Vertebrata, taken from Colbert's *Evolution of the Vertebrates*. This gives some idea of the detail that is to be found in the taxonomic analysis of a large, well-studied group, such as the vertebrates. Also, note here the taxa that have been introduced in addition to the more traditional Linnaean ones, e.g., Subclass, Infraclass, Cohort, etc.

Table I–3. The Phylum Cordata (Colbert, 1969).

SUBPHYLUM HEMICHORDATA: *Balanoglossus*

SUBPHYLUM CEPHALOCHORDATA: *Amphioxus*

SUBPHYLUM UROCHORDATA: sea squirts

SUBPHYLUM VERTEBRATA: the vertebrates

CLASS AGNATHA: jawless vertebrates; ostracoderms, lampreys, and hagfishes.
  SUBCLASS MONORHINA: * a dorsal nostril between the eyes
    ORDER CEPHALASPIDA: * amored ostracoderms, with a flattened head shield
    ORDER ANASPIDA: * small, armored, deep-bodied ostracoderms
  SUBCLASS DIPLORHINA: * no dorsal nostril
    ORDER PTERASPIDA: * ostracoderms with a head shield of large plates
    ORDER COELOLEPIDA: * unarmored ostracoderms, denticles covering body
  SUBCLASS CYCLOSTOMATA: modern jawless fishes
    ORDER MYXINIFORMES: hagfishes
    ORDER PETROMYZONTIFORMES: lampreys

CLASS PLACODERMI: * early jawed fishes, mostly heavily armored
    ORDER ARTHRODIRA: * armored fishes with jointed necks
    ORDER PYCTODONTIDA: * small, armored placoderms
    ORDER PETALICHTHYIDA: * armored fishes related to arthrodires
    ORDER ANTIARCHI: * small placoderms with jointed, movable, pectoral spines
    ORDER RHENANIDA: * "skatelike" placoderms
    ORDER PHYLLOLEPIDA: * flattened, heavily plated placoderms
    ORDER PALAEOSPONDYLOIDEA: * *Palaeospondylus,* an enigmatic form

CLASS ACANTHODII: * spiny fishes
    ORDER CHMATIFORMES: * primitive acanthodians
    ORDER ISCHNACANTHIFORMES: * specialized forms, with reduced spines
    ORDER ACANTHODIFORMES: * persisting acanthodians

CLASS CHONDRICHTHYES: cartilaginous fishes, the broad category of sharks
  SUBCLASS ELASMOBRANCHII: sharks
    ORDER CLASDOSELACHIFORMES: * ancestral sharks
    ORDER PLEURACANTHIFORMES: * early freshwater sharks
    ORDER HETERODONTIFORMES: archaic sharks
    ORDER HEXANCHIFORMES: primitive sharks
    ORDER LAMNIFORMES: modern sharks
      SUBORDER GALEOIDEI: the majority of living sharks
      SUBORDER SQUALOIDEI: dogfishes and related forms
      SUBORDER BATOIDEI: skates and rays
  SUBCLASS HOLOCEPHALI: sharklike fishes
    ORDER CHIMAERIFORMES: chimaeras or ratfishes

CLASS OSTEICHTHYES: bony fishes
SUBCLASS ACTINOPTERYGII: ray-finned fishes
INFRACLASS CHONDROSTEI: primitive ray-finned fishes
ORDER PALAEONISCIFORMES: * ancestral ray-finned fishes
ORDER POLYPTERIFORMES: surviving palaeonischid types, living in Africa
ORDER ACIPENSERIFORMES: sturgeons and paddlefishes and their extinct relatives
INFRACLASS HOLOSTEI: intermediate ray-finned fishes
ORDER SEMIONOTIFORMES: * early holosteans
ORDER PYCNODONTIFORMES: * deep-bodied holosteans
ORDER AMIIFORMES: the central holostean group; the modern bowfin
ORDER ASPIDORHYNCHIFORMES: * heavily scaled, elongated holosteans
INTRACLASS TELEOSTEI: advanced ray-finned fishes
SUPERORDER PHOLIDOPHOROMORPHA: * transitional from holosteans to teleosts
ORDER PHOLIDOPHORIFORMES: * ancestors of teleosts
SUPERORDER LEPTOLEPIDOMORPHA: * the beginning of teleost radiation
ORDER LEPTOLEPIFORMES: * generalized teleosts
SUPERORDER ELOPOMORPHA: primitive, varied teleosts
ORDER ELOPIFORMES: ancestors of the tarpons
ORDER ANGUILLIFORMES: eels
ORDER NOTOCANTHIFORMES: certain deep-sea fishes
SUPERORDER CLUPEOMORPHA: persistently primitive teleosts
ORDER CLUPEIFORMES: herrings and their relatives
SUPERORDER OSTEOGLOSSOMORPHA: primitive Cretaceous teleosts and their descendants
ORDER OSTEOGLOSSIFORMES: primitive, tropical freshwater fishes
ORDER MORMYRIFORMES: snouted freshwater fish of Africa
SUPERORDER PROTACANTHOPTERYGII: basically primitive, but progressive teleosts
ORDER SALMONIFORMES: salmon and trout
ORDER CTENOTHRISSIFORMES: * Cretaceous ancestors of spiny-finned teleosts
ORDER GONORHYNCHIFORMES: milk fishes and their relatives
SUPERORDER OSTARIOPHYSI: a majority of freshwater teleosts
ORDER CYPRINIFORMES: characins, minnows, carp
ORDER SILURIFORMES: catfishes
SUPERORDER PARACANTHOPTERYGII: advanced teleosts, paralleling the acanthopterygians
ORDER POLYMIXIIFORMES: beardfishes
ORDER PERCOPSIFORMES: pirate perch and freshwater relatives
ORDER GADIFORMES: cod, haddock
ORDER BATRACHOIDIFORMES: toadfishes
ORDER LOPHIIFORMES: anglers
ORDER GOBIESOCIFORMES: clingfishes
SUPERORDER ATHERINOMORPHA: varied teleosts
ORDER ATHERINIFORMES: flying fishes, killifishes
SUPERORDER ACANTHOPTERYGII: spiny teleosts, a majority of teleosts
ORDER BERYCIFORMES: primitive acanthopterygians, squirrel fishes
ORDER ZEIFORMES: John Dory and other tropical teleosts
ORDER LAMPRIDIFORMES: moon fish
ORDER GASTEROSTEIFORMES: sticklebacks, seahorses
ORDER CHANNIFORMES: the snakehead, *Channa*
ORDER SYNBRANCHIFORMES: tropical coastal fishes
ORDER SCORPAENIFORMES: sculpins, sea robins
ORDER DACTYLOPTERIFORMES: large-finned Oriental fishes
ORDER PEGASIFORMES: tropical armored teleosts
ORDER PERCIFORMES: the majority of spiny teleosts
ORDER PLEURONECTIFORMES: flatfishes
ORDER TETRAODONTIFORMES: plectognath fishes
SUBCLASS SARCOPTERYGII: lobe-finned, air breathing fishes
ORDER CROSSOPTERYGII: progressive air breathing fishes
SUBORDER RHIPIDISTIA: * ancestors of the amphibians
SUPERFAMILY OSTEOLEPIFORMES: * leading toward the amphibians
SUPERFAMILY POROLEPIFORMES: * primitive and aberrant forms
SUBORDER COELACANTHINI: predominantly marine crossopts; the surviving *Latimeria*
ORDER DIPNOI: lungfishes

CLASS AMPHIBIA: amphibians, the earliest tetrapods or land-living vertebrates
SUBCLASS LABYRINTHODONTIA: * late Paleozoic and Triassic solid-skulled amphibians with complex vertebrae
ORDER ICHTHYOSTEGALIA: * ancestral labyrinthodonts
ORDER TEMNOSPONDYLI: * late Paleozoic and Triassic labyrinthodonts
SUBORDER RHACHITOMI: * the culmination of labyrinthodont evolution
SUBORDER STEREOSPONDYLI: * large, aquatic labyrinthodonts
SUBORDER PLAGIOSAURIA: * flat, broad-skulled labyrinthodonts
ORDER ANTHRACOSAURIA: * late Paleozoic labyrinthodonts, evolving toward reptiles
SUBORDER SCHIZOMERI: * very early anthracosaurs
SUBORDER DIPLOMERI: * primitive anthracosaurs
SUBORDER EMBOLOMERI: * typical anthracosaurs
SUBORDER SEYMOURIAMORPHA: * intermediate between labyrinthodonts and reptiles
SUBCLASS LEPOSPONDYLI: * late Paleozoic amphibians with spool-shaped vertebrae
ORDER NECTRIDIA: * varied lepospondyls, including diplocaulids
ORDER AISTOPODA: * ancient, limbless lepospondyls
ORDER MICROSAURIA: * small, early lepospondyls
SUBCLASS LISSAMPHIBIA: the modern amphibians
INFRACLASS SALIENTIA: frogs, toads, and their ancestors
ORDER PROANURA: * ancestors of frogs and toads
ORDER ANURA: frogs and toads
INFRACLASS CAUDATA: tailed lissamphibians
ORDER URODELA: salamanders and newts
ORDER APODA: coecilians; limbless, tropical amphibians

CLASS REPTILIA: reptiles; scaled or armored tetrapods reproducing by the amniote egg
SUBCLASS ANAPSIDA: reptiles with solid skull roof
ORDER COTYLOSAURIA: * stem reptiles
SUBORDER GEPHYROSTEGOMORPHA: * early, primitive cotylosaurs

Suborder Captorhinomorpha: * generally small, carnivorous cotylosaurs
Suborder Procolophonia: * small, specialized cotylosaurs
Suborder Diadectomorpha: * large herbivores
Order Chelonia: turtles
Suborder Proganochelydia: * ancestral turtles
Suborder Amphichelydia: * primitive turtles
Suborder Pleurodira: * side-neck turtles
Suborder Cryptodira: * vertical-neck turtles
Suborder ?Eunotosauria: * *Eunotosaurus,* a doubtful turtle ancestor
Order Mesosauria: * ancient aquatic reptiles
Subclass Synapsida: * the mammal-like reptiles, with a skull opening below the postorbital-squamosal bones
Order Pelycosauria: * early mammal-like reptiles
Suborder Ophiacodontia: * primitive pelycosaurs
Suborder Sphenacodontia: * carnivorous pelycosaurs
Suborder Edaphosauria: * herbivorous pelycosaurs
Order Therapsida: * varied advanced mammal-like reptiles
Suborder Anomodontia: * small to large herbivorous therapsids
Infraorder Dromasauria: * small, generalized types
Infraorder Dinocephalia: * large, massive herbivores
Infraorder Venyukoviamorpha: * large, partially beaked herbivores
Infraorder Dicynodontia: * beaked therapsids, often with tusks
Suborder Phthinosuchia: * primitive therapsids
Suborder Theriodontia: * advanced, carnivorous therapsids
Infraorder Gorgonopsia: * primitive theriodonts
Infraorder Bauriamorpha: * specialized theriodonts
Infraorder Cynodontia: * late advanced theriodonts
Infraorder Tritylodontia: * highly adapted theriodonts
Infraorder Ictidosauria: * theriodonts close to mammals
Subclass Euryapsida: * generally marine reptiles, with a skull opening above the postorbital-squamosal bones
Order Protorosauria: * ancestral land-living euryapsids
Order Sauropterygia: * marine euryapsids
Suborder Nothosauria: * small, primitive sauropterygians
Suborder Plesiosauria: * large advanced sauropterygians
Superfamily Pistosauroidea: * ancestral plesiosaurs
Superfamily Pliosauroidea: * short-necked plesiosaurs
Superfamily Plesiosauroidea: * long-necked plesiosaurs
Order Placodontia: * mollusk-eating euryapsids
Order Ichthyosauria: * ocean-living reptiles of fishlike form
Subclass Diapsida: the ruling reptiles, with two skull openings separated by the postorbital-squamosal bones
Infraclass Lepidosauria: primitive diapsids and their direct descendants
Order Eosuchia: * ancestral lepidosaurs
Order Rhynchocephalia: beaked lepidosaurs; the modern tuatera
Order Squamata: lizards and snakes
Suborder Lacertilia: lizards
Suborder Ophidia: snakes
Infraclass Archosauria: advanced diapsids
Order Thecodontia: * ancestral archosaurians
Suborder Proterosuchia: * early, primitive thecodonts
Suborder Pseudosuchia: * varied thecodonts
Suborder Aetosauria: * heavily, armored thecodonts
Suborder Phytosauria: * aquatic, crocodilelike thecodonts
Order Crocodilia: crocodilians
Suborder Protosuchia: * ancestral crocodilians
Suborder Mesosuchia: * Mesozoic crocodilians
Suborder Sebecosuchia: * aberrant crocodilians
Suborder Eusuchia: * modern crocodilians; gavials, crocodiles, alligators
Order Saurischia: * saurischian dinosaurs
Suborder Palaeopoda: * ancestral saurischians
Infraorder Palaeosauriscia: * Triassic carnivores
Infraorder Plateosauria: * Triassic herbivores
Suborder Sauropoda: * gigantic, swamp-dwelling dinosaurs
Suborder Theropoda: * carnivorous saurischians
Infraorder Coelurosauria: * small to medium-sized carnivores
Infraorder Carnosauria: * large and gigantic carnivores
Order Ornithischia: * ornithischian dinosaurs
Suborder Ornithopoda: * duck-billed and related dinosaurs
Suborder Stegosauria: * plated dinosaurs
Suborder Ankylosauria: * armored dinosaurs
Suborder Ceratopsia: * horned dinosaurs
Order Pterosauria: * flying reptiles
Suborder Rhamphorhynchoidea: * primitive pterosaurs
Suborder Pterodactyloidea: * advanced pterosaurs

Class Aves: birds
Subclass Archaeornithes: * Jurassic toothed birds
Order Archaeopterygiformes: * *Archaeopteryx*
Subclass Neornithes: true birds
Order Hesperornithiformes: * *Hesperornis* and its allies
Order Ichthyornithiformes: * *Ichthyornis* and related genera
Order Tinamiformes: tinamous
Order Struthioniformes: ostriches
Order Rheiformes: rheas
Order Casuariiformes: cassowaries, emus
Order Aepyornithiformes: * elephant birds
Order Dinornithiformes: * moas
Order Apterygiformes: kiwis
Order Gaviiformes: loons
Order Podicipediformes: grebes
Order Procellariiformes: albatrosses, petrels
Order Sphenisciformes: penguins
Order Pelecaniformes: pelicans, frigate birds
Order Ciconiiformes: herons and storks
Order Anseritormes: ducks, geese, swans
Order Falconitormes: vultures, hawks, falcons, eagles
Order Galliformes: grouse; quail, turkeys, pheasants
Order Gruiformes: cranes, rails, limpkins, phororhacids
Order Diatrymiformes: * *Diatryma* and related genera
Order Charadriiformes: shore birds, gulls, auks
Order Columbiformes: pigeons, doves, the dodo
Order Psittaciformes: lories, parrots, macaws
Order Cuculiformes: cuckoos, roadrunners

Order Strigiformes: owls
Order Caprimulgiformes: goatsuckers
Order Apodiformes: swifts, hummingbirds
Order Coliiformes: colies
Order Trogoniformes: trogons
Order Coraciiformes: kingfishes, rollers, hoopoes, hornbills
Order Piciformes: barbets, toucans, woodpeckers
Order Passeriformes: perching birds; flycatchers, ovenbirds, lyrebirds, songbirds

Class Mammalia: mammals; tetrapods with hair, that suckle the young
Subclass Eotheria: * very primitive Triassic and Jurassic mammals
Order Docodonta: * docodonts
Order Triconodonta: * triconodonts
Subclass Prototheria: egg-laying mammals
Order Monotremata: the recent platypus and echidna
Subclass Allotheria: * a long line of early mammals
Order Multituberculata: * early mammals with multicuspid teeth
Suborder Plagiaulacoidea: * primitive multituberculates
Suborder Taeniolabidoidea: * large persisting multituberculates
Suborder Ptilodontoidea: * small forms with shearing teeth
Subclass Theria: most of the mammals
Infraclass Pantotheria: * the first of the therians
Order Eupantotheria: * ancestors of marsupials and placentals
Order Symmetrodonta: * symmetrodonts
Infraclass Metatheria: pouched mammals
Superorder Marsupialia: marsupials
Order Marsupicarnivora: opossums, carnivorous marsupials
Order Paucituberculata: opossum-rats and related fossil forms
Order Peramelina: bandicoots
Order Diprotodonta: phalangers, koala, wombat, diprotodonts, kangaroos, and wallabies
Infraclass Eutheria: placental mammals
Order Insectivora: insectivores, the most primitive placentals
Suborder Proteutheria: * primitive insectivores
Suborder Macroscelidea: elephant shrews
Suborder Dilambdodonta: moles, hedgehogs, shrews
Suborder Zalambdodonta: tenrecs, the golden mole
Order Dermoptera: the colugo and fossil relatives
Order Taeniodonta: * taeniodonts
Order Tillodontia: * tillodonts
Order Chiroptera: bats
Suborder Microchiroptera: the insectivorous bats
Suborder Megachiroptera: the fruit-eating bats
Order Primates: lemurs, tarsiers, monkeys, apes, men
Suborder Plesiadapoidea: * ancestral primates
Suborder Lemuroidea: lemurs
Suborder Tarsioidea: tarsiers
Suborder Platyrrhini: New World monkeys
Suborder Catarrhini: Old World monkeys, apes, men
Superfamily Parapithecoidea: * ancestral catarrhines
Superfamily Ceropithecoidea: Old World monkeys
Superfamily Hominoidea: apes and men
Order Edentata: New World edentates
Suborder Palaeanodonta: * ancestral edentates
Suborder Cingulata: armadillos and glyptodonts
Suborder Pilosa: sloths, ground sloths, anteaters
Order Rodentia: rodents
Suborder Protrogomorpha: * primitive rodents
Suborder Sciuromorpha: squirrels
Suborder Caviomorpha: South American rodents
Suborder Myomorpha: mice and rats and their relatives
Suborder Castorimorpha: beavers and their relatives
Suborder Theridomyomorpha: * certain European fossil rodents
Suborder Hystricomorpha: Old World porcupines
Suborder Thryonomyomorpha: cane and rock "rats"
Suborder Ctenodactylomorpha: gunis and related forms
Order Lagomorpha: hares, rabbits and pikas
Order Cetacea: porpoises and whales
Suborder Archaeoceti: * ancestral whales
Suborder Odontoceti: porpoises, dolphins, toothed whales
Suborder Mysticeti: whalebone whales
Order Creodonta: * ancient carnivorous placentals
Suborder Deltatheridia: * early creodonts
Suborder Hyaenodontia: * varied and persisting creodonts
Order Carnivora: the modern carnivorous placentals
Suborder Fissipedia: land-living carnivores
Superfamily Miacoidea: * ancestral fissipeds
Superfamily Canoidea: dogs, bears, pandas, raccoons, mustelids
Superfamily Feloidea: civets, hyenas, cats
Suborder Pinnipedia: marine carnivores; seals, sea lions, walruses
Order Condylarthra: * ancestral hoofed mammals or ungulates
Order Amblypoda: * primitive ungulates
Suborders Pantodonta * and Dinocerata: * large, early Tertiary ungulates
Order Tubulidentata: aardvarks
Order Pholidota: pangolins
Order Xenungulata: * certain ancient ungulates of South Africa
Order Pyrotheria: * very large South American ungulates
Order Notoungulata: * the most varied of the South American ungulates
Suborder Notioprogonia: * primitive notoungulates
Suborder Toxodontia: * large specialized notoungulates
Suborder Typotheria: * small, rabbitlike notoungulates
Suborder Hegetotheria: * small notoungulates
Order Astrapotheria: * large, possibly amphibious South American ungulates.
Order Litopterna: * "camel-like" and "horse-like" South America ungulates.
Order Penssodactyla: odd toed hoofed mammals
Suborder Hippomorpha: horses, titanotheres
Superfamily Equoidea: horses
Superfamily Brontotherioidea: * titanotheres
Suborder Ancylopoda: * chalicotheres

Suborder Ceratomorpha: rhinoceroses, and tapirs
Superfamily Tapiroidea: tapirs
Superfamily Rhinocerotoidea: rhinoceroses
Order Artiodactyla: even toed hoofed mammals
Suborder Paleodonta: * early artiodactyls
Superfamily Dichobunoidea: * ancestral artiodactyls
Superfamily Entelodontoidea: * entelodonts
Suborder Suina: piglike artiodactyls
Superfamily Suoidea: pigs and peccaries
Superfamily Anthracotherioidea: * anthracotheres
Superfamily Hippopotamoidea: hippopotamuses
Suborder Ancodonta: * ancodonts
Superfamily Cainotherioidea: * caenotheres
Superfamily Merycoidodontoidea: * oreodonts
Suborder Tylopoda: camels and llamas
Suborder Ruminantia: advanced artiodactyls
Superfamily Traguloidea: tragulids and their relatives
Superfamily Cervoidea: deer
Superfamily Giraffoidea: okapis and giraffes
Superfamily Bovoidea: antelopes and cattle
Order Proboscidea: elephants and their progenitors
Suborder Moeritherioidea: * ancestral proboscideans
Suborder Dinotherioidea: * dinotheres
Suborder Barytherioidea: * barytheres
Suborder Euelphantoidea: mastodonts and elephants
Superfamily Gomphotherioidea: * long-jawed mastodonts and their descendants
Superfamily Mastodontoidea: * crested toothed mastodonts
Superfamily Elephantoidea: stegodonts, mammoths, elephants
Order Sirenia: sea cows
Order Desmostylia: desmostylids; large marine waders
Order Hyracoidea: hyraxes
Order Embrithopoda: * *Arsinoitherium,* a gigantic rhinoceroslike mammal.

* An asterisk after a taxon indicates that its members are extinct.

References

Blackwelder, R. E., 1967. *Taxonomy: A Text and Reference Book*. Wiley, New York.

Bold, H., 1973. *The Morphology of Plants*. 3rd ed. Harper & Row, New York.

Colbert, E. H.: *Evolution of the Vertebrates: A History of Backboned Animals through Time*. 2nd ed. Wiley, 1969.

Fernald, M. L., 1950. *Gray's Manual of Botany*. 8th ed. Van Nostrand Reinhold, New York.

Grassé, P. P., 1952–1977. *Traité de Zoologie*. Vols. I–XVII. Masson et Cie, Paris.

Hyman, L. H., 1940–1961. *The Invertebrates,* vols. 1–6. McGraw-Hill, New York.

Keeton, W. T., 1980. *Biological Science*. 3rd ed. Norton, New York.

Larousse Encyclopedia of Animal Life, 1967. McGraw-Hill, New York.

Mayr, E., 1969. *Principles of Systematic Zoology*. McGraw-Hill, New York.

Sneath, P.H.A., and R. R. Sokal, 1973. *Numerical Taxonomy*. W. H. Freeman, San Francisco.

Whittaker, R. H., 1969. New concepts of kingdoms of organisms. *Science 163:* 150–160.

Whittaker, R. H., and L. Margulis, 1978. Protist classification and the kingdoms of organisms. *BioSystems 10:* 3–18.

II. ANSWERS TO PROBLEMS

Problems from Chapter 5

1.

Population	Frequency of gene A
I	0.2
II	0.7
III	0.4
IV	0.6

2. (a) Frequency of $L^m = 0.92$. Frequency of $L^n = 0.08$.
   (b) No. Note that calculating the frequency of L^m as p from the value of $\sqrt{p^2}$ is equal to $\sqrt{305/361} = \sqrt{0.845} = 0.92$. Similarly, the frequency of L^n as q is equal to 0.105. Since $p + q = 1$ when a Hardy-Weinberg equilibrium is present, and here $p + q = 1.025$, it appears that the population is close to but not at an equilibrium. This is confirmed by calculating the expected frequency of the heterozygote from $2\,pq$. At equilibrium this would be 0.193, but the observed value is 0.144. This difference between expected and observed values also indicates that the population is not in a Hardy-Weinberg equilibrium.
   (c) The proportion of progeny derived from L^mL^n mothers is 0.146. This answer is obtained as follows:

Matings where the mother has L^mL^n genotype ♀ × ♂	Frequency of matings
$L^mL^n \times L^nL^n$	$0.144 \times 0.845 = 0.123$
$L^mL^n \times L^nL^m$	$0.144 \times 0.144 = 0.021$
$L^mL^n \times L^mL^m$	$0.144 \times 0.011 = 0.002$
Total	0.146

   (d) The proportion of L^mL^n children from L^mL^m fathers is 0.070. The answer is obtained as follows:

Matings of L^mL^n fathers where L^mL^n progeny are possible ♂ × ♀	Frequency of matings	Frequency of Progeny L^mL^m	L^mL^n	L^nL^n
$L^mL^n \times L^mL^n$	$0.845 \times 0.144 = 0.122$	0.061	0.061	–
$L^mL^n \times L^nL^n$	$0.845 \times 0.011 = 0.009$	–	0.009	–
			0.070	

3. (a) The alleles initially present and their frequencies of occurrence would remain unchanged. Changes in the environment would matter because such changes would, in all probability, affect the selection pressures acting on the population, and this

would eliminate the Hardy-Weinberg equilibrium.

(b) Yes, because selection acting on one locus would affect all loci linked to that locus.

(c) If we assume that the new genes also met the conditions for a Hardy-Weinberg equilibrium, the new equilibrium will start the moment the new migrations cease and will then be expressed in the next generation and perpetuated from there.

4. (a) The frequency of taster × taster marriages is 0.41. This is determined by calculating the frequency of a taster in this population, i.e., 728/1138 or 0.64, and then multiplying 0.64×0.64, which is the frequency of the marriage in question.

(b) The ratio of taster to non-taster children in the offspring of parents who are tasters is 6.1 : 1.0. This is calculated as follows:
The frequencies of the *TT* and *Tt* parents is p^2 and $2\,pq$, respectively. These values are determined from calculating q.
$q = \sqrt{410/1138} = \sqrt{0.36} = 0.6$
and therefore
$p = 1 - q = 0.4$
$p^2 = 0.16$
and
$2pq = 0.48$

Marriages between taster parents ♀ ♂	Frequencies of marriages
TT × *TT*	$0.16 \times 0.16 = 0.026$
TT × *Tt*	$0.16 \times 0.48 = 0.077$
Tt × *TT*	$0.48 \times 0.16 = 0.077$
Tt × *Tt*	$0.48 \times 0.48 = 0.230$
	0.410

N.B. Compare the total frequency of taster × taster marriages, i.e., 0.41, to the answer obtained by a different method in 4(a).

Progeny from taster × taster marriages

TT	*Tt*	*tt*
0.026	–	–
0.039	0.038	–
0.038	0.039	–
0.058	0.115	0.058
0.161	0.192	0.058

the total of *TT* and *Tt* is 0.353, and 0.353 : 0.058 is 6.1 : 1.0

(c) The ratio of taster to non-taster children from marriages between taster and non-taster parents is 1.7:1.0. This is determined as follows.

Marriages between taster and non-taster parents ♀ ♂	Frequencies of marriages
TT × *tt*	$0.16 \times 0.36 = 0.058$
Tt × *tt*	$0.48 \times 0.36 = 0.173$
tt × *TT*	$0.36 \times 0.16 = 0.058$
tt × *Tt*	$0.36 \times 0.48 = 0.173$

Progeny from taster × non-taster marriages

TT	*Tt*	*tt*
–	0.058	—
–	0.086	0.087
–	0.058	—
–	0.087	0.086
	0.289	0.173

5. (a) Populations I and II are in Hardy-Weinberg equilibrium, but population III is not.
For population III:

$$\text{frequency of } A' = p = \frac{85 + 85 + 50}{400} = 0.55$$

$$\text{frequency of } A' = q = \frac{65 + 65 + 50}{400} = 0.45$$

With these values of p and q, the expected numbers of the three genotypes, i.e., p^2, $2\,pq$, and q^2 are

AA	*AA'*	*A'A'*
0.302×200	0.495×200	0.202×200
$= 60$	$= 99$	$= 40$

These expected values deviate significantly from the observed numbers.

(b) There are fewer heterozygotes than expected.

(c) If we assume that combining the populations has not changed the conditions for a Hardy-Weinberg equilibrium, the equilibrium will be reached in the next generation.
The frequencies of the three genotypes are:

AA	*AA'*	*A'A'*
$p^2 = 0.302$	$2\,pq = 0.495$	$q^2 = 0.202$

Problems from Chapter 6

1. 10^2 or 100 newly mutated (mutant) genes. This is calculated from $10^7 \times 10^{-5} = 10^2$, which is the population size multiplied by the mutation rate.
2. No, because one kind of assortative mating can be inbreeding, e.g., brother × sister matings.
3. Actual size = 493
   Breeding size = 52 + 405 = 457
   Effective size = $N_e = \frac{4N_f N_m}{N_f + N_m} = \frac{4(405)(52)}{(405+52)}$
   $= \frac{84240}{457} = 183.93$
4. When the loss of the allele in question, due to selection, equals the rate of appearance of the allele due to mutation, then there will be an equilibrium. Hence (loss due to selection) = (appearance due to mutation) or
   $spq^2/(1-sq^2) = \mu p$
   $sq^2(1-1)/(1-sq^2) = \mu(1-q)$
   $3q^2/(1-sq^2) = \mu$
   $sq^2 = \mu(1-q^2)$
   $q^2 = \mu(1-sq^2)/s$
   But here $s = 1$ and q^2 is very low; therefore, sq^2 is negligible and so
   $q^2 = \mu/s$
   $q = \sqrt{\mu/s}$
   and since $s = 1$,
   $q = \sqrt{\mu}$

This is at equilibrium so q can be expressed as the square root of the mutation rate.

$$sp(1-p)^2 = \nu q$$
$$sp(1-p)(1-p) = \nu(1-p)$$
$$sp(1-p) = \nu$$
$$p(1-p) = \nu/s$$

but $1-p$ can be treated as 1 because p is very small as a result of the allele being a dominant lethal. Hence,
$p = \mathrm{V}/s$
and
$s = 1$

therefore
$p = \mathrm{v}$

The frequency of the allele for retinoblastoma, at equilibrium $\hat{p}$, can be expressed as the rate of mutation to that allele, if we realize that when the allele is expressed in the development of a new individual it is lethal. Hence, this allele really only persists as a new mutant in a gamete.

6. In infantile amaurotic idiocy, the frequency of the homozygote q^2 is equal to μ (since $q = \sqrt{\mu}$). These occasional homozygotes can arise from the offspring from two heterozygotes. In retinoblastoma, $p^2 = \nu^2$, theoretically. The homozygous individuals can only arise if a sperm carrying the mutant allele fertilizes an egg carrying the mutant allele. Practically speaking, this is so rare $(3 \times 10^{-6})^2$, that we will not expect to find it.

Indices

ORGANISM INDEX

SUBJECT/AUTHOR INDEX

Definitions are located on pages identified by italic type

SOURCES FOR FIGURES

PART 1
The Bettmann Archive.

CHAPTER ONE
Fig. 1-1: B. Doonan and R. Bloodgood; **Fig. 1-2: B and C.** Redrawn from W.G. Whaley et al., *Journal of Biophysical and Biochemical Cytology.,* vol. 5, 1959, by permission; **Fig. 1-3: A.** © L.V. Bergman & Assoc., Inc. **B.** Tseh An Chan, New Jersey Agricultural Experiment Station. **C.** © L.V. Bergman & Assoc., Inc.; **Fig. 1-4:** Redrawn from *Animals without Backbones* by R. Buchsbaum, 2nd ed., by permission of The University of Chicago Press. Copyright © 1976; **Fig. 1-5:** Redrawn from E.O. Dodson and P. Dodson, *Evolution: Process and Product,* 2nd ed. D. Van Nostrand/ Reinhold Co., 1976. Fig. 3-14, by permission; **Fig. 1-6:** Reproduced from E.D. Hanson, *Animal Diversity,* 2nd ed., Prentice-Hall, Inc., 1964, by permission; **Fig. 1-8: A.** Redrawn from *Animals Without Backbones* by R. Buchsbaum, 2nd ed., by permission of The University of Chicago, Press. Copyright © 1976. **B.** E.D. Hanson; **Figs. 1-9 and 1-10;** E.D. Hanson; **Fig. 1-11:** © L.V. Bergman & Assoc., Inc; **Figs. 1-12, 1-13, and 1-14:** Redrawn from *Animals Without Backbones,* by R. Buchsbaum 2nd ed., by permission of The University of Chicago Press. Copyright © 1962; **Figs. 1-15, 1-17A, B, 1-18, 1-20, 1-21:** E.D. Hanson; **Fig. 1-16:** Redrawn from H.S. Jennings, *Behavior of the Lower Organisms,* Indiana University Press, copyright © 1976, by permission.

CHAPTER TWO
Fig. 2-1: Reproduced from E.D. Hanson, *Aniaml Diversity,* 2nd ed., Prentice-Hall, Inc., 1964, by permission; **Fig. 2-2:** From *Darwin's Finches* by D. Lack. Copyright © 1953 by Scientific American, Inc. All rights reserved, by permission; **Fig. 2-3: A through E.** From *The Vertebrate Body* by A.S. Romer and T.S. Parsons, 5th ed., copyright © 1977, Saunders, by permission; **Fig. 2-4:** The American Museum of Natural History; **Fig. 2-5:** From *Evolution,* 3rd ed., by J.M. Savage. Copyright © 1977 by Holt, Rinehart and Winston. Reprinted by permission of Holt, Rinehart and Winston; **Box 2-1:** The American Museum of Natural History; **Box 2-2: A.** The American Museum of Natural History. **B.** National Library of Medicine. **C.** From *Darwin's Finches* by D. Lack. Copyright © 1953 by Scientific American, Inc. All rights reserved, by permission; **Box 2-3: A.** George Eastman House. **B.** The New York Public Library.

CHAPTER THREE
Fig. 3-1: From *Heredity, Evolution, and Society,* 2nd ed., by I.M. Lerner and W.J. Libby. W.H. Freeman and Company. Copyright © 1976, by permission; **Figs. 3-3, 3-5, 3-6, and 3-7:** From *Genetics,* 2nd ed., by M.W. Strickberger. Macmillan Publishing Company, Inc., by permission; **Fig. 3-8:** From M. Smith, *American Scientist,* 67:58, 1979. Fig. 2, by permission; **Fig. 3-10:** Redrawn by permission from M.W. Strickberger (1976) *Genetics,* 2nd ed., Macmillan Publishing Co., Inc., New York.

PART 2
Above, The Bettmann Archieve, below, from G. Hartwig, *The Polar and Tropical Worlds,* 1874. Courtesy J. Berry.

CHAPTER FOUR
Fig. 4-2: B. From *Heredity, Evolution, and Society,* 2nd ed., by I.M. Lerner and W.J. Libby. W.H. Freeman and Company. Copyright © 1976, by permission; **Fig. 4-4:** From *Evolution,* 3rd ed., by J.M. Savage. Copyright © 1977 by Holt, Rinehart and Winston. Reprinted by permission of Holt, Rinehart and Winston; **Fig. 4-5:** After H.S. Jennings, 1909; **Fig. 4-6:** After H.S. Jennings, 1916; **Fig. 4-7:** E.D. Hanson; **Fig. 4-9: A.** M.E.B. Joyner. **B.** Carolina Biological Supply; **Figs. 4-10 and 4-11:** Reproduced by permission from W.H. Brown, *The Plant Kingdom,* Ginn and Co. (Xerox Corporation). © Copyright 1935, William H. Brown; **Box 4-6:** National Science Foundation.

CHAPTER FIVE
Fig. 5-2: Courtesy H.B.D. Kettlewell; **Fig. 5-3:** Redrawn with permission of Macmillan Publishing Company, Inc., from *The Life of Bacteria* by W.D. Stansfield. Copyright © 1977; **Figs. 5-6 and 5-7:** Redrawn from *Population Genetics* by C.C. Li, with permission of the University of Chicago Press, Copyright © 1955; **Fig. 5-9:** Redrawn with permission of Worth Publishers, New York, from *Biology,* 2nd ed., by H. Curtis. Copyright 1975. p. 875; **Box 5-1: A.** National Library of Medicine. **B.** Fisher Memorial Trust, Gonville and Caius College, Cambridge. Photo: A.C. Barrington-Brown. **C.** Courtesy University of Chicago Press, publisher of *Evolution and the Genetics of Populations* by S. Wright. Photo: Lewellyn Studio.

CHAPTER SIX
Fig. 6-1: *Co-Evolution Quarterly* No. 17, Spring 1978, p. 47; **Figs. 6-3 and 6-5:** From *Evolution* by T. Dobzhansky, F.J. Ayala, G.L. Stebbins, and J.W. Valentine. W.H. Freeman and Company. Copyright © 1976, by permission.

PART 3
From C. Darwin, *Journal of Researches,* New York, 1896. Courtesy E.A. Christenson.

CHAPTER SEVEN
Fig. 7-2: Reproduced by permission from G.G. Simpson, *The Meaning of Evolution.* Yale University Press, New Haven, Conn., 1960; **Fig. 7-4:** After E. Mayr, *Animal Species and Evolution,* Belknap Press of Harvard University Press, Cambridge, Mass., 1963, by permission; **Fig. 7-5:** Redrawn with permission of Macmillan Publishing Company, Inc., from *The Life of Bacteria* by W.D. Stansfield. Copyright © 1977; **Fig. 7-7:** R.M. Tullar.

CHAPTER EIGHT
Figs. 8-1 and 8-2: From *Processes of Organic Evolution,* 3rd ed., by G.L. Stebbins, © 1977. Reprinted by permission of Prentice-Hall, Inc., Englewood Cliffs, N.J.; **Fig. 8-4:** From *Adaptation,* 2nd ed., by B. Wallace and A.M. Srb, © 1964. Reprinted by permission of Prentice-Hall, Inc., Englewood Cliffs, N.J.; **Fig. 8-5: A and B.** After A.C. Wilson, 1974; **Fig. 8-6:** G. LeFevre; **Fig. 8-7:** From *Evolution* by T. Dobzhansky, F.J. Ayla, G.L. Stebbins, and J.W. Valentine. W.H. Freeman and Company. Copyright © 1977, by permission; **Figs. 8-8 and 8-9:** From *Chromosomes, Giant Molecules and Evolution,* by B. Wallace, W.H. Norton, Inc., copyright © 1966, by permission; **Box 8-1: A through G.** From *The Story of Pollination* by B. J. D. Meeuse, Copyright © 1961 by John Wiley and Sons, Inc., by permission; **H.** Reproduced from T. Delevoryas, *Plant Diversification,* 2nd ed., Holt, Rinehart and Winston, copyright 1966, 1967. After various sources, by permission; **Box 8-2: A, B, and C.** From *Chromosomes, Giant Molecules and Evolution* by B. Wallace, W.H. Norton, Inc., copyright © 1966, by permission.

CHAPTER NINE
Fig. 9-2: From E. Haeckel, *Evolution,* 1866; **Fig. 9-4:** From A. Romer, *Man and the Vertebrates,* University of Chicago Press, 1941, by permission; **Fig. 9-7: A.** J.E. McCosker, Steinhart Aquarium. **B.** Wards Natural Science Establishment; **Fig. 9-8:** S.C. Bisserôt; **Fig. 9-9: B.** From a drawing after Simpson and Beck, 1965, in *Evolution* by T. Dobzhansky, F.J. Ayala, G.L. Stebbins, and J.W. Valentine. W.H. Freeman and Company, Copyright © 1977, by permission; **Fig. 9-11:** Redrawn from M. Moran in "Evolution" in *Encyclopaedia Britannica,* 15th ed. (1974), by permission; **Fig. 9-12:** From *Phylogenetic Systematics* by W. Henning. Copyright © 1966 by University of Illinois Press, by permission; **Fig. 9-14:** Reproduced by permission from A.C. Wilson, "Evolutionary Importance of Gene Regulation," Stadler Symposium, v.7, 1975; **Box 9-1: B.** From *Molecular Genetics: An Advanced Treatise,* Part 2, J.H. Taylor (ed.), Academic Press, 1963-1968, by permission.

PART 4
Reproduced by permission from I.W. and V.G. Sherman, *Biology, A Human Approach,* Oxford University Press, New York, 1979, Fig. 24-12.

CHAPTER TEN
Fig. 10-2: S. Fox et al.; **Fig. 10-3:** L. Meszoly, Harvard University, by permission; **Fig. 10-4: A.** E.S. Barghoorn. **B.** D. Eicher. **C and D.** J. Robertson. **E.** The American Museum of Natural History; **Fig. 10-5:** The Smithsonian Institution; **Fig. 10-6: A(1 and 2).** R.S. Solecki. **B.** By permission Robert Harding Associates; **Fig. 10-7:** From *The Study of Botany* by Adams, Baker, and Allen, © 1970, Addison-Wesley Publishing Company, Inc. . Reprinted with permission; **Fig. 10-10:** J.W. Schopf; **Fig. 10-11:** S. Fox, Institute

for Cellular and Molecular Evolution, University of Miami; **Fig. 10-14: A and B.** The American Museum of Natural History. **C and D.** J. Ostrom, *American Scientist,* 67, 1979; **Fig. 10-15:** The Peabody Museum of Natural History, Yale University; **Box 10-1:** National Library of Medicine; **Fig. 10-17:** Adapted from C. Ponnamperuma (1972) *The Origins of Life,* Thames and Hudson, Ltd. London, by permission; **Fig. 10-18:** Redrawn by permission from M. Petterson (1978) *Journal of Social Biological Structures.* Pages 201-206, Fig. 1; **Fig. 10-19:** Reproduced from E.D. Hanson, *Animal Diversity,* 3rd ed., Prentice-Hall, Inc., 1964, by permission.

CHAPTER ELEVEN

Fig. 11-1: © L.V. Bergman & Associates, Inc.; **Fig. 11-2:** Redrawn from *The Life of Bacteria* by K.V. Thimann. Copyright 1955, Macmillan Publishing Company, Inc, by permission; **Fig. 11-3:** Redrawn by permission from H. Morowitz and M.E. Tourtelotte, *Scientific American,* 1962; **Fig. 11-5:** From *Traité de Zoologie,* vol. I, Part I by P.P. Grasse. Masson et Cie., Paris, by permission; **Fig. 11-16 and Box 11-3:** Reproduced by permission from W.H. Brown, *The Plant Kingdom,* Ginn and Co. (Xerox Corporation). © Copyright 1935, William H. Brown; **Fig. 11-10:** Reproduced from G.N. Calkins, *The Protozoa,* Columbia University Press, 1901, by permission; **Fig. 11-11:** From *Evolution* by T. Dobzhansky, F.J. Ayala, G.L. Stebbins, and J.W. Valentine. W.H. Freeman and Company. Copyright © 1976, by permission.

CHAPTER TWELVE

Figs. 12-1, 12-3, 12-5, 12-6, 12-7, 12-9, 12-12, 12-16, 12-20, A through D: Reproduced by permission from W.H. Brown, *The Plant Kingdom,* Ginn and Co. (Xerox Corporation). © Copyright 1935, William H. Brown; **Fig. 12-2:** Adapted from *Plant Diversity: An Evolutionary Approach* by R.F. Scagel, R.J. Bandoni, G.E. Rouse, W.B. Schofield, J.R. Stein, and T.M.C. Taylor. © 1965, 1969 by Wadsworth Publishing Company, Inc., Belmont, California 94002. Reprinted by permission of the publisher; **Fig. 12-4:** From H.C. Bold, *Morphology of Plants,* 3rd ed., Harper and Row Publishers, Inc., New York, 1973, by permission; **Figs. 12-8 and 12-10:** Reproduced from T. Delevoryas, *Plant Diversification,* 2nd ed., Holt, Rinehart and Winston, copyright 1966, 1977. After various sources, by permission; **Fig. 12-11:** After A.J. Eames and L.H. MacDaniels, *An Introduction to Plant Anatomy,* 2nd ed., McGraw-Hill, 1947, by permission; **Fig. 12-13: A and B.** E.D. Hanson; **Fig. 12-14:** Field Museum of Natural History, Chicago; **Fig. 12-15: A.** From H.N. Andrews, *Ann. Missouri Bot. Garden,* v. 61, p. 186, 1974, by permission. **B.** Trustees of The British Museum. **C.** Carolina Biological Supply. **D.** © A. Rokach, N.Y. Botanical Garden; **Fig. 12-17: A and B.** M.E.B. Joyner; **Fig. 12-18:** Reproduced from T. Delevoryas, *Plant Diversification,* 2nd ed., Holt, Rinehart and Winston, copyright 1966, 1977. After various sources, by permission; **Fig. 12-19:** E.D. Hanson; **Box 12-1: A and B.** Reproduced by permission from W.H. Brown, *The Plant Kingdom,* Ginn and Co. (Xerox Corporation). © Copyright 1935, William H. Brown. **C and D.** From C.B. Beck, ed., *Origin and Early Evolution of Angiosperms,* © 1976, Columbia University Press, by permission.

CHAPTER THIRTEEN

Figs. 13-2 and 13-10: From T.I. Storer et al., *General Zoology,* 6th ed., copyright 1979 by permission McGraw-Hill Book Company; **Fig. 13-13:** E.D. Hanson; **Fig. 13-14: A.** After Hyman, 1951. **B.** After Gurney, 1942. **C.** From Segrove, 1951. *Q. Jl. Microsc. Sci,* 82; **Fig. 13-15: A.** After Hyman, 1951. **B.** After Gurney, 1942. **C.** Redrawn from *Animals Without Backbones,* 2nd ed. by R. Buchsbaum, by permission of The University of Chicago Press, 1962; **Fig. 13-17: A.** E.D. Hanson. **B.** From *Living Marine Molluscs* by C.M. Yonge and T.E. Thompson, published by Collins, London 1976, by permission.

CHAPTER FOURTEEN

Fig. 14-1: From T.I. Storer et al., *General Zoology,* 6th ed., copyright 1979, by permission McGraw-Hill Book Company; **Fig. 14-3: A.** Reproduced from Q. Bone, *The Origin of Chordates,* 2nd ed., Oxford University Press, 1979, by permission. **B.** The British Museum; **Figs. 14-4, 14-7, 14-11, 14-13, 14-14, and 14-21:** Adapted by permission from a drawing by Lois M. Darling © 1969, in *Evolution of the Vertebrates,* 2nd ed., by E.H. Colbert. John Wiley & Sons, 1969. **Fig. 14-5: A through D.** From J.R. Norman, *A History of Fishes,* published by Ernest Benn Ltd., copyright 1931, by permission; **Figs. 14-6 and 14-9:** From J.Z. Young, *The Life of Vertebrates,* 2nd ed. © Oxford University Press 1962, published by Oxford University Press, by permission; **Fig. 14-8:** From Williston, *Osteology of the Reptiles,* published by Harvard University Press. Reprinted with permission © 1925; **Fig. 14-10:** From *The Evolution of the Vertebrates* by E.H. Colbert, 2nd ed. by permission John Wiley & Sons, Inc.; **Fig. 14-12:** Redrawn after A.S. Romer, *The Vertebrate Body,* W.B. Saunders Company, Philadelphia, 1949; **Fig. 14-20: A through E.** From *The Life of Vertebrates* by J.Z. Young, 2nd ed. © Oxford University Press 1962, published by Oxford University Press, by permission; **Fig. 14-24:** Robert Harding Associates; **Fig. 14-25:** After Jerison, 1976; **Box 14-2: A.** R.F. Sison. © National Geographic Society. **B.** From M.D. Leakey, *Olduvai Gorge* Vol. 3. Published by Cambridge University Press, © 1972. Reproduced by permission of Cambridge University Press.